AF665378

# Spectroscopic and Diffraction Techniques in Interfacial Electrochemistry

# NATO ASI Series

**Advanced Science Institutes Series**

*A Series presenting the results of activities sponsored by the NATO Science Committee, which aims at the dissemination of advanced scientific and technological knowledge, with a view to strengthening links between scientific communities.*

The Series is published by an international board of publishers in conjunction with the NATO Scientific Affairs Division

| | |
|---|---|
| **A Life Sciences**<br>**B Physics** | Plenum Publishing Corporation<br>London and New York |
| **C Mathematical and Physical Sciences**<br>**D Behavioural and Social Sciences**<br>**E Applied Sciences** | Kluwer Academic Publishers<br>Dordrecht, Boston and London |
| **F Computer and Systems Sciences**<br>**G Ecological Sciences**<br>**H Cell Biology** | Springer-Verlag<br>Berlin, Heidelberg, New York, London, Paris and Tokyo |

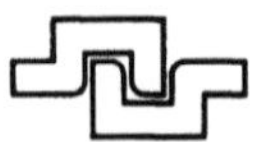

**Series C: Mathematical and Physical Sciences - Vol. 320**

# Spectroscopic and Diffraction Techniques in Interfacial Electrochemistry

edited by

C. Gutiérrez

Instituto de Química Física "Rocasolano", C.S.I.C.,
Madrid, Spain

and

C. Melendres

Argonne National Laboratory,
Argonne, Illinois, U.S.A.

Springer Science+Business Media, B.V.

Proceedings of the NATO Advanced Study Institute on
Spectrocopic and Diffraction Techniques in Interfacial Electrochemistry
Puerto de la Cruz, Tenerife, Spain
July 3–15, 1988

**Library of Congress Cataloging-in-Publication Data**

NATO Advanced Study Institute on Spectroscopic and Diffraction Techniques in Interfacial Electrochemistry (1988 : Puerto de la Cruz, Canary Islands)
Spectroscopic and diffraction techniques in interfacial electrochemistry / edited by C. Gutiérrez and C. Melendres.
p. cm. -- (NATO ASI series. Series C, Mathematical and physical sciences ; no. 320)
"Proceedings of the NATO Advanced Study Institute on Spectroscopic and Diffraction Techniques in Interfacial Electrochemistry, Puerto de la Cruz, Tenerife, Spain, 3-15 July 1988."
Includes bibliographical references and index.
ISBN 978-94-010-5684-7 ISBN 978-94-011-3782-9 (eBook)
DOI 10.1007/978-94-011-3782-9
1. Electrochemistry--Congresses. 2. Surface chemistry--Congresses. 3. Spectrum analysis--Congresses. 4. Diffraction--Congresses. I. Gutiérrez, C. (Claudio), 1942- . II. Melendres, C. (Carlos), 1939- . III. Title. IV. Title: Interfacial electrochemistry. V. Series.
QD551.N38 1988
541.3'7--dc20 90-5353

ISBN 978-94-010-5684-7

*Printed on acid-free paper*

CONTENTS

## PREFACE

Electrochemistry is one of the oldest branches of Physical Chemistry. Having its foundations in the work of Faraday, Arrhenius and others, it evolved from the study of transport in electrolyte solutions to that of electrode kinetics. Kinetic methods are inherently unable to identify unequivocally the species involved in a reaction. Therefore, beginning in the 70s many spectroscopic and diffraction techniques were applied to the study of the electrode–electrolyte interface, in order to identify intermediary reaction species, and even the spatial arrangement of atoms or molecules at the interface. In order to disseminate these techniques, a NATO Advanced Sutdy Institute was held at Puerto de la Cruz, Tenerife (Canary Islands, Spain) from July 2 to 15, 1988. The Institute consisted of tutorial type lecutures, poster sessions, and round–table discussions. It was attended by over 65 participants from NATO–member countries, and others from Argentina and Japan.

In the present volume most of the lectures presented at the Institute have been collected. At least one chapter is devoted to each technique. Emphasis has been made on case studies, rather than theory, which can be found in textbooks and other publications. Our purpose in this book is to help the electrochemists uninitiated in spectroscopic methods to decide which techniques would be suitable for application to their particular problems.

We thank all the lecturers who contributed to this volume, and even those UHPs (Unrepentant Habitual Procrastinators) who did not in spite of our urgings to do so. We greatly appreciate the many hours they spent in preparing their lectures and in after–hours discussions with the participants. We owe a great deal of gratitude to our local co–director, Prof. Arévalo, for his excellent organization of the local arrangements.

It was our pleasure to have as company participants from so many lands. Finally, we thank the Scientific Affairs Division of NATO for their financial support.

C. Gutiérrez and C. A. Melendres

## LIST OF PARTICIPANTS

ALBARELLI, Michael J.
Baker Laboratory, Department of Chemistry, Cornell University, Ithaca, NY 14853, U.S.A.

ANDERSON, Mark R.
Department of Chemistry, University of Utah, Salt Lake City, UT 84112, U.S.A.

ANDREU, Rafael.
Depto. de Quimica Física, Fac. de Química, Universidad de Sevilla, 41012-Sevilla. Spain.

BEREKET, Gözen.
Dept. of Chemistry, Faculty of Arts and Science, University of Anatolia, Turkey.

CASADO GIMENEZ, Juan.
Dept. of Physical Chemistry, Universidad Autónoma de Barcelona, Bellaterra, Barcelona, Spain.

CELDRAN, Raquel.
Universidad Autónoma de Madrid, Facultad de Ciencias, Depto. Química, C-II, 28049 Madrid, Spain.

CROUIGNEAU, Pierre.
Laboratoire de Chimie I, UER Sciences, Université de Poitiers, 86022 Poitiers Cedex, France.

DA SILVA, Rui.
Faculdade de Engenharia, Universidade do Porto, 4000 Porto, Portugal

EPPEL, Steve
Department of Physics, Case Western Reserve University, Cleveland, OH 44106, U.S.A.

ERBOLUKBAS, Aysen.
Istanbul Universitesi Fen Fakültesi Fizik Bölümü, Vezneciler Istanbul, Turkey.

ESKIKAYA, Tulay.
Chemistry Dept., Istanbul Technical University, 82060 Istanbul, Turkey

FERNANDES, Joáo.
Instituto Superior Técnico, Lab. Electroquímica,
Universidade Técnica de Lisboa, 1096 Lisboa Codex, Portugal.

FERREIRA, Mario G.S.
Depto. de Eng. Química, Instituto Superior Técnico, Universidade
Técnica de Lisboa, 1096 Lisboa Codex, Portugal.

FONSECA, Ines.
Dept. of Chem., Faculty of Science of Lisbon, CECUL, R. da Escola Politecnica
58-1293 Lisboa Codex, Portugal.

FONTANESI, Claudio.
University of Modena, Dept. of Chemistry, Via Campi 183, 41100 Modena,
Italy.

GIORDANO, Cristina.
INIFTA, Casilla de Correo 16, Sucursal 4, (1900) La Plata, Argentina.

GOMEZ, Mercedes
Depto. de Química, Facultad de Ciencias, C-II, Universidad Autónoma de
Madrid, 28049 Madrid, Spain.

GOSLICH, Roland.
Insttitut für Solarenergieforschung GmbH, Sokelant Straße 5,
D-3000 Hannover 1, West Germany.

HANSEN, Galen J.
Physics Dept., UMC 4415, Utah State University, Logan, UT 84322. USA.

JAEGERMANN, Wolfram.
Hahn-Meitner Institut, Glienicker Str. 100, D-1000 Berlin 39,
West Germany.

JEHOULET.
Laboratoire des Surfaces et Interfaces, 13 rue de Toul, 59046 Lille Cedex,
France.

JIMENEZ FERRER, M. Isabel.
Dpto. Física Aplicada, Facultad de Ciencias,
Ciudad Universitaria de Cantoblanco, 28049 Madrid. Spain.

JUSTO, Maria Julia.
L.N.E.T.I., Az. Lameiros, Est. Paco Lumiar, 1600 Lisboa, Portugal.

KAUTEK, W.M.
Laboratory for Chemical Surface Technologies, Federal Insitute for Materials
Research, Unter den Eichen 87, D-1000 Berlin 45, West Germany

KESSEL, Robert.
Inst. of Physical Chemistry, Universität Düsseldorf, D-4000 Düsseldorf, Universitätsstrasse 1, Gebäude 26.32, West Germany.

LARRAMONA, Gerardo
Instituto de Química Física "Rocasolano", CSIC, Serrano, 119, 28006-Madrid, Spain.

LAUERMANN, Iver.
Institut für Solarenergieforschung, SokelantstraBe 5, D-3000 Hannover 1, West Germany.

MAKKY, Wagih.
Dept. of Electrical Engineering, University of Illinois, Chicago, IL 60680, U.S.A.

MARASSI, Roberto.
Dept. of Chemistry, University of Camerino, Via S. Agostino 1, Italy.

MARCO SANZ, Jose F.
Instituto de Química Física "Rocasolano", CSIC, Serrano, 119, 28006-Madrid, Spain.

MARTIN NEVSKAIA, Daniela.
Dpto. Física Aplicada, Facultad de Ciencias, Ciudad Universitaria de Cantoblanco, 28049 Madrid, Spain.

McMAHON, John J.
Dept. of Chemistry, Fordham University, Bronx, N.Y. 10458, U.S.A.

MEISSNER, Dieter.
Institut für Solarenergiesforschung, Sokelantstrasse 5, D-3000 Hannover 1, West Germany.

MONTEIRO, Fernando J.
Depto. de Engenharia Metalúrgica, Universidade do Porto, Rua dos Bragas, 4099 Porto Codex, Portugal.

MüLLER JEVENOIS, Carlos.
Dpto. Química Física, Universidad de Barcelona, Marti i Franques 1, 08028 Barcelona, Spain.

OLLIER, Nathalie.
Laboratoire d'Electrochimie Interfaciale (LEI), CNRS,
1, Place Aristide Briand, 92195 Meudon Principal Cedex, France.

OZSAVASCI, Cengiz.
Gazi Universitesi, Fen-Edebiyat Fakültesi, Kimya Bölümü,
06500 Teknikokullar, Ankara, Turkey.

PALITEIRO, Carlos.
Department of Chemistry, University of Coimbra, 3049-Coimbra, Portugal.

PEÑA, Maria Jesús.
Depto. de Química, Facultad de Ciencias, C-II, Universidad Autónoma de Madrid, 28049-Madrid, Spain.

PEREIRA, M. Isabel S.
CECUL, Fac. de Ciencias, R. Da Escola Politécnica, 1200 Lisboa, Portugal.
PHILPOTT, M.R.
IBM Almaden Research Center, 650 Harry Road, San Jose, CA 95120-6099, U.S.A.

PIAZZA, Giuseppe.
Dipartimento Chimico "G. Ciamician", Via Selmi, 2, 40127 Bologna, Italy.

PIMENTA, Gervasio.
ICEN-LNETI, Depto. de Quimica, P-2686 Sacavém Codex, Portugal.

RADZISZEWSKI, J. G.
Center for Structure and Reactivity, Department of Chemistry, The University of Texas at Austin, Austin, TX 78712-1167, U.S.A.

RANGEL, Carmen Mireya.
LNETI, Department of Materials Technology, Corrosion and Protection, 22 Paco do Lumiar, 1699 Lisboa Cedex, Portugal.

RIMMASCH, Jörk.
Institut für Solarenergieforschung, SokelantstraBe 5, D-3000 Hannover 1. West Germany.

ROBINSON, Jeanne M.
Dept. of Chemistry, University of Oregon, Eugene, OR 97403, U.S.A.

SA, Carlos M.
Depto. de Engenharia Metalúrgica, Universidade do Porto, Rua dos Bragas, 4099 Porto Codex, Portugal.

SANTOS LOPES, Maria Irene.
CECUL, Dep. Química, Faculdade de Ciencias, Rua da Escola Politécnica, 1200 Lisboa, Portugal.

SANZ, Fausto.
Dpto. Química Física, Fac. Química, Universidad de Barcelona, Av. Diagonal 647, 08028 Barcelona, Spain.

SARAC, A. Sezai
Technical University of Istanbul, Departmant of Chemistry, Maslak-Istanbul, Turkey

SHANNON, Curtis.
Department of Chemistry, University of Texas, Austin, TX 78712, U.S.A.

TADJEDDINE, Abderrahmane.
LEI du CNRS, 1 Pl. Aristide Briand, F 92195 Meudon Ppl. Cedex, France.

YOSHIKAWA, Kozo.
Takasago R & D Center, Mitsubishi Heavy Industries, Ltd., 2-1-1 Shinhama, Arai-Cho, Takasago, Hyogo Pref., 676 Japan.

## LIST OF CONTRIBUTORS

ABERDAM, Daniel.
Lab. de Spectrometrie Physique. Université Joseph Fourier Grenoble I, BP87, F38402 St Martin d'Hères, France.

ARVIA, Alejandro J.
Instituto de Investigaciones Físicoquímicas Teóricas y Aplicadas (INIFTA), Casilla de Correo 16, Sucursal 4, 1900-La Plata, Argentina.

BEDEN, Bernard.
Lab. de Chimie I- Electrochimie et Interactions, U.A. au C.N.R.S. 350, Université de Poitiers, 86022 Poitiers, France.

BOCKRIS, John O'M.
Department of Chemistry, Texas A & M University, College Station, Texas 77843, U.S.A.

CHANG, Richard K.
Section of Applied Physics and Center for Laser Diagnostics, Yale University, New Haven, CT 06520, U.S.A.

GUTIERREZ, Claudio
Instituto de Química Física "Rocasolano", Consejo Superior de Investigaciones Científicas, Serrano, 119, 28006-Madrid, Spain.

KOLB, Dieter M.
Fritz-Haber-Institut der Max-Planck-Gesellschaft, D-1000 Berlin 33, West Germany.

KÖTZ, Rüdiger.
ASEA Brown Boveri Corporate Research Center, CH 5405 Baden, Switzerland.

MELENDRES, Carlos.
Materials Science and Chemical Technology Division, Argonne National Laboratory, Argonne, IL 60439, U.S.A.

OCKO, Benjamin
Brookhaven National Laboratory, Bldg. 510B, Upton, N.Y. 11973, U.S.A.

PLIETH, Waldfried J.
Institut für Physikalische und Theoretische Chemie, Freie Universität Berlin, Takustrasse 3, D-1000 Berlin 33, West Germany.

ROBINSON, James.
Department of Physics, University of Warwick, Coventry CV4 7AL, United Kingdom.

SHEN, Y R.
Department of Physics, University of California, Berkeley, CA 94720,U.S.A.

VALLET, Catherine E.
Chemistry Division, Oak Ridge National Laboratory, Oak Ridge, TN 37831, U.S.A.

## LIST OF LECTURERS WHO FAILED TO SEND THEIR MANUSCRIPTS

BEWICK, Alan.
Department of Chemistry, University of Southampton, Southampton SO9 5NH, United Kingdom.

CAHAN, Boris D.
Case Center for Electrochemical Studies and Department of Chemistry, Case Western Reserve University, Cleveland, OH 44106, U.S.A.

HOFFMAN, Richard W.
Department of Physics, Case Western Reserve University, Cleveland, OH 44106, U.S.A.

SASS, Jürgen K.
Dept. of Surface Reactions, Fritz-Haber-Institut der Max-Planck-Gesellschaft, Faradayweg 4-6, D-1000 Berlin 33/Dahlem, West Germany.

# THE ADVANCING FRONTIER IN THE KNOWLEDGE OF THE STRUCTURE OF INTERPHASES

J. O'M. BOCKRIS AND ANUNCIA GONZALEZ-MARTIN
*Surface Electrochemistry Laboratory*
*Texas A&M University*
*College Station, Texas 77843*

**ABSTRACT.** This is a comprehensive document about the structure of the double layer in electrochemical systems, which involves mainly the metal/ and semiconductor/solution interphases, as well as metal/molten-salt interphases. It begins with an account of the classical method of investigating interfaces, which appears to be no longer applicable. Among topics discussed are the different models of water structure at the double layer, the various potentials in the interphasial region, and ab-initio calculations.

## 1. Introduction

It is firstly necessary to define "interphase" and distinguish it from "interface". The latter is a surface in contact with another phase. The interphase refers to the region between the two phases.

The region of interphasial studies is larger than that indicated by the content of the usual paper about "the double layer". Thus the interphase may extend to several hundred, or even thousand Å (e.g., in dilute electrolytic solutions).

The most studied systems are boundaries between metals and aqueous electrolytic solutions. In addition, the dominating experimental base concerns experiments made with mercury, because with this metal it is possible to easily determine the interfacial tension. Thus Hg played a special role in forming the basis of present models. Among the most studied of these interfaces, after the mercury/aqueous solution system, is that in the mercury/non-aqueous solution, then in the noble metal/ aqueous solution, the metal/molten salt interphase, the semiconductor/ aqueous solution interphase, the insulator/solution interphase, and finally interfaces involving colloids and moving boundaries. Large numbers of other interphases exist which have been little studied. The most important of these is the biological cell/solution interface.

Study of the double layer goes back to Helmholtz, 1879 [1]. This worker pictured the situation between the metal and an aqueous solution as consisting of two planar sheets of charge, one sheet due to an excess charge on the metal, the other one (standing a few Å away into



*C. Gutiérrez and C. Melendres (eds.),*
*Spectroscopic and Diffraction Techniques in Interfacial Electrochemistry*, 1–54.

the solution) due to the ions which would be the partners of those excess charges. As these ions were pictured as a planar sheet, the situation was called a <u>double</u> layer, as shown in Fig. 1.

The base of double layer studies until the 1970's was the remarkable contribution made in the 19th century by W. Gibbs, "the electrocapillary thermodynamics". However, Gibbs' original theory was obscure, and it was not until 1953 that Parsons and Devanathan [2] produced a theory with which most scientists in the field could live.[1]

These original thermodynamic contributions were widely applied by D. Grahame [3]. He made a seminal contribution to the work of the double layer by his many measurements of capacitance as a function of potential for simple, basic systems involving mercury and solutions containing the salts of Group 1A and Group 2A in the periodic table.

On this basis, the so-called "Grahame's theory" was evolved, and the ideas in it taken as applicable to other systems. In this theory, cations are supposed to remain hydrated upon adsorption , and always to be separated from the metal by at least one hydration layer.[2] Grahame developed the idea that anions manage to come into direct contact with the metal and get bound to it in a way which is more than that which would be required by simple electrostatic equivalence. Thus, Grahame's postulate about cation adsorption enabled measurements (involving mercury) of the Gibbsian surface excess of anions to be resolved into two parts, that of the diffuse layer, and that of contact (or "specific") adsorption. Thus cations were regarded as all in the diffuse layer (Fig. 2). Therefore, it was thought that for cations it would be possible to utilize Gouy's theory (which had been earlier suggested for the diffuse layer), and directly from this, deduce the potential of the Gouy-Helmholtz plane, $\phi_{2\text{-}b}$ (Fig. 3).

Thus, thermodynamically valid experimental determination of both the total anion surface excess and the metal charge could be used to discriminate between the cation and anion concentration in the diffuse layer as well as the concentration of adsorbed anions in the inner layer ($\phi_{2\text{-}b}$ was obtained from the knowledge of the <u>cation</u> concentration. The anion concentration in the diffuse layer is, then,

---

[1]The principal difficulty was the application of reversible thermodynamic reasoning to situations in which one of the interfaces was "polarizable" (this term means "no electrons can pass across the interface"). The situation was overcome by Parsons and Devanathan [2]. In their deduction, the correct adjustments of electrons and concentrations were made <u>around</u> the cell, to keep the interfaces in equilibrium, without involving a crossing of the polarizable interface.

[2]Whether cations are separated by one hydrated layer (their own hydration shell), or whether they are separated by this and by another hydrated layer (that on the electrode) is a point which remains undecided. However, it seems difficult to obtain a fitting model consistent with a value of 16 $\mu F\ cm^{-2}$ for the double layer capacitance on the negative side of the p.z.c., unless there is both a layer of water <u>and</u> a hydration layer between the electrode and the ion [4].

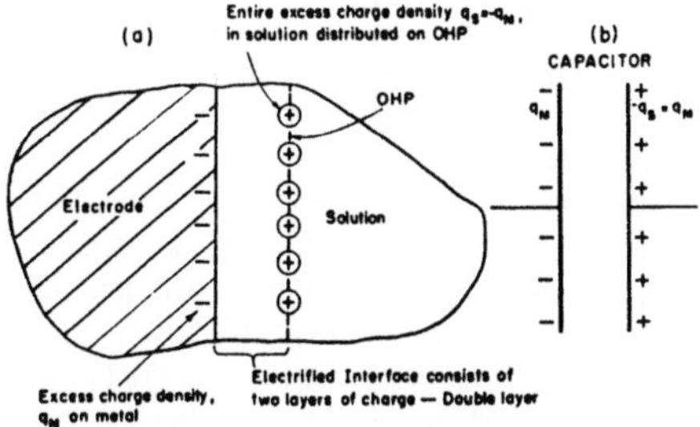


Fig. 1. (a) Helmholtz model of the double layer. (b) A parallel-plate capacitor, the electrical equivalent of the double layer. OHP: outer Helmholtz plane; $q^s$ and $q^m$: excess charge in solution and metal, respectively.

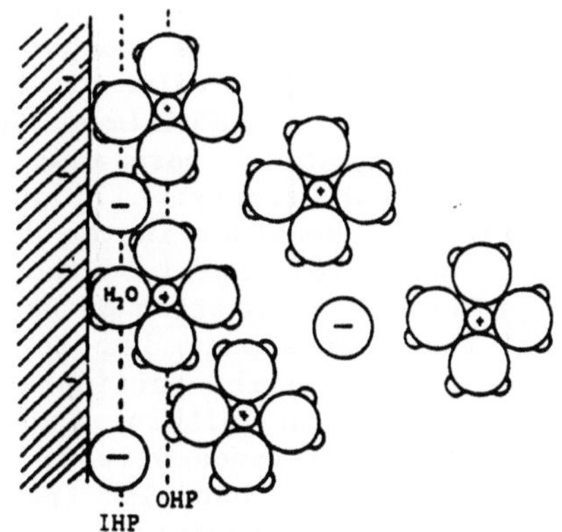


Fig. 2. Grahame model of the interfacial region in the immediate vicinity of the electrode. IHP: inner Helmholtz plane.

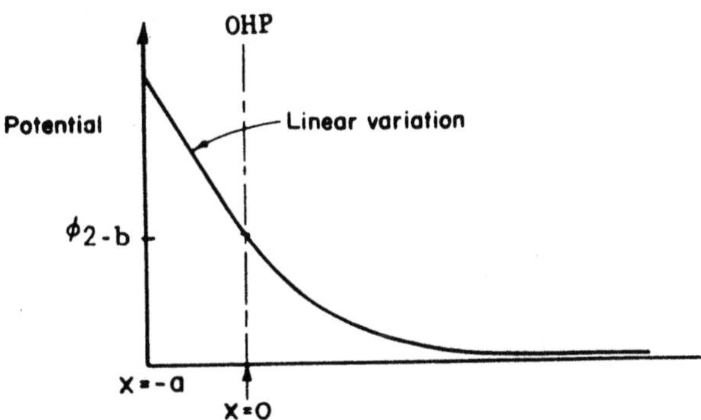


Fig. 3. Schematic representation of the potential ($\phi$) as a function of the distance from the electrode, x, as described by the Stern model.

also determinable from Gouy's model by use of the knowledge of $\phi_{2-b}$). Therefore, distinction between ions which sit in the outer Helmholtz plane, in the inner Helmholtz plane, and in the diffuse layer could be made using Grahame's model (Fig. 4).

However, Grahame's approach to study of the interphasial region was not easily carried over to those involving solids. Thus,

a. Measurement of the interfacial tension at the solid/solution interphase involved great difficulties (see below).

b. The capacitance of the double layer (or polarizable) region for the mercury/solution interface had been extensively measured by Grahame [3]. This capacitance was found to be 16 $\mu$F cm$^{-2}$ for a wide variety of alkaline metal and alkaline earth salts, particularly halides.

Because mercury is a liquid, it had been assumed between around 1950 and 1970 that this capacitance must represent the true double layer capacitance of any metal. Therefore, any observed changes in the value of this capacity would be connected with the difference of real surface area per geometric area which could be found in solids.

Frumkin et al [5] tested this hypothesis by measuring the capacitance of the liquid-gallium/solution interface. Such capacitance, at the double layer region, should have identical value as that for Hg. However, it was found to be different (at the minimum point, 2-3 $\mu$F cm$^{-2}$ greater) (Fig. 5) [6]. Some suggestions were made to explain this difference. For example, Frumkin and Grigoryev [7] suggested it was due to a difference of the dielectric constant in the double layer for Ga and Hg. On the other hand, Conway [8] explained the difference as due to stronger specific adsorption and orientation of water dipoles in Ga than in Hg. Nevertheless, the discrepancy led to increasing reservations concerning the Stern-Grahame model, particularly when a contribution from the Thomas-Fermi length within the metal is considered. Thus, if the double layer capacitance were to consist of contributions not only from the solution side, but also from the metal side, then the Stern-Grahame model was indeed incomplete.

At the present time, two new trends are visible in the fundamental investigations of interfacial structures. It may be that these will lead to revised model of the double layer in the coming decades. They are:

a. The use of direct spectroscopic methods for determining the structure of the interface (see the paper by J. O'M. Bockris and M. Gamboa in this symposium); and

b. The application of quantum mechanical methods to the interface with the resulting accounting of the electron overlap potential difference at the interface [9,10].

Among areas which need attention are not only the basic solid/solution interphases, but particularly the interphases involving biological systems. Then, the object of this article is to "catch the double layer on the wing" in the 1988 view of the situation at the present time.

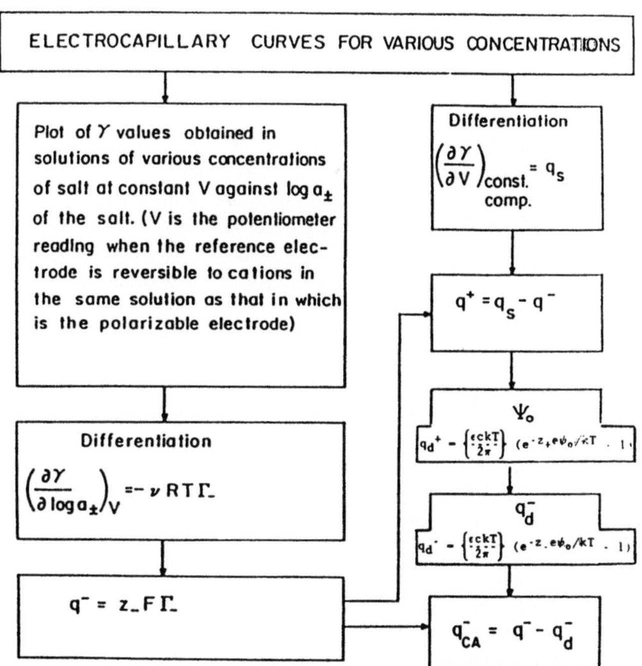


Fig. 4. Grahame model to calculate the contact-adsorbed charge density, $q_{ca}$. $q_d^-$ and $q_d^+$ are the anionic and cationic charges in the diffuse layer, respectively. $\gamma$: surface tension; $\Gamma_-$: anion surface excess; $\Psi o$: potential at OHP; $z_+$ and $z_-$ are the charge on the cation and anion, respectively. $a_\pm^2 = a_+a_-$, where $a_+$ and $a_-$ are the cationic and anionic activity, respectively.

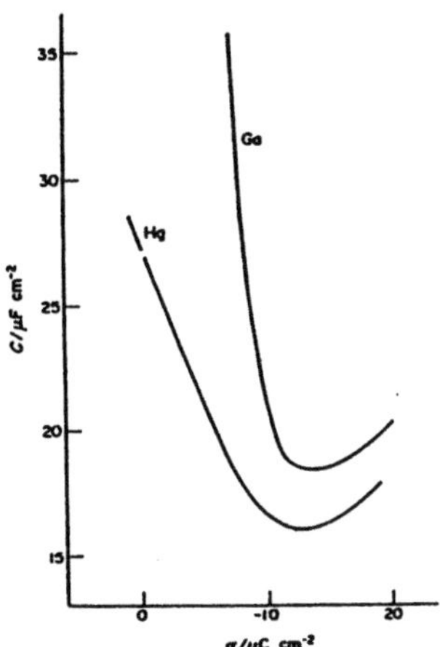


Fig. 5. Differential capacity of liquid-Hg and liquid-Ga electrodes in 0.5 M $Na_2SO_4$.

## 2. The Fading of the Thermodynamic Approach to Investigation of Interphases

The thermodynamic method presents difficulties in application to the solid/solution interface. These arise as follows:

a. There are several methods of attempting to obtain the interfacial tension of solid/solution interfaces in a direct way. For example, in order to determine the dependence of the surface tension, $\gamma$, on the electrode potential, Beck [11] proposed a method based on the change in the length of a stretched metal tape with changing potential. On the other hand, Fredlein and Bockris [12,13] measured the dependence on potential of the bending of a thin glass strip metallized on one side and fastened at one end (Fig. 6) [14]. The number of lines in a diffraction grating attached to the strip passing by the laser spot measured the degree of bending. The dependence of the "differential surface tension" against potential was obtained (Fig. 7) [12].

For a liquid-metal/solution interface, a change of potential or concentration can be assumed to result in a thermodynamically reversible change in the surface tension, leading to a change in area of the meniscus. However, this may not necessarily be so with the solid/solution interface because of the influence of stress.

Thus, at an infinitely long time, a change of, e.g., concentration in the solution or electrode potential should give rise to a change in the area of the interface, which would be dependent only upon the equilibrium value of the surface free energy. In solids, the relaxation of the interatomic forces may take a considerable time. Thus, the reaction due to a change in potential may be caused not only by the change of surface tension, but also by the unrelaxed surface stress.

However, this difficulty may not be as bad as it is sometimes made out to be. Thus, Couchman and Davidson [15] concluded that the correction term for the effect of stress on surface tension would be negligible. The position remains at the present unresolved.

b. It is possible to approach the surface tension of any interface by a double integration of the capacity with respect to potential because of the equation:

$$C = \left(\frac{\delta^2\gamma}{\delta\phi^2}\right) \tag{1}$$

There are three difficulties on the application of this equation to determine $\gamma$:

b.1. It is often difficult to find a region in the capacity-potential relationship in which the double layer capacity is the only one present (pseudo-capacities often come in). On the other hand, it may be necessary to work at frequencies sufficiently high so that the double layer capacity predominates, but low enough to avoid inductance effects. Such difficulties obtrude at semiconductor/solution interfaces, where the double layer capacity is often much bigger (in series mode) than the space charge capacity, so that the former is not seen at any rate at relatively low frequencies.

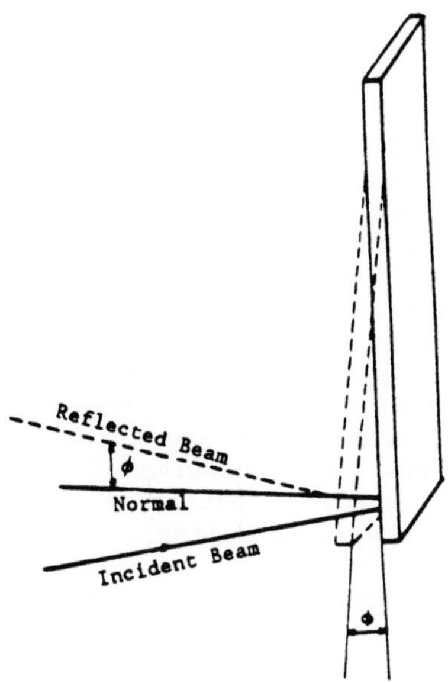


Fig. 6. Measurement of the surface tension of a solid-metal electrode.

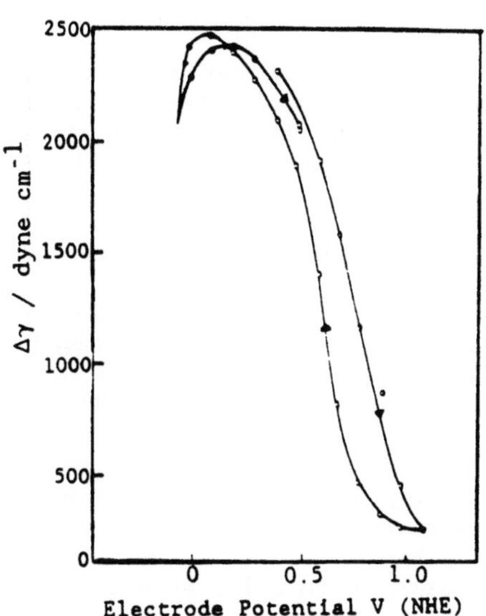


Fig. 7. Change of the differential tension, Δγ, with potential for a gold electrode in 0.1 M KCl solution.

b.2. Two constants are needed in the integration of Eq. (1). One constant involves the potential of zero charge, p.z.c., easily found; but the other involves the $\gamma_{max}$ at the p.z.c. At the first sight, it seems that the integration constants, involved in the evaluation of Eq. (1), would cancel, thus enabling a determination of the surface excess to be made at any potential from capacitance measurements. However, this is not necessarily the case because the p.z.c. and $\gamma_{max}$ are often concentration-dependent. Therefore, the surface excess of the p.z.c. will be concentration-dependent.

b.3. The Grahame's assumptions. It is much less acceptable to assume that cations do not undergo contact adsorption on solids, including semiconductors, than it is on Hg. It may be that a direct test of Grahame's assumptions, with regard to the non-specificity of cation adsorption, could be made by radiotracer determinations, of, for example, a number of alkali cations as a function of potential. The identity of their behavior with respect to potential variation would check the applicability of Grahame's view to solids.

Last of all, there is a certain timeliness in all approaches. The thermodynamic approach to the structure of interfaces, belongs historically to the pre-1970 period. The reasoning involves subtle concepts and implicit assumptions which reduces its clear applicability, particularly to solids.

Instead of trying to make the older Gibbsian analysis work for solids, it is better to graduate to more direct spectroscopic approaches.

## 3. Water Structure at the Interface

In the 1920's, Frumkin [16] mentioned that water might contribute with a dipole potential to the potential drop at the interface, However, considerations of the water structure at the interface were avoided in the use of thermodynamic concepts in much of the Russian work about interface up to 1980's.

In the first molecular approach taken to adsorption, that by J. A. V. Butler [17], no structure was discussed for the water molecules at the interface.

Bockris and Potter [18] introduced a contribution of oriented water dipoles to electrode kinetic phenomena. They portrayed the double layer as having water molecules all oriented in the same direction.

J. T. Law [19] recorded measurements of the surface tension of mercury in contact with water vapor as a function of temperature. He interpreted his measurements in terms of the entropy of the surface layer. He concluded that the entropy of the water layer was consistent only with dimerization of the water molecules. Kemball [20] reached the same conclusion.

A theory was suggested by Watts-Tobin [21], as well as by Mott and Watts-Tobin [22], who proposed a two-state water model (Fig. 8) [21]. The importance of this model is that it seemed to give rise to an explanation of the capacitance hump, which occurs on the positive side of the p.z.c. (Fig. 9) [23]. They explained the hump as due to the

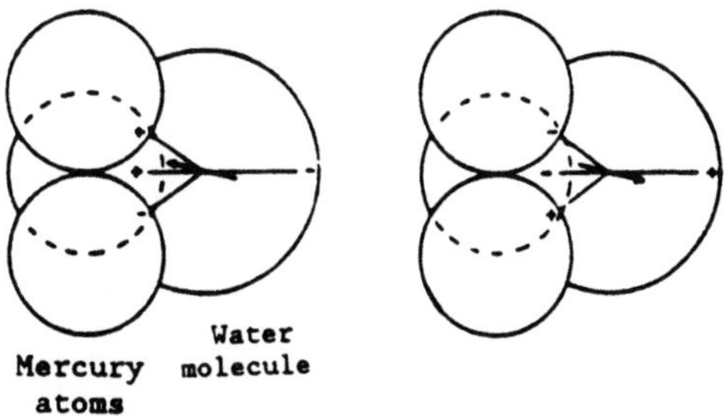


Fig. 8. Original Watts-Tobin model of the "up" and "down" water molecules at the interface.

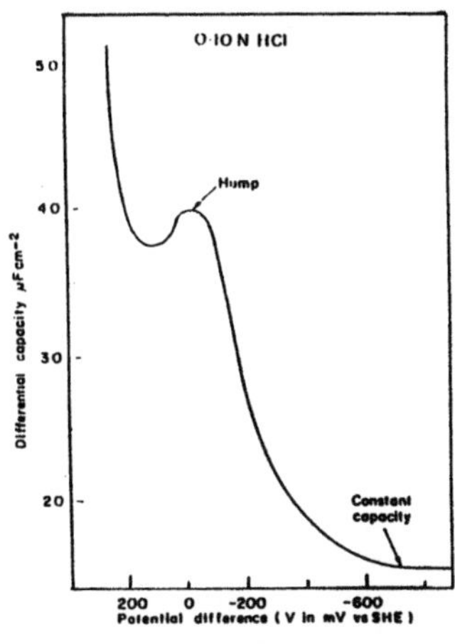


Fig. 9. The experimental capacity potential curve snowing the constant-capacity region

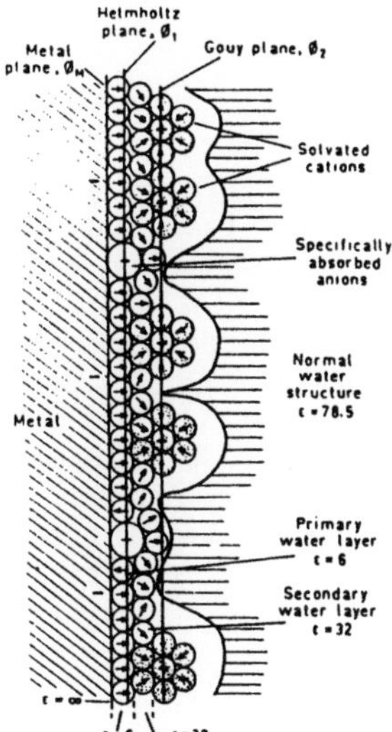


Fig. 10. The BDM model of the double layer

dielectric properties of water: there would be an increase in the effective dielectric constant of the water monolayer in the inner layer, marking the point where the average dipole moment of the water in the layer changes sign. Thus, the hump would be due to the high dielectric constant of water when the field in the inner region is not so strong as to give saturation.

Thus, according to Mott and Watts-Tobin, the structure of water was highly oriented in one direction at the extremes of potential, and reaches a minimum degree of order near the p.z.c. It follows that the capacitance should pass through a maximum when the field vanishes in the inner region. Based on this model, Mott and Watts-Tobin predicted that the hump may not be at the electrocapillarity maximum, or point of zero charge, p.z.c. In the original Mott and Watts-Tobin paper, it is stated that specific adsorption may change the position of the hump to the anodic side of the p.z.c. On the other hand, in the absence of non-symmetrical situation of the dipole moment in water, the water molecules will change their orientation at the p.z.c., which is consistent with the Mott and Watts-Tobin model.

Bockris, Devanathan, and Muller (BDM) [4] put forward an independent and different theory of water molecules in the double layer (Fig. 10). The BDM theory, independently deduced, has two important differences from Watts-Tobin's model. These are the following:

a. The water capacitance is supposed to be in series (and not in parallel) with the ionic contribution to the capacitance, so that:

$$1/C_{DL} = 1/C_{ion} + 1/C_{dipole} \tag{2}$$

where $C_{DL}$ is the double layer capacitance, $C_{ion}$ is the ionic contribution to the capacitance, and $C_{dipole}$ is the water contribution[3]. This equation is derived from the equation due to Lange and Mischenko [24]:

$$\Delta\phi = \Delta\psi + \Delta\chi \tag{3}$$

where $\Delta\phi$ is the Galvani or inner potential difference at the interface, $\Delta\psi$ is the outer potential difference which arises from the free charges at the interface, and $\Delta\chi$ is the dipole potential difference at the interface.

The introduction of Eq. (2) is, of course, an important result. Applied to the Watts-Tobin's model, it implies that when the dipole capacitance reaches a maximum near the p.z.c., its effect upon the observed capacitance of the double layer would be minimal (see Sec. 4).

b. BDM allowed for an interaction between nearest neighbors among water molecules. The effect of this is great, leading to a considerable increase in the capacitance of the double layer due to water molecules.

---

[3]For a parallel mode:

$$C_{DL} = C_{ion} + C_{dipole} \tag{2a}$$

Further, this indicates that the $C_{dipole}$ is less likely to show up at the p.z.c. because the $C_{dipole}$ is in series with the $C_{ion}$.

The value of the $C_{dipole}$ near the p.z.c. depends considerably upon the constant of interaction between the water molecules. In the original BDM paper, the capacitance due to dipoles was considered too high (and thus their effect too low) to be the cause of the capacitance hump (see Sec. 4). The BDM position has been developed by Bockris and Habib [25] who took into account the formation of other types of water molecules in the double layer, in particular dimers, which were supposed not to contribute to the orientational properties of water (see below). Thus, the potential difference due to dipoles is reduced. Calculations corresponding to this model for the dipole capacitance showed that it is even higher than that calculated by BDM. Therefore, in the series mode, the contribution of the dipole capacitance to the overall situation of the observed capacitance would be smaller than that calculated by BDM. Thus, the Bockris and Habib model supports the conclusion from BDM that the dipole capacitance does not affect the overall observed capacitance, in contrast to the results of the Watts-Tobin model.

On the other hand, Damaskin and Frumkin [26] unaware of these conditions, modified the Watts-Tobin model. They proposed that there are chemisorbed water dipoles oriented with the oxygen towards the metal surface, as well as water clusters (which are pictured as associated molecules having a small dipole moment. Therefore, they are oriented either with the electrode field or against it). Fig. 11 depicts this view of the water structure at the interface [27]. This model can be made to reproduce the observed shapes of the capacity-potential curves (Fig. 12) [26], but only if the chemisorbed water molecules are surprisingly assumed to vary exponentially in capacity with potential. Parsons [28] pointed out other defects of this model. For example, the parameters related to chemisorbed water seem to be arbitrary, and the model is unable to predict the capacity rise at negative charges. In addition, the concentration of chemisorbed water molecules is allowed to increase freely with the electrode charge without requiring any change in the concentration of water in the clusters.

In order to overcome these drawbacks, Parsons [28] suggested a four-state water model, where the four states are the two positions of individual water molecules and the two positions of the individual clusters (dipole oriented with the electrode field or against it) (Fig. 13) [27]. In addition, different interaction energies exist between each species (the clusters and free molecules) with the metal and the bulk liquid. By adjustment of the six parameters involved, this model may be made to predict not only capacity curves having shapes similar to those of the experimental ones for Hg, but also satisfactory proportions between minimum and maximum values (Fig. 14) [28]. Also, a capacity rise at negative charges is predicted. This is explained as reorientation of free molecules combined with clusters disruption. However, considering Parsons' principal equation for $C_{DL}$, a total of six adjustable parameters are used. Thus, the model loses credibility.

Among the difficulties of these two last models are the following:

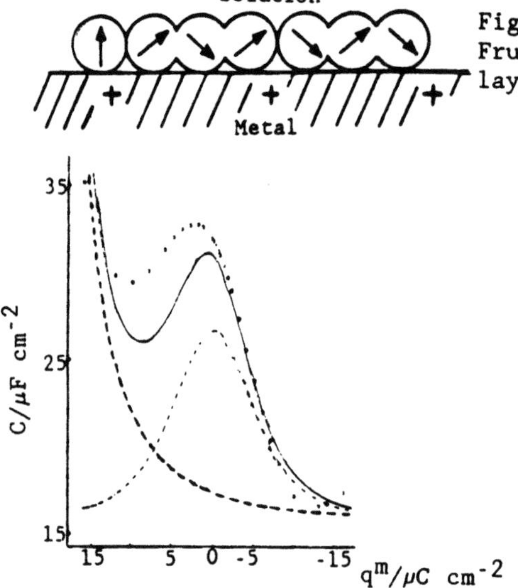


Fig. 11. The Damaskin and Frumkin model of the double layer.

Fig. 12. Experimental and calculated inner-layer capacity-charge curve for Hg in contact with aqueous solution of not specifically adsorbed electrolyte at 0° C. Capacity associated with clusters (-··-), with chemisorbed molecules (---), and total capacity (——), according to the Damaskin and Frumkin model. (o) Experimental points.

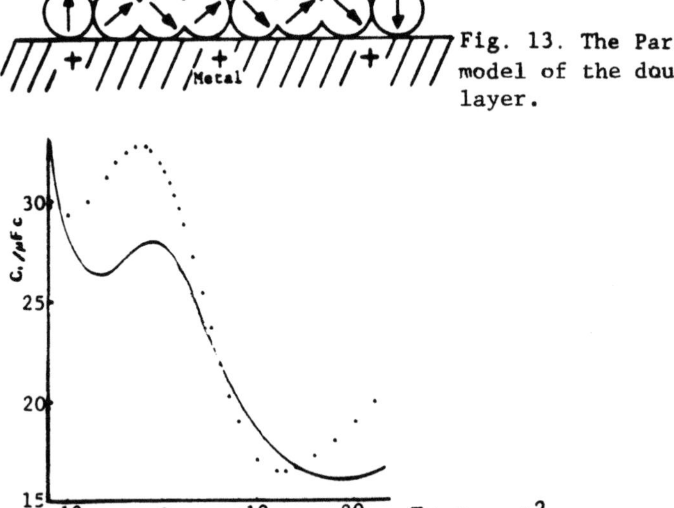


Fig. 13. The Parsons model of the double layer.

Fig. 14. Experimental (Δ) and calculated (——) inner-layer capacity-charge curve for Hg in contact with aqueous solution of not specifically adsorbed electrolyte at 0° C. Theoretical curve according to the Parsons model.

a. It is difficult to discern clearly what it is meant by "clusters". Do these associated molecules turn as one group or turn as individuals? The net dipole moment of a cluster, particularly for more than four water molecules, would be expected to be very small. On the other hand, rotation of an entire cluster would involve a substantial entropy of configuration.

b. Use of many arbitrary adjusted parameters to obtain agreement with experiment. For example, Habib [29] showed that an irrational polynomial (to determine coverage, $\theta$, as a function of potential, V):

$$\theta = A + BV + CV^2 + DV^3 + \ldots \tag{4}$$

could be made to account well with details of $\theta$-V curves for organic adsorption.

Bockris and Habib [25] introduced a three-state water model (Fig. 15) in which they started from the idea of Law [19] that a third component on the surface of the electrode would be a dimer.

Bockris and Habib made a number of unpublished calculations of the energy of monomers, dimers, trimers and tetramers. They found that the total energy (involving the polymer formation as well as different metal/water and water/water interactions) was a minimum for the case of the dimer (the calculation was made by attempting to equilibrate the water molecules on the electrode with the first layer of molecules in the solution). For this reason, as well as from the result obtained by Law [19] and Kemball [20], they concluded that the most likely structure in the double layer was a mixture of free water molecules oriented with the electric field and against it ("up" and "down" water molecules) together with dimers. The calculated ratio was about 3 dimers to 1 up (or down) water molecule[4].

Later, Bewick and Kunimatsu [30] made a potential modulation approach to IR measurements on water on single crystals of platinum, and found 5 peaks which were in the range given by the O-H frequency (Fig. 16). By comparing this work with the work of Luck [31], who studied clusters of water molecules using matrix methods, Bewick and Kunimatsu concluded that these peaks represent various polymer groups (dimer, trimer, etc.) on the surface of the electrode.

By using FTIR, Habib and Bockris [32] made a comprehensive

---

[4]In a discussion of the Bockris and Habib model, Giudelli [27] has misunderstood the equilibria being considered. Bockris and Habib considered all the equilibria of water in the double layer but adduced reasons for that with the solution to control the situation. On this basis, the equilibrium in their calculation is the appropriate one. Giudelli evidently thought the equilibrium on the electrode dominates. He goes on to give an elaborate development of a similar model to that of Bockris and Habib (three-state). However, his calculation contains three adjustable parameters and is aimed largely at explaining capacitance humps. He does not calculate the entropy-charge relation, nor the hump in terms of specific adsorption. The major question if the series calculation is viable remains open.

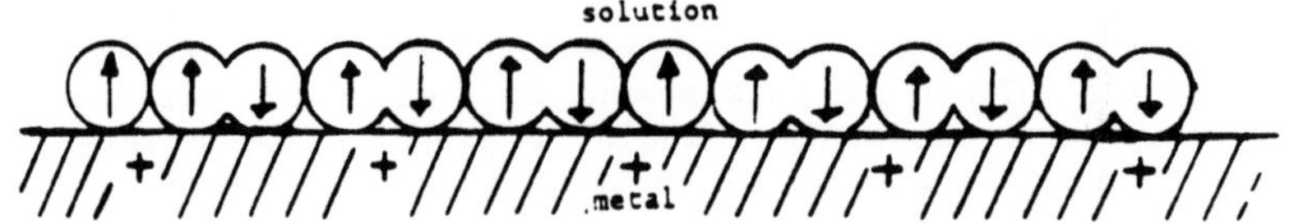


Fig. 15.- The three-state water model of the double layer (Bockris and Habib model).

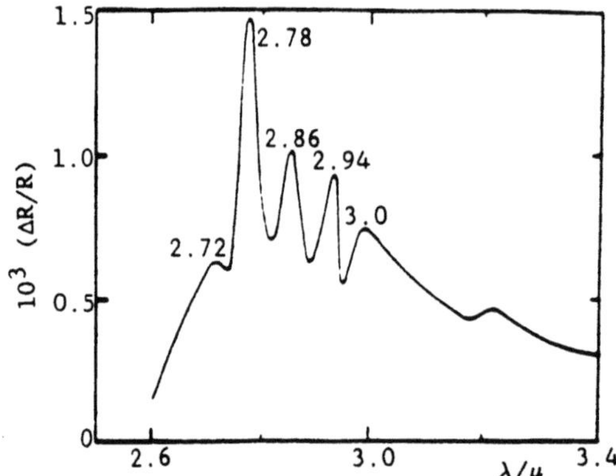


Fig. 16. Spectrum from Pt/1 M $H_2SO_4$ in the O-H adsorption region. Modulation from 0.4 to 0.8 V at 8.5 Hz. The peaks at 2.72, 2.78, 2.86, 2.94, and 3.0 correspond to the signal from monomer, dimer, trimer, tetramer, and polymer water molecules, respectively.

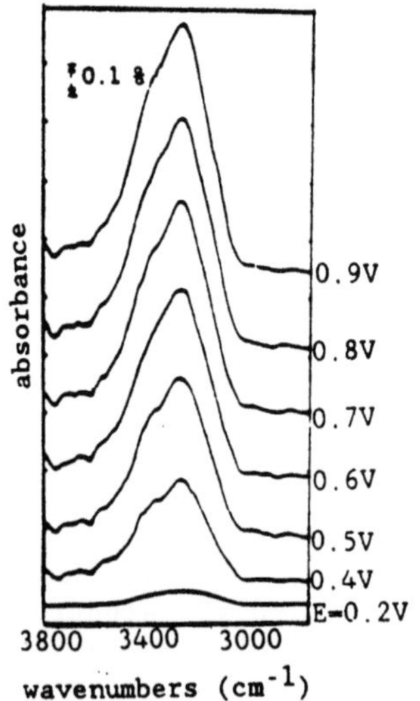


Fig. 17. Differential IR spectra expressed in percent absorbance of O-H vibration in adsorbed water molecules on Pt in a solution of 50% $D_2O$ + 0.1 M NaF at various potentials. Each spectra is the difference between the spectra at the mentioned potential and that at 0.1 V (reference potential).

analysis of the dependence of the O-H signal in water upon potential. Using a NaF solution and a Pt electrode, they obtained the results given in Fig. 17, where the peak for OH is shown as a function of potential. It is seen that there is an increase of the peak height as the potential becomes more positive, away from the p.z.c. for Pt. Fig. 18 shows that there is a similar peak for OD, verifying that the main peak measured is indeed that for OH.[5] Finally, the interpretation to which Bockris and Habib came was the following: the minimum spectroscopic interaction in the infrared of the OH frequency is at the p.z.c. This indicates that the water molecules are "lying down", i.e., their vectors are not interacting with the oncoming light beam. As the potential of the electrode becomes more positive, the theories developed above would suggest that the water molecules begin to stand up (at least, a fraction of those which are not dimers) and thus interact more with the oncoming light beam. Thus, the observed peak height vs. potential graph, as shown in Fig. 19, is in excellent qualitative agreement with this idea. Quantitative agreement is also good, although certainly it does not distinguish between one form of the water standing-up model and another.

Thus, spectroscopic measurements do confirm for the first time that there is a "standing-up" form of water on the surface and that it does vary with potential. What the spectroscopic measurement has not done is distinguish between some kind of polymer on the surface and the standing up waters, i.e., it is not yet possible to get a quantitative analysis of the fraction of waters in the free form.

## 4. Models for the Capacitance Hump

One of the more consistent aspects of the phenomenology of the metal/solution double layer is the existence of a capacitance hump a few milivolts positive to the p.z.c. (Figs. 9, 20 and 21; Table I) [3,23]. The occurrence of this capacitance maximum is not only found for Hg, but also for a number of solid materials (Fig. 22 [33]). In addition, the hump decreases with increase of temperature and is exceeded on the positive branch by a rising capacitance (Fig. 23) [21].

The first attempt to explain the hump was given by Mott and Watts-Tobin [21,22]. They regarded the total capacitance of the interface as consisting of two capacitances added together. The first was supposed to be due to the ions consisting of the double layer - the usual parallel plate capacitance. The second capacitance component was connected with the change of the potential contribution of the water dipoles with the charge on the metal [21].

Watts-Tobin pictured the water molecules as relatively saturated far from the p.z.c. However, near the p.z.c., the water molecules were

---

[5]The much lesser area under the OD peak, for observations on Pt in the solution containing 50% $D_2O$, arises because of the weaker degree of hydrogen bonding for bonds involving D as compared with the similar bonds involving H.

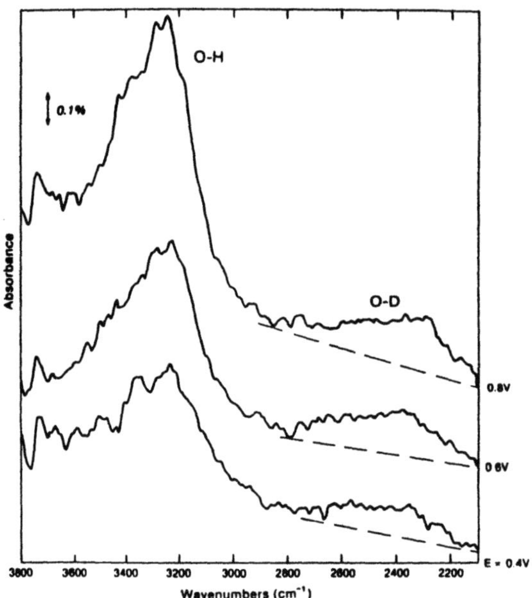


Fig. 18. O-H and O-D vibrational bands at various potentials.

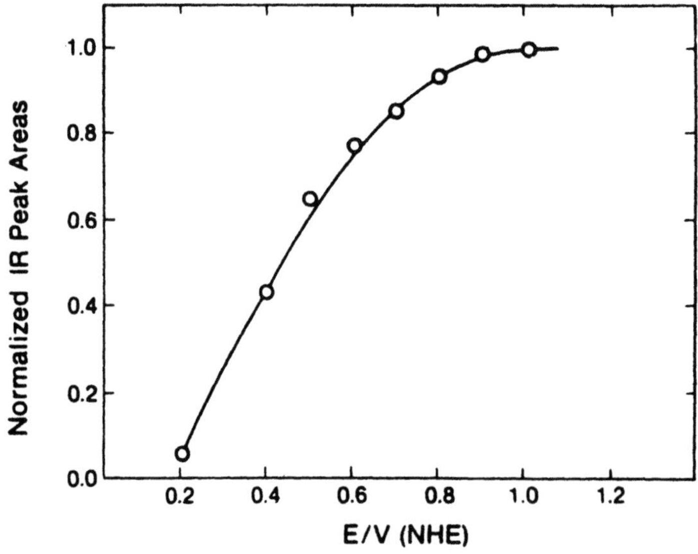


Fig. 19. IR peak areas (normalized to 1 at the maximum for O-H vibration) in adsorbed water molecules on Pt as a function of potential.

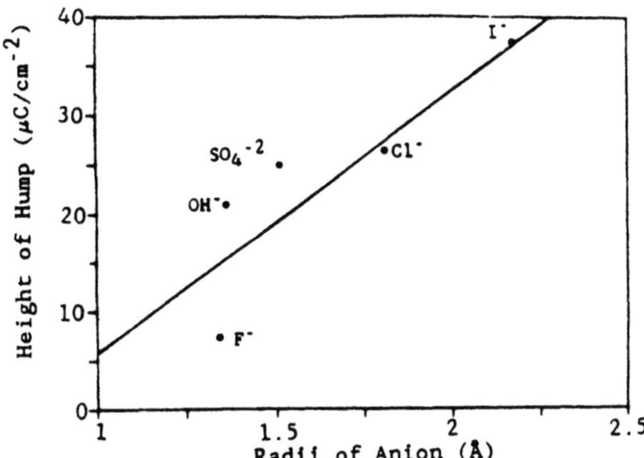


Fig. 20. Relation between the height of the hump and ion size for Hg/aqueous solution interfaces. Concentration: 1 M.

**Table I. Hump potential (relative to the p.z.c.) for aqueous solution at different concentrations.**

| Salt | Concentration (M) | Potential (Volts) |
|---|---|---|
| NaF | 0.01 | 0.40 |
| | 0.1 | 0.28 |
| NaCl | 0.01 | 0.24 |
| | 0.1 | 0.16 |
| | 1 | 0.08 |
| $NaNO_3$ | 0.1 | 0.02 |
| | 1 | 0.00 |
| $HClO_4$ | 0.001 | 0.28 |
| | 0.01 | 0.16 |
| | 0.1 | 0.06 |
| | 1 | -0.04 |

**Table II. Height of the hump and anion radii for aqueous solutions at different concentrations.**

| Anion | Radii (Å) | Concentration (M) | Height of the Hump ($\mu C/cm^2$) |
|---|---|---|---|
| $F^-$ | 1.34 | 0.1 | 3.8 |
| | | 1 | 7.4 |
| $Cl^-$ | 1.81 | 0.01 | 17.6 |
| | | 0.1 | 21.8 |
| | | 1 | 26.4 |
| $I^-$ | 2.17 | 1 | 37.5 |
| $OH^-$ | 1.36 | 1 | 21.0 |
| $NO_3^-$ | 1.23 | 0.1 | 13.5 |
| | | 1 | 14.0 |
| $ClO_4^-$ | 1.48 | 0.001 | 7.2 |
| | | 0.01 | 10.3 |
| | | 1 | 12.8 |

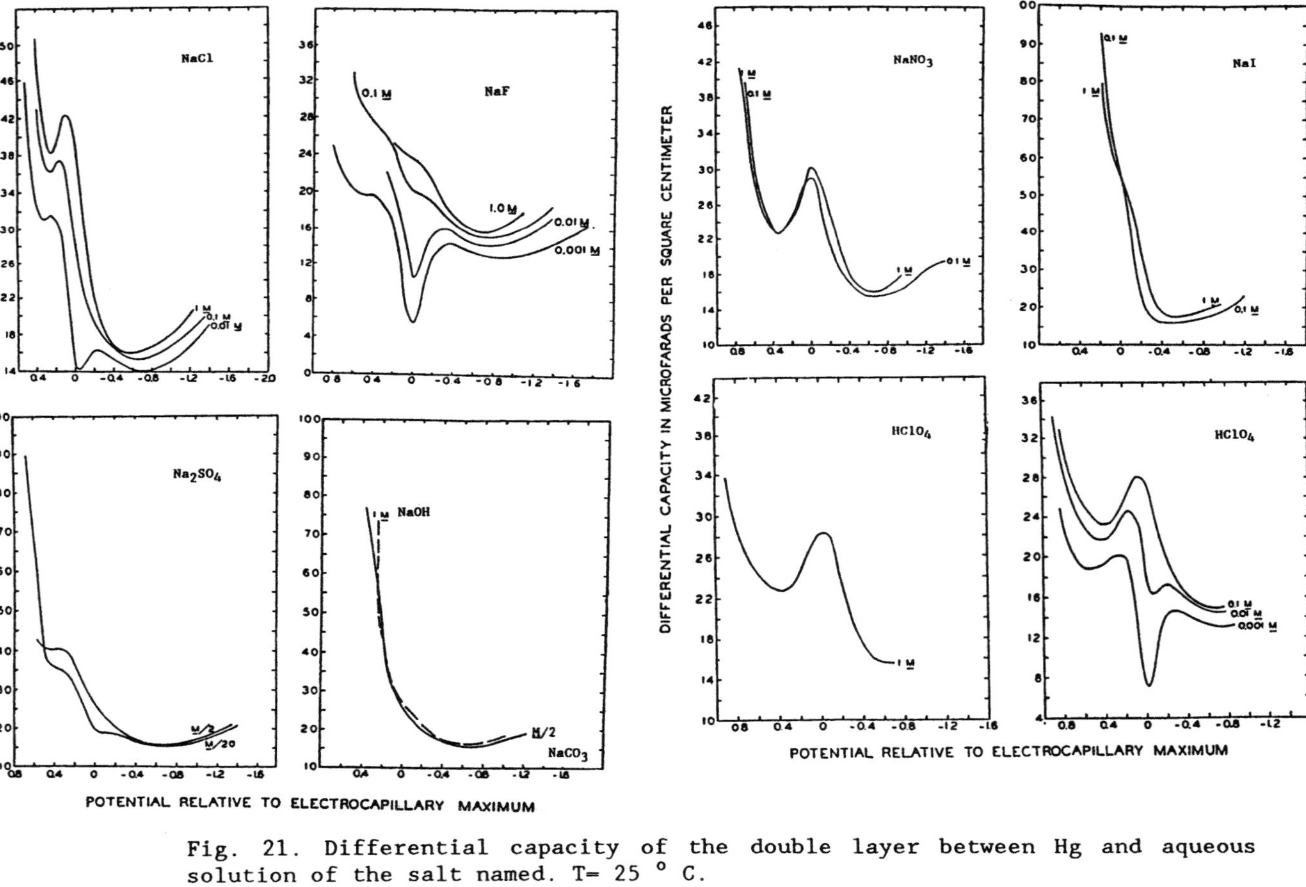


Fig. 21. Differential capacity of the double layer between Hg and aqueous solution of the salt named. T= 25 ° C.

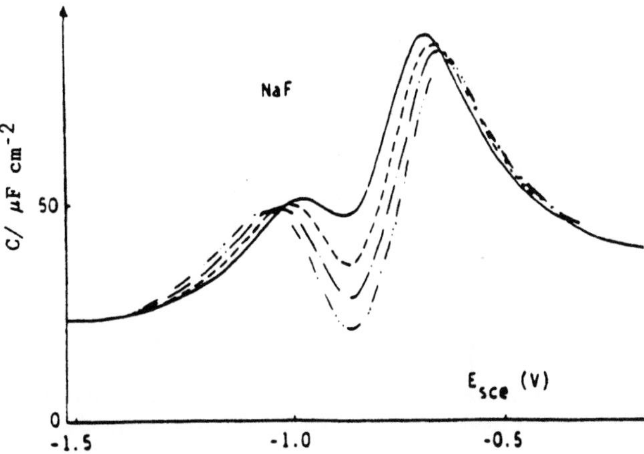


Fig. 22. Capacitance curves for the (100) face of silver in aqueous solutions of $KPF_6$ for different concentrations of electrolyte: (——)100, (---)40, (-·-)20, (-····-)10. Sweep rate = 5mV $s^{-1}$. Frequency = 20 Hz.

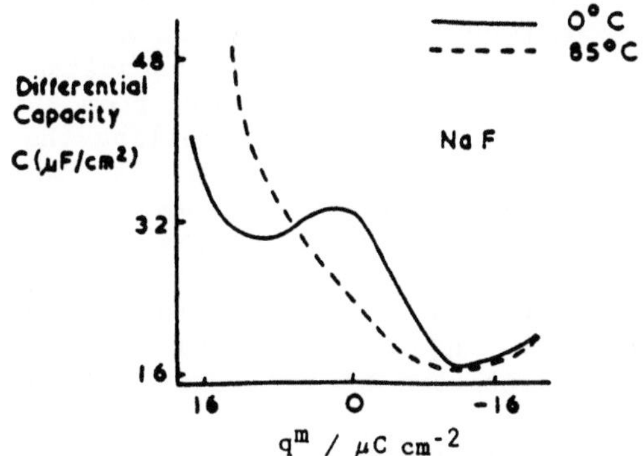


Fig. 23. Capacity values for the double layer between Hg and NaF at two different temperatures.

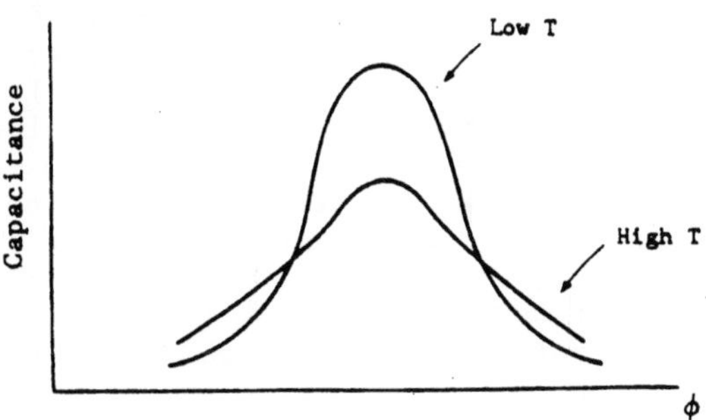


Fig. 24. Calculated capacitance of a monolayer of water as a function of the potential according to Mott and Watts-Tobin model.

no longer firmly adsorbed in either direction, but became mixed in their orientation. This mixture would be analogous to an unsaturated dielectric, and therefore, higher than the dielectric experienced for the inner layer at extremes of positive and negative charge. Thus, this picture of the water molecules gave rise to a maximum at the p.z.c. for the capacitance curve (Fig. 24).

The model first stated implies, of course, that the water capacitance is in parallel with the ionic capacitance, and most workers have found this rather difficult to accept. Thus, differentiation of the well-known Lange-Mischenko equation, (Eq. 3), with respect to the metal charge, $q^m$, would suggest that the observed capacitance of the double layer is related to the ionic and the dipole component in terms of the reciprocals (Eq. 2). Therefore, a model for the capacitances in series is suggested. Although the Watts-Tobin model was accepted by many at first, in recent times most workers accept a series arrangement of the two capacities.

In this case, the original Watts-Tobin capacitance (which is supposed to maximize at the p.z.c.) does not make a reasonable account of the facts. Thus, only if the capacitance grows to a maximum at the p.z.c., and is added to the ionic capacitance (in a parallel mode), it gives a maximum for the observed capacitance. On the other hand, if one follows the Watts-Tobin argument, the capacitance of the double layer becomes numerically almost equal to the water capacitance at charges far from the p.z.c. (for this gets smaller away from the p.z.c., and thus becomes the most important component controlling the double layer for a series mode). However, it is well known that this is not the case, i.e., that the double layer capacitance remains relatively constant in the region of 16 $\mu$C cm$^{-2}$ for Hg on the negative branch, and does not decrease towards very low values on either side of the p.z.c., as the Watts Tobin model applied to a series would suggest.In addition, if one subtracts the dipole capacitance (in a series arrangement), because of the reversed orientation of the water molecules, then the values for $C_{DL}$ become negative.

In 1963, an independent treatment of the capacitance due to water molecules was published by Bockris, Devanathan and Muller (BDM) [4] shortly after the publication of the Mott and Watts-Tobin paper. The BDM theory, which took into consideration lateral interactions between water molecules, predicts a minimum at the p.z.c. for the water capacitance contribution (Fig. 25) (cf. the maximum suggested for the Watts-Tobin model). Nevertheless, the contribution of $C_{dipole}$ to $C_{DL}$ was neglected in the BDM theory because of the high value calculated for the $C_{dipole}$ (even at the p.z.c.) after taking into account the interaction between water molecules. In terms of the three-state water model of Bockris and Habib [25], where only about one third of the molecules are monomers, $C_{dipole}$ at p.z.c. further increases.

Both theories, Mott and Watts-Tobin and BDM, begin with the same equation for the dipole potential difference due to the solvent molecules, $g^s{}_{dipole}$. That is:

$$g^s{}_{dipole} = \frac{4\pi N\mu R}{\epsilon} \tag{5}$$

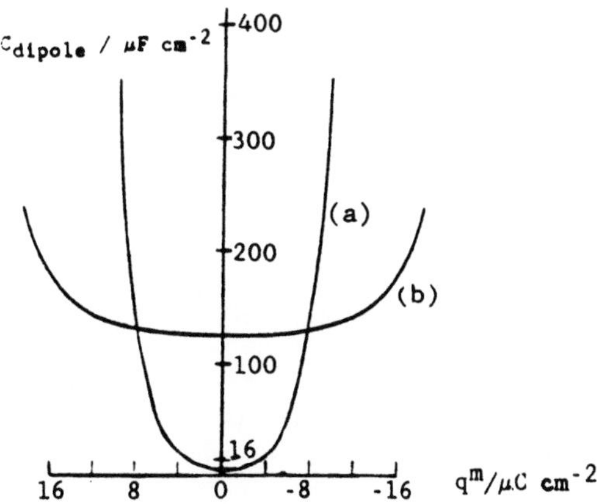


Fig. 25. Capacitance due to water dipoles, $C_{dipole}$, according to the BDM model. (a) without lateral interaction; (b) with lateral interactions. Energy of interaction is 0.5 kT; coordination number is 8.

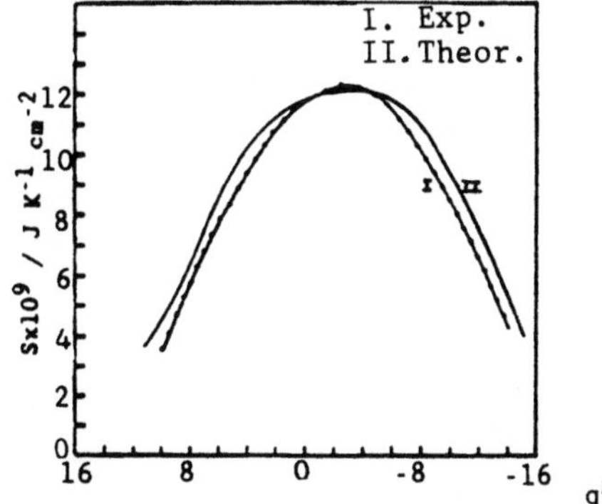


Fig. 26. The solvent excess entropy as a function of the electrode charge. Experimental curve for 0.1 M NaCl solution.

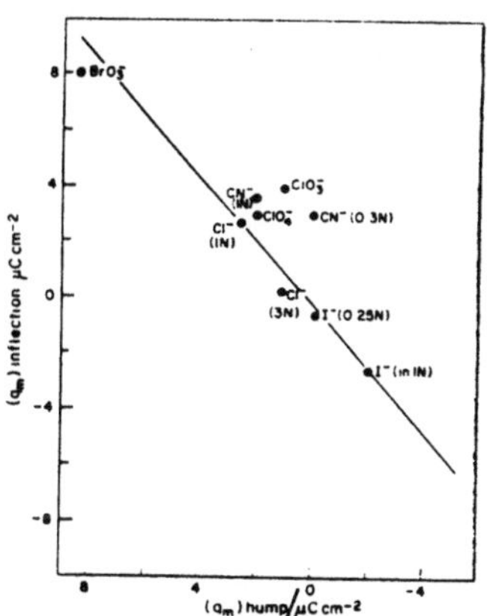

Fig. 27. The location of the hump on the C versus $q^m$ curve; comparison between theory (BDM) and experiments.

where N is the number of water molecules on the electrode surface per $cm^2$, $\epsilon$ is the dielectric constant of the adsorbed water layer, and $\mu$ is the effective dipole moment of a monomer. R is the key point for the development of these theories. For Watts-Tobin, R is given by[6]:

$$R = \tanh \frac{\mu X}{kT} \tag{6}$$

where X is the electric field, k is the Boltzman constant, and T is the temperature. On the other hand, the BDM model considers the potential difference at the interface to consist of the contribution of the Volta p.d., $\Delta\psi$, which is added to the potential difference due to dipole p.d. at the surfaces (Eq. 3). In addition, in the BDM model, the strong interaction between neighboring water molecules was allowed, and gave rise to the equation:

$$R = \tanh \left\{ \frac{\mu X}{kT} - \frac{UcR}{kT} \right\} \tag{7}$$

where U is the lateral interaction energy of a dipole with its nearest neighbor and c is the coordination number (number of adsorbed water molecules that surrounded one adsorbed water molecule). At a first view, the BDM model should give rise to the Mott and Watts-Tobin model when U is equal to zero. However, this is not the case (compare Figs. 24 and 25). This is because of the equation used to calculate X. Watts-Tobin took a simplistic equation for X:

$$X = \frac{\phi}{d} \tag{8}$$

where $\phi$ is the potential drop from the metal to the solution, and d is the width of the inner region. Using Eq. (8) in Eq. (6), and considering the surface charge, $q^m$, expressed as:

$$q^m = \frac{\epsilon}{4\pi d}\phi + \frac{N\mu}{d}\tanh \left\{ \frac{\mu\phi}{kTd} \right\} \tag{9}$$

Thus,

$$C = \frac{dq}{d\phi} = \frac{\epsilon}{4\pi d} + \frac{N\mu^2}{kTd^2} \operatorname{sech}^2 \left\{ \frac{\mu\phi}{kTd} \right\} \tag{10}$$

and

$$C_{dipole} = \frac{N\mu^2}{kTd^2} \operatorname{sech}^2 \left\{ \frac{\mu\phi}{kTd} \right\} \tag{11}$$

which predicts a maximum at the p.z.c. (Fig. 24).

---

[6]Watts-Tobin introduced Eq. 6 without any reasonable explanation. On the other hand, in the BDM model, Eq. 7 is obtained by considering not only the equilibrium condition for the water molecule adsorption, but also the ratio between the net number of water molecules contributing to the dipole p.d., and the total number of water molecules, N, standing on the metal surface [34].

On the other hand, BDM expressed X as:

$$X = \frac{4\pi q^m}{\epsilon} \tag{12}$$

Use of Eq. (12) in Eqs. (7) and (5) gives (for U equal to zero):

$$g^s{}_{dipole} = \frac{4\pi\mu N}{\epsilon} \tanh\left\{\frac{\mu 4\pi q^m}{\epsilon kT}\right\} \tag{13}$$

Thus,

$$\frac{1}{C_{dipole}} = \frac{dg^s{}_{dipole}}{dq^m} = \frac{16\pi^2\mu^2 N}{kT\epsilon^2} \operatorname{sech}^2\left\{\frac{\mu 4\pi q^m}{\epsilon kT}\right\} \tag{14}$$

Therefore,

$$C_{dipole} = \frac{kT\epsilon^2}{16\pi^2\mu^2 N} \cosh^2\left\{\frac{\mu 4\pi q^m}{\epsilon kT}\right\} \tag{15}$$

which predicts a minimum of $C_{dipole}$ at the p.z.c. Eq. (14) is the same expression as obtained from the BDM model for U equal to zero (Fig. 25). This minimum of the $C_{dipole}$ at the p.z.c. could have an important contribution to $C_{DL}$ (for a series mode). However, in the above, lateral interactions have not been considered. When BDM worked with those interactions (Eq. (7)) the value of $C_{dipole}$ was too high (even at the p.z.c.) to contribute to $C_{DL}$. Thus, the turning over of water molecules cannot be regarded as occurring entirely under the field of the electrode acting upon a series of individual dipoles. The dipoles themselves interact electrically, in a fashion which will tend to maintain the "up" and "down" positions. This prevents the repulsive situation arising whereby the molecules are all oriented in the same direction.

Again, it is likely to be bonding between the water molecules in a more complex structure (Sec. 3), and these hydrogen bondings will tend to slow down the turning over of the water molecule. Qualitatively, this means that as $q^m$ changes, $g^m{}_{dipole}$ will be much smaller than it would be according to the original approximation, so the $C_{dipole}$ would be bigger[7] (Eq. 14). Thus, because of the reciprocal addition (Eq. 2), the effect on the capacitance is smaller than had been thought to be the case for the original Watts-Tobin model (passing through a maximum at the p.z.c. and the two capacitances in parallel), or the BDM version of this in which the dipole capacitance passes through a minimum at the p.z.c. Therefore, it appears that the Watts-Tobin model of the hump in

---

[7]Correspondingly, as $q^m$ approaches zero, $\Delta\psi$ will change greatly for small changes of $q^m$. Therefore, $dq^m/d\Delta\psi$ will be small near the p.z.c. However, when $q^m$ is large, change of $q^m$ will make little difference in $\Delta\psi$, as all the molecules will tend to be oriented in one direction. Hence, at extremes of $q^m$, $dq^m/d\Delta\psi$ will be large, as all the water molecules will tend to be oriented in one direction. Hence, at extremes of $q^m$, $C_{dipole}$ will be large and, in the series arrangement, of no importance.

terms of water capacitance is not tenable.

Moreover, the introduction of polymeric forms of water into the water structure, would dilute the monomers, and consequently, there would be a decrease of the dipole p.d. due to the solvent molecules which gives rise to a further increase in the $C_{dipole}$.

Most of the present workers on the double layer continue to implicitly support water rotation models. One of the facts which seems inconsistent with a water rotation model is the position of the maximum of the entropy of water molecules in the double layer. If the hump is due to a maximum disorder of the water molecules on the surface, the position of the hump should coincide with the maximum entropy of the water molecules at the interface. Bockris and Habib [34] calculated the solvent excess entropy as a function of the electrode charge, finding a good agreement with the experimental results (Fig. 26). It is remarkable that the maximum of entropy is in the cathodic side and not in the anodic side of the curve where the capacitance hump is always found. This is an important result because this favors models of capacitance humps which are not dependent upon water molecules orientation[8].

There is an alternate theory of the hump, suggested by BDM, which is founded qualitatively on the fact that the height of the hump depends upon the degree of specific adsorption. Previously to this theory, Devanathan [35] had related the measured differential capacitance to $dq_{ca}/dq^m$, the rate of charge changes with the charge on the metal, at which specifically adsorbed ions enter in the compact layer. Thus, the development of this theory was carried out by considering the effect of specific adsorption on the increase of the capacity that would result from the variation of specific adsorbed charge on the double layer as a function of $q^m$. The model is based on the following experimental tendencies [4]:

a. The height of the hump increases with the radius of the anion and with the ionic concentration (Table II and Fig. 20) [3].

b. The charge at which the hump appears (or in other words, the hump potential relative to the p.z.c.) decreases (becomes less positive) with the increase of ion size and concentration (Table I).

These general tendencies can be observed in Fig. 21, which shows some of the result obtained by Grahame [3].

Mott and Watts-Tobin [22] attempted an explanation of the fact that the hump is more marked in the presence of specific adsorption. Their qualitative explanation is that the effective dipole moment of water molecules in the Helmholtz layer is larger when this layer is wider (which is an effect of the specific adsorption).

The experimental facts (Tables I and II, and Figs. 20 and 21) on the anionic properties are difficult to reconcile with a water rotation model. On the other hand, according to the BDM model [4], a dependence of the magnitude of the hump on the anion is consistent with the interpretation of the hump as due to the variation of the specific

---

[8]To make the Watts-Tobin model work, one has to believe both in a parallel arrangement of the water and ion capacitances and in Eq. (8).

adsorption with metal charge.

In addition, if one plots the charge on the surface as a function of charge on the metal, one indeed obtains an inflection. It is easy to show that this inflection should correspond to the capacitance hump, according to the specific adsorption theory of the hump. Thus, a test arises for the specific adsorption theory of the hump, and is shown in Fig. 27. It seems to be difficult to deny that, for the ions shown, the charge on the metal where the humps occurs is near to that at which the inflection of the $q_{ca}$-$q^m$ graph occurs. It is this inflection which is supposed to indicate the capacitance hump on the specific adsorption model.

A more complete version of the BDM theory was worked out by Bockris and Habib [36]. They utilized the single imaging model which was developed after showing that multiple imaging models were not numerically significant as compared with the single imaging ones (this depends on the fact that the boundary between the Helmholtz and Gouy layer is not a sharp one). Thus, the Bockris and Habib model on the capacitance hump differs importantly from the BDM model because it takes into account not only the repulsion between anions on the surface (the proposed physical reason for the hump), but also the attraction which would occur at higher coverage between anions, due to dispersion interactions. Therefore, it successfully predicted the observed increase of the capacitance on the positive side of the hump, leading to an increase of the capacitance at higher anodic potentials.

Thus, the anion adsorption interpretation seems to be favored by these qualitative correlations to anion properties and quantitative correlations to the properties of inflections of humps. Therefore, the mechanism of humps seems difficult to explain in terms of water molecules so long as one adheres to the series model for the water molecule capacitance and that of the ions (Table III).

Nevertheless, there are difficulties in accepting the specific adsorption theory of the hump. The main one is that the hump exists for NaF-containing systems. At first, it seemed that this fact made the specific adsorption theory untenable because $F^-$ ion was thought not to adsorb specifically. However, it is now known that the $F^-$ ion adsorbs specifically, although not till about +6 $\mu C\ cm^{-2}$, according to both the Grahame mode of calculation and measurements. However, one must now consider that the Grahame's stance is questionable (particularly, in respect to the application of Gouy's theory for the calculation of the charge in the diffuse layer at concentrations far above those at which such models work for activity coefficient calculations).

### 5. Potentials at Interphases

Physical electrochemists are often embarrassed by the fact that, when asked the numerical value of the potential to which they refer, they admit that numerical values of most of the potentials discussed cannot be known by measurement. A cell potential is the most frequently used potential in electrochemistry and the only one known with thermodynamic soundness. However, some others can be known by means of either

Table III. Comparison of two mechanism of the hump.

| FACT | WATER ROTATION MODEL | SPECIFIC ADSORPTION MODEL |
|---|---|---|
| The entropy of water molecules in the double layer passes through a maximum on the negative side of the p.z.c. | Theory would predict a maximum on the positive side. | Consistent with an anion-caused positive charge for the position of the hump. |
| Hump increases with concentration of anion. | Interpretation unclear. | Follows clearly from the model. |
| Height of hump increases with radius. | Not immediately interpretible in this model. | Follows obviously from model. |
| Hump occurs always positive to the pzc even for solid materials such as silver and gold. | One might expect a much greater variation of the position of the hump because water would bind differently to the various metals. | Specific adsorption follows at once from the model. |
| Probe molecule pyrazine finds maximum disorder of water at $-3\mu$ coul $cm^{-2}$. | Inconsistent with water as the origin of the hump (on the positive side). | Fact is consistent with the anion-caused hump. |
| Charge in hump is same as charge for inflection point ($q_{ca}$-$q^m$ relation) | Inconsistent. | Consistent. |
| At potentials anodic to the hump, C increases rapidly. | Does not follow for model. | Follows for model (if dispersion force is included). |

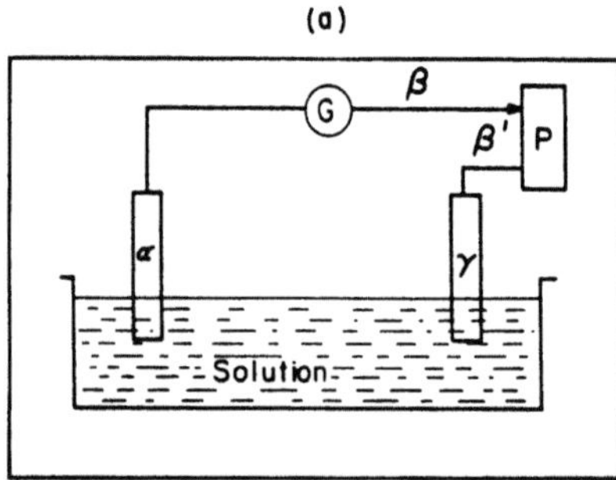


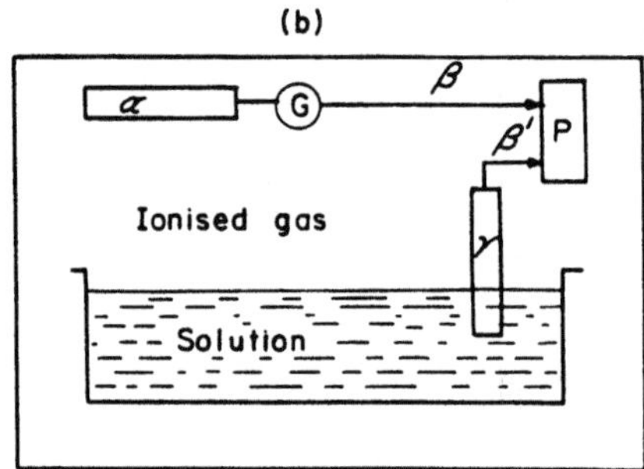


Fig. 28. Measurement of the Volta potential, ${}^{m}\Delta^{s}\psi$. (a) Electrode inside the solution. (b) Electrode outside the solution. G: galvanostat; P: potentiostat. $\alpha$, $\beta$, and $\gamma$ represent different phases. $\beta$ and $\beta'$ are made of the same material but are at different potentials.

experiments or (more doubtfully) by means of calculations involving models.

## 5.1. THE OUTER POTENTIAL OR THE VOLTA POTENTIAL.

The outer potential or Volta potential of a phase is the electrostatic work done by bringing a charge from infinity to a point just outside the surface, far enough away from it so that the image forces do not have a significant effect upon the potential.

The difference in outer potential between two chemically different phases is measurable (since it involves the work required to move a test charge through vacuum from one phase to the other). Thus the value of the outer potential of a phase may be found if the potential of the test charge at infinite distance from the phase is taken as zero.

Bockris and Reddy [23] describe a method (originally by Klein and Lange [37]) to determine the difference in outer potential between a electrode and solution, ${}^{m}\Delta^{s}\psi$. Two measurements of a cell potential are required: one is taken when the electrode is in contact with the solution under totally polarizable conditions. The other one is recorded when the electrode is separated from the solution by a space filled with an ionizing unreactive gas a low pressures (Fig. 28). The theory [23] shows that ${}^{m}\Delta^{s}\psi$ is the difference of these potentials.

## 5.2. POTENTIALS OF ZERO CHARGE, p.z.c.

This is the most commonly measured potential at an electrode/solution interface, apart from the thermodynamic cell potentials of classical electrochemistry. The principal advantage of the p.z.c. is that it removes the contribution to the potential at the interface due to charge on the electrode. At the interphase:

$$ {}^{m}\Delta^{s}\phi = {}^{m}\Delta^{s}\psi + {}^{m}\Delta^{s}\chi \tag{16} $$

where ${}^{m}\Delta^{s}\phi$ is the inner (or Galvani) potential difference and ${}^{m}\Delta^{s}\chi$ is the difference of the dipole and Thomas-Fermi potential.

At the p.z.c. the charge on the metal is zero ($q^{m}=0$). Therefore, ${}^{m}\Delta^{s}\psi$ is equal to zero (${}^{m}\Delta^{s}\psi = 0$). Thus,

$$ {}^{m}\Delta^{s}\phi = {}^{m}\Delta^{s}\chi \tag{17} $$

## 5.3. THE SO-CALLED ABSOLUTE POTENTIAL

Consider the electrochemical cell shown in Fig. 29 [38]. $V_m$ is the potential exerted by the potentiometer to cause zero current to flow; m and m′ are of the same metal as the test electrode, but differ in potential by the potential exerted by the potentiometer.

For zero current:

$$ \bar{\mu}_e^{m} - \bar{\mu}_e^{m'} + nFV_{cell} = 0 \tag{18} $$

where $\bar{\mu}_e^{m}$ and $\bar{\mu}_e^{m'}$ are the electrochemical potentials at m and m′,

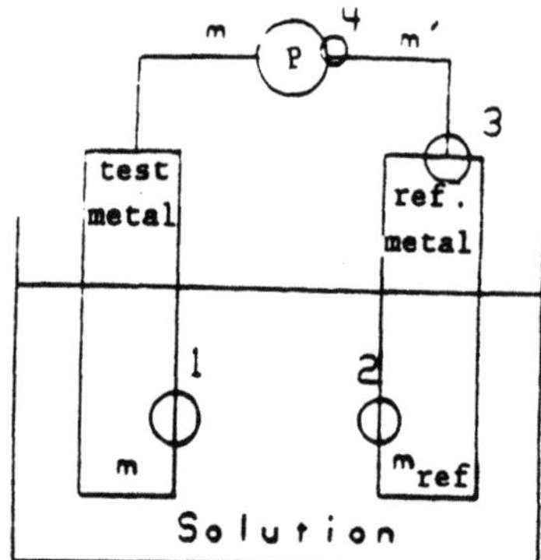


Fig. 29. Schematic diagram of the electrochemical cell showing various interfaces involved in the measurement of the electrode potential.

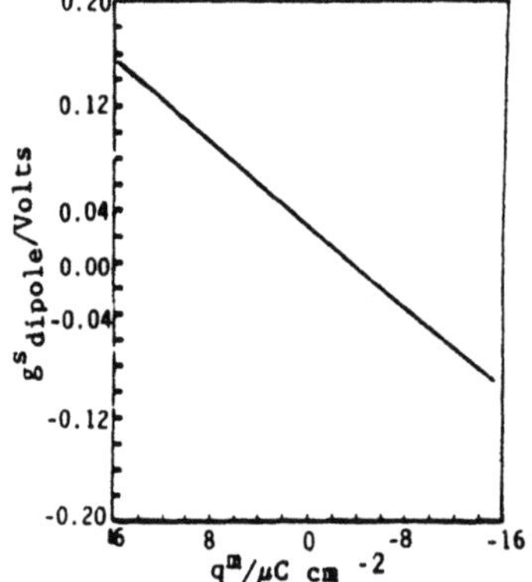


Fig. 30. The surface potential of water at the Hg/solution interface as a function of the electrode charge.

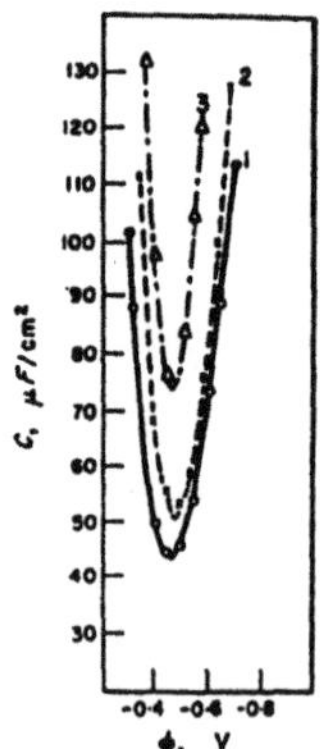


Fig. 31. Capacitance curves of a Pb-electrode/molten salt. 1, NaCl, 820 °C; 2, NaBr, 800 °C; 3, NaI, 800 °C.

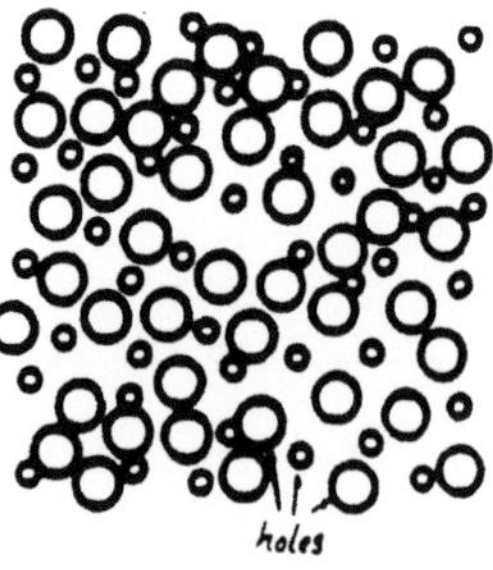


Fig. 32. Probable structure of the molten salt.

respectively; F is the Faradaic constant, and n is the number of transferred electrons. Because m and m' are of the same metal:

$$\mu_e^m = \mu_e^{m'} \tag{19}$$

where $\mu_e^m$ is the chemical potential of an electron in the phase m. On the other hand:

$$\bar{\mu}_e^m = \mu_e^m - nF\phi^m \tag{20}$$

where $\phi^m$ is the inner (or Galvani) potential at m. A similar relation to Eq. (20) holds for m'. Combining Eqs. (18)-(20) one has:

$$V_{cell} = {}^{m}\Delta^{m'}\phi \tag{21}$$

Thus for the cell in Fig. 29:

$${}^{m}\Delta^{s}\phi - {}^{m'}\Delta^{ref}\phi + {}^{s}\Delta^{ref}\phi = V_{cell} \tag{22}$$

where ${}^{m}\Delta^{s}\phi$, ${}^{m'}\Delta^{ref}\phi$, and ${}^{ref}\Delta^{s}\phi$ are the Galvani potential at the m/solution interface, at the m'/reference electrode interface, and at the reference electrode/solution interface, respectively.

At the equilibrium between the phase m' and the reference electrode (interface 3), one has:

$$\bar{\mu}_e^{m'} = \bar{\mu}_e^{ref} \tag{23}$$

Using Eq.'s (18) and (19), one obtains:

$$nF\,{}^{ref}\Delta^{m'}\phi = \mu_e^{ref} - \mu_e^m \tag{24}$$

and using Eq. (24) in Eq. (22), one obtains:

$$nFV_{cell} = (nF\,{}^{m}\Delta^{s}\phi - \mu_e^m) - (nF\,{}^{ref}\Delta^{s}\phi - \mu_e^{ref}) \tag{25}$$

Thus, it is possible to conceive the potential of a cell as consisting of two series of contributions: one characteristic of the metal as the test electrode, and the other one characteristic of the reference electrode. It seems reasonable to regard _each of the bracketed terms as corresponding to the concept of "electrode potential"_.

There are some workers who regard the identification of the quantities in the brackets in Eq. (25) as ill-named. Thus, in the minds of many physical chemists, the "electrode potential" is identified with the Galvani potential difference, ${}^{m}\Delta^{s}\phi$, (the actual electrical potential difference across the interphase involving a solution). The use of Eq. (25) to represent "the electrode potential" mixes up two quantities, a Galvani potential difference together with a _chemical_ potential, and this raises semantic discussion.[9]*

Thus, considering Eq. (25), it is possible to work out a series of so-called absolute or vacuum scale electrode potentials, $({}^m\Delta^s\phi - \mu_e{}^m/nF)$, if it is possible to obtain a numerical value for the last term, $({}^{ref}\Delta^s\phi - \mu_e{}^{ref}/nF)$ for a given reference electrode (i.e. the hydrogen reference electrode) (see discussion of this quantity below). Nevertheless, the "absolute electrode potential" obtained in this way, may physically represent rather complex quantities.

It has been pointed out by Frumkin and Damaskin [40] that one could add a constant K to each part of the Eq. (25) without consequences:

$$nFV_{cell} = (nF{}^m\Delta^s\phi - \mu_e{}^m + K) - (nF{}^{ref}\Delta^s u - \mu_e{}^{ref} + K) \quad (26)$$

Thus, if the reference electrode is at the vacuum scale:

$$nFE_{(vac\ scale)} = nF{}^m\Delta^s\phi - \mu_e{}^m + K \quad (27)$$

Kanevsky [41] suggested that K may be equal to $F\phi^s$. This assumption would lead to the following relation:

$$nFE_{(vac\ scale)} = nF\phi^m - \mu_e{}^m = -\bar{\mu}_e{}^m \quad (28)$$

Khan et al [38] argued that the addition of K seems arbitrary because no potential term was left out in the derivation of Eq. (25). In addition, there is no stated reason which justifies K being taken as $F\phi^s$. This last point was also criticized by Trassati [42] because the choice of K as $F\phi^s$ reduces the electrode potential of the metal to the electrochemical potential of the metal (Eq. (28)). Therefore, the metal solution interface is not explicitly involved.

## 5.4. THE SO-CALLED VACUUM SCALE (OR ABSOLUTE) HYDROGEN ELECTRODE POTENTIAL

Bockris and Argade [39] made a deduction of the value of the quantity $(F{}^{ref}\Delta^s\phi - \mu_e{}^{ref})$ for the hydrogen reference electrode. This quantity is used for the evaluation of the so-called absolute electrode potential of the last section. They used two different paths to obtain the result, and the final values are quite close. One of the ways is based on thermodynamic relations of the potential drops between the different components of the cell. The other one is based on the Born-Haber cycle for the hydrogen reaction.

---

[9]On the other hand, Bockris and Argade [39] clearly worked out the quantity ${}^m\Delta^s\phi$, and identified this with the idea of the absolute potential.

*For a more complete analysis of these various terms and the meaning of potential, absolute potential, electrode potential, etc., see Khan, Kainthla, and Bockris [38].

The first path was developed as follow. At the p.z.c., the Galvani potential difference at the metal/solution interface, ${}^{m}\Delta^{s}\phi$, is expressed as Eq. (17). ${}^{m}\Delta^{s}\chi$ can be expressed in terms of the surface potentials of the free phases (in contact with vacuum: $\chi^{m}$ for the metal; and $\chi^{s}$ for the solution), as well as an interaction term, $\delta\chi$. The last term takes into account the changes in the orientation of the solvent dipoles and the electron overlap at the metal surface when the two phases are brought in contact which each other. Thus,

$$ {}^{m}\Delta^{s}\chi = \chi^{m} - \chi^{s} + \delta\chi \qquad (29) $$

On the other hand, at the p.z.c. [39],

$$ -\mu_{e}{}^{m} + F\chi^{m} = -\alpha^{m} = \Phi^{m} \qquad (30) $$

where $\alpha^{m}$ is the real potential of the electron in m and $\Phi^{m}$ is the work function. Using Eqs. (29) and (30) in Eq. (25) at the p.z.c., one obtains that [39]:

$$ [V_{cell}]_{q=0} = \Phi^{m}/F - \chi^{s} + \delta\chi - ({}^{ref}\Delta^{s}\phi - \mu_{e}{}^{ref}/F) \qquad (31) $$

Thus,

$$ -({}^{ref}\Delta^{s}\phi - \mu_{e}{}^{ref}) = [V_{cell}]_{q=0} - \Phi^{m}/F - \chi^{s} + \delta\chi \qquad (32) $$

It is interesting to see that $({}^{ref}\Delta^{s}\Phi - \mu_{e}{}^{ref})$ is independent on the material of the reference electrode and does not influence the value of the p.z.c. [39].

Choosing the NHE as the reference electrode, the reversible reaction is the following:

$$ H_2 \rightleftharpoons 2H^{+} + 2e \qquad (33) $$

Under standard conditions,

$$ \frac{1}{2}\mu_{H_2}{}^{o} = \bar{\mu}_{H^+}{}^{o} + \bar{\mu}_{e}{}^{ref} \qquad (34) $$

Thus,

$$ \frac{1}{2}\mu_{H_2}{}^{o} = \mu_{H^+}{}^{o} + F\phi^{s} + \mu_{e}{}^{ref} - F\phi^{ref} \qquad (35) $$

Therefore,

$$ -({}^{ref}\Delta^{s}\phi - \mu_{e}{}^{ref}/F) = (\frac{1}{2}\mu_{H_2}{}^{o} - \mu_{H^+}{}^{o})/F \qquad (36) $$

where $\mu_{H}$ and $\mu_{H^{+o}}$ are independent on the metal. Eq. (36) is consistent with the independence of the "absolute hydrogen electrode potential" upon the material used for the reference electrode.

The value of the potential of the reference electrode, (${}^{ref}\Delta^{s}\phi - \mu_e{}^{ref}/F$), was found by using the values of $[V_{cell}]_{q=0}$, $\Phi^m$, and $\delta\chi$ for Hg in Eq. (32). This is the estimated value of NHE as the reference electrode (Eq. (36)), and its calculated value was 4.76 V [39].

This value was compared with the one obtained when a thermodynamic cycle is use for the hydrogen reaction [39]:

$$\begin{array}{ccc} H_{(g)} & \xrightarrow{\ I\ } & H^+_{(g)} + e_{(g)} \\ \Delta G_d \uparrow & & \downarrow \Delta G_{h,H^+} \quad \downarrow \mu_e \\ \frac{1}{2}H_{2(g)} & \xrightarrow{{}^{ref}\Delta^{s}\phi F} & H^+_{(aq)} + e_{(ref)} \end{array} \tag{37}$$

where $\Delta G_d$, I, and $\Delta G_{h,H^+}$ are the free energy of dissociation, ionization and hydration, respectively. Thus,

$$F\,{}^{ref}\Delta^{s}\phi = \Delta G_d + I + \Delta G_{h,H^+} + \mu_e{}^{ref} \tag{38}$$

Therefore,

$$-(F\,{}^{ref}\Delta^{s}\phi - \mu_e{}^{ref}) = -(\Delta G_d + I + \Delta G_{h,H^+}) \tag{39}$$

From Eq. (39), a value of 4.78 V was found for (${}^{ref}\Delta^{s}\phi - \mu_e{}^{ref}/F$) [39].

Trassati [42,43] modified this numerical value and suggested it to be 4.31 V based on a revised value of the free energy of hydration of the proton and chemical potential of the metal. Many electrochemists recognize some uncertainty in the value of the quantity concerned. That is (cf. Eq. (25)):

$$E_{NHS} = E_{(vac\ scale)} - 4.5 \tag{40}$$

where $E_{NHS}$ is the cell potential respect to the hydrogen reference electrode.

## 5.5. THE DEDUCTION OF OTHER POTENTIALS IN THE INTERPHASE

According to Guggenheim [44] the experimental determination (as distinct from calculation) of absolute potentials at interfaces is impossible. However, it <u>is</u> possible to calculate interesting potentials pertinent to interphases, such as the following:

5.5.1. *The Galvani Potential Difference*, ${}^{m}\Delta^{s}\phi$. Bockris and Argade [39] estimated the absolute potential across a $Na/Na^+$ standard reversible electrode. From Eq. (31):

$$^{Na}\Delta^{s}\phi_{ref} = -[V_{cell}]_{q=0} + \Phi^m/F - \chi^s + \delta\chi + \mu_e{}^{Na}/F \tag{41}$$

Using $[V_{cell}]_{q=0}$, $\Phi^m$, $\delta\chi$ for Hg, $\chi^s$ as 0.2 V, and the Fermi energy of sodium as -3.2 eV, the estimated value of ${}^{Na}\Delta^{s}\phi_{ref}$ is -1.24 V.

5.5.2. *Calculation of the surface potential difference,* $^m\Delta^s\chi$. The surface potential difference, $^m\Delta^s\chi$, is equal to $^m\Delta^s\phi$ at the p.z.c. (Eq. 17). Therefore, if the value of $^m\Delta^s\phi$ is known at the p.z.c., $^m\Delta^s\chi$ can be calculated.

5.5.3. *Dipole Potential Difference Due to Solvent Molecules,* $g^s{}_{dipole}$. An estimate of the dipole potential difference can be made for a given potential for a given model of the interface. According to Bockris and Habib's three-state water model [25], only monomers, which appear to cover one third of the electrode surface, contribute to the dipole potential difference. Thus, the potential contribution by the water molecules, $g^s{}_{dipole}$, is given by Eq. (5).

On the other hand, the Bockirs and Habib model introduced a modification of the BDM model. This was done by considering the difference in the standard free-energy change associated with the adsorption of "up" water molecules, $\Delta G^c\uparrow$, and that energy of adsorption for the "down" water molecules, $\Delta G^c\downarrow$. Thus, with this modification, R is given by:

$$R = \frac{\exp[(-\Delta G^c\uparrow + b)/kT] - \exp[(-\Delta G^c\downarrow - b)/kT]}{\exp[(-\Delta G^c\uparrow + b)/kT] + \exp[(-\Delta G^c\downarrow -b)/kT]} \tag{42}$$

where

$$b = \frac{\mu X}{kT} - \frac{UcR}{kT} \tag{43}$$

(for the meaning of these terms, see Sec. 4)

Fig. 30 shows a plot of $g^s{}_{dipole}$ as a function of the electrode charge (with one third monomers assumed).

5.5.4. *Calculation of the Electron Overlap Potential,* $g^m{}_{dipole}$. The surface potential difference, $^m\Delta^s\chi$, is given by:

$$^m\Delta^s\chi = g^m{}_{dipole} - g^s{}_{dipole} \tag{44}$$

where $g^m{}_{dipole}$ is the electron overlap potential at the metal solution interface [45]. Eq. (44) can be use to estimate $g^m{}_{dipole}$ at the p.z.c., where $^m\Delta^s\chi$ can be calculated (Sec. 5.5.2), as well as $g^s{}_{dipole}$ (Sec. 5.5.3). Thus, using Eq. (44) in Eq. (25) at the p.z.c., one obtains [45]:

$$[g^m{}_{dipole}]_{q=0} = [V_{cell}]_{q=0} + [g^s{}_{dipole}]_{q=0} + \mu_e{}^m/F - \mu_e{}^{ref}/F + {}^{ref}\Delta^s\phi \tag{45}$$

Bockris and Habib [45] calculated $[g^m{}_{dipole}]_{q=0}$ for two metals, Cd and Zn. They used the values of $[g^s{}_{dipole}]_{q=0}$ previously calculated for both Cd/solution interface and for Zn/solution interface by Habib [28]. The chemical potential of Cd and Zn were taken from the work of Paasch et al [46]. The quantity $({}^{ref}\Delta^s\phi - \mu_e{}^{ref}/F)$ was already calculated [39,42]. The values for the p.z.c.'s were taken from the work of Trasatti [47]. Table IV shows the calculated values of the electron overlap potentials $g^m{}_{dipole}$. The estimation of $g^m{}_{dipole}$ might involve an uncertainty of as much as ± 0.3 V.

Table IV. Calculated values of the electron overlap potential, $g^m{}_{dipole}$, at metal/solution interfaces.

| Metals | $-\mu_e^m$/ eV | $V_{q=0}$ (NHE)/ V | $[g^s(\text{dipole})]_{q=0}$/ V | $[g^m(\text{dipole})]_{q=0}$/ V | $\Phi^m$/ eV | $\chi^m$/ V | $\delta\chi^m$/ V | $\Delta_s^m\phi_{q=0}$/ V |
|---|---|---|---|---|---|---|---|---|
| Cd | 2.57 | −0.72 | 0.06 | 1.08 | 4.10 | 1.53 | −0.45 | 1.02 |
| Zn | 1.97 | −0.63 | 0.07 | 1.78 | 4.12 | 2.15 | −0.37 | 1.71 |

On the other hand, $g^m{}_{dipole}$ is related to the surface potential of the metal in contact with vacuum, $\chi^m$, by the following relation [45]:

$$g^m{}_{dipole} = \chi^m + \delta\chi^m \quad (46)$$

From the theory of the work function, one obtains that [45]:

$$F\chi^m = \Phi^m + \mu_e{}^m \quad (47)$$

Therefore, $\delta\chi^m$ can be calculated from Eq. (46) where $\chi^m$ can be obtained from Eq. (47). Table IV contains the value of $\delta\chi^m$ for two metals, as given by reference [45].

Alternatively, Goodisman [10] and others (Shmickler, Badiali, etc.) have calculated the electron overlap potential from quantum mechanical principles. One can begin such considerations by applying the uncertainty principle to the situation of the electron in the interface. Thus, the uncertainty in position, Δx, would be given by:

$$\Delta x = h/p \quad (48)$$

where h is Planck's constant, p is the momentum of the electron arising from the interface. The order of magnitude of the kinetic energy of the electron should be a few electron volts. Taking it as 1 eV, the value of Δx is 3.5 Å. If the double layer field is $10^7$ V per cm, then:

$$g^m{}_{dipole} = 0.35 \text{ V} \quad (49)$$

This roughly estimated value, a pre-quantal calculation, can be compared with the modelistic result obtained by Bockris and Habib [45] (Table IV), of about 1 V.

Goodisman [10] carried out a simple quantum mechanical calculation to find a surface potential at the metal/vacuum interface. On the other hand, the self-consistent jellium model, which takes into account both coulombic and non-coulombic forces, appears to be more realistic than the hard wall model for electrons at the metal surface [10]. In the jellium model, the ion cores are represented by a semi-infinite background, instead of a hard wall. However, calculations of the electronic profile are more complex [10]. According to this model, a value of 4.8 V is obtained for $\chi^m$. This value is higher than the phenomenologically derived values of Bockris and Habib (Table IV). According to Trasatti [42], the calculated jellium values for $\chi^m$ are the upper limits of such calculations.

### 5.6. SUMMARY CONCERNING CALCULATIONS OF POTENTIALS IN THE DOUBLE LAYER

As it is evident from the development of this section, few of the calculations of the potential differences in interfaces are model-free. The wide spread values for $g^m{}_{dipole}$ illustrate the point. The values will change with time. Furthermore, some of the data which is needed to calculate the potentials is missing, as, for example, detailed knowledge of the Fermi energy for many metals and other materials.

The situation here is that "some ideas may be obtained of the numerical values of the various potentials in the double layer".

## 6. The Metal/Molten-Salt Interface

It might be thought at first that such an interface would be experimentally more difficult than the aqueous solutions interfaces, (particularly the mercury/solution interface), but this is far from true. In fact, both the experimental behavior and the theory are simpler.

This is partly because there are no solvent molecules; there is no $H^+$ or $OH^-$ discharge, nor tendency toward oxide film formation. Therefore, the potential range over which one may expect polarizable behavior is increased.

The typical behavior given by the inverted parabola is shown in Fig. 31. It is observable that the minimum occur at the p.z.c. [49]. The graph shown suggests that the situation at the metal/molten-salt interface corresponds, in a sense, to a diffuse layer. This is because there is a degree reminiscent of the minimum which one obtains in aqueous solutions near the p.c.z.

At first sight, it is difficult to see a molten salt in a diffuse layer model, because one tends to think of the salt as particularly concentrated. However, this may not be a reasonable view. Bockris and Richards [50] pointed out that during the melting of salts (e.g., sodium chloride), a typical volume increase is about 25%. On the other hand, the average internuclear distance decreases. Then, it is only possible, to think of a large amount of free space in molten salts. Thus, based on these facts, Bockris and Richards developed a version of the theory of holes in liquids. On this basis, they were able to calculate adequately, without the use of adjustable parameters, the compressibility and free volume of a number of salts. It is therefore possible to see the molten salt as having a structure of the kind shown in Fig. 32. Thus, it would be possible for the insertion of an electrode into a molten-salt to attract a larger number of cations than anions into a certain area around the plate by means of a non-equilibrium filling of holes to favor one type of ion than the other.

This model was treated by Sotnikov and Esin [51]. They took a Boltzman distribution of such vacancies. Preferential filling of the vacancies by cations (and/or anions, depending upon the charge on the electrode) was taken as giving rise to the surface excess on the solution side. Further, they provided an equation for the capacitance as a function of potential. Here, agreement with experiment was good and without adjustable parameters.

On the other hand, Monte Carlo methods of computer simulation have been used to study structures in molten salts. Woodcock and Singer [52] were able to predict thermodynamic properties by the use of a Born-Mayer-Huggin interionic potential. McDonald and Singer [53] concluded that the simulations gave rise to large and persistent voids in the structure, as suggested in the Bockris and Richards model (1957) [50].

## 7. The Semiconductor/Solution Interface

In the semiconductor, the concentration of carriers is typically in the order of $10^{16}$ $cm^{-3}$, whereas in a metal it is about $10^{23}$ $cm^{-3}$. Thus, a metal is brimming over with electrons, whereas in the semiconductor the concentration of electrons is some $10^5$-$10^8$ times less. Here electrons and holes may be regarded in the way of cations and anions in the solution rather than electrons in the metal. For this reason, the distribution of the electrons and holes is governed by laws which are similar to the distribution of ions in solution, which was first derived by Gouy for charges outside an electrode.

These ideas have given rise to the concept of a Schottky "barrier" inside the semiconductor (Fig. 33). "Barrier" refers to the theory of the distribution of electrons and holes in semiconductors which was worked out (in analogy to Gouy's theory) by Green and Seiwartz and others in the 1950's [54].

The first application of the semiconductor principles to the semiconductor/solution interface was made by two physicists in the Bell Telephone Laboratory, Garrett and Brattain [55].

It is a feature of the work on the semiconductor/solution interface that largely <u>physicists</u> have made contributions in respect to phenomena connected with photoelectrochemistry, particularly during the 1970's. Most of them have treated the semiconductor/solution interface as if it were between the semiconductor and the metal. This has given rise to certain misunderstandings.

The earliest paper in which electrochemical kinetics principles[10] were applied to semiconductors is that of Green [56,57]. Green pointed

---

[10]Garret and Brattain [55] have the distinction of having discussed the semiconductor/solution interface in a published paper for the first time. However, their discussion was entirely Nernstian, i.e., not modelistic. Thus, they did not deal either in concepts of interfacial properties nor in kinetics.

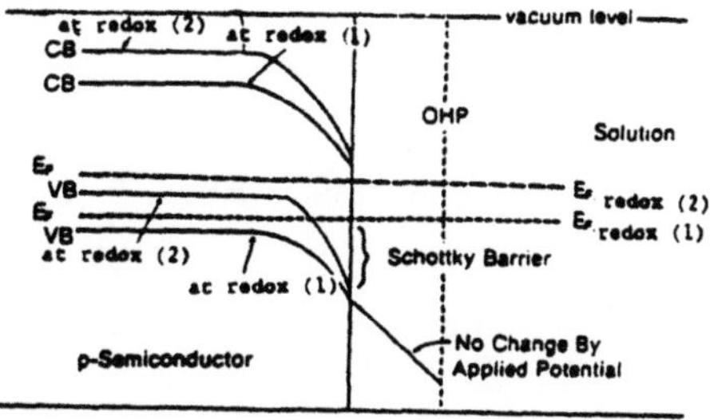


Fig. 33. Schottky barrier model at the semiconductor/solution interface using different redox couples, redox (1) and redox (2).

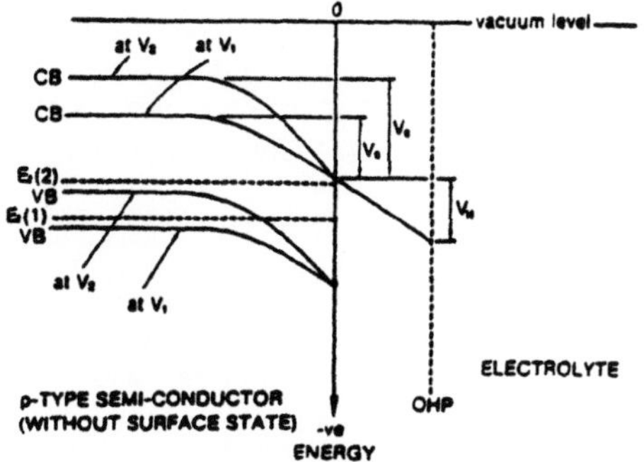


Fig. 34. Schematic diagram of a p-type semiconductor/solution interface at two applied potentials, $V_1$ and $V_2$. $V_s$ and $V_H$ are the p.d. in the space charge region and Helmholtz region, respectively.

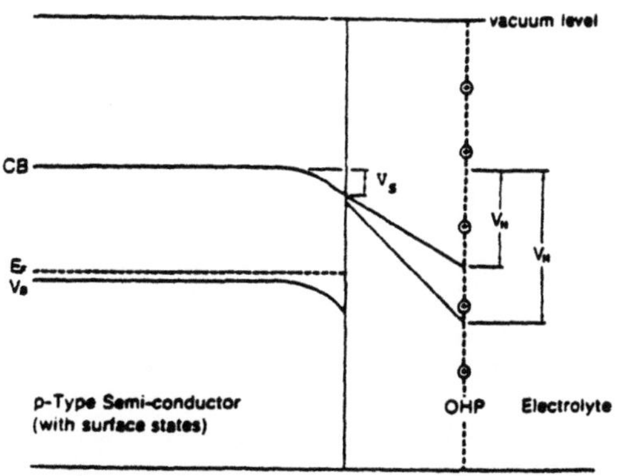


Fig. 35. Schematic diagram of a p-type semiconductor/solution interface in the presence of high density of surface states. $V_s$ and $V_H$ are the p.d. in the space charge region and Helmholtz region, respectively.

out that the entire situation with regards to the so-called Schottky barrier theory of semiconductors applied to electrolytic solutions would change if the surface states were sufficiently high in concentration. Figs. 34 and 35 show the semiconductor/solution interface in the absence and in the presence of surface states, respectively.

It is possible to work out the potential difference in the Schottky barrier as a function of surface states and to contrast this with the potential difference in the double layer. This is shown in Fig. 36 for the p-Si/aqueous solution under illumination [58]. Therefore, there is a full analogy to the situation between the part played by surface states at the semiconductor/solution interface and the part played by specifically adsorbed ions in the metal/solution interface.

It seems likely, in fact, that all semiconductor/solution interfaces contain a high degree of surface states. The normal surface states which are present in semiconductor/vacuum interfaces are probably expunged in the electrolytic situations because water will combine with the dangling bonds from the semiconductor atoms, and these will no longer exist. On the other hand, water will form a complete layer on the semiconductor/solution interface and may give rise to a large number of surface states. In addition, specifically adsorbed ions may be the origin of surface states. Also, reactions leading to hydrogen or oxygen on the semiconductor surface will increase the surface state concentration.

There are clearly two kinds of surface states:

a. The first kind are essentially traps. Electrons and holes which attempt to pass through the interface to the solution may be trapped at the surface states, recombine with the carrier of opposite sign, and therefore diminish, e.g., the efficiency of photoelectrochemical processes (Fig. 37).

b. On the other hand, the second kind refers to "good surface states" which exist in the band gap, and thus increase the likelihood of electron transfer between electrode and solution. Thus the possible efficiency of some photoelectrochemical processes is increased (Fig. 38).

There are several methods of determining surface state concentration at the semiconductor/solution interface. However, the actual measurements of surface states at this interface is not satisfactory at this time. The methods are the following:

a. The Impedance Technique [59]. One may determine the impedance of the interface, make various equivalent circuits, attempt to evaluate the capacitance due to the surface states, and evaluate the number of surface states from this capacitance as a function of potential. The procedure is highly dependent upon obtaining the correct equivalent circuit (Fig. 39) [60]. This may involve uncertainties among various alternatives which reproduce experimental behavior with nearly equal facility.

b. Another method includes Sub-Band Gap Photocurrent [61]. Qualitatively, it is clear that insofar as current passes when the frequency of the light is less than the band gap, excitation must be

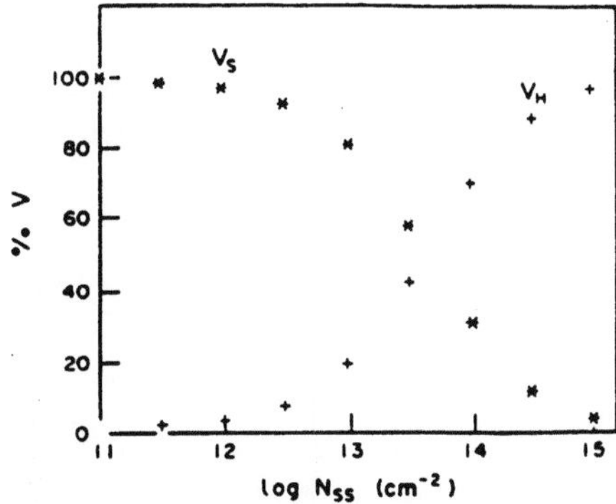


Fig. 36. Relative potential drop in the space charge region, $V_s$, and Helmholtz region, $V_H$, as a function of surface state density, $N_{ss}$, for a p-Si/1 M $H_2SO_4$ interface under illumination.

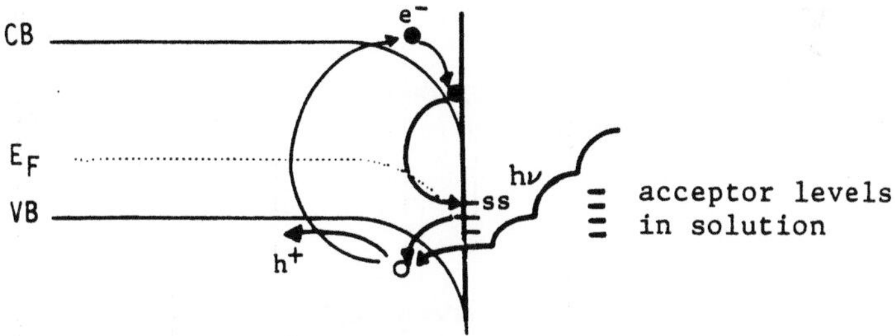


Fig. 37. Schematic representation of "bad surface states" (ss) for a p-type semiconductor/solution interface. Electrons are excited by light which energy is higher than the band-gap energy, $E_g$ ($h\nu > E_g$). Once the electron is in the conduction band (CB), it may go to a surface state instead of to an acceptor level in solution. From the surface state, the electron may recombine with a hole.

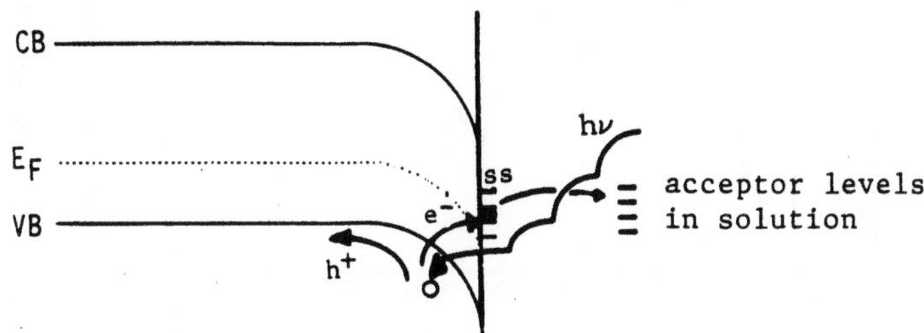


Fig. 38. Schematic representation of "good surface states" (ss) for a p-type semiconductor/solution interface. Electrons are excited by sub-band gap light, ($h\nu < E_g$, where $E_g$ is the band-gap energy) to a surface state, and finally reach an acceptor level in solution (thus, electrochemical reaction takes place).

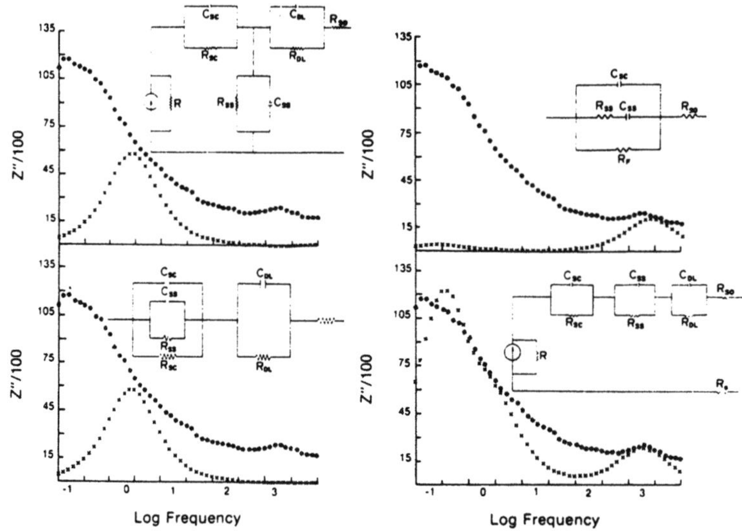


Fig. 39. Stimulated imaginary impedance-frequency plot (x) for the photo-electrochemical reduction of $CO_2$ on CdTe. (*) experimental values.

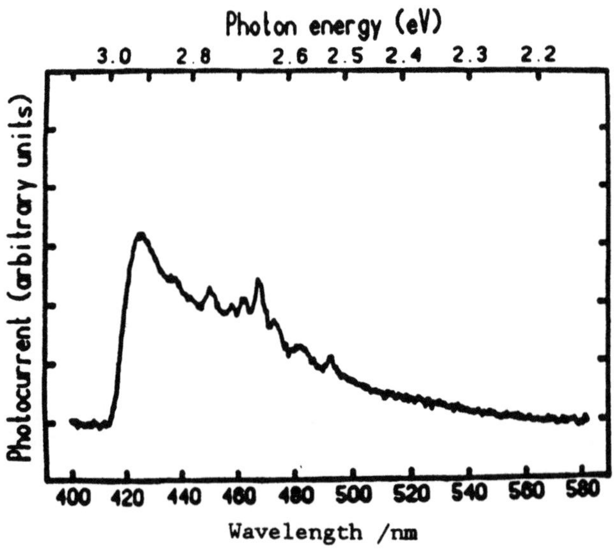


Fig. 40. The sub-band gap photocurrent response of a $TiO_2$ sample in 1 M KOH at 0 V (SCE).

occurring at surface states in the band gap (Fig. 40).

c. The Sub-Band Gap Capacitance [62]. This last method involves measurements of the capacitance using sub-band gap light. Changes in the capacitance values are related to the filling or emptying of surface states (Fig. 41). A manipulation of this method may give rise to a measurement of surface states.

The situation in 1988 is that there has been little comparison among these various methods for the same system, so the determination of surface states (rather analogously against the determination of specific adsorption at solid/solution interfaces) is uncertain.

In spite of the absence of consistent determinations of surface states at semiconductor/solution interfaces, it is possible to make certain conclusions concerning the frequency of occurrence of a high density of surface states at these interfaces. Thus

a. All semiconductors are highly dependent, in respect to their photo-electrochemical kinetics, upon the condition of the surface. This fact would hardly be so if the kinetics of the electrode reactions concerned were associated with internal energy barriers.

b. The value of the Tafel slope for the rate-determining step of electron transfer to ions in solution (or viceversa) is different if it is determined by the Schottky barrier than if it is determined by the Helmholtz double layer [57,63].

Thus, Schottky barrier rate determining gives rise to the following equation:

$$d\ (\log i)/dV = RT/F \qquad (50)$$

Helmholtz barrier rate determining gives,

$$d\ (\log i/dV) = 2RT/F \qquad (51)$$

There is no doubt that in most cases 2RT/F is the observed slope [63,64]. However this is obscured in many cases by the fact that most of the time the top part of the i-V curve (where the current is near its limiting current value) is simply observed. Thus, the interesting and informative Tafel region is avoided.

Table V gives a series of Tafel slopes for hydrogen evolution; they are much nearer to 2RT/F than to RT/F [64].

c. When the concentration of surface states is determined for situations involving hydrogen or oxygen evolution, the concentrations found are usually high, e.g., $10^{13}$ - $10^{14}$ $cm^{-2}$. Thus, Chandrasekaran et al [58] studied the p-Si/aqueous solution interface under illumination where the Faradaic reaction involved was the hydrogen evolution. They used the Impedance Technique and determined the dependence of surface state density on the electrode potential (Fig. 42). They concluded that adsorbed ions induce surface states.

The introduction of adequate methods for the determination of surface states is the rate determining step in progress towards knowledge of the structure of the semiconductor/solution interface. On the other hand, direct spectroscopic methods of investigation, particularly FTIR, are needed.

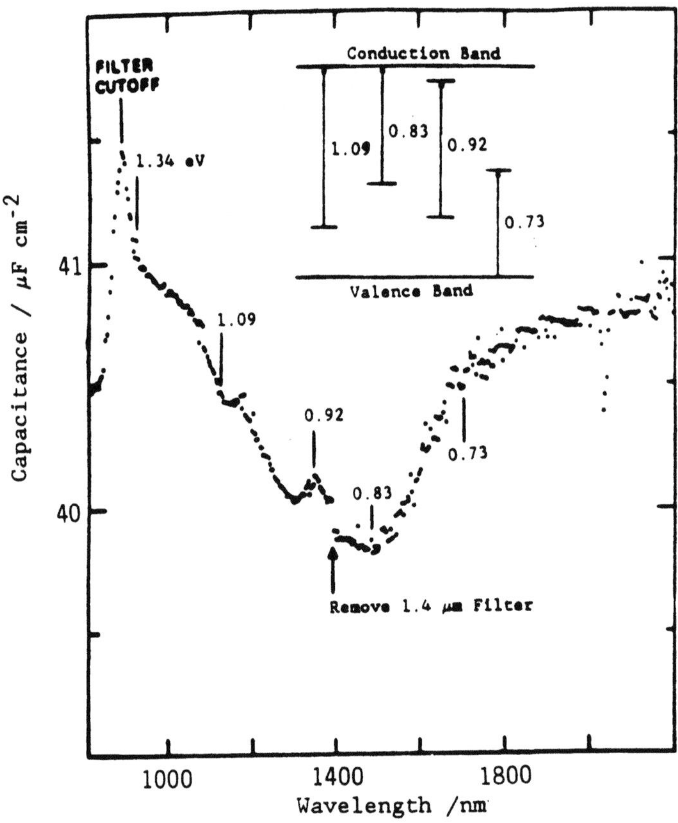


Fig. 41. The sub-band gap photocapacitance response of a n-GaAs sample in 0.1 M tetraethylammonium perchlorate/acetonitrile solution. Frequency: 5 Hz.

Table V. Tafel slope for $H_2$ evolution on semiconductor in the presence of deposited metal.

| Electrode | Tafel Slope (V decade $^{-1}$) |
|---|---|
| p-Si/Pt * | 0.10 |
| p-InP/Pt * | 0.11 |
| p-MoSe3/Pt * | 0.21 |
| n-Si/Pt | 0.17 |
| n-Si/Co | 0.13 |
| n-Si/Ni | 0.11 |
| n-Si/Au | 0.11 |
| n.Si/Pb | 0.11 |
| n-Si/Cd | 0.21 |

*photoelectrochemical reaction

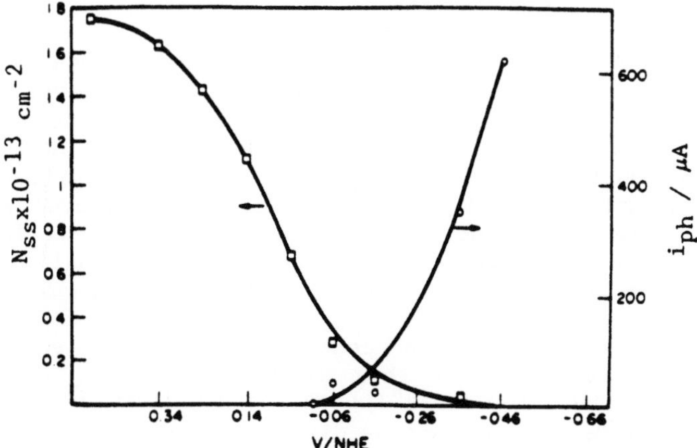


Fig. 42. Surface state density, $N_{ss}$, as a function of electrode potential for the p-Si/aqueous $H_2SO_4$ interface under illumination.

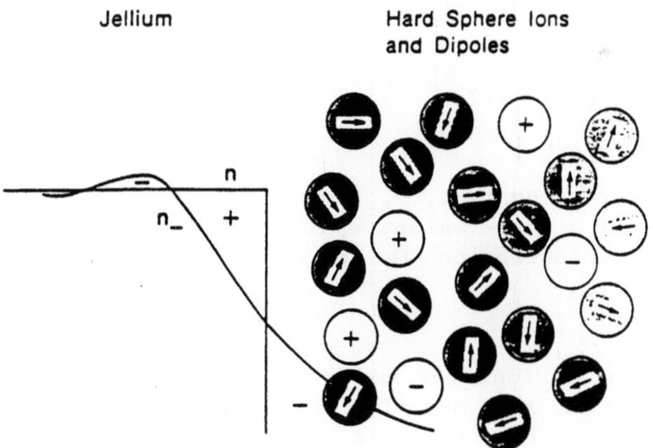


Fig. 43. Ion-dipole jellium model of the double layer.

## 8. Solutions to the Poisson Equation

The normal solutions of the Poisson equation are well known. They form the backbone of the classical theory of the double layer and are applied in the theory of colloidal stability where the solutions are so low in concentration that the most important part of the double layer is the diffuse double layer. On the other hand, this diffuse layer theory is frequently applied by workers in double layer problems regardless of questions about the dependence of its validity upon concentration to solutions where the concentration is usually not less than $10^{-3}$ M and may reach (e.g. in specific adsorption calculations) as high as 5 M.

The Poisson equation is a continuum equation, and may be reasonably applied so long as this condition can be maintained. Therefore, the electrode "should see" the surrounding ions in the form of smoothed-out charge-density and not as discrete charges. On this basis, there are arguments which suggest that the range of concentration in the applicability of the Gouy theory is severely limited. The principal argument concerns graininess. At low concentrations, the effective thickness of the ionic cloud, $\kappa^{-1}$, is large, and a number of ions exists within the region between the electrode surface and the $\kappa^{-1}$ thickness. In this case, the "smoothed-out cloud of charge" (the continuum condition) is valid. However, when the ionic concentration increases, $\kappa^{-1}$ declines faster than the average distance between particles in solution ,l. Here, the electrode "sees" discrete charges and not the smoothed-out cloud of charge. Eventually, the volume of solution between the electrode and the thickness of the diffuse layer contains no ions. It seems reasonable to take this as a limiting condition for the beginning of graininess and breakdown of the continuum condition.

Thus [23],

$$\kappa^{-1} = \left\{ \frac{\epsilon kT}{8\pi N_a z^2 e^2 C} \right\}^{1/2} \qquad (52)$$

where $\epsilon$ is the dielectric constant, $N_a$ is the Avogadro number, z is the charge of the ion (for a z:z- valent electrolyte), e is the electron charge, and C is the concentration. On the other hand,

$$l = \left\{ \frac{1000}{fN_a C} \right\}^{1/3} \qquad (53)$$

where f is the number of ions per molecule of electrolyte. Table VI [65] contains values of $\kappa^{-1}$ and l for different electrolytes at different concentrations. However, this argument is not exact because the distance with which one is comparing the thickness of the double layer is calculated under equilibrium conditions.

Direct tests of the diffuse double layer are difficult to make. Specific adsorption makes such tests more difficult. One test used by Grahame [3] is to take measurements of the capacitance of the double layer near to the p.z.c. Here, the diffuse layer capacitance becomes small, and it dominates the double layer capacitance because of the addition of reciprocals (Eq. (2)). However, this test seems to be an

Table VI. Calculated effective thickness of the diffuse layer, $\kappa^{-1}$, and average distance between ions, l, for different electrolytes.

| Concen-tration(M) | Type of Salt (z:z') 1:1 | | 1:2 | | 2:2 | | 1:3 | |
|---|---|---|---|---|---|---|---|---|
| | $\kappa^{-1}$(Å) | l (Å) | $\kappa^{-1}$(Å) | l (Å) | $\kappa^{-1}$(Å) | l (Å) | $\kappa^{-1}$(Å) | l (Å) |
| $10^{-4}$ | 304 | 203 | 176 | 203 | 152 | 203 | 124 | 203 |
| $10^{-3}$ | 96 | 94 | 56 | 94 | 48.1 | 94 | 39.3 | 94 |
| $10^{-2}$ | 30.4 | 43.6 | 18 | 43.6 | 15.2 | 43.6 | 12.4 | 43.6 |
| $10^{-1}$ | 9.6 | 20.3 | 56 | 20.3 | 4.81 | 20.3 | 3.93 | 20.3 |
| $10^{0}$ | 3.04 | 9.4 | 1.76 | 9.4 | 1.52 | 9.4 | 1.24 | 9.4 |
| $10^{1}$ | 0.96 | 4.36 | 0.56 | 4.36 | 0.48 | 4.36 | 0.39 | 4.36 |

undemanding one because of the concentration at which it was carried out (ten times less than the concentration at which the diffuse layer theory is applied in calculations connected with specific adsorption determination [66]).

Joshi and Parsons [67] and Parsons and Trasatti [68] carried out examination of the dependence of the charge in the diffuse layer on the metal charge. They assumed that there is no significant degree of specific adsorption in the concentrations considered. They found that the experimental diffuse charge seems to be only slightly greater than that predicted from the theory. Again the concentrations used were smaller than those in which Gouy's theory is applied [66].

In summary, diffuse layer theory is applicable only up to concentration lower than those at which it is usually applied in double layer work. This limiting validity of the diffuse layer theory is particularly important because it is just in the higher regions of concentration where specific adsorption takes place. The suggested invalidity of the Gouy theory above 0.01 M may mean that specific adsorption calculations based on it need revision.

## 9. Ab-Initio Calculations in the Double Layer.

In the approaches to the double layer which we have described hitherto, a _model_ has initially been supposed and one finds out the degree to which it fits experiments. Such an approach is perfectly satisfactory -it is the essence of the scientific approach- as long as it is done with complete honesty and with the use of no more than one adjustable parameter. Thus, the parameter adjustments will make most models fit if the number of parameters (3-4) are sufficient.[11]

In this view, some ab-initio calculations (no model assumed) are attractive. Starting with the work of Lang and Kohn [48,69] and then following along the line of Goodisman [10], Badiali [70], Blum [71], Schmickler [9] and Henderson [9,69,72], a new series of papers in the double layer have been written during the 1980's. These papers are characterized by complex mathematical treatments, and this may be all to the good, although in some instances naive statements and conclusions result. The work involves the use of a large number of acronyms. Hitherto, it applies at the p.z.c.

Among the difficulties which have to be borne in appreciating this recent approach, is that its authors have not understood the relation of the model to their names. Thus, the union of the Helmholtz theory with that of Gouy which was made by Stern (which gives radically different values from those obtained on Gouy's model) is not referred

[11] Of course it is obviously ideal to make the comparison calculation with no adjustable parameters. However, the application of a single adjustable parameter to locate the calculation upon a given scale is generally necessary. Two parameter equations may also provide some valid tests, but when the number of adjustable parameters exceeds two, the agreement between theory and experiment begins to decline.

to by the accepted title (Stern model), but all previous models are called "Gouy's theory" or modifications of Gouy's theory.

The radical modification of the Gouy's theory which has been carried out by these authors is the integration of the charge-distance relationship not up to the electrode surface, but to a boundary which represents ions in contact with the double layer. Thus a theory has been produced which, in form, resembles closely the Stern theory in that it has, basically, two terms: one representing the capacitance due to the first layer of ions; the second representing the space charge out into the solution. Solvent effects are taken into account in a simplistic way, i.e., the solvent distribution is taken Watts-Tobian only. There is no allowance for the extremely strong interactions which occurs between water molecules (as a result of both the electrostatic interaction and the hydrogen bonding interactions) although such interactions alter qualitatively the predicted influence of the solvent on double layer properties. The most important contribution which these theories have made is to introduce into the double layer theory a term, its bases were calculated by Bockris and Habib [45], i.e., the $g^m{}_{dipole}$ term. They have not only calculated this term in a more quantum mechanical way than that which has been calculated by others, but also differentiated it with respect to charge, and therefore obtained a capacitance which may be called here the Thomas-Fermi capacitance. Fig. 43 shows the model of the double layer by Schmickler and Henderson [9].

This contribution is of value because it explains the fact that the capacitance of gallium is different from that of mercury. This fact is presumably due to the different contributions in the Thomas-Fermi region. The water layer structure in contact with the electrode is clearly left open to speculation.

Qualitatively speaking, when electrons approach the surface of a metal positively charged, the electrons pass through nuclei at the surface of the metal and form a charge cloud <u>outside</u> the electrode surface in the solution. However, if the electrode is charged negatively, the electrons would tend to underlap the electrode surface, i.e., the approach to the surface would be repulsive because that is where the excess charge lies.

This brings up the interesting question of how the water layer is affected by $g^m{}_{dipole}$. One would expect it to be affected greatly for positively charged metals, and not at all for negatively charged metals. Hitherto, it has been assumed that water molecules lie parallel to the surface, and at the p.z.c., they "stand up" on the anodic side. It has also been <u>assumed</u> that when the excess charge is negative, the water molecules will have a parallel or symmetrical behavior to that on the positive side. This may not be so if the $g^m{}_{dipole}$ theory is correct, and this gives rise then to the interesting speculation on the behavior of water molecules in respect to the "standing up" property on both sides of the p.z.c. The extension of the Habib and Bockris FTIR determinations is necessary [32].

On the other hand, there are plenty of phenomena which have already been examined which suggest that a situation outside the potential of zero charge is symmetrical (i.e. the adsorption of simple alcohols on Hg). Here, the variation of the coverage with potential is determined by competition between alcohol and water dipoles; the adsorption of methanol as a function of potential is symmetrical around the p.z.c. (Fig. 44) [73]. However, this does not seem consistent with the concept of overlap and underlap of a significant electron cloud.

Thus, the $g^{m}_{dipole}$ may have a lesser magnitude than has been suggested by the approximate theory now extant.

## 10. The Role of Ex-Situ Methods

The use of high vacuum to electrochemical systems with consequent electron spectroscopy was introduced by Baker, Revie and Bockris [74, 75][12]. The application of ex-situ electron spectroscopy to the study of surface films (e.g., passive layers) is undoubtedly great. Thus, XPS, Auger, SIMS and Rutherford backscattering was applied by Murphy, Pou and Bockris to studies of the passive layer on iron [78-80].

The advantages of the ex-situ method of approach are not only the variety of methods which can be used (so long as it is possible to put a vacuum on a sample), but also the greater accuracy which can be obtained by using methods which have long been in surface chemical investigations of the metal/vacuum interface. An element of controversy and doubt enters when such methods are applied to examine the surface of metals which have been in contact with solution and contain chemisorbed quantities remaining after vacuum removal of the solution.

Thus:

a) Preparation of sample. If the sample is to be prepared by evacuation and evaporation of the bulk water from the surface, then the original film must be a thin one. Supposing that it is 1 $\mu$ in thickness, then the number of moles which are in a layer above a 1 $cm^2$ electrode is $10^{-4}$x$10^{-3}$ M. If the largest acceptable deposit of material from the solution is 1% of the surface, then the molarity of the solution must be less than $10^{-4}$, and this seems to present a difficulty.

b) Removal of water. If the measurements are to be made at room temperature, water molecules will no longer be at the surface since they have been lost from the double layer. Thus, if the water is removed, a principal constituent which influences the presence of organic materials on the surface of electrodes (and influences the presence of ions there) is gone. So that, whatever results are obtained may not be relevant to the electrochemical situation. These theories relating the dependence of the adsorption upon potential, in which water is a primary determinant, meet difficulties in being evaluated by means of vacuum techniques.

Insofar as the material is strongly chemisorbed, the dependence on water will grow less and, in the limit, results relevant to the electrochemical situation will be obtained.

---

[12]Hubbard, who has continued publishing almost exclusively in this area, published his first paper in 1976 [76]. However, a preliminary abstract of work to be done was published by Hubbard in 1973 [77].

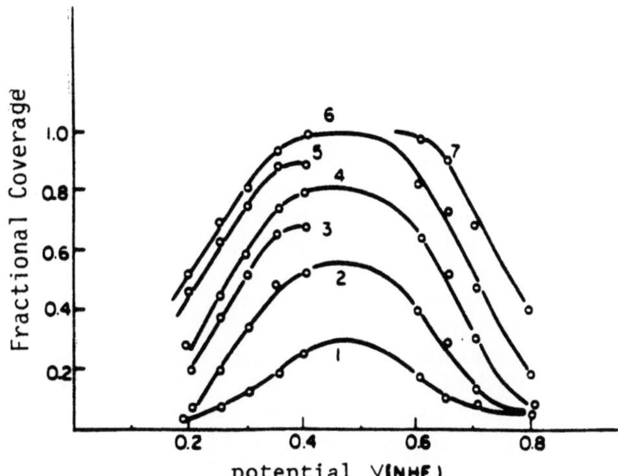


Fig. 44. Dependence of methanol adsorption on the electrode potential in 1 N $H_2SO_4$ at various methanol concentrations: (1) $10^{-3}$ M; (2) $10^{-2}$ M; (3) $5x10^{-2}$ M; (4) $10^{-1}$ M; (5) $5x10^{-1}$ M; (6) 1 M; (7) 5 M.

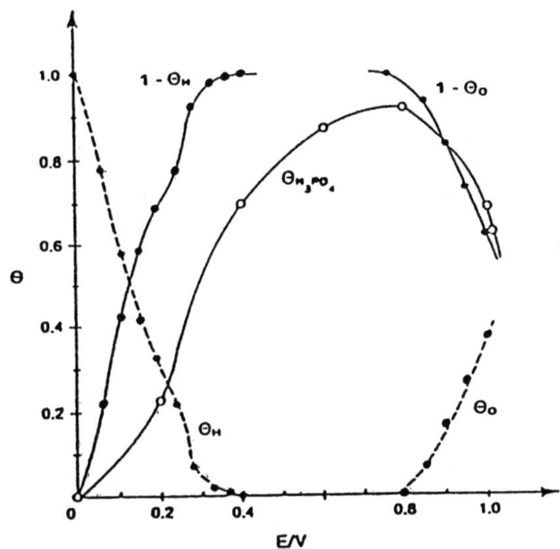


Fig. 45. Hydrogen adsorption, oxide layer formation, and phosphoric acid adsorption on a Pt electrode as a function of potential at 298 K.

c) Loss of potential control? The electrode is potentiostated to prepare the area which is later to be looked in vacuum. This means that the total potential between the metal and the reference electrode is constant and, in the absence of an IR drop, this is a satisfactory control. However, if the IR drop is significant, the potential which is kept constant is that of the electrode together with the IR drop. As the film of this solution is evaporated, the IR drop increases, and the potential controlling the adsorption changes.

4) Special treatment. One of the techniques which has been used by Bockris and McHardy [81] is to jet liquid air onto the electrode at a given condition, remove the electrode frozen in an ice layer, and immerse this in a canister of liquid air. The electrode can then be transported to the appropriate spectroscopic device, and the water sublimed in vacuum. The method is of interest in looking at oxide films or the constitution of materials which would be stable under these conditions, which will not change because of the application of a vacuum.

The ex-situ methods do have some part in the examination of thin layers of oxide films, and perhaps in contributing towards knowledge of the structure of the double layer. However, it seems that there will always be ambiguities in the relevance of the knowledge obtained. Thus, Wieckowski [82] reported a comparison of the results of radiotracer adsorption experiments with those arising from vacuum examinations. For example, Laane and Hubbard [83] used UHV technique to find the maximum surface concentration of adsorbed $I^-$ species. The result was comparable with that one obtained by Khidifov et al [84] using the radiotracer technique. However, quantitative agreement between the solution [85] and vacuum [86] work is poor for the studies of $CN^-$ adsorption on platinum.

A good example of a situation where ex-situ methods would be likely to give the wrong results is given by the recent work of Zelenay et al on the adsorption of phosphoric acid from solution onto platinum [87]. Fig. 45 shows the correlation between the coverage-potential diagram of the phosphoric acid and the hydrogen and oxygen present on the surface [87]. It is seen that at negative and positive potentials, the adsorption of phosphoric acid declines with the increase of hydrogen and oxygen, respectively. Thus, phosphoric acid adsorption in these regions is clearly dependent upon the presence of hydrogen and oxygen which have been electrolytically deposited on the solution (the deposition depends upon an exact potential being maintained between the electrode and solution). In the vacuum situation, hydrogen and oxygen would not be there; therefore, phosphoric acid, controlled by the presence of the hydrogen and oxygen, would adsorb differently in their absence.

## 11. The First Twelve Conclusions

1. The thermodynamic approach to the double layer is probably exhausted. Its past triumphs are under a cloud. It should be replaced by direct in-situ techniques.

2. The data accrued through potential sweep determinations of adsorption on noble metals is limited because of uncertainties in respect to partial charge transfer. The method is difficult to apply to non-noble metals because of co-dissolution. Although a good tool for a preliminary scan, i-V curves are often difficult to interpret, and should be replaced by direct in-situ spectroscopic methods as tools for double layer investigations.

3. A number of quantities which are referred to in the double layer as though determinable exist at present on an uncertain basis. Some of those quantities are the following: specific adsorption, surface states, the p.z.c. at solid-metals, and the flat band potential at semiconductors under most practical conditions. Correspondingly, most of the named potentials in the double layer (e.g. the Galvani potential difference) are only calculated and not determined - the values contain model assumptions.

4. A central difficulty is the lack of clarity concerning the solution concentration in which the Gouy theory can be applied with confidence.

5. An important aspect of double layer determinations lies in the entropy of the solvent. Measurements on liquid metals such as gallium are to be encouraged.

6. Organic adsorption measurements in the double layer are for the most part governed by steady state kinetics rather than equilibrium adsorption. For those cases which can be considered true equilibrium adsorption situations, the measurement of the adsorption as a function of temperature is valuable because it gives rise to information concerning the entropy of the adsorbed entities. Thus, this would increase the basis of model building.

7. The introduction of $g^{m}_{dipole}$ under considerations of the structure of the double layer is a new step and valuable. However, it brings into question basic aspects of the double layer. One of them is the extent to which the solvent or specific adsorption concentration may be affected by the "electron cloud" outside the solid phase.

8. The structure of the semiconductor/solution interface has been looked at simplistically. In most practical cases involving electrolytes, it is probable that a considerable Helmholtz component affects the potential difference.

9. In building up models for the interface, there is no doubt that helpful information can be obtained with experiments in vacuum. However, such information can seldom be direct because the double layer without water is not comparable with the double layer with water.

10. The introduction of direct methods into the investigation of the double layer is the most important element in the advancing experimental front. The spectroscopic examination of the solvent may be mentioned. The use of apparatus which can determine intermolecular forces in the regions below 10 Å, and the scanning tunneling microscope, are examples of devices with enormous power for the near future.

11. The effect of the insulator/aqueous interface has been little investigated. Much more attention should be devoted to this area, as well as those involving biological/solution interfaces, including

protein adsorption and interfaces between the mitochondrial and the surrounding solution.

12. In a general way, the future indicates a bringing together of the concepts of surface chemistry and surface electrochemistry, where lesser distinction will exist between each area. As most of the practical surface systems involve an interface with a solvent, most surface chemistry is electrochemistry.

## 12. References


1.- H. L. von Helmholtz, *Wied. Ann.*, **7**, 337 (1879).
2.- R. Parsons and M. A. V. Devanathan, *Trans. Faraday Soc.*, **49**, 404 (1950).
3.- D. C. Grahame, *Chem. Rev.*, **41**, 441 (1947).
4.- J. O'M. Bockris, M. A. V. Devanathan, and K. Müller, *Proc. Roy. Soc. (London)*, **A274**, 55 (1963).
5.- A. Frumkin, N. Polianovskaya, N. Griforyer and I. Gagotskaya, *Electrochim. Acta*, **10**, 793 (1965).
6.- L. Doubova, A. De Battist, and S. Trasatti, *Electrochim. Acta*, **31**, 881 (1986).
7.- A. Frumkin and B. Grigoryev, *Elektrokhimiya*, **8**, 412 (1972).
8.- B. E. Conway, in *Atomistic of Fracture*, R. M. Latanisum and J. R. Pickens, Eds. (Plenum Press, New York, 1981) p. 497.
9.- W. Schmickler and D. Henderson, *Prog. Surf. Sci.*, **22**, 323 (1986).
10.- J. Goodisman, *Electrochemsitry: Theoretical Foundations. Quantum and Statistical Mechanics, Thermodynamics, The Solid State*. (John Wiley & Sons, New York, 1987). Ch. 3.
11.- T. Beck, *J. Phys. Chem.*, **73**, 466 (1969).
12.- R.A. Fredlein, A. Damjanovic and J. O'M. Bockris, *Surf. Sci.* **25**, 261 (1971).
13.- R. A. Fredlein and J. O'M. Bockris, *Surf. Sci.*, **46** ,641 (1974).
14.- J. O'M. Bockris, in *Electrochemistry: the "Past Thirty and the Next Thirty Years"* H. Bloom and F. Gutmann, Eds. (Plenum Press, New York, 1977), Ch. 28.
15.- P. R. Couchman and C. R. Davidson, *J. Electroanal. Chem.*, **85**, 407 (1977).
16.- A. Frumkin, *Z. Phys.*, **35**, 792 (1926).
17.- J. A. V. Butler, *Proc. Roy. Soc.*, **A122**, 399 (1929).
18.- J. O'M. Bockris and E. C. Potter, *J. Chem. Phys.*, **20**, 614 (1952).
19.- J. T. Law, *Ph. D. Thesis*, Royal College of Science, London S. W. 7 (1951).
20.- C. Kemball, *Proc. Roy. Soc.*, **A190**, 117 (1947).
21.- R. J. Watts-Tobin, *Phil Mag.*, **6**, 133 (1961).
22.- N. F. Mott and R. J. Watts-Tobin, *Electrochim. Acta*, **4**, 79 (1961).
23.- J. O'M. Bockris and A. K. N. Reddy, *Modern Electrochemistry*, Plenum Press, Vol.2, Ch. 7.
24.- E. Lange and K. P. Mischenko, *Z. Phys. Chem.*, **A149**, 1 (1930).
25.- J. O'M. Bockris and M. A. Habib, *Electrochim. Acta*, **22**, 41 (1977).
26.- B. Damaskin and A. Frumkin, *Electrochim. Acta*, **19**, 173 (1974).
27.- R. Guidelli, in *Trends in Interfacial Electrochemistry*, NATO ASI

Ser., Ser. C, A. F. Silva, Ed. (D. Reidel Pub. Co., Holland, 1986) Vol. 149, p. 387.
28.- R. Parsons, *J. Electroanal. Chem. and Interfacial Electrochem.* **59**, 229 (1975).
29.- M. A. Habib, *Ph. D. Dissertation*, Flinders University (1975).
30.- A. Bewick and K. Kunimatsu, *Surf. Sci.*, **101**, 131 (1980).
31.- W. A. P. Luck, in *Water, a Comprehensive Treatise*, Vol 2, Ed. F. Franks (Plenum, New York, 1972) Ch. 4.
32.- M. A. Habib and J. O'M. Bockris, *Langmuir*, **2**, 288 (1986).
33.- A. Hamelin, in *Modern Aspects of Electrochemistry*, B.E. Conway and J.O'M. Bockris, Eds. (Plenum Press, New York, 1974) Vol. 16, Ch.1.
34.- J. O'M. Bockris and M. A. Habib, *J. Electroanal. Chem.*, **65**, 473 (1975).
35.- M .A .V. Devanathan, *Trans. Faraday Soc.*, **50**, 373 (1954).
36.- J. O'M. Bockris and M. A. Habib, *J. Res. Inst. Catalysis*, **23**, 47 (1975).
37.- Klein and E. Lange, *Z. Elektrochem.*, **43**, 570 (1937).
38.- S. U. M. Khan, R. C. Kainthla, and J. O'M. Bockris, *J. Phys. Chem.*, **91**, 5974 (1987).
39.- J. O'M. Bockris and S. D. Argade, *J. Chem. Phys.*, **49**, 5133 (1968).
40.- A. Frumkin and B. Damaskin, *J. Electroanal. Chem. Interfacial Electrochem.*, **66**, 150 (1975).
41.- E. Kanevsky, *Zh. Fiz. Khim.*, **22**, 1397 (1948).
42.- S. Trassati in *Comprehensive Treatise of Electrochemistry*, J. O'M. Bockris, B. E. Conway, and E. Yeager, Eds.; (Plenum Press, New York, 1980). Vol 1, Ch. 2.
43.- S. Trasatti, *J. Electroanal. Chem.*, **52**, 313 (1974).
44.- E. A. Guggenhein, *J. Phys. Chem.*, **33**, 842 (1929).
45.- J. O'M. Bockris and M. A. Habib, *J. Electroanal. Chem.*, **68**, 367 (1976).
46.- G. Paasch, H. Eschrig and W. John, *Phys. Stat. Sol.*, **51**, 283 (1972).
47.- S. Trasatti, *J. Electroanal. Chem.*, **33**, 351 (1971).
48.- N. D. Lang and W. Kohn, *Phys. Rev. B1*, 4555 (1970).
49.- E. A. Urshe, N.G. Bukun, D.I. Leinis and A.N. Frumnon, *Electrochim. Acta*, **9**, 431 (1964).
50.- J. O'M. Bockris and N. E. Richards, *Proc. Roy. Soc.*, **A241**, 44 (1957).
51.- A. I. Sotnikov and O. A. Esin, *Physical Chemistry and Electrochemistry of Molten Salts and Slags*, Proc. of Third All-Soviet Conferences Khimiya, Leningrade (1968).
52.- L. V. Woodcodk and K. Singer, *J. Phys. Chem. Solids*, **31**, 1199 (1973).
53.- I. R. McDonald and K. Singer, *Chemistry in Britain*, **9**, 54 (1973).
54.- R. Seiwartz and M. Green, *J. Appl. Phys.*, **29**, 1034 (1958).
55.- W. H. Brattain and C. G. B. Garrett, *Beel Syst. Tech. J.*, **34**, 129 (1955).
56.- M. Green, *J. Chem. Phys.*, **31**, 200 (1959).
57.- M. Green, in *Modern Aspects of Electrochemistry*, J. O'M. Bockris and B. E. Conway, Eds. (Butterworkrths & Co (Pub.), London, 1959) Vol 2, Ch. 5.

58.- K. Chandrasekaran, R. C. Kainthla, and J. O'M. Bockris, *Electrochim. Acta,* **33**, 327 (1988).
59.- S. R. Morrison, in *Electrochemistry at Semiconductor and Oxidized Metal Electrodes* (Plenum Press, New York, 1980) Ch. 2.
60.- K. Chandrasekaran and J. O'M. Bockris, *Electrochim. Acta,* **32**, 1393 (1987).
61.- M. A. Butler, M. Abramovich, F. Decker, and J. F. Juliao, *J. Electrochem. Soc.*, **128**, 200 (1981).
62.- R. Haak and D. Tench, *J. Electrochem. Soc.*, **131**, 275 (1984).
63.- J. O'M. Bockris and S. U. M. Khan, *J. Electrochem. Soc.*, **132** 2648 (1985).
64.- J. O'M. Bockris, M. Szklarczyk, A. Q. Contractor, and U. M. Khan, *Int. J. Hydrogen Energy,* **9**, 741 (1984).
65.- J. O'M. Bockris and A. K. N. Reddy, *Modern Electrochemistry,* Plenum Press, Vol.2, Ch. 3, p. 199.
66.- M. A. Habib and J. O'M. Bockris in *Comprehensive Treatise of Electrochemsitry,* J. O'M. Bockris, B. E. Conway and E. Yeager, Eds. (Plenum Press, New York, 1980) Vol 1, Ch. 4.
67.- K. M. Joshi and R. Parsons, *Electrochim. Acta,* **4**, 129 (1961).
68.- R. Parsons and S. Trasatti, *Trans. Faraday Soc.*, **65**, 3314 (1969).
69.- N. D. Lang and W. Kohn, *Phys. Rev.* **B3**, 1215 (1971).
70.- J. P. Badiali, *Electrochim. Acta,* **31**, 149 (1986).
71.- L. Blum, D. Henderson, and R. Parsons, *J. Electroanal. Chem.*, **161**, 389 (1984).
72.- D. Hendersons, in *Trends in Interfacial Electrochemistry,* A. F. Silva, Ed., (Reidel, Dordrecht, Hollland, 1986). p. 183.
73.- Vasilyer and Bagotzky, *"Fuel Cells: their Electrochemical Kinetics"*, Vasilyer and Bagotzky, Eds. (Tram. Consultants Bureau, New York, 1966).
74.- R. W. Reive, J. O'M. Bockris and B. G. Baker, *Surf. Sci.*, **52**, 664 (1975).
75.- R. W. Reive, B. G. Baker, and J. O'M. Bockris, *J. Electrochem. Soc.*, **122**, 1460 (1975).
76.- A. T. Hubbard, *J. Electroanal. Chem.* **69**, 313 (1976).
77.- A. T. Hubbard, *Crit. Rev. Anal. Chem.*, **3**, 201 (1973).
78.- O. J. Murphy, J. O'M. Bockris, T. E. Pou, D. L. Cocke, and G. Sparrow, *J. Electrochem. Soc.*, **129**, 2149 (1982).
79.- T. E. Pou, O. J. Murphy, V. Young, and J. O'M. Bockris, *J. Electrochem. Soc.*, **131**, 1244 (1984).
80.- O. J. Murphy, T. E. Pou, J. O'M. Bockris, and L. L. Tungsen, *J. Electrochem. Soc.*, **131**, 2785 (1984).
81.- J. McHardy, *Ph.D. Thesis*, University of Pennsylvania (1970).
82.- A. Wieckowski, *private communication*, 1988.
83.- R. F. Laane and A. T. Hubbard, *J. Phys. Chem.*, **79**, 808 (1975).
84.- Sh. Sh. Khidirov and V.E. Kazarinov, *Elektrokhimiya,* **5**, 616 (1969).
85.- G. Horanyi and E. M. Rizmayer, *J. Electroanal. Chem.*, **215**, 369 (1986).
86.- J. L. Stickney, S. D. Rosasco, G. N. Salaita, and A. T. Hubbard, *Langmuir,* **1**, 1 (1985).
87.- P. Zelenay, M.A. Habib and J.O'M. Bockris, *Langmuir,* **2**, 393 (1986).

# SOME RECENT SPECTROSCOPIC APPROACHES TO THE SOLID-SOLUTION INTERFACE

J. O'M. BOCKRIS and M. GAMBOA-ALDECO
*Surface Electrochemistry Laboratory*
*Texas A&M University*
*College Station, TX 77843*

A brief account is given of the different spectroscopic approaches used to study the solid-solution interphase. It gives information concerning the structure of the interphase on noble metals,on iron and on numerous semiconductors in contact with aqueous and non-aqueous solutions. It represents the first comprenhensive document which treats structure at the interface as a function of potential and concentration in solution.

## 1. Introduction.

### 1.1 VARIOUS IN-SITU METHODS OF INVESTIGATION.

In-situ methods of looking at electrodes in solution were commenced by Blomgren and Bockris in a paper in 1960 in which, for the first time, they showed *in-situ radiotracer* measurements of potential-dependent adsorption[1].

In-situ optical methods were introduced into electrochemistry by Reddy, Devanathan and Bockris in 1963[2,3]. The method at that time was called "*chronoellipsometry*" because an ellipsometer was used simply to signify the formation of a film at a certain time. *Ellipsometry in electrochemistry* with potential control was described by Reddy and Bockris in 1964[4].

Determination of adsorption by the *measurement of the resistance of thin layers* was introduced by Bockris, Cahan and Stoner in 1969,[5] and the first *study of plasmon in electrochemistry* was carried out by Peik and Bockris in 1971.[6] In the same year, Bockris, Damjanovic and Fredlein introduced the *bending beam method* of determining surface tension on solid-solution interfaces[7]. In 1975, Bockris and O'Grady published the first application of Mössbauer *spectroscopy to electrochemistry* where they discribe the amorphous character of passive films[8].

*Ex-situ methods in high vacuum conditions* with transfer from solution to high vacuum in the same equipment were introduced by Revie, Baker, and Bockris in 1975[9]. The impression that these methods were



*C. Gutiérrez and C. Melendres (eds.),*
*Spectroscopic and Diffraction Techniques in Interfacial Electrochemistry,* 55–102.

first reported by Hubbard has some validity in the fact that an abstract[10] was published by Hubbard about his work before the Revie et al.[9] paper of 1975. However, Hubbard's first paper on ex-situ electrochemistry was not published until 1976[11].

Ellipsometry was developed extensively during the 1960's, particularly for work on passive layers[2-4,12-15]. It was then applied in 1969 by Chiu and Genshaw[16] to the measurement of the adsorption of ions in submonolayer amounts. A complete solution to the equations for films to obtain the three unknowns (thickness, refractive index and absorption coefficient) was developed by Paik and Bockris[17], who included the reflection of the parallel component of light as the third variable.

The potentialities of the ellipsometric method seem great, so long as it is developed into the IR and utilized for adsorption measurements, because the time response of the ellipsometric parameters is less than those of present FTIR work (see below). Hitherto, however, *IR ellipsometry* for metal/solution interphases has not been developed extensively[18].

IR spectra at electrodes were first mentioned by Hansen, Kuwana and Osteryoung in 1966[19]. However, these early attempts involved total internal reflection methods, and are thought to yield information which may have included data from the solution near the surface.

Table I shows a summary of the first-paper contributions to *in-situ and ex-situ* methods for the investigation of the double layer.

The IR methods have been greatly improved in recent years. The introduction of *Fourier Transform Spectroscopy*[20,21] and *Potential Modulated External Reflectance IR Spectroscopy*[22] have proved to be useful in the application to electrochemical systems.

Later development of the method depends much upon a theorem due to Greenler[23]. This theorem gives rise to a unique surface selection rule arising from the physics of reflection of radiation from highly conducting surfaces. Fig. 1 shows the reaction of both perpendicular (s-) and parallel (p-) polarized light upon striking a surface. Thus, in the lower left diagram of fig. 1, the vectors arising from the s-polarized light cancel in respect to the information obtained at the surface, i.e., the s-polarized light does not interact with the material on the surface. On the other hand, p-polarized light, as shown in the right bottom diagram of fig. 1, results in vectors which add, so information from both, the solution and the electrode is retained.

The upper part of fig. 1 shows the phase shift as a function of the angle of incidence. Thus, the large change in phase shift between the p- and s- modes of light occurs only at angles of incidence which are above about 60$^{o}$, and particularly, above 80$^{o}$. Thus, if the reflected beam is arranged so that the perpendicular mode of polarization (containing information from the solution) is subtracted from the parallel mode (containing information from the solution and from the surface), the result, looked at in a simple way (see below) should be information from the surface only.

Another important consequence due to this reflection rule is that only those fundamental vibrations of adsorbed species which have a finite value of the dipole derivative perpendicular to the surface are

Table I. Contributions to Methods for the investigation of the Double Layer.

| | | |
|---|---|---|
| In-Situ Radiotracer | E. Blomgren; J.O'M. Bockris | *Nature* 186(1960)305 |
| Ellipsometry | J.O'M. Bockris; M.A.V. Devanathan; A.K.N. Reddy | *J. Electroanal. Chem.* 6(1963)61 |
| Resistance of Thin Layers | J.O'M. Bockris; B.D. Cahan; G. Stoner | *Chem. Inst.* 1(1969)273 |
| Mössbauer Spectroscopy | W.E. O'Grady; J.O'M. Bockris | *Chem.Phys.Lett.* 5(1970)116<br>*Surf. Sci.* 38(1973)249 |
| Absolute Ellipsometric Measurements | W. Paik; J.O'M. Bockris | *Surf. Sci.* 28(1971)61 |
| Plasmon Spectroscopy | W. Peik; J.O'M. Bockris | *Surf. Sci.* 27(1971)191 |
| The Bending Beam Method | A. Fredlein; A. Damjanovic; J.O'M. Bockris | *Surf. Sci.* 25(1971)261 |
| In-Situ SERS (Surface Enhanced Raman Scattering) | M. Fleischmann, P.J. Hendra; A.J. McQuillan | *Chem. Phys. Lett.* 26(1974)163 |
| Auger Electron Spectroscopy (u.h.v. method) | R.W. Revie; B.G. Baker; J.O'M. Bockris | *Surf. Sci.* 52(1975)664 |
| In-Situ IR Spectroscopy | S. Bewick; K. Kunimatsu; S. Pons | *Electrochim. Acta* 25(1980)465 |
| In-Situ EXAFS (Extended X-Ray Absorption Fine Structure) | G.G. Long; J. Kruger; D.R. Black; M. Kuriyama | *J. Electroanal. Chem.* 150(1983)603 |

**REFLECTED LIGHT**

**PHASE CHANGE OF REFLECTED LIGHT DEPENDS UPON**
1. **ANGLE OF INCIDENCE**
2. **STATE OF POLARIZATION**

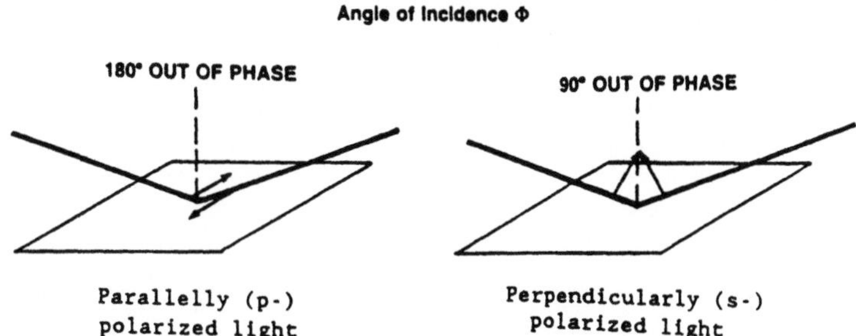


Fig. 1 Phase change of p- and s- polarized light upon reflection on a surface.

able to undergo interaction with the radiation. This means that it is possible to deduce the orientation of species relative to the electrode surface from the spectra. For example, the infrared band for CO is expected to be stronger for the perpendicular orientation than for the parallel orientation (fig. 2)[24].

In the laboratory at Texas A&M University, two methods have been used, one of which is original to the laboratory. The first one, SNIFTIRS (Subtractively Normalized Interfacial Fourier Transform Infrared Spectroscopy)[24,25] consists of obtaining a spectrum using the parallel mode of polarization of light at a designated potential at which results are wished to be obtained (adsorption potential) and subtracting from it the spectrum under the same intensity of the parallel polarized light by taking it at a potential where there is independent evidence that no adsorption exists (reference potential). Finally the obtained spectrum is normalized with respect to the intensity at the reference potential (c.f. fig. 3). Thus,[1]

$$A = -\frac{(I_p)_V - (I_p)_{Ref}}{(I_p)_{Ref}} \qquad (1)$$

where A is the absorbance, $(Ip)_V$ is the reflected light intensity of parallel polarized light at the chosen potential and $(Ip)_{Ref}$ is the reflected light intensity at the reference potential at which adsorption is taken to be negligible.

Although such a method has been used extensively in the laboratory[24,26,28], a far better method is to use PM-FTIRRAS (Polarization Modulation Fourier Transform Infrared Reflection-Absorption Spectroscopy)[22,29]. In this method, the basic idea is to follow Greenler theorem and modulate the modes of polarization of light, subtracting the intensity of the reflected perpendicular from that of the parallel polarized light. It is to be stressed that the resultant of the substraction contains information predominantly from the surface only when the angle of incidence is a high one (see the upper diagram on fig. 1).

Thus, in a measurement using FTIRRAS, one obtains a parallel beam, $I_p$, and a perpendicular beam, $I_s$, after reflection from the electrode, and then evaluates the absorbance. Hitherto, the presentation has followed the understanding of Greenler's theorem which had been applied to interphases which involve solids and gases. At such interphases, all the absorption can be attributed to changes at the metal/gas interphase. Nevertheless in the electrochemical situation, when a metal or a semiconductor is in contact with a <u>solution</u> further changes may occur, e.g. variation of the concentration in the region close to the electrode, or when using aqueous solutions, strong IR absorbance by the

---

[1]Absorbance=$\alpha cx$, where $\alpha$ is the absorption coefficient, x is the path length and c is the concentration of the solute. Then, $I=I_o\exp(-\alpha cx)$, $\Rightarrow I-I_o=I_o(\exp(-\alpha cx) - 1)$. For small absorbances,

$$(I-I_o)/I_o = (1-\alpha cx)-1 \Rightarrow \text{Absorbance} = -(I-I_o)/I_o = \alpha cx$$

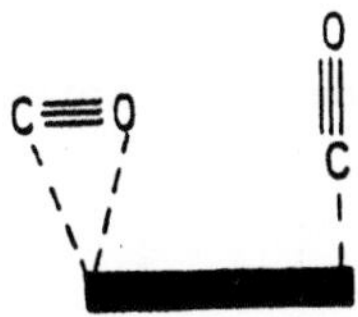


Fig. 2 Possible orientations for CO adsorbed at a surface[20].

**SNIFTIRS**

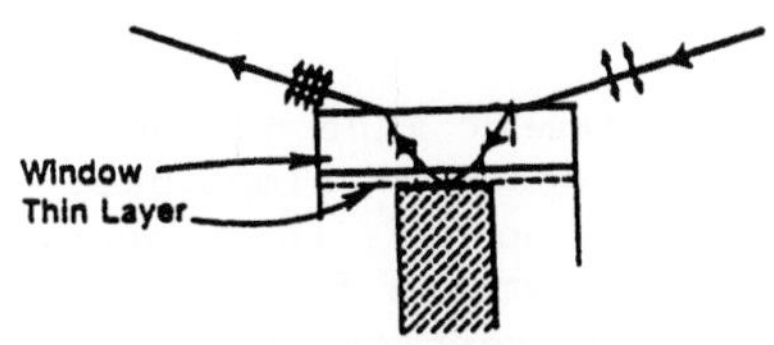


**Reflectance apparatus for the SNIFTIRS techniques.**

$$\text{Absorbance} \approx -\frac{\Delta R}{R} \approx -\frac{(I_p)_V - (I_p)_R}{(I_p)_R}$$

Fig. 3 Subtractively Normalized Interfacial Fourier Transform Infrared Spectroscopy.

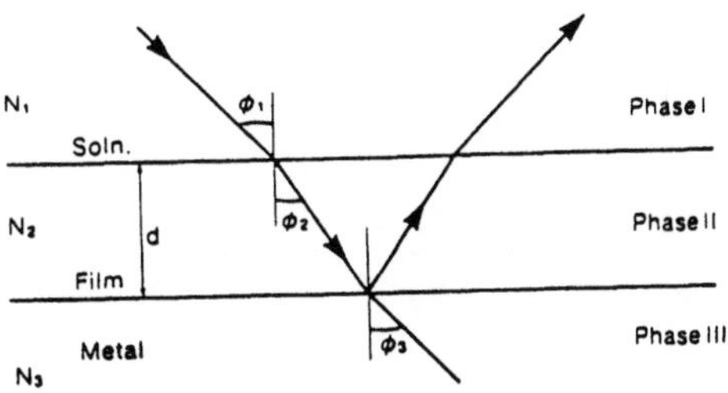


Fig. 4 Reflection at the three-phase interphase. $d$=film thickness; $\emptyset_1$= angle of incidence; $\emptyset_2$ and $\emptyset_3$ are the angles of refraction in the film and substrate, respectively.

water molecules. To avoid this, therefore, a fiducial reading is taken, similar to that with SNIFTIRS, and subtracted from the expression for the measurement taken at the potential V,[2]

$$A = -2\left[\frac{I_p - I_s}{I_p + I_s}\right]_V + 2\left[\frac{I_p - I_s}{I_p + I_s}\right]_{Ref} \tag{2}$$

Potential modulation methods (to be distinguished from SNIFTIRS)are less desirable because ionic substances, in particular organic materials, take substantial times to adsorb and to desorb. For example, in the case of the adsorption of benzene from 1 M $H_2SO_4$, when V = 0.1 V (NHE), and $\theta_{eq}$ is 0.15, the time for reaching equilibrium is in the region of minutes[30]. Thus, the relaxation time for adsorption under diffusion control is given by[30] $\tau = K\delta/D$, where K is the equilibrium constant for the equilibrium between adsorbed material and the solution, D is the diffusion coefficient, and $\tau$ is the relaxation time (K is in seconds, $\delta$ in cm, and D in $cm^2\ sec^{-1}$).

## 1.2. BASIC THEORY OF SURFACE INFORMATION FROM FTIR.

It has been noted above that the statement made of the FTIR method in polarization modulation was based upon a simplistic version of Greenler's theorem. Thus, investigation of the theory about the dependence of the reflected beam upon distance from the electrode shows that the difference of the parallel and vertical beam does not drop to zero immediately, but there is some residual intensity even up to $\lambda/2$ from the electrode[29]. Because the adsorption region on the electrode is less than 10 Å thick, this extra information from the solution, remaining after the subtraction, is a danger to the validity of the method and must be quantitatively analyzed.

Chandrasekaran and Bockris[29] made a calculation of the value of the absorbance separately for parallel and vertical modes of polarization applied as a function of distance to the electrode surface. The following account is essentially theirs.

Fresnel coefficients for an interface for perpendicular and parallel polarized light for two-phase systems are given by[29] (cf. fig. 4)

$$r_{\perp}(1,2) = \frac{N_1 \cos \Phi_1 - N_2 \cos \Phi_2}{N_1 \cos \Phi_1 + N_2 \cos \Phi_2} \tag{3}$$

$$r_{\parallel}(1,2) = \frac{N_2 \cos \Phi_1 - N_1 \cos \Phi_2}{N_2 \cos \Phi_1 + N_1 \cos \Phi_2} \tag{4}$$

where $N_1$ and $N_2$ are the complex refractive indices at phase 1 (solution) and phase 2 (metal), $\Phi_1$ is the angle of incidence and $\Phi_2$ is the angle the light makes with respect to the normal inside the metal

---

[2] Since $I \sim I_o$, $A \sim (I_o - I)/I = 2(I_o - I)/2I = 2(I_o - I)/(I_o + I) = -2(I - I_o)/(I + I_o)$. Considering that $I = I_p$ and $I_o = I_s$, then $A = -2(I_p - I_s)/(I_p + I_s)$.

(fig. 4). Similarly, for a three-phase system (fig. 4), the equivalent Fresnel coefficients are[31]

$$r_{\perp}(1,2,3) = \frac{r_{\perp}(1,2) + r_{\perp}(2,3)e^{-2i\beta}}{1 + r_{\perp}(1,2)r_{\perp}(2,3)e^{-2i\beta}} \tag{5}$$

$$r_{\parallel}(1,2,3) = \frac{r_{\parallel}(1,2) + r_{\parallel}(2,3)e^{-2i\beta}}{1 + r_{\parallel}(1,2)r_{\parallel}(2,3)e^{-2i\beta}} \tag{6}$$

where $\beta$ is the change in phase of the beam during its travel through the phase 2 of thickness $d$;

$$\beta = 2\pi N_2(d \cos \Phi_2)/\lambda \tag{7}$$

and where $\lambda$ is the wavelength.

For a three-phase system consisting of solution, adsorbed layer and metal substrate, $N_1$, $N_2$ and $N_3$ are the complex refractive indices of the solution, thin film and the metal, respectively. In the absence of a thin film, $\beta$ is zero. Then

$$r_{\perp}(1,3) = \frac{r_{\perp}(1,2) + r_{\perp}(2,3)}{1 + r_{\perp}(1,2)r_{\perp}(2,3)} \tag{8}$$

$$r_{\parallel}(1,3) = \frac{r_{\parallel}(1,2) + r_{\parallel}(2,3)}{1 + r_{\parallel}(1,2)r_{\parallel}(2,3)} \tag{9}$$

Reflectivity R of a three-phase system can be calculated from[31]

$$R_{\perp}(1,2,3) = |r_{\perp}(1,2,3)|^2 \tag{10}$$

$$R_{\parallel}(1,2,3) = |r_{\parallel}(1,2,3)|^2 \tag{11}$$

and reflectivity in the absence of film is

$$R_{\perp}(1,3) = |r_{\perp}(1,3)|^2 \tag{12}$$

$$R_{\parallel}(1,3) = |r_{\parallel}(1,3)|^2 \tag{13}$$

The ratio of reflectivity with and without film can be calculated from r(1,2,3) and r(1,3)[31],

$$\frac{R_{\perp}(d)}{R_{\perp}(0)} = \left|\frac{r_{\perp}(1,2,3)}{r_{\perp}(1,3)}\right|^2 \tag{14}$$

$$\frac{R_{\parallel}(d)}{R_{\parallel}(0)} = \left|\frac{r_{\parallel}(1,2,3)}{r_{\parallel}(1,3)}\right|^2 \tag{15}$$

where R(d) is the reflectivity in the presence of the film thickness d

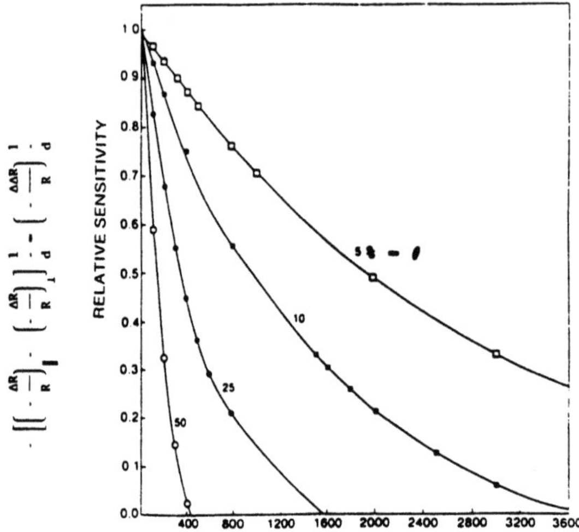


Fig. 5 Relative absorbance as a function of thickness of the film. Relative absorbance=$(-\Delta\Delta R/R)/d$. Values normalized to unity at the surface.

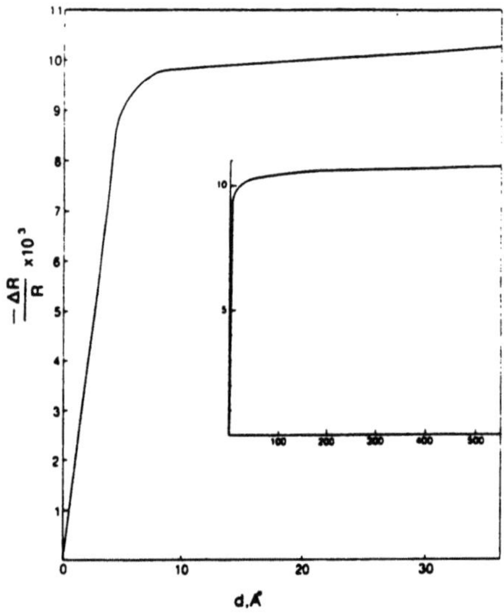


Fig. 6 Absorbance for the surface and solution species as a function of distance at a surface coverage of 0.5.

and R(0) is the reflectivity in the absence of a film. The absorbance[3] of the thin film for a given *d* can be calculated for the two kinds of polarizations,

$$\left(-\frac{\Delta R}{R}\right)_{\perp} = -\frac{R_{\perp}(d) - R_{\perp}(0)}{R_{\perp}(0)} = -\left[\frac{R_{\perp}(d)}{R_{\perp}(0)} - 1\right] \tag{16}$$

$$\left(-\frac{\Delta R}{R}\right)_{\parallel} = -\frac{R_{\parallel}(d) - R_{\parallel}(0)}{R_{\parallel}(0)} = -\left[\frac{R_{\parallel}(d)}{R_{\parallel}(0)} - 1\right] \tag{17}$$

Thus, equations 16 and 17 show the absorbance of thin films, and it is possible therefore theoretically to calculate this absorbance from a knowledge of the thickness of the layer, the complex refractive indices, and the angle of incidence for the perpendicular and parallel polarized light. The modes of obtaining these particular quantities are given by the authors[29].

The absorbance as calculated from equations 16 and 17 above are shown for several coverages of $CO_2$ in fig. 7, as a function of the distance from the electrode for an angle of incidence of 87°[29]. The absorbance of the perpendicularly polarized light (not shown in fig. 7) is about $10^2$ times smaller than that of parallelly polarized light. In fig. 5, a plot is made of

$$-\left[\left(-\frac{\Delta R}{R}\right)_{\parallel} - \left(-\frac{\Delta R}{R}\right)_{\perp}\right]\frac{1}{d} = \left(-\frac{\Delta\Delta R}{R}\right)\frac{1}{d} \tag{18}$$

This "relative sensitivity" is a function of distance, and the values are normalized to unity at the surface. In a general way it is seen that the substantial contributions to the relative sensitivity exists particularly for low coverages, even at 2000 Å from the electrode. The solution contribution to the absorbance at various coverages is shown in Table II. Thus, the surface contribution can be as low as 44% at 0.1 coverage.

The substantial absorbance which is shown in fig. 5 looks less threatening when one takes into account the concentration difference between the surface and the film. The final values are plotted in fig. 6 (it is these values which give rise to the surface contribution of the table). Nevertheless, these values have sufficient information from the solution to justify the use of the fiducial subtraction (c.f. eq. 2), which wipes out those contributions which are not directly from the surface.

## 1.3. SUPPORTING EVIDENCE FOR THE SURFACE CHARACTER OF THE INFORMATION.

(1) In the measurements made for the difference spectrum of ammonium ions at 0.1 M adsorbed on p-GaP in acetonitrile containing 0.5 M $LiClO_4$

---

[3] Absorbance $= -\log(R/R_{ref}) = -\log[(R_{ref} + R - R_{ref})/R_{ref}] = -\log[1 + \Delta R/R_{ref}]$. Since $\Delta R/R_{ref} \ll 1$, then Absorbance $= -\Delta R/R_{ref}$[27].

Table II. Solution contribution to the absorbance at various coverages before the subtraction of the spectrum at reference potential.

| Coverage | Absorbance | |
|---|---|---|
| | Surface contribution (%) | Solution contribution (%) |
| 0.50 | 97.5 | 2.5 |
| 0.25 | 85.6 | 14.4 |
| 0.10 | 44.5 | 54.5 |

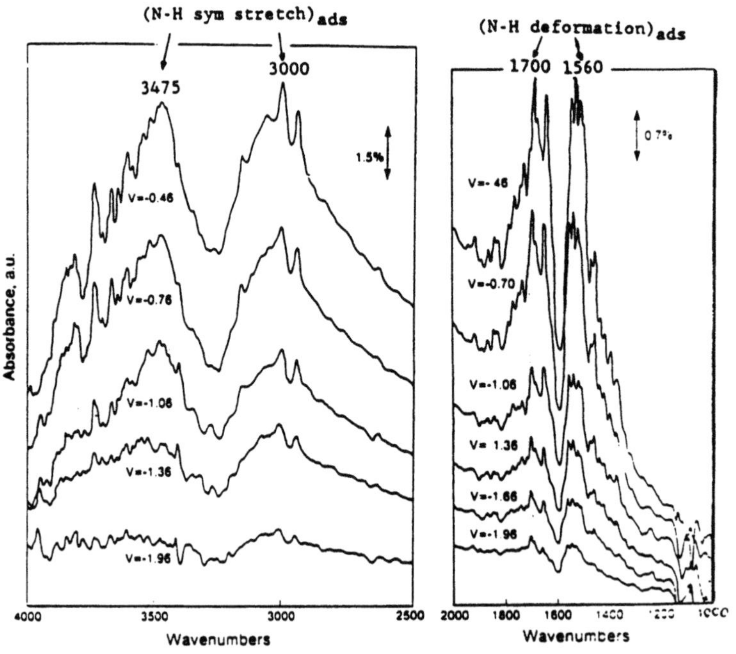


Fig. 7 Absorption spectrum of ammonium ion on GaP in acetonitrile.

[29], there are two peaks of equal intensity corresponding to the N-H symmetric stretching vibrations centered at 3475 and 3000 $cm^{-1}$, and two more peaks of equal intensity in the region 1500-1800 $cm^{-1}$ corresponding to N-H deformation vibrations (fig. 9) (c.f. Section 9c).

If the measurement represented information from the solution, one would obtain six peaks, four for the adsorbed species (those described before),and two for the solution species[29]: this is not observed.

(2) The dependence of the peak areas of many species upon potential in a rational way is, of course, an indication of the surface nature of the information (e.g. fig. 13, 32, 38).

(3) The work of Nichols and Bewick[32] on FTIR spectra obtained from hydrogen on single crystal platinum, as well as on other noble metals, varies reasonably with the potential[4], and this variation is in fair quantitative accordance with the prediction of the variation of surface coverage with potential[33].

(4) These specific examples, however, have recently been added to in a remarkable way by the work of Scharifker et al.[34] who measured the adsorption of sulfate ions on platinum. When accidental poisoning of the solution with copper took place under conditions in which the copper became deposited on the platinum electrode, the authors found a radical change in the dependance of sulfate adsorption on potential: the new behavior corresponded to the results obtained independently by radiotracer measurements on copper[35] (fig.8). This evidence appears to provide a simple and direct proof of the surface nature of the FTIR information, at least for sulfate ions on metals.

In summary, the FTIR technique is capable of determining the variation of adsorbed hydrogen, adsorbed inorganic ions, and adsorbed organic radicals. However, absolute determinations without calibration by another method is more difficult but will be discussed below.

## 2. Potential-Dependent Water Orientation: An In-Situ Spectroscopic Study.

Most of the studies of water at interfaces have been theoretical[36-38]. Few measurements have been made of the interface from the point of view of the spectrum of water. Bewick and Kunimatsu[22] concluded that several polymers of water existed on the surface of platinum but more recently they have modified this view to one in which the main species present are monomers and dimers[40]. Recently, a paper was published by Bockris and Habib[41] in which special attention was given to an attempt to obtain information concerning the orientation of water molecules on the surface.

Fig. 9 shows the differential IR spectra (SNIFTIRS) expressed in

---

[4]The work of Bockris and Carbajal[33] in the spectra obtained for hydrogen on polycrystalline iron shows a variation of potential as it corresponds to the expected one for the mechanism of proton discharge followed by hydrogen recombination. A similar work by Bewick and Nichols[32] for hydrogen on Pt presents parallel evidence.

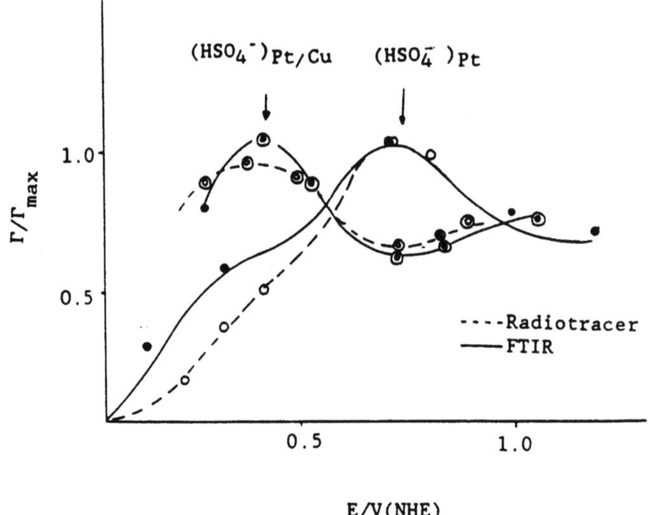


Fig. 8 Relative surface coverage for sulfate ion adsorption as a function of potential. FTIR and Radiotracer[35] results are shown. The integrated absorption peak was the stretching symmetric vibration of $HSO_4^-$.

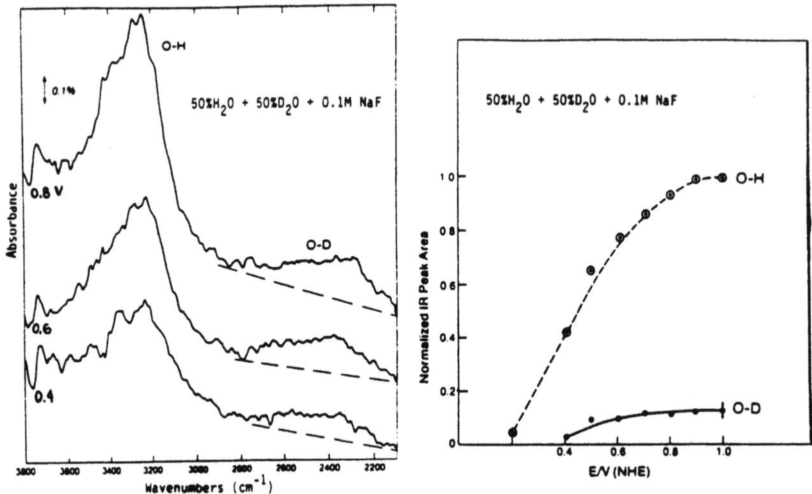


Fig. 9 O-H and O-D vibration bands at various potentials. Differential IR spectra. Reference potential is 0.1V (NHE).

Fig. 10 Comparison of O-H and O-D vibrational peak areas as a function of potential.

percent absorbance of the O-H and O-D stretching vibration frequency for adsorbed water molecules on platinum in a solution of 50% $H_2O$ + 50% $D_2O$ at various potentials with respect to the spectrum at 0.1 V (NHE) in 1 M NaF. The areas under the IR peaks were normalized to unity at the maximum for the O-H vibration and are shown for adsorbed water on platinum as a function of potential in fig. 10. Here, the normalized IR peak for O-D is also shown.

The selection rule for IR discussed before establishes that the interaction of a dipole oscillator near the reflecting metal surface with incident radiation will be different depending upon what is the orientation of the dipole relative to the surface[23]. Thus, at 0.1 V (NHE) the water molecules on or near the surface are evidently oriented parallel to the surface, i.e. both hydrogen and oxygen atoms in a water molecule lie on the surface and the average dipole derivative perpendicular to the surface is zero. This is evidenced by the minimum energy configuration of adsorbed water molecules, which corresponds to an alignment of dipoles in the plane to the two-dimensional array, i.e. parallel to the surface plane, as shown by Parsons and Reeves[42].

As the potential is increased anodically, the oxygen end of the water dipole becomes more oriented towards the surface. This results in a net dipole-moment increase, resulting in a corresponding enlargement of the intensity of the IR absorption band.

The calculated degree of orientation of adsorbed water molecules, ($N\uparrow/N_T$), using a version of the BDM model of 1963[38] with correction for dimer formation on the surface[43], is compared with experiment in fig. 11. This degree of orientation of the water molecules with their oxygen end towards the surface may be calculated as a function of the electric field in the double layer, X, and lateral interaction energy, U, by

$$\frac{N\uparrow}{N_T} = (1-\theta\uparrow)\ \exp\ [-(\Delta G_c^\circ\uparrow-\mu X + Uc\theta\uparrow)/kT] \tag{19}$$

where $N_T$ is the total number of sites containing adsorbed water, $\mu$ is the dipole moment of water, c is the coordination number, $\theta\uparrow$ is the fractional coverage by the "up" water molecules, $\Delta G^\circ\uparrow$ is the chemical interaction energy of the dipole with the metal, and X is the double layer field given by $X = (4\pi q_M/\epsilon)$ where $q_M$ is the charge density on the electrode and $\epsilon$ is the dielectric constant of the adsorbed water layer.

An obvious question to ask is whether the zero intensity situation does correspond to the pzc on Pt, as would be expected in the model. Determinations of the pzc at different pH's have been made by Gileadi, Argade and Bockris[44]. The value found at pH 7 is about 0.18 V (NHE), and this is in very reasonable agreement with the value of 0.2 seen in the diagrams, i.e., the model fits the results to an unusual degree. Fig. 10 also shows that the peak areas due to O-D vibration are about 4-8 times less than that due to O-H vibration. This difference in vibrational intensity between O-D and O-H vibration in the adsorbed OH or OD containing species, may be explained in terms of their difference in hydrogen bonding, because the intensity of the O-H or O-D vibration in water depends very critically on the extent of hydrogen bonding to the immediate environment. A similar behavior was found for species in solution[45,46].

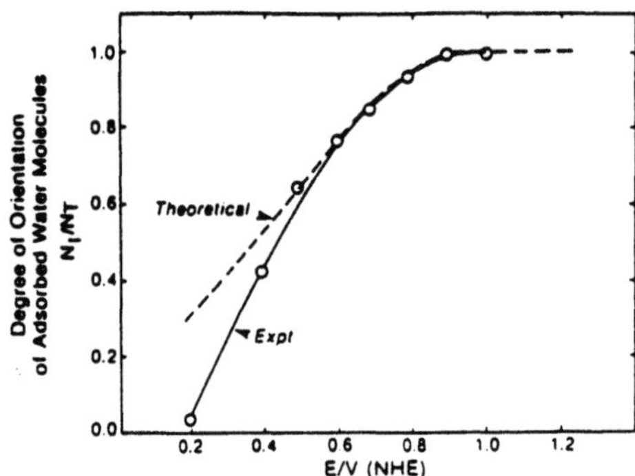


Fig. 11 Normalized IR peak areas for O-H vibration in adsorbed water molecules on Pt as a function of potential, compared with the degree of orientiation calculated theoretically. ($\Delta G_{C\uparrow}/kT$=-0.1 and Uc/kT=2.5)

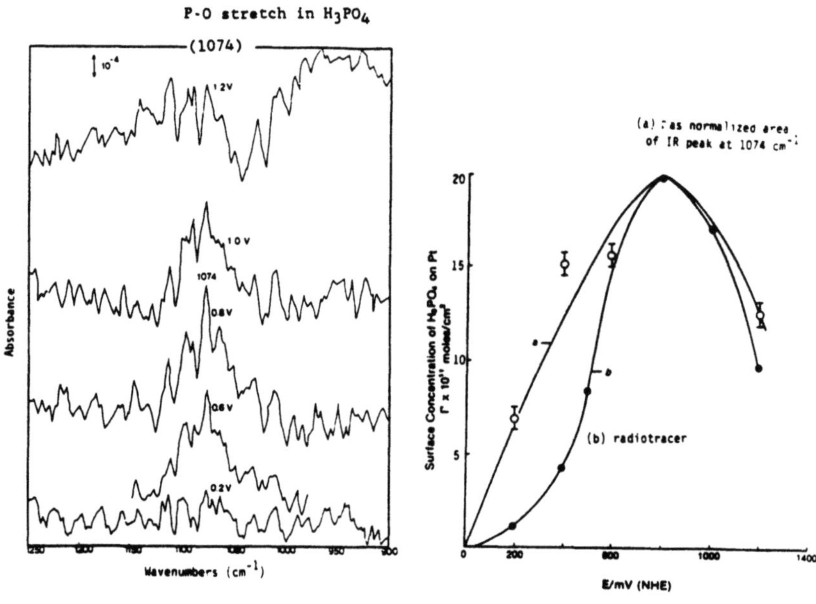


Fig. 12 Differential IR spectra of $H_3PO_4$ species adsorbed on Pt. 1M $HClO_4$+6.6x$10^{-3}$M $H_3PO_4$.

Fig. 13 Surface concentration, Γ, of $H_3PO_4$ on Pt as a function of potential.

## 3. Spectroscopic Evidence Concerning the Adsorption of Phosphoric Acid and Trifluoromethane Sulfonic Acid at the Solid-Solution Interface.

The progress of fuel cell research has received considerable impetus during the last ten years from the discovery[47] that certain acids, particularly those containing fluoride ions, have a greatly enhanced effect upon the rate of production of oxygen at a given overpotential. Therefore they would seem favorable to use as acids in fuel cell operation.

The problem at issue here was to find out adsorption characteristics of the two solvent species[48,49]. Thus, it was suspected that, for phosphoric acid in which oxygen reduction was so much slower than in trifluoromethane sulfonic acid, a high degree of coverage had given rise to blocking of the surface. The corresponding anion coverage from the TFMSA would be expected to be much smaller.

In this particular research[48], an FTIR study of the adsorption from aqueous solution of phosphoric acid onto Pt and Au electrodes at various concentrations and potentials was carried out. The spectra measured were of parallel polarized IR radiation, and the method used was the SNIFTIRS approach.

Fig. 12 shows the situation for the differential spectra using p-polarized light and $6.6 \times 10^{-3}$ M $H_3PO_4$ at different potentials, using the spectra at 0.0 V (NHE) as the reference. The normalized areas of these peaks as a function of potential are shown in fig. 13, where the results of radiotracer measurements are also shown. As was discussed before, these results represent the surface concentration of phosphoric acid on Pt.

It is of interest to note that the species present in solution for phosphoric acid at 25°C are given by the dissociation reactions[50]:

$$\begin{aligned} H_3PO_4 &= H^+ + H_2PO_4^- & K_1 &= 10^{-2.15} \\ H_2PO_4^- &= H^+ + HPO_4^{=} & K_2 &= 10^{-7.2} \\ HPO_4^{=} &= H^+ + PO_4^{\equiv} & K_3 &= 10^{-12.3} \end{aligned} \tag{20}$$

where $K_1$, $K_2$ and $K_3$ are the respective dissociation constants at room temperature. Due to the presence of 1 M $HClO_4$ (base electrolyte), about 1% of the $H_3PO_4$ dissociates to give $H_2PO_4^-$ ions at room temperature, and about 99% remains as undissociated acid.

In solution, $H_3PO_4$ has a spectral band 900-1100 $cm^{-1}$ due to the P-O stretching vibration[51]. The differential spectra of the platinum electrode surface at different potentials with $H_3PO_4$ in solution are shown to have a major peak at 1074 $cm^{-1}$, which is in the region of the P-O stretching frequency. This peak (fig. 12) first increases with increase of anodic potential, reaches a maximum, and then decreases with increase of potential. Were it due to the P-O vibration in the $H_2PO_4^-$ anion, it would be difficult to explain the reduction in the height of the peak as the potential becomes more anodic (c.f. fig. 13). Thus, it seems likely that this peak comes from the P-O vibration in the $H_3PO_4$ molecule.

Fig. 14 shows the adsorption isotherm for $H_3PO_4$ which varies linearly with the logarithm of the concentration at constant potential,

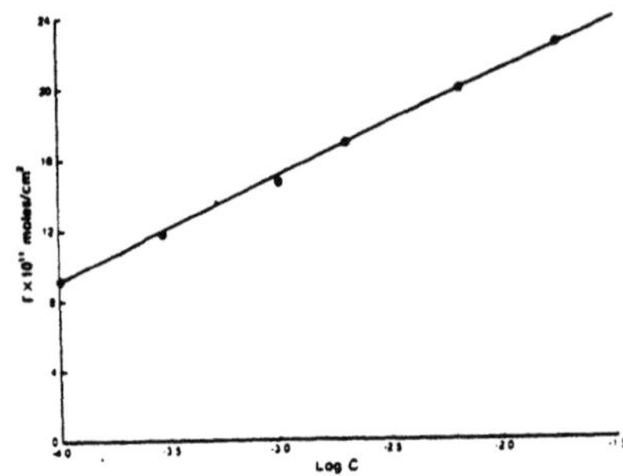


Fig. 14 Variation of the surface concentration of $H_3PO_4$ on Pt with the logarithm of the concentration of $H_3PO_4$ in solution.

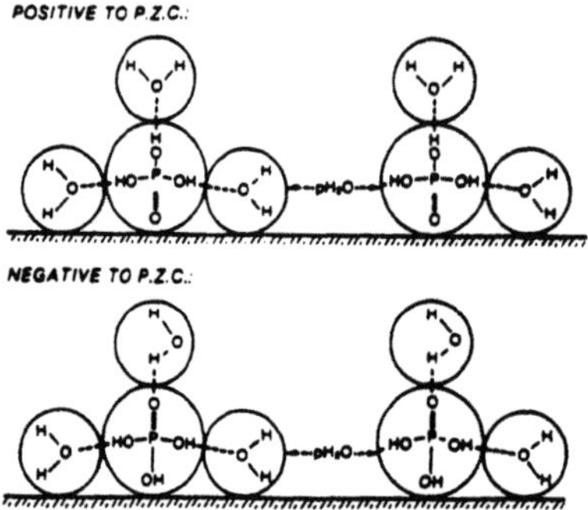


Fig. 15 Schematic model of adsorbed $H_3PO_4$ molecules separated from each other by an adsorbed water molecule.

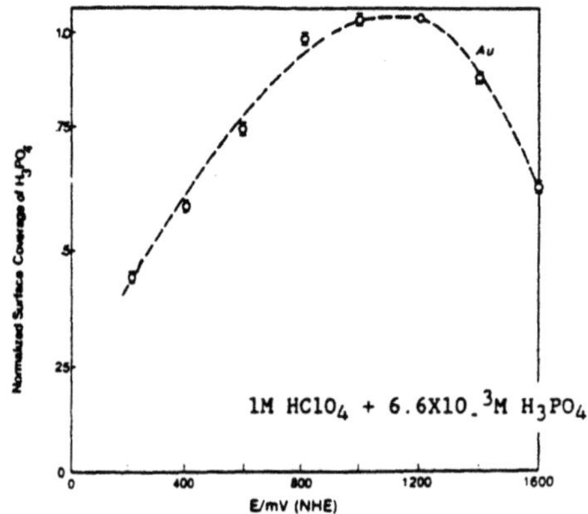


Fig. 16 Areas of IR peaks at 1074 $cm^{-1}$ on Au normalized to 1 at maximum as a function of the elctrode potential.

i.e., $f(\theta) = A \log Kc$, where A is a constant, $\theta$ is the fraction of the surface covered, c is the concentration of $H_3PO_4$ in solution and K the adsorption equilibrium constant. At 800 mV (NHE), the surface concentration is $20\times10^{-11}$ moles $cm^{-2}$ (from radiotracer measurements) [49] and if the area occupied by a single phosphoric acid molecule is 12 $Å^2$ based on the molecular $H_3PO_4$ molecule, the maximum fraction of the surface coverage is only 0.15. This coverage is somewhat low for the Temkin type of isotherm observed. It is more probable that the adsorbed species is laid down on platinum at sites with water molecules in between, as shown in fig. 15. Under these circumstances, the real area occupied by each $H_3PO_4$ molecule will be larger and the coverage will be closer to 0.4, well in the Tempkin region.

The question arises why the coverage decreases at high potentials, as shown in fig. 13. The two factors which may cause such decrease are the displacement of the $H_3PO_4$ by the platinum oxide and/or electrostatic interactions with water molecules.

The measurement on gold (fig. 16) showed that $\theta$ starts decreasing from a higher potential of about 1.1 V, and this is the potential of the gold oxide formation. On the other hand, the symmetry of the $\theta$-V curve seems to indicate the displacement of $H_3PO_4$ from the surface by water molecules.

Similar measurements to those made for phosphoric acid were made for trifluoromethane sulfonic acid, and for both these acids at temperatures up to 100°C[49]. It was found that there was a great difference in the time constant for the adsorption. For phosphoric acid this was about 3 minutes, but for the sulfonic acid less than 1 minute.

The differential spectra of adsorbed $CF_3SO_3^-$ are shown in fig. 17. The wide absorption band in the region from 1400-950 $cm^{-1}$ consists of several peaks. The peak observed at 1400 $cm^{-1}$ is ascribed to the asymmetric stretching vibration of the S-O bond, and is slightly shifted towards lower frequencies as compared with the region 1405-1420 reported in aqueous solution for $CF_3SO_3^-$[52]. In respect to the peaks observed at 1295 and 1155 $cm^{-1}$, they are assigned to the stretching frequencies of the C-F bond, and are within about 10-15 wave numbers of those observed in aqueous solution. The peaks shown in fig. 16 at 1045 $cm^{-1}$ are typical of a symmetric stretching vibration of the S-O bond in $CF_3SO_3^-$, which occurs for the non-adsorbed ion at 1032-1040 $cm^{-1}$[52].

The adsorption isotherm for $CF_3SO_3^-$ at 0.8 V using the radiotracer technique is shown in fig. 18a, and the corresponding results with $H_3PO_4$ are shown in fig. 18b. Comparing with various isotherm plots, it was found that the Bockris-Swinkels isotherm[53], namely,

$$\frac{\theta}{(1-\theta)^n}\,\frac{[\theta+n(1-\theta)]^{n-1}}{n^n} = \frac{C_i}{C_w}K \qquad (21)$$

applies well to the $CF_3SO_3^-$ data. In this equation, n is the number of water molecules displaced by one molecule of adsorbing solute; $C_i$ and $C_w$ is the concentration in moles per liter of the solute and solvent, respectively; and K is the equilibrium constant for adsorption. A modified equation of this same isotherm[49], namely,

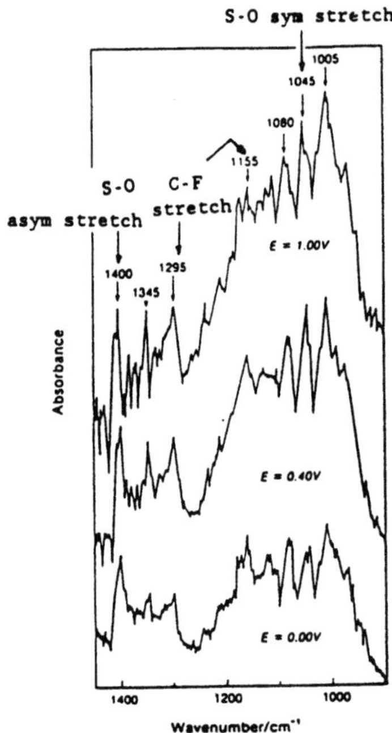


Fig. 17 Differential spectra of adsorbed $CF_3SO_3^-$ on Pt at various potentials. $2x10^{-2}$M $CF_3SO_3^-$; T=298°K

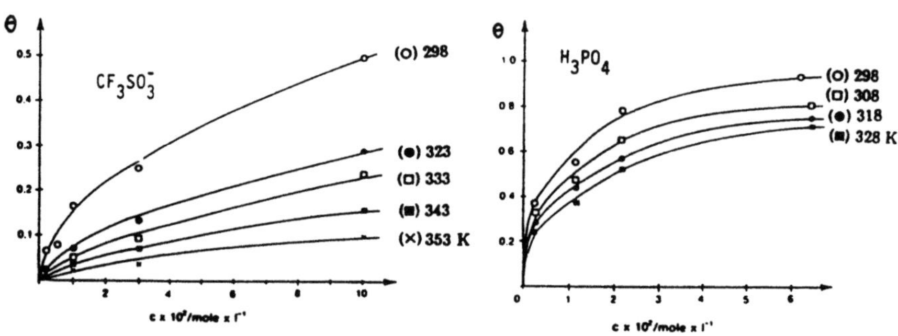


Fig. 18 Adsorption as a function of concentration at various temperatures. E=0.8V (NHE).

$$\frac{\theta}{(1-\theta)^n}\frac{[\theta+n(1-\theta)]^{n-1}}{n^n}\exp\left(-\frac{A}{RT}\theta^3\right)=\frac{K}{Cw}c_i \tag{22}$$

gave a good agreement for the $H_3PO_4$ data. In this equation, A is a term based upon lateral dispersive interaction between adsorbate molecules [54] ($A=[(3/4\pi^3)(Nh\nu/\epsilon o)] = 3.9$ kJ mol$^{-1}$, where $\epsilon_0$ is the electronic dielectric constant of water and $\nu$ is a vibration frequency in the adsorbed molecule). The values found for n were 3 for both $CF_3SO_3^-$ and $H_3PO_4$. A reversible adsorption was established for both substances. The plots of $\Delta G^o_{ads}/T$ as a function of 1/T are shown in fig. 19 for $CF_3SO_3^-$ and for phosphoric acid.

The frequencies of individual motions of $CF_3SO_3^-$ in solution and in the adsorbed state are shown in Table III[52]. Only a shift in the C-F vibration of the adsorbed species with respect to that in solution was observed. It may be that this increase in energy of the adsorbed state arises from the weaker hydrogen bonding of the fluorides to water when the molecule is adsorbed.

Little attention has been paid to the question of choosing an appropriate standard state in measurement of adsorption from solution[55]. In the case of the Bockris-Swinkels isotherm[53], the expression for chemical potential becomes[48]

$$\mu_i = \mu_i^\circ + RT\ln[f(\theta)/f(\theta^\circ)] \tag{23}$$

where

$$f(\theta) = \frac{\theta}{(1-\theta)^n}\frac{[\theta+n(1-\theta)]^{n-1}}{n^n} \tag{24}$$

and $f(\theta^\circ)$ represents the corresponding term with $\theta=\theta^\circ$ and

$$\mu_i^\circ = (\mu_i^\circ)_I + RT\ln f(\theta^\circ). \tag{25}$$

Thus, the most convenient standard state of $\theta^\circ$ is calculated by solving the equation

$$f(\theta^\circ) = 1 \tag{26}$$

and under these circumstances, $\theta^\circ$ in the present case, was found to be 0.582.

In respect to the potential-dependent adsorption of phosphoric acid, extensive tests were made to find out if this adsorption could be explained in terms of competition with water or in terms of competition with oxygen and hydrogen[48]. The hydrogen adsorption and oxide layer formation is shown in fig. 20. Considerable influence is clearly exerted by hydrogen at low and at high potential, but the middle part of the phosphoric acid curve shows up well in an analysis of water competition, as shown in fig. 21.

The potential dependence of the $CF_3SO_3^-$ adsorption was analyzed using the adsorption isotherm for ions given by Bockris, Devanathan and Muller[38]. From this equation, it was found that the value of $\epsilon$, the dielectric constant for lateral interaction, which fit the best was ca. 50. Fig. 22 shows a model which is consistent with this value.

Fronm the data concerned it is possible to make an analysis of the entropy. Calculation of the entropy of $CF_3SO_3^-$ and $H_3PO_4$, as well as for water in both dissolved and adsorbed states, gives rise to the

Table III. Frequencies of individual motions observed in $CF_3SO_3^-$ in solution [52] and in adsorbed state.

| $CF_3SO_3^-$ in soln,* wavenumber/ $cm^{-1}$ | adsorbed $CF_3SO_3^-$, wavenumber/ $cm^{-1}$ | shift upon adsorption, wavenumber/ $cm^{-1}$ | assignment* |
|---|---|---|---|
| ca. 1405 | 1400 | ca. −5 | $\nu_{as}$(S⩦O) |
| 1257–1280 | 1295 | 15–38 | $\nu_{C-F}$ |
| 1156 | 1155 | none | $\nu_{C-F}$ |
| 1032–1040 | 1045 | ca. 5 | $\nu_{sym}$(S⩦O) |

Table IV. Results of Calculation of models of Adsorption (T=298°K).

| | | $CF_3SO_3^-$ | | $H_3PO_4$ | |
|---|---|---|---|---|---|
| | | total surface entropy/ J $mol^{-1}$ $K^{-1}$ (cal $mol^{-1}$ $K^{-1}$) | $\Delta S_{ads}$/J $mol^{-1}$ $K^{-1}$ (cal $mol^{-1}$ $K^{-1}$) | total surface entropy/ J $mol^{-1}$ $K^{-1}$ (cal $mol^{-1}$ $K^{-1}$) | $\Delta S_{ads}$/J $mol^{-1}$ (cal $mol^{-1}$ $K^{-}$ |
| 2D translation + restricted rotation | communal translation | 140 (33.4) | −115 (−27.6) | 148 (35.4) | −92 (−22.0) |
| | translation within one "cell" | 131 (31.4) | −90 (−21.4) | 140 (33.4) | −68 (−16.2) |
| 2D translation, no rotation | communal translation | 125 (30.0) | −130 (−31.0) | 130 (31.2) | −110 (−26.2) |
| | translation within one "cell" | 117 (28.0) | −105 (−25.2) | 122 (29.2) | −85 (−20.4) |
| no translation, restricted rotation | | 14 (3.4) | −241 (−57.6) | 26 (6.2) | −214 (−51.2) |
| no translation, no rotation | | 8 (2.0) | −247 (−59.0) | 8 (2.0) | −232 (−55.4) |
| experiment | | | −97 ± 10 (−23.2 ± 2.4) | | −22.0 ± 59 (−52.6 ± 14.1) |

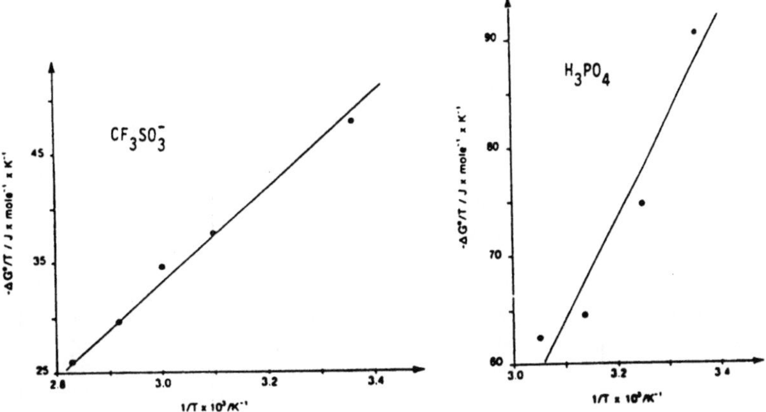


Fig. 19 Temperature dependence of the standard free energy of adsorption on Pt at 0.80V (NHE).

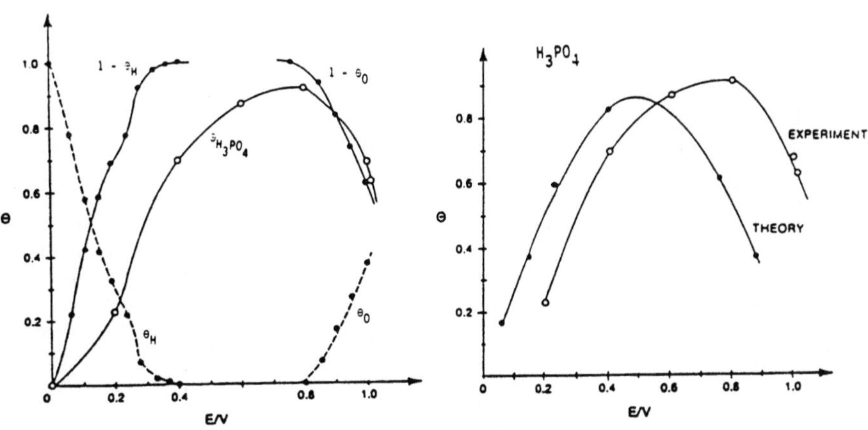


Fig. 20 Hydrogen adsorption, oxide layer formation and phosphoric acid adsorption on a Pt electrode as a function of potential. T=298°K.

Fig. 21 Potential dependence of $H_3PO_4$ adsorption. $6.2x10^{-2}$M $H_3PO_4$, T=298°K.

total change in entropy upon adsorption of solute accompanied by water displacement, and these values are summarized in Table IV for various models. The results are quite unambiguous. For $CF_3SO_3^-$, the best agreement is reached with the model of the surface in which the adsorbed entity retains two degrees of translational freedom and has a restricted rotation within one cell. The phosphoric acid indication is radically different with respect to the fact that the optimal agreement is obtained for a model in which the adsorbed entity has no translational degree of freedom, but some restricted rotation.

This study is a good example which shows the advantages of having some spectroscopic information about the positioning of the molecules, which, combined with the coverage of surfaces, allows us to obtain an improved knowledge of the adsorption of the ions.

The work also has remarkable influence upon the picture for a fuel cell situation because it shows that, although the phosphoric acid is much more adsorbed on the surface than the trifluoromethane sulfonic acid at room temperature (thus giving credence to the fact that phosphoric acid is more "blocking") (fig. 23) when the corresponding pictures are made at a higher temperature by extrapolating, using the isotherm, it is seen that the acids come to about the same adsorption (c.f. fig. 24)[51].

## 4. Studies Concerning the Surface of Iron.

It is interesting to note that the number of methods which are available for investigation of electrical processing declines upon leaving the noble metals. For an active metal such as iron, it is not possible to use the potential sweep method owing to the co-dissolution of iron in the anodic region and film formation at still higher potentials.

Much work has been done on the constitution of passive layers. It has been shown by means of Mössbauer spectroscopy[8], Auger spectroscopy[57], XPS[58], Rutherford backscattering[59], and other similar techniques that water is an essential part of the passive film.

In the studies reported here[60], the major impetus was to examine the spectroscopy of the iron-oxygen bond by means of spectroscopic ellipsometry, and also to try to obtain information from XPS concerning the presence of $OH^-$ and the correlation between this ion (which must arise from water) and the growing presence of $Fe^{III}$ as the oxidation of the iron proceeds. The extinction coefficient of the spectra of the passive layer of iron for different potentials is shown in fig. 25; and fig. 26 shows the corresponding XPS data for the $OH^-/O^{2-}$ and the $Fe^{2+}/Fe^{3+}$ spectra.

As far as the spectroscopic data are concerned, two peaks were identified. Molecular orbital calculations[61] and data on the absorption spectra of $\alpha$-$Fe_2O_3$ and of thick passive layers on iron[62] show that the peak at 2.3 eV (5250 Å) can be assigned to the ferric ion in the passive layer. The assignment of the 2.05 eV (6000 Å) was less straight forward due to the lack of relevant FeO and $Fe_3O_4$ spectra. However in the work of Schlegel[63] on $Fe_3O_4$ a small non-identified

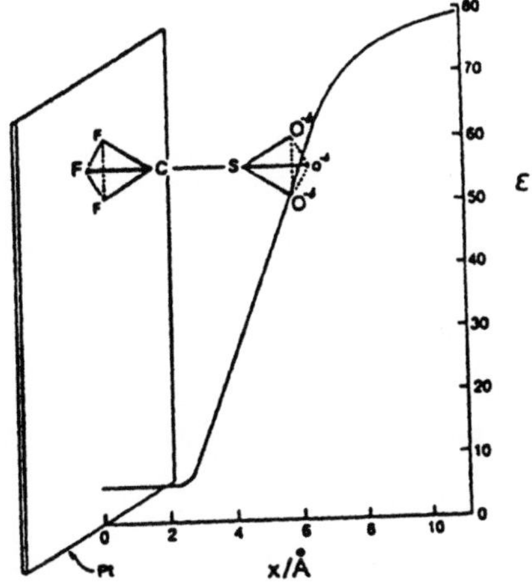


Fig. 22 Schematic diagram of adsorbed $CF_3SO_3^-$ in the region of the distance dependent dielectric constant in the double layer.

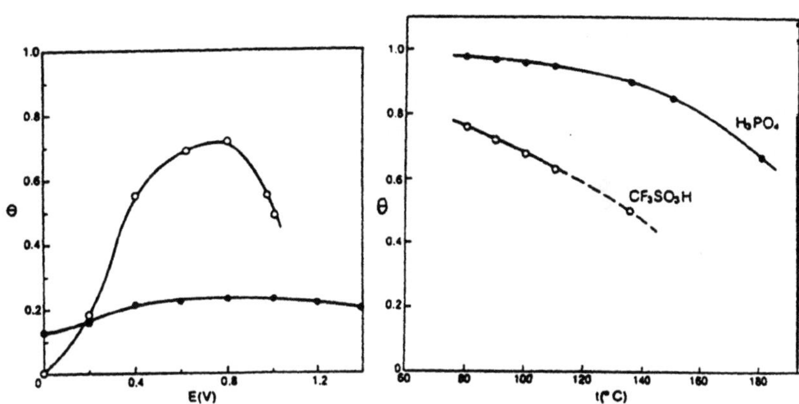


Fig. 23 A comparison of adsorption of $CF_3SO_3^-$ and $H_3PO_4$ on Pt as a function of potential.

Fig. 24 Extrapolation of adsorption data of 5.0M TFMSA and 98% $H_3PO_4$ in the high temperature region.

peak at c.a. 2.0 eV is present. As this peak was not observed in the absorption spectrum of $\alpha$-$Fe_2O_3$, it seemed reasonable to assign the peak at 2.05 eV to the $Fe^{II}$ ion. Fig. 25 also shows that in the potential region less than 0.2 V/NHE, there are significant changes in peak height, width, and intensity with potential. Doubtless these results arise from structural changes induced by hydration and oxidation of the passive layer.

The decrease of the peak at 2.05 eV with respect to the peak at 2.3 eV when the potential becomes more positive can be attributed to a change in ferrous to ferric without significant changes in the degree of hydration. At more anodic potentials, c.a. 0.0 V/NHE, the intensity of the spectrum declines by a factor of 5 and only small changes in the relative intensities of the two peaks are observed. The important changes in the spectrum of the $Fe^{3+}$ ion may be associated with an increase in hydration of the passive layer as the potential becomes more positive. Finally, at 0.2 V/NHE, the peak at 2.05 eV disappears, and the peak at 2.3 eV decreases slightly corresponding to the absence of $Fe^{2+}$ ions. Again the increasing hydration of the passive layer (c.f. fig. 26) with the corresponding decrease in the extinction coefficient, is the cause of a considerable decrease in lattice distortion following the $Fe^{2+} \rightarrow Fe^{3+}$ conversion in the oxide film. Fig. 26 shows XPS data on $Fe^{2+}/Fe^{3+}$ and is consistent with a rapid decline of this ratio of potentials above 0 V(NHE).

The data on XPS (fig.29) are somewhat scattered in respect to the points. Nevertheless it seems reasonable to make a straight line rising from left to right through the $OH^-/O^{2-}$ graph as a function of potential going in the anodic direction, and then this will correspond to the decrease of the $Fe^{2+}/Fe^{3+}$ line which correspondingly declines: the hydration increases as the passive layer develops.

These studies are rare in that they exemplify ellipsometric spectroscopy. At the moment, our ellipsometric machine is being developed in favor of in-situ infrared use, and we will then hope to be able to make a corresponding comparison of the spectra obtained from the FTIR measurements to those obtained from corresponding, but different, ellipsometric techniques.

## 5. Breakdown of Passive Layers.

Although the spectroscopic methods which have been utilized for the study of the iron oxide film are very appropriate, there is added need for radiotracer measurements in studying the breakdown of passive layers.

In earlier spectroscopic work, it was shown by Revie, Baker and Bockris[64] that entry of the chloride ion in contact with the passive layer gives rise to a decrease of the water content of the film. There is evidence, therefore, that the mechanism of depassivation is some kind of displacement of water by the chloride ion. It is needed to obtain some more definite information which would confirm or deny the hypothesis. For this reason, the radiotracer determination of the activity of the chloride ion was obtained as a function of time and its

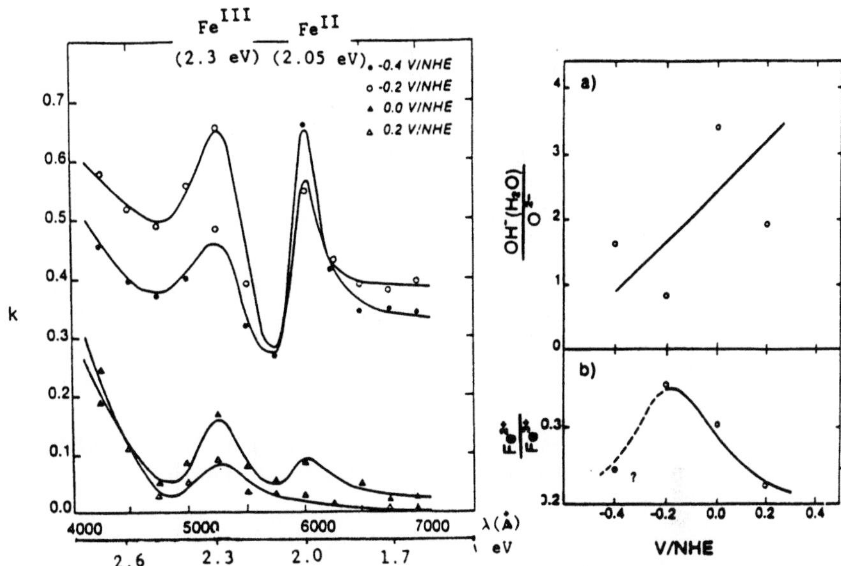


Fig.25 Extinction coefficient spectra, k, of iron for different potentials. Borate buffer, pH=8.4.

Fig. 26 XPS data. Potential relationship of the passive film of a) $OH^-(H_2O)/O^{2-}$, and b) $Fe^{2+}/Fe^{3+}$.

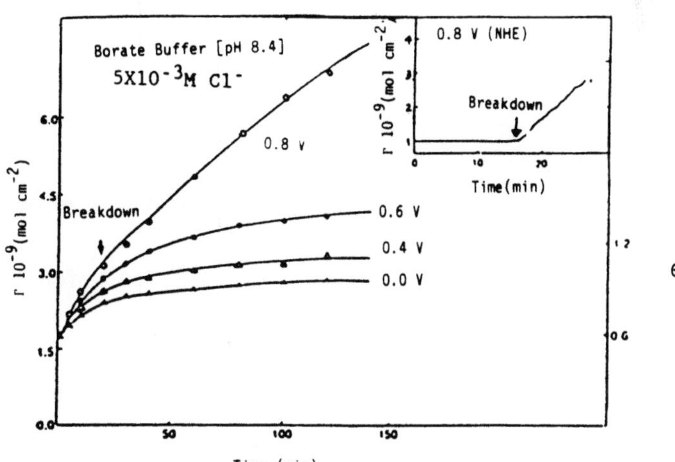


Fig. 27 Total "surface" concentration of $Cl^-$ at different potentials as a function of time. $5x10^{-3}M\ Cl^-$.

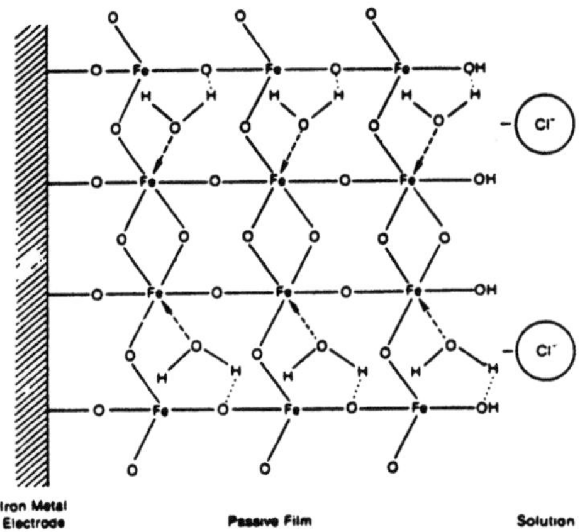


Fig. 28 Schematic representation of chloride penetration channels in terms of the hydrated polymeric oxide model of passivity.

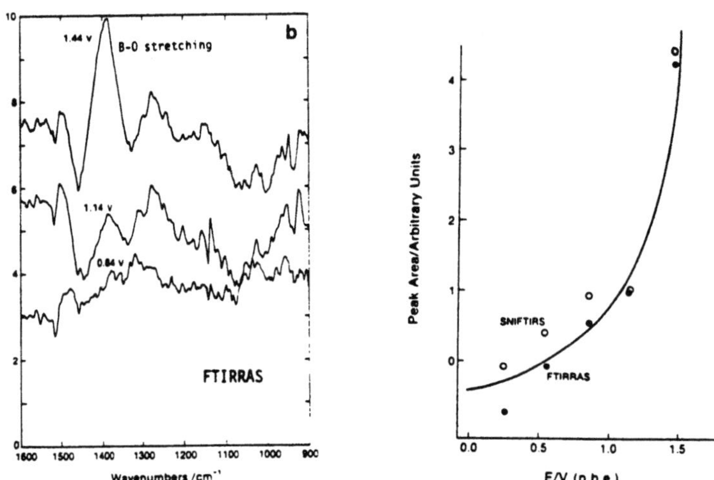


Fig. 29 Difference spectra of the electrode/solution interface at different potentials.

Fig. 30 Integrated band (B-O stretching) intensity as a function of potential.

Table V.ESCA analysis of surface films on Fe, grown from borate solution,

| E (V(NHE)) | Angle (deg) | % boron |
|---|---|---|
| 0.0 | 90 | 3.5 |
| 0.3 | 90 | 2.5 |
| 0.9 | 90 | 2.0 |
| 0.9 | 15 | 12.0 |

diffusion through the passive layer[65]. The essential result is shown in fig. 27. Here, one sees the total surface concentration, Γ, at different potentials as a function of time up to 150 minutes after the contact of the radiotracer chloride ion with the passive layer. Up to 0.8 V/NHE the variation with time is rather standard, and similar to a Langmuir-type behavior. However, at 0.8 V/NHE a different behavior begins, and a greatly enhanced activity of the chloride ion in the oxide is observed. Somewhat after this (see insert in fig. 27), the breakdown of the passive layer occurs, as indicated by a great increase in the conductivity. Thus, this breakdown takes place clearly as a result of, and not as a cause of, the sudden increase of the total surface concentration of chloride ions in iron, and this clearly is consistent with the model given for the exchange mechanism of water by the chloride ion[59] (c.f. fig. 28).

## 6. The Adsorption of Borate Ions on Passive Films.

Most of the work carried out on the passivity of iron concerns neutral solutions, and the buffer usually used is a borate buffer. It is obvious that one should find out to what extent the borate ion is adsorbed upon the passive layer, and whether this has a part in making the layer passive.

In this case[66], both SNIFTIRS and FTIRRAS spectra of 0.15M $H_3BO_3$ + 0.0375M $Na_2B_4O_7$ solution (pH = 8.5) were recorded at different potentials (fig. 29). Also fig. 30 shows the integrated band at 1400 $cm^{-1}$ frequency corresponding to the out of phase B-O stretching band[67]. It is interesting to note that when sulfate ions are adsorbed on the surface, no spectroscopic evidence of borate adsorption was obtained[66]. This together with the reduction of the potential range in which iron is passive suggests that the borate ions themselves may have some effect on passivity. However, fig. 30 shows that significant concentration of $BO_3^{3-}$ in the film does not begin until the potential of this is 0.4 V positive with respect to the beginning of passivation. Thus, borate ions adsorbed at the interface cannot be a main stabilizing factor for the passive films even though it is present in the film at low concentration as shown in Table V (ESCA analysis).

## 7. The Adsorption of Intermediate Radicals and Inhibitors.

### 7.1 STUDY OF THE ADSORPTION OF THIOCYANATE ON ACTIVE AND PASSIVATED IRON.

The study of the adsorption of thiocyanate on active and passivated iron was carried out both with radiotracer and with FTIR spectroscopy [68]. The thiocyanate peaks are shown in fig. 31. The corresponding signal as a function of potential, both from radiotracer and FTIR studies, is shown in fig. 32.

It was found that the concentration dependence of adsorption for bare iron and passive iron gave rise to plots which agreed with the

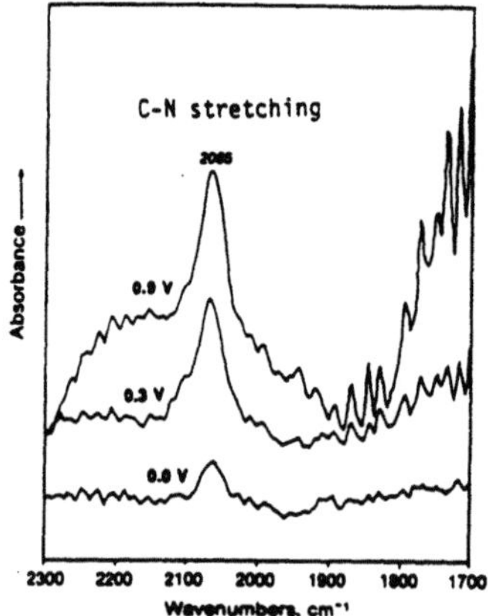


Fig. 31 SNIFTIRS spectra of thiocyanate adsorption on passive iron.

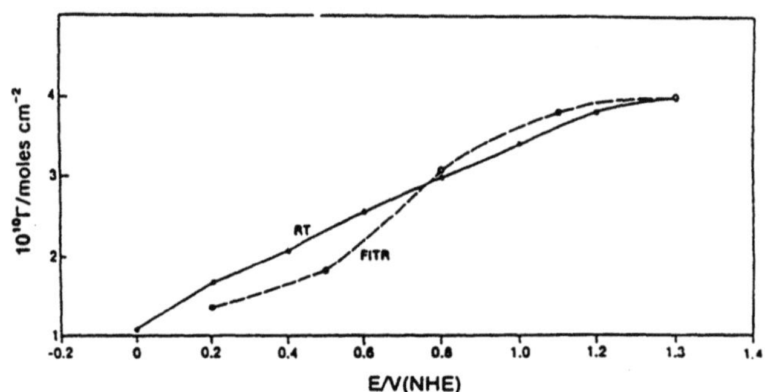


Fig. 32 Comparison of the potential variation of the integrated absorption intensity due to the adsorption of thiocyanate with the potential variation of the corresponding radiotracer signal.

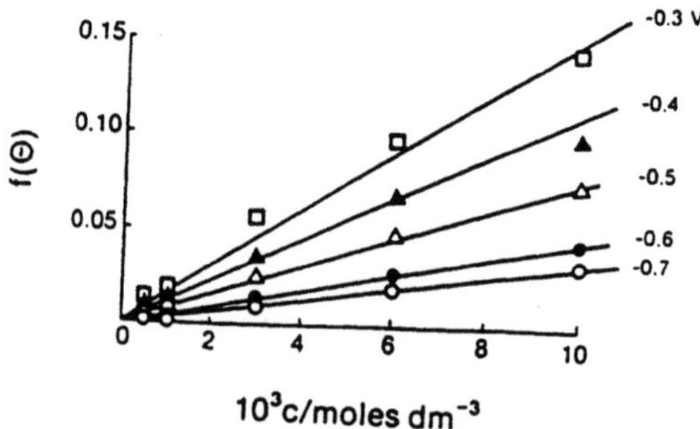


Fig. 33 f(Θ) vs. C for the adsorption of thiocyanate on iron at different potentials, V(NHE).

Bockris-Swinkels isotherm[53] when two moles of solvent are displaced per mole of $SCN^-$ adsorbed (n=2):

$$f(\theta) = \frac{\theta}{(1-\theta)^n} \frac{[\theta + n(1-\theta)]^{n-1}}{n^n} = \frac{K}{c_w} c_i \qquad (27)$$

The corresponding plots for $f(\theta)$ vs. concentration are shown in fig. 33. In terms of single imaging of the charge in the electrode (BDM model[38]), the adsorption isotherm (eq. 27) can be expressed as

$$f(\theta) = \frac{2r_i F}{q_{max}} \frac{c_i}{c_w} \exp\{- \frac{\Delta G_c^o}{RT} + \frac{4\pi e_o (x_2 - x_1)}{\epsilon kT} q_m - [\frac{\pi^{7/2} e_o 2 r_i^2}{4\epsilon kT} (\frac{q_{max} N_A}{F})^{3/2}] \theta^{3/2}\} \qquad (28)$$

where $\Delta G_c^o$ includes the coverage-independent chemical interactions, the term $4\pi e_o(x_2-x_1)q_M/(\epsilon kT)$ takes into account the metal-ion coulombic interactions, and the third term in the exponent includes the interionic coulombic repulsions. $q_M=e_o/(4r_i^2)$ represents the maximum amount of contact-adsorbed-charge per unit area, $x_2$ and $x_1$ the distances of the OHP and IHP planes, respectively and the rest of the variables have their usual meaning. An excellent agreement was obtained for the plot of [ln C - ln $f(\theta)$] vs. $\theta^{3/2}$ (c.f. fig. 34). The assumptions involved considered a unit charge still upon the thiocyanate ion, and the dielectric constant of a surface layer equal to 10. From the slope the value of $r_i$ was consistent with the radius of the thiocyanate ion ($1.6 \times 10^{-8}$cm)[69]. Standard free energies of adsorption were found to be -1.26 kcal mole$^{-1}$ on the bare iron, but on the passive iron only 0.7 k cal mole$^{-1}$.

## 7.2 HYDROGEN.

The classical intermediate in electrochemical reactions is, of course, the hydrogen atom, which has had so much treatment in electrochemical kinetic literature, beginning in the 1950's.

One of the principal criteria of the pathway of the electrochemical hydrogen evolution reaction is the state of adsorption, whether it is high and near to full coverage, or low, below 10%. It is easy to determine such matters on noble metals by coulometric methods, but on metals such as iron in which there is perhaps the greatest interest in knowing the coverage of hydrogen, the coulometric methods are difficult to apply because of the co-dissolution of the metal[70,71].

In the work recently carried out by Bockris, Carbajal, Scharifker and Chandrasekaran[33] on the determination of the coverage of the iron surface by hydrogen, this problem was approach by means of the FTIR spectroscopy technique. The absorption spectrum of iron in borate buffer solution at -0.9 V/NHE is shown in fig. 35.

There are no recorded determinations of the frequency of vibration for hydrogen adsorbed on iron. However, Barclay[72] discussed this question and took the Fe-H vibration from inorganic complexes in solution[73,74] and the average value obtained was $1885 \pm 25$ cm$^{-1}$.

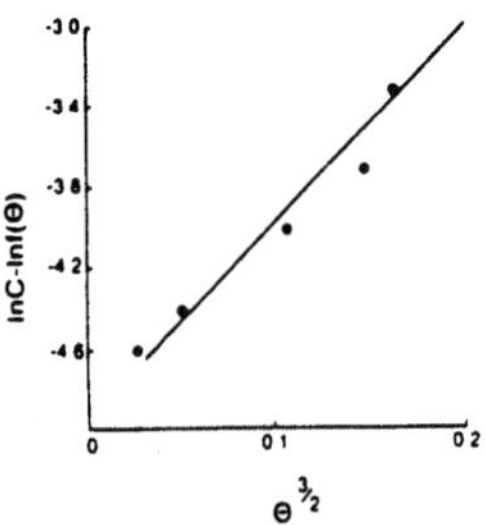


Fig. 34 Plot of [ln c - ln f(Θ)] vs. [$\Theta^{3/2}$] for the adsorption of thiocyanate on active iron at -0.6 V.

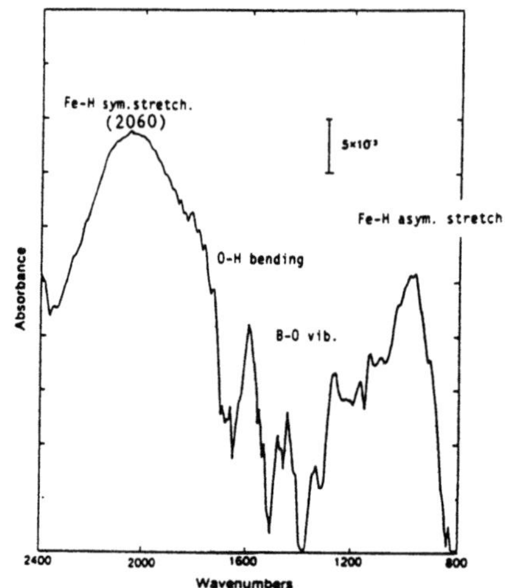


Fig. 35 Differential IR spectra in abosrbance of Fe-H vibration at -0.9V (NHE) in borate buffer solution.

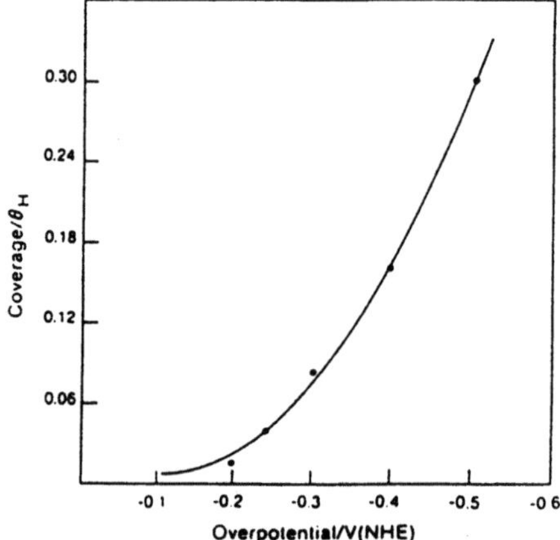


Fig. 36 Coverage-Overpotential relation for H adsorbed on Fe corrected for reference potential.

Various pieces of evidence[75,76] suggest that it is reasonable to consider this frequency range as extending to cover the 2060 $cm^{-1}$ peak for the adsorbed material. A further confirmation of the identity of the wave numbers observed in fig. 35 as being associated with adsorbed H (2060 $cm^{-1}$) was the experiment carried out with $D_2O$ instead of $H_2O$. In this case an absorbance maximum appears at 1485 $cm^{-1}$. It is well known that the ratio of the frequencies of the hydrogen isotope is given by the square root of the inverse ratio of their masses, i.e.,

$$(\nu_1/\nu_2) = \sqrt{\mu_2/\mu_1} = \sqrt{2/1} = 1.41 \tag{29}$$

The ratio frequencies found for the case of $H_2O/D_2O$ was 2060/1485=1.39, which agrees reasonably well. The rest of the peaks were directly identified as shown in fig. 35[66,77,78].

An attempt to estimate roughly the surface coverage was made by calculating the extinction coefficient from the IR peak intensity. It is known that[79]

$$\frac{8\pi^2}{3}\,\frac{m_e \nu c}{he^2}\,|\mu|^2 = \frac{4m_e c\epsilon_o A}{N} \tag{30}$$

where $m_e$ is the mass of the electron, $\nu$ is the wave number of the adsorption peak, $\mu$ is the transition dipole moment, A is the absorptivity, and the rest of the terms have their usual meaning. The absorptivity A can be related to the absorption coefficient at a given frequency, $\alpha$, by $A = \int_{\nu_1}^{\nu_2} \alpha_1 d\nu$ where $\nu_1$ and $\nu_2$ represent the breadth of the absorption band. The value of the transition dipole moment is not known for Fe-H. It is a quantity which generally has to be calculated quantum mechanically. Determination of this quantity has been made for the Si-H bond for which the value of $\nu$ is 2100 $cm^{-1}$. In the calculations for the transition dipole moment of Si-H, the value obtained was 1.33 Debye/Å[80]. Using a value of 1.5 Å for the appropriate Fe-H bond length[69], one obtains a value of $\mu_{Fe-H}$ = 1.99D. Using this value of $\mu$ in eq. 30, one obtains A (the integrated absorption coefficient). From the expression

$$A = \int_{\nu_1}^{\nu_2} \alpha d\nu \sim \alpha_{mean}(1/2)c(\nu_2 - \nu_1) \tag{31}$$

$\alpha_{mean}$ can be obtained. $\nu_2 - \nu_1 = 500$ $cm^{-1}$ (cf. fig. 35), therefore $\alpha_{mean} = 8 \times 10^6$ $cm^2mol^{-1}$. The concentration of adsorbed species at the interface was obtained by consideration of the extinction coefficient, $\epsilon = \alpha_{mean}/2.303$, and the optical path through the adsorbate, $d=2l/\cos\theta$, where $l$ is the Fe-H bond length ($1.5 \times 10^{-8}$ cm) and $\theta$ the angle of incidence of the IR radiation.

The experimentally measured absorbance, $a$, at -0.9V, was $1.8 \times 10^{-2}$, at an angle of incidence, $\theta$, of 40°. Thus, from $a = \epsilon cd$, where c is the concentration of absorbing molecules, c is found to be 0.15 mol $cm^{-3}$. The value of $\Gamma$ can be correspondingly obtained from the value of the Fe-H distance, $l$, i.e., $\Gamma = cl = 2 \times 3 \times 10^{-9}$ mol $cm^{-2}$. Since $\Gamma_{max} = 2.7 \times 10^{-9}$ mol $cm^{-2}$ and the roughness factor of the surface is 3,

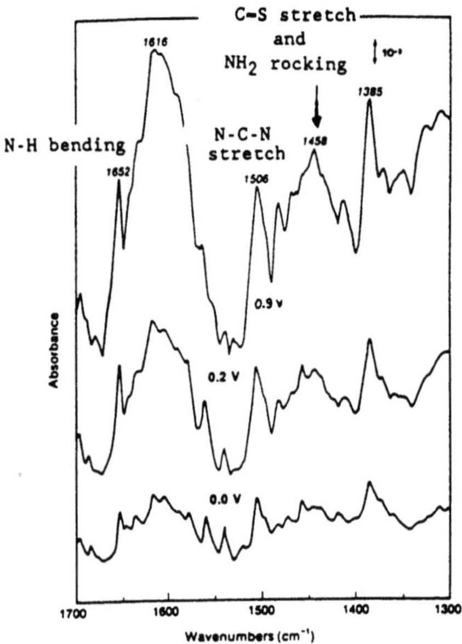


Fig. 37 SNIFTIRS spectra of thiourea adsorbed on passive iron at different potentials.

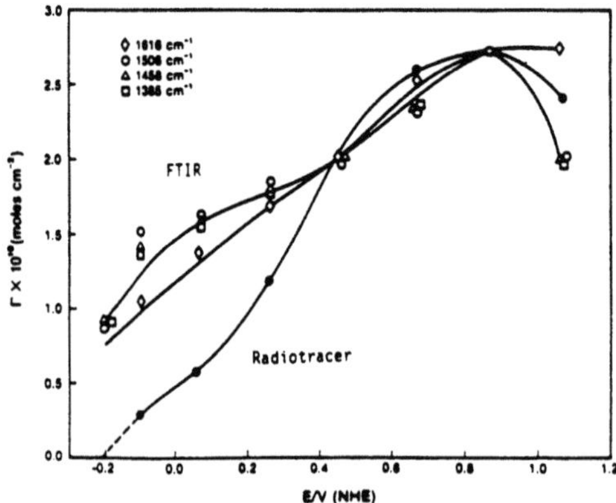


Fig. 38 Integrated adsorption intensities of spectral bands corresponding to the adsorption of thiourea on passive iron, as a function of electrode potential. Radiotracer results are also shown.

then $\theta = \Gamma/3\Gamma_{max} = 0.28$.

However, the values for the hydrogen coverage must be corrected by the fact that at the corrosion potential there is a $\theta$ value of 0.03[81] which has to be added to the original values. The corrected coverage as a function of overpotential is shown in fig. 36.

An important matter concerns the situation in respect to the slope of the $\theta$-$\eta$ line, which is shown in fig. 36. Plotting this logarithmically, it shows a slope of F/6RT, the expected value for a couple discharge combination mechanism being F/4RT[82]. (However, the Tafel slope for hydrogen evolution is 3RT/F, and this corresponds exactly to a 6RT/F slope to d(log $\theta$)/dV, as observed.) Similar results as these for iron have been recently obtained by Nichols and Bewick for single crystals of platinum and polycrystals of platinum, rhodium and ruthenium[32].

## 7.3 THE ADSORPTION OF THIOUREA ON IRON

The adsorption of thiourea on passivated iron in aqueous solutions has been investigated, both with radiotracers and with Fourier Transform Infrared Spectroscopy[68]. The techniques agreed well.

The SNIFTIRS spectra of thiourea on passivated iron is shown in fig. 37. The potential dependence of the thiourea on passivated iron as obtained from the integrated area of the absorption peaks is compared with the results obtained from the radiotracer technique and is shown in fig. 38.

The peak at 1652 $cm^{-1}$ and the band at 1660 $cm^{-1}$ belong to the N-H scissoring (bending) vibration[83]. The peak at 1506 $cm^{-1}$ corresponds to the N-C-N stretching vibration[83]. In aqueous solution, this peak is observed to occur at 1420$cm^{-1}$, but is shifted to 1506 $cm^{-1}$ on adsorption due to the increase of the double bond character of the C-N bond, resulting from adsorption with the sulfur atom towards the metal surface[84]. The peak at 1458 $cm^{-1}$ corresponds to the $NH_2$ rocking, and the C=S stretching[83], which seemed to have about the same frequency. The isotherm which fits is the Bockris-Swinkels isotherm[53].

## 7.4 ADSORPTION, ORIENTATION AND POLYMERIZATION OF 1-OCTYNE-3-OL ON IRON.

The initial adsorption, structural transition, and polymeric growth of 1-octyne-3-ol on iron electrodes in sulfuric acid solution have been examined by using time-resolved in-situ automatic ellipsometry[85,86].

The adsorption of 1-octyne-3ol on iron in $10^{-2}$M $H_2SO_4$ at -0.4 V(NHE) for different concentrations of octynol in the solution was monitored in respect to thickness as a function of time (fig. 39). At low concentrations of 1-octyne-3ol two plateaus with different time constants can be distinguished. The first plateau characterized by fast adsorption (~2 min) attains its maximum thickness of about 2.2 Å at a concentration of $10^{-4}$M, followed by a concentration-dependent transition and a second plateau with a maximum thickness of 3.9 Å at the same concentration. When the concentration becomes greater than $10^{-3}$M the two plateaus are reduced to two inflection points after which continuous growth of the film occurs.

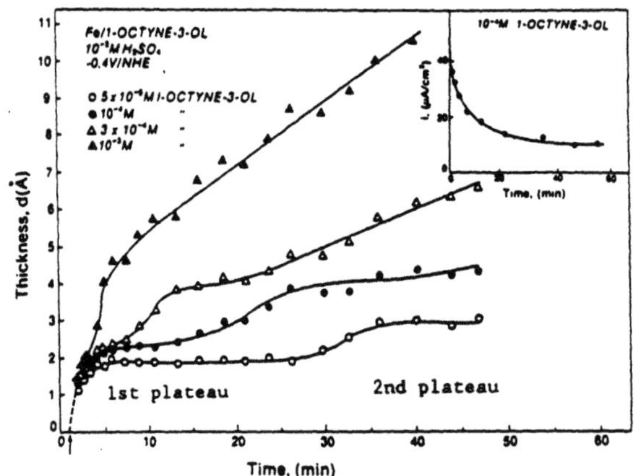


Fig. 39 Thckness-time relationships for adsorption of 1-octyne-3ol on iron in 0.01M $H_2SO_4$.

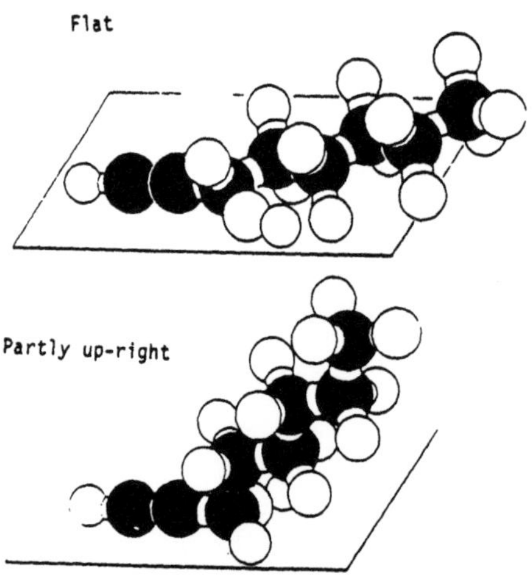


Fig.40 Adsorption of 1-octyne-3ol on iron in flat and partly up-right orientations.

Thus, from the thickness-time relationship (fig. 39) it is clear that there is a first adsorption followed by a transition, and it is proposed that this transition is shown in fig. 40, where the octynol lies first flat (first plateau) (c.f. fig. 40I), and secondly in an upright orientation (second plateau) (c.f. fig.40II). These different orientations of adsorbed octynol are supported by the estimated thickness of close-packed monolayers in orientations I and II, determined from covalent radii. In orientation I (2 Å), the ethylenic bond parallel to the surface, with the oxygen atom toward the surface, provides the most stable configuration at low concentration and/or short time. However, the position of carbon atom 2 and 3 of the acetylenic link is liable in respect to rotation, so that the change from form I to form II can occur. In the configuration II (3.9 Å) the molecule is partly standing up and gives rise to a greater thickness of the layer. Similar changes have been observed in the case of several organics adsorbed on platinum[87]. Upon overcrowding, they rise and adsorb in a position which creates more room.

Chloride ion exhibits a pronounced effect on the adsorption and reorientation of 1-octyne-3-ol on iron from acid solution[86]. The adsorption inhibition of octynol by $Cl^-$ gives rise to thinner adsorption layers, suggesting strong competitive adsorption mechanisms. The orientational transition is shifted towards longer times with a stronger adsorption of octynol in the upright orientation. The rate of growth of the polymer layers is unaffected by the presence of chloride ions.

## 8. Organic Adsorption on Semiconductors.

Polarization modulation Fourier Transform Infrared Reflection Absorption Spectroscopy (FTIRRAS) and Subtractively Normalized Interfacial Fourier Transform Infrared Spectroscopy (SNIFTIRS) have been applied to the adsorption of ions and neutral molecules on p-silicon and p-gallium phosphide in aqueous as well as non-aqueous solution[88]. The potential and concentration dependence were observed.

a)Adsorption of Thiourea on p-Silicon. In fig. 41 the spectra for thiourea on silicon is observed. The spectrum is similar to the absorption spectrum of thiourea on passivated iron[89]. A broad adsorption peak at 1600 $cm^{-1}$ is prominent, and it was attributed to the N-H bending vibration [83].

Adsorption of neutral molecules as a function of bias potential usually passes through a maximum[90]. When the free energies of adsorption of water in the various positions on the surface are the same, the adsorption maxima of the neutral molecules can be associated with the pzc. But for most interfaces, the free energy of adsorption of water is different for orientations taken up at various potentials. Hence, the adsorption maximum does not coincide with the pzc. The later was determined to be 0.14 V(NHE) for the silicon-aqueous interface containing 0.8M NaF. Thus, on silicon, adsorption of thiourea does not pass through a maximum (c.f. fig. 42), but continues to increase with increase of anodic potential. The behavior can be understood if adsorbed thiourea is regarded as a Zwitterion in which the $NH_2$ groups

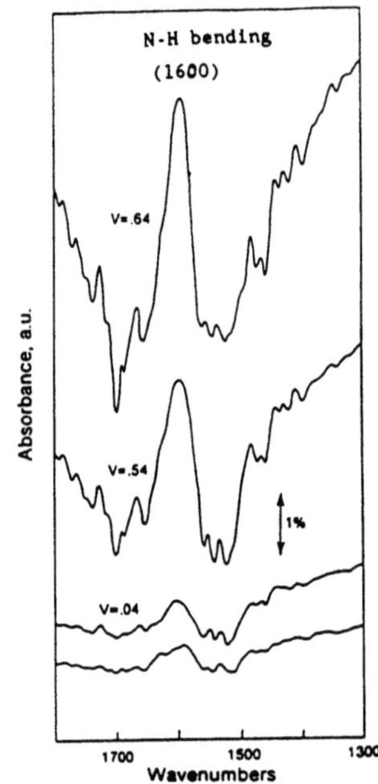


Fig. 41 The adsorption spectrum of thiourea adsorbed on Silicon. 25mM thiourea, 0.8M NaF, aq. solution.

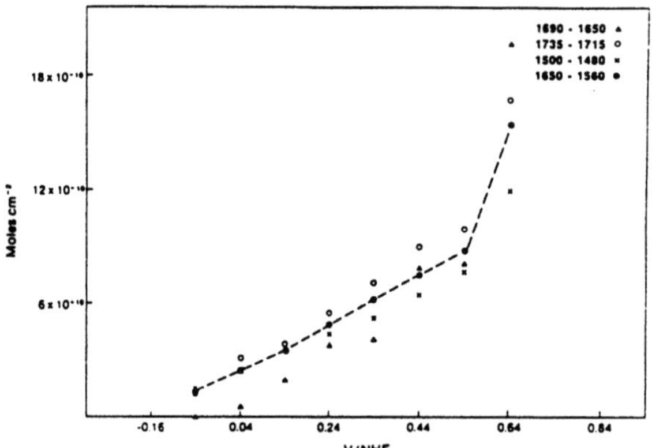


Fig. 42 Surface concentration Θ, of thiourea as a function of potential at a silicon-aqueous 0.8M NaF interface. Thiourea is 25 mM.

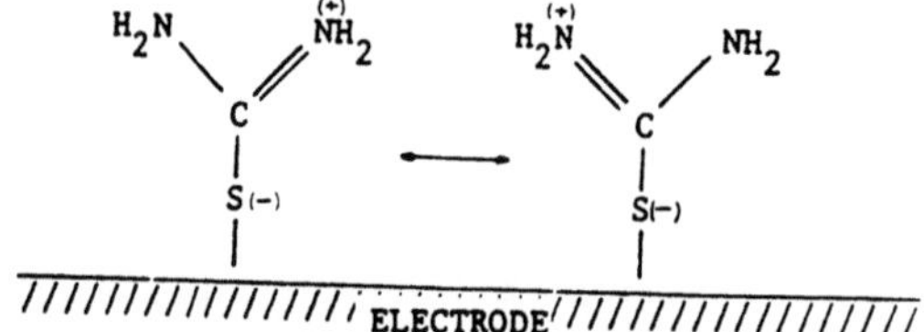


Fig. 43 Possible structure of adsorbed thiourea on Silicon.

are in the diffuse layer:

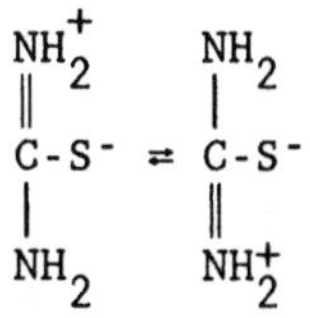

$$\begin{matrix} NH_2^+ & & NH_2 \\ \| & & | \\ C\text{-}S^- & \rightleftharpoons & C\text{-}S^- \\ | & & \| \\ NH_2 & & NH_2^+ \end{matrix}$$

The model shown in fig. 43 illustrates this. The distance between silicon surface and amino group nitrogen is about 5.4 Å, about the expected thickness of the double layer. Thus, the partially positively charged amino groups would lie at the Helmholtz-Gouy boundary. If adsorption occurs as shown in fig. 43, it would increase in the anodic direction in a way similar to the adsorption of anions. This model is supported by the fact that the N-C-N stretching peak is shifted to higher frequencies, indicating an increased double bond nature of C-N after adsorption.

b) Acetonitrile and Acetate adsorption on GaP. A remarkable spectrum for acetonitrile adsorbed on GaP-acetonitrile interface was obtained and is shown in fig. 44. A sharp and strong absorbance peak corresponding to a C≡N vibration is seen at 2250 $cm^{-1}$. SNIFTIRS spectrum for acetate adsorbed on GaP in acetonitrile containing 0.8M $LiClO_4$ and 0.1 M sodium acetate are shown in fig. 45. The absorption maximum at 1593 $cm^{-1}$ is attributed to C=O stretching vibration of the carboxylate group[78]. At sufficiently cathodic potentials, the value of $(I_p - I_s)/(I_p + I_s)$ becomes indeterminate. This condition arises because of the very high degree of adsorption of acetonitrile under sufficiently cathodic conditions. However, at potentials positive to +0.24 V NHE the spectrum can be recorded. The maximum of the CN frequency is shifted by 30 $cm^{-1}$ to higher frequencies compared to those measured in solution. This is consistent with the surface adsorbed nature of the species giving the 2250 $cm^{-1}$ signal.

Adsorption of acetonitrile increases towards more positive potentials on the platinum surface, whereas here increased adsorption is noticed towards less positive potentials (fig. 46). Such differences in behavior can be due to the difference in the pzc between the platinum and gallium phosphide systems, and/or due to a stronger adsorption of acetate ions compared with that of tetrafluoroborate used by Pons et al.[25].

c) Adsorption of Ammonium Ions on GaP. Ammonium ions adsorbed on GaP show two broad peaks centered around 3475 and 3000 $cm^{-1}$, corresponding to the N-H symmetric stretching (fig. 9)[91]. Two peaks of equal intensity were observed for N-H deformation: one at 1700 and another at 1560 $cm^{-1}$. Two different N-H groups are involved. From the equal intensities of the two peaks, it is evident that each peak corresponds to two different N-H deformation vibrations.

Three orientations of the adsorbed ammonium ion are possible as shown in fig. 47. Unlike the ammonium ion in homogeneous solutions, the adsorbed ammonium is not a symmetrical tetrahedron. Two different N-H groups are present for the adsorbed ammonium ion. The peak at 1560 $cm^{-1}$ is attributed to the N-H bond projecting towards the solution as it lies close to the ammonium ion absorption in homogeneous solution.

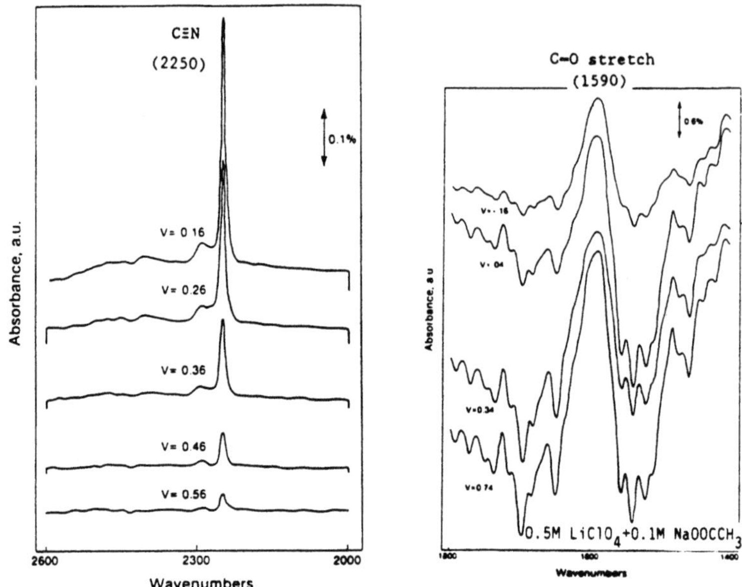


Fig. 44 Absorption spectrum of acetonitrile adsorbed on Ga-P acetonitrile interface. 0.5M $LiClO_4$+0.1M $NaOOCCH_3$.

Fig. 45 SNIFTIRS absorption spectrum of acetate on GaP.

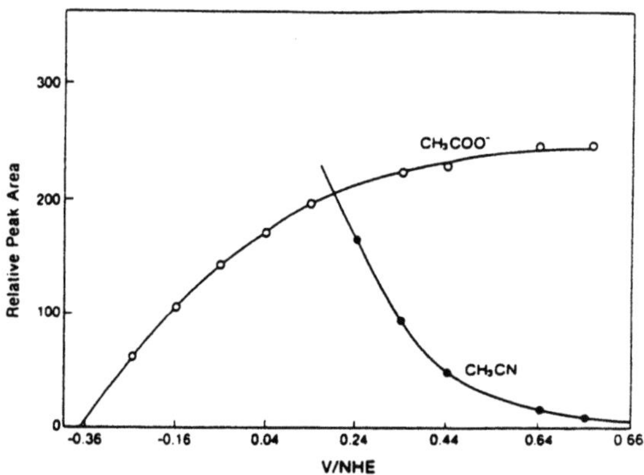


Fig. 46 Relative adsorption of acetate and acetonitrile on GaP as a function of bias potential.(0.1M acetate.)

The absorption maximum at 1700 $cm^{-1}$ is attributed to the N-H bonds associated with the adsorbed hydrogen atom of the ammonium ion[91]. Since the intensities of these two peaks are equal at all bias potentials, it is proposed that the ammonium ion adsorption occurs through the two hydrogen atoms, as shown in fig. 47. For the other two orientations (Figs. 47a and 47c), the intensity ratios should be 3:1 and 1:3. Symmetric stretching vibrations of the adsorbed ammonium ions also indicate a similar behavior. Thus, such spectroscopic examinations give rise to structural conclusions concerning adsorbed entities.

## 9. The In-Situ Spectroscopic Investigation of Adsorbed Intermediate Radicals and Electrochemical Reactions: $CO_2^-$ on Pt.

Adsorption of carbon dioxide, carbon dioxide anion radicals, and acetonitrile were studied on platinum in acetonitrile using Polarization Modulation Fourier Transform Infrared Reflection Absorption Spectroscopy[29]. The mechanism of the $CO_2$ reduction is well known to follow the pathway[92]

$$CO_{2aq} = CO_{2ads} \tag{32}$$

$$CO_{2ads} + e^- = CO^-_{2ads} \tag{33}$$

$$CO^-_{2ads} + H_2O + e^- \rightarrow HCOO^- + OH^- \tag{34}$$

The proposed mechanism involves adsorbed $CO_2$ and adsorbed $CO_2^-$. However, there has been no direct evidence for the presence of the adsorbed $CO_2^-$ radical, and it was this that we tried to find from our spectroscopic investigations. Particular care was taken to avoid residual water, and the solution was pre-electrolyzed in a nitrogen atmosphere for 2 hours to remove it. The final water content was estimated to be as much as 0.01%. Carbon dioxide was bubbled through the solution for 40 minutes, and the cell then sealed.

The spectra reported here were obtained by subtracting from the spectra at various potentials, the reference spectrum at 0.0 V(NHE). The difference spectrum of surface adsorbed species on Pt in acetonitrile at -1.2 V(NHE) is shown in fig. 48. The peaks pointing downward show a decrease in surface concentration. Three peaks pointing downward and one peak pointing upward in the region 2500-1500 $cm^{-1}$ are seen in fig. 48. There is a broad maximum around 1680 $cm^{-1}$. Correspondingly, a sharp peak is observed at 2342 $cm^{-1}$, and may be identified with adsorbed $CO_2$[93] (c.f. fig. 48). Absorbance corresponding to the C-N stretching vibration of the adsorbed acetonitrile[91] at 2250 $cm^{-1}$ and adsorbed $CH_3CN$ at various electrode potentials are shown in fig. 50.

The absorption maximum of 1680 $cm^{-1}$ is attributed to the adsorbed $CO_2^-$ radical. The IR spectrum of this radical has been recorded at -190°, and it is a sharp maximum at 1671 $cm^{-1}$[94]. The broadening of the spectrum (full width, half maximum 40 $cm^{-1}$) indicates that the radical is adsorbed on the surface of the electrode. The 1680 $cm^{-1}$ peak cannot be due to water on the surface of the electrode, as the water peak appears at 1590 $cm^{-1}$. The potential dependence of the two

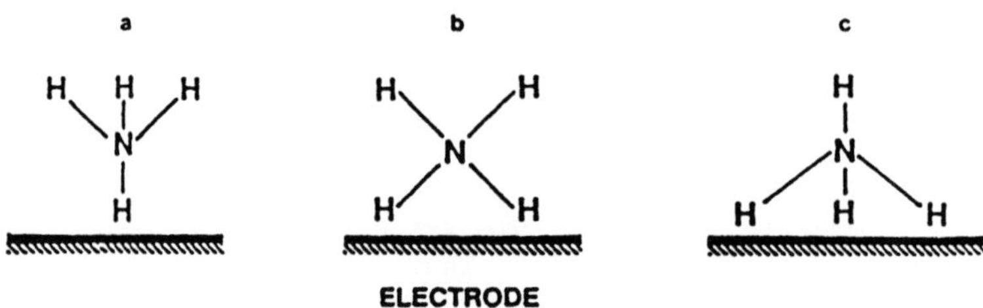


Fig. 47 Possible structure of adsorbed ammonium on GaP.

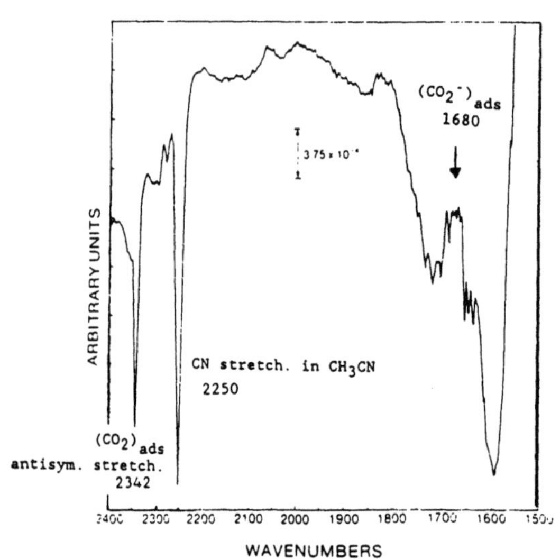


Fig. 48 The differential spectrum of surface adsorbed species on Pt in acetonitrile at -1.2V (NHE).

opposite, i.e., the intensity at 1680 $cm^{-1}$ peak increases towards more cathodic potentials whereas the intensity of the 1590 peak decreases.

The integrated areas of the peaks between 1650 and 1700 $cm^{-1}$ at several bias potentials are shown in fig. 49. The adsorption of the anion $CO_2^-$ increases in the cathodic direction. If anion adsorption occurred from the bulk of the solution, decrease in adsorption coverage would be expected when the potential is moved in this direction. Thus, the increase in anion concentration at cathodic potentials may be explained if $CO_2^-$ is formed as an intermediate radical; from eqs. (32)-(34),

$$\theta_{CO_2} = K_1 P_{CO_2} \tag{35}$$

and with (33) in equilibrium

$$\theta_{CO_2^-} = K_2 K_1 P_{CO_2} \exp(-VF/RT), \tag{36}$$

where V is the electrode potential, $K_1$ and $K_2$ are equilibrium constants for (32) and (33), respectively. Thus, $\theta_{CO_2^-}$ increases with increasing negative value of V (fig. 49). With (34) as rate determining step, the current density, i, becomes,

$$i = 2FK_3\theta_{CO_2}\exp(-\beta VF/RT), \tag{37}$$

$$= 2FK_1K_2K_3P_{CO_2}\exp[-(1+\beta)VF/RT]. \tag{38}$$

Thus it is possible to obtain information on the rate-controlling step by identifying the radical and following the surface coverage with potential.

The carbon dioxide adsorbed on platinum in acetonitrile has a maximum of 2342 $cm^{-1}$. The IR spectrum of $CO_2$ in the gas phase shows a maximum at 2349 $cm^{-1}$[93]. Since the frequency shifts are very small, $CO_2$ is probably physically adsorbed. It has been reported that adsorption of $CO_2$ in platinum at anodic potential involves chemisorbed CO[95], but no adsorbed CO was observed at the cathodic potentials on platinum and acetonitrile.

If the $CO_2$ were adsorbed parallel to the surface of the electrode, asymmetric stretching vibrations would not be IR-active, for the electric field vector of the parallel polarized light is zero in the plane parallel to the surface of the electrode. Since adsorbed $CO_2$ absorbs at 2340 $cm^{-1}$, which is attributed to the antisymmetric stretching vibration, it was proposed that adsorption of $CO_2$ occurs through one oxygen atom, and the other oxygen atom is projecting towards the solution, as shown in fig. 51. Symmetric stretching vibrations of $CO_2$ are not IR-active as the transition dipole moment for this symmetric stretching is zero. Bending vibrations for $CO_2$ occur at 667 $cm^{-1}$, which is beyond the range of the instrument detection.

The relative concentration of $CO_2$ decreases in the cathodic direction (c.f. fig. 52). It is possible that decrease in concentration of $CO_2$ may be due to the reduction of $CO_2$ to give $CO_2^-$, which is absorbed at 1680 $cm^{-1}$. The potential of zero charge of platinum in acetonitrile has been determined to be +0.3 V NHE[96]. Concentration of

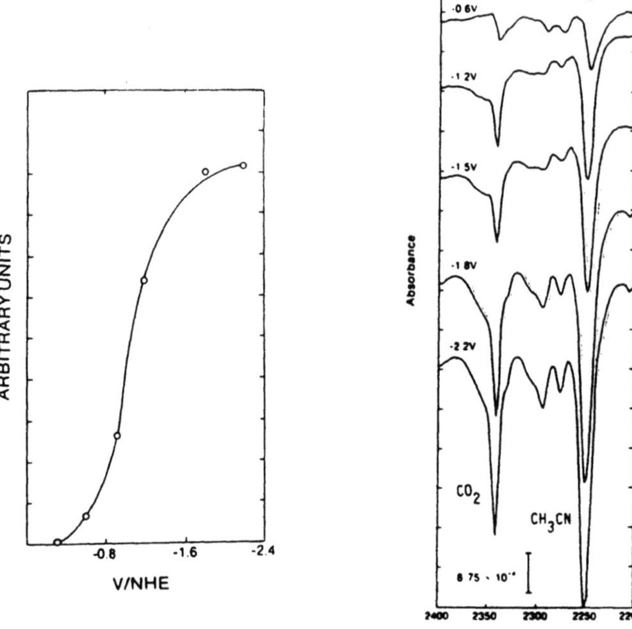


Fig. 49 Relative concentration of adsorbed $CO_2^-$ on Pt.

Fig. 50 Absorption spectra of adsorbed $CO_2$ and $CH_3CN$ on Pt in acetonitrile containing 0.4M $LiClO_4$.

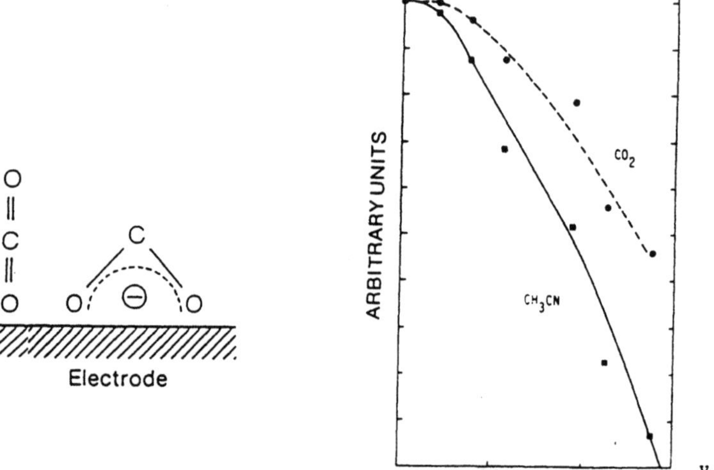


Fig. 51 Possible structures of adsorbed CO on Pt.

Fig. 52 Relative concentration of adsorbed $CO_2$ and $CH_3CN$

adsorbed $CO_2$ should decrease on either side of the potential of zero charge so the present results are consistent with this fact.

## 10. Summary.

The present paper is a review and summary of the work carried out at the Surface Electrochemistry Laboratory of Texas A&M University from about 1984. It gives information concerning the structure of the interface on noble metals, on iron and on several semiconductors in contact with aqueous and non-aqueous solutions. It represents the first comprehensive document which treats the structure at the interface as a function of potential and concentration in solution, obtained by spectroscopic means in electrochemistry.

Firstly, it is shown that the Fourier Transform Infrared Spectroscopy does work in respect to giving information at the surfaces[29,32-34], a point about which doubt had been expressed. This confirmation comes not only from one direct test which occurred from a coincidental happening[34], but also by mathematical analysis using Fresnel equations of the difference between the reflectance intensity of the parallel and vertical planes of polarization as a function of distance out into the solution[29].

It is shown that the method gives rise to very basic information concerning water structure of the interface[41], establishing for the first time, that the orientation of water which has been claimed by theorists over the last 20 years[36-39] does have some reality: water "lies down" at the pzc, and stands up as the potential of the electrode becomes more positive (perhaps also as it becomes more negative)[41].

Then, in a comparative study of the adsorption of phosphoric acid and trifluoromethane sulfonic acid anion[48,49], it is shown that the occupancy of the surface, as a function of potential, differs radically between the two substances at low temperatures, supporting the older view that phosphoric acid blocking of the interface was the origin of the much faster oxygen reduction obtained in the presence of the sulfonic acid[56]; but that this view disappears when the temperature is sufficiently increased so that, when the $\theta_{acid}$ is compared at conditions which might pertain to fuel cell use, the coverage is about the same. The work, carried out as a function of temperature, also leads to the formation of a number of isotherms which in turn has led to entropy determinations and indication of the movements of these materials on the surface.

Some of the studies have been carried out upon iron, and here ellipsometric spectroscopy has been used to obtain information upon the types of oxide which are formed in the various stages of passivation of iron, and the variation of these passive layers in respect to the degree of water which is in them[60]: water increases with oxidation of the passive film to the ferric state. Borate ions are adsorbed to the extent of a few percent on the surface of the passive layer, but, under conditions in which the iron is passivating, the presence of borate seems negligible[66].

The examination of inhibitors on iron has also been carried out with the aid both of ellipsometry and of FTIR. For example, thiocyanate [64] has been examined in respect to its isotherm under electrochemical conditions, and the single imaging isotherm with anion repulsion, deduced originally by Bockris, Devanathan and Müller[38], has been shown to apply well to the results in a layer of dielectric constant of 10. The inhibitor thiourea[68] has been examined spectroscopically with FTIR on iron, giving rise to the conclusion that the molecule adsorbes in such a way that the sulfur atom is attached to the iron and the $NH_2$ groups point towards the solution. Octynol, an inhibitor which is particularly effective in strong acid solutions in protecting iron, has been subject to spectroscopic examination, especially with respect to ellipsometry[85,86]. It has been shown that when the octynol first adsorbs upon the surface, it is in a "lying down" position, but as time increases and the concentration upon time on the surface of iron increases, it stands up. Further, it forms multilayers on the iron surface.

Furthermore, one of the greater advances which has been made is the beginning of the examination of intermediate radicals of inorganic and organic reactions in surfaces. Thus, the Fe-H bond has been identified at 2066 $cm^{-1}$, and this identity has been checked by measuring the O-H and O-D stretch frequency which have the appropriate ratios[34]. The scissoring motion also has given rise to a spectrum which corresponds to the spectrum earlier observed by EELS work[97]. This work is remarkable in the fact that it also has given rise to a rough estimate of the coverage of hydrogen on iron (by interpolating the transition dipole moment of Fe-H among a number of other known transition dipole moments); and by a proof of the correctness of the results and the fact that they show an appropriate variation of $\theta_H$ with respect to potential (i.e., one that corresponds to theory for a couple discharge mechanism of hydrogen on iron).

The $CO_2$ reduction mechanism is one which has had a great deal of investigation in the past 20 years[92].In many of these studies the suggested sequence is supposed to be one involving $CO_2^-$, although this ion has never been actually identified[92]. In the present investigation, a remarkable advance seems to have been made in respect to the fact that a frequency has been found in the literature for the $CO_2^-$ anion, and this frequency was indeed discovered in the spectrum during the reductin of $CO_2$ on platinum. The $\theta_{CO_2^-}$ increases in coverage with increasing <u>negative</u> potential, and this then leads to an example of the use of spectroscopy in mechanism determination because the result is remarkable enough to have given rise to marked questioning. It turns out to be uniquely explained by the existence of the following mechanism:

$$CO_{2aq} = CO_{2ads} \tag{32}$$

$$CO_{2ads} + e^- = CO^-_{2ads} \tag{33}$$

$$CO^-_{2ads} + H_2O + e^- \rightarrow HCOO^- + OH^- \tag{34}$$

Another example of mechanistic information arising from the use of FTIR in organic reaction stems from the examination of the oxidation of methanol. Here, the prevailing opinion has been that the mechanism involves adsorbed CO, but it has never been clear whether there are two forms of this or not. Indeed, two forms of CO have been found on the surface of platinum, and their variation with potential has been obtained. It has been shown that the sum of these (as observed by FTIR) follows precisely the radiotracer and (to a lesser degree) the potential sweep examination of CO on the surface of platinum in methanol oxidation. This is another example of how FTIR determinations can add to the determination of mechanisms on the surface of electrodes[29,92].

Lastly, a number of experiments has been carried out on the adsorption of ions and organic molecules on semiconductors. This is another example of the versatility of the FTIR method, where such observations could not be made by the use of methods such as potentiostatic sweep measurements, due to several problems as the co-dissolution of the semiconductor[88]. Correspondingly, the acetate ions have been shown to be adsorbed on gallium phosphide, and their adsorption as a function of potential in acetonitrile has been obtained; with special reference to the competition as a function of potential, the acetate increases with positive potential, displacing the acetonitrile. The study of the adsorption of ammonium on GaP has given rise to a clear determination of its structure (c.f. fig. 47).

The conclusions to the study are that FTIR is probably the leading method in electrochemistry at this time for knowledge of surface species. It appears to be applicable under all circumstances, and even, perhaps, to give a very rough indication of the amount of material on the surface, orientation and structure of the adsorbed entity. It is possible to obtain information on intermediates in reactions, and these may help in determining a path in rate-determining step. On the other hand, ellipsometric spectroscopy, particularly in the UV visible (but also potentially the IR) may be a coming tool because of the high speed of the measurement.

The background of all these methods is the radiotracer method, and it must not be neglected because it is now some 28 years old. Its major advantage -when it can be applied- is that it gives an accurate rendition of the total coverage by means of the isotope. On the other hand, and if this information is combined with other data (i.e. potential sweep for noble metals and FTIR and ellipsometry for non-noble substrates), understanding of the structures and perhaps of the mechanisms may be obtained.

The three methods reported here are being joined in our laboratory at this moment by the STM (Scanning Tunneling Microscope) and by an approach to the use of EXAFS(Extended X-Ray Absorption Fine Structure).

## 11. References.


1. E. Blomgren and J. O'M. Bockris, *Nature* **186** (1960) 305.
2. J. O'M. Bockris, M.A.V. Devanathan and A.K.N. Reddy, *J.Electroanal. Chem.* **6** (1963) 61.
3. J. O'M. Bockris, M.A.V. Devanathan and A.K.N. Reddy, *Proc.Roy.Soc.*

**A279** (1964) 327.
4. A.K.N. Reddy and J. O'M. Bockris, *Proceedings of the Symposium on the Ellipsometer and its use in the Measurements of Surface and Thin Films, Bureau of Standards* (1964) 229.
5. J. O'M. Bockris, B.D. Cahan and G. Stoner, *Chem. Inst.***1** (1969) 273.
6. W. Peik and J. O'M. Bockris *Surf. Sci.***27** (1971) 191. Aslo, W. Peik, M.A. Genshaw and J. O'M. Bockris, *J. Phys. Chem.***74** (1970) 4266.
7. A. Fredlein, A Damjanovic and J.O'M Bockris, *Surf Sci* **25** (1971) 261
8. W.E. O'Grady and J. O'M. Bockris, *Surf. Sci.***10** (1975) 245. Also the short note *Chem. Phys. Letters* **5** (1970) 116.
9. R. W. Revie, J. O'M. Bockris and R. G. Baker, *Surface Sci.* **52** (1975) 664; and *J.Electrochem.Soc.* **122** (1975) 1460.
10. A.T. Hubbard, Crit. *Rev. Anal. Chem.* **3** (1973) 201.
11. R.M. Ishikawa and A.T. Hubbard, *J.Electroanal. Chem.* **69** (1976) 317.
12. J. Kruger and J.P. Calvert, *J. Electrochem. Soc.* **114** (1967) 43.
13. A.K.N. Reddy, M.G.B. Rao and J. O'M. Bockris, *J.Chem.Phys.* **42** (1965) 2246.
14. A.K.N. Reddy, M.G.B. Rao and J. O'M. Bockris, *J.Electrochem. Soc.* **113** (1966) 1133.
15. A.K.N. Reddy, M.A. Genshaw, J.O'M. Bockris, *J. Chem. Phys.* **48** **(1968) 671.**
16. Y. Chiu and M. A. Genshaw, *J. Phys. Chem.* **73** (1969) 3571.
17. W. Paik and J. O'M. Bockris, *Surface Sci.* **28** (1971) 61.
18. M.J. Dignam and M.D. Baker, *J. Vac. Sci. Technol.* **21** (1982) 80.
19. W. Hansen, W.W.T. Kuwana and R. A. Osteryoung, *Anal. Chem.* **38** (1966) 1809.
20. H. Neugebauer, G. Nauer, N. Brinda-Konopik and G. Gidaly, *J. Electroanal. Chem.* **122** (1981) 381.
21. D. L. Dubois and J. A. Turner, *J.Amer.Chem.Soc.* **104** (1982) 4989.
22. S.Bewick, K.Kunimatsu and B.S.Pons, *Electrochim. Acta* **25** (1980) 465
23. R. G. Greenler, *J.Chem.Phys.* **44** (1969) 310; and *J.Chem.Phys.* **50** (1969) 1963.
24. S. Pons, J. K. Foley, J. Russell, M. Seversen in *Modern Aspects of Electrochemistry* (Bockris-Conway-White, Eds., Plenum Press: New York, 1986), **Vol. 17**, p. 223.
25. A. Bewick, S. Pons in *Advances in IR and Raman Spectroscopy* (Clark-Hester, Eds., Wiley Heyden: London, 1984) **Vol. 12**, p. 1.
26. M.A. Habib and J. O'M. Bockris, *J.Electroanal.Chem.* **180** (1984) 287.
27. M. A. Habib and J. O'M. Bockris, *J.Electrochem.Soc.* **132** (1985) 108.
28. M. A. Habib and J. O'M. Bockris, *Langmuir* **2** (1986) 388.
29. K. Chandrasekaran and J. O'M. Bockris, *Surface Sci.* **185** (1987) 495.
30. A.K.N. Reddy in *Electrosorption* (Gileadi Ed., Plenum Press: New York, 1967), p. 53.
31. K. Schroder and R. Memming, *Ber. Bungenges Phys Chem* **89** (1985) 385.
32. R.J. Nichols and A. Bewick, *J. Electroanal. Chem.* **243** (1988) 445.
33. J. O'M. Bockris, J. L. Carbajal, B. Scharifker and K. Chandrasekaran, *J.Electrochem.Soc.* **134** (1987) 1957.
34. B. Scharifker, K. Chandrasekaran, M. Gamboa-Aldeco, P. Zelenay, and J. O'M. Bockris, to be published in *Electrochim. Acta.*
35. G. Horanyi, *J.Electroanal.Chem.* **31** (1971) 95.
36. S. Trasatti in *Comprehensive Treatise of Electrochemistry*

(Bockris-Conway-Yeager Eds., Plenum Press: New York, 1984) **Vol. 1.**
37. M. A. Habib in *Modern Aspects of Electrochemistry* (Bockris-Conway-White, Eds., Plenum Press: New York, 1986) **Vol. 2.**
38. J. O'M. Bockris, M.A.V. Devanathan and K. Muller, *Proc. Roy. Soc.* **274** (1963) 55.
39. A. Bewick, *J.Electroanal.Chem.* **150** (1983) 481.
40. A. Bewick, *NATO ASI Series on In-Situ Methods*, Tenerife, 1988.
41. M. A. Habib and J. O'M. Bockris, *Langmuir* **2** (1986) 389.
42. R. Parsons and R. Reeves, *J.Electroanal.Chem.* **123** (1981) 141.
43. J. O'M. Bockris and M. A. Habib, Electrochim. Acta **22** (1976) 41.
44. E. Gileadi, S.D. Argadi and J.O'M. Bockris, *J.Phys. Chem.* **70** (1966) 2044.
45. D. Eisenberg and W. Kauzmann in *The Structure and Properties of Water* (Oxford University Press: London, 1965).
46. K. Kretzschmer, J. Sass, A. M. Bradshaw and S. Holloway, *Surf. Sci.* **115** (1982) 183.
47. A. J. Appleby and B.S. Baker, *J.Electrochem.Soc.* **125** (1978) 404.
48. M. A. Habib and J. O'M. Bockris, *J.Electrochem.Soc.* **132** 1985 108.
49. P. Zelenay, M. Habib, J. O'M. Bockris, *Langmuir* **2**, 1986, 393.
50. D. D. Perrin, *Ionization Constants of Inorganic Acids and bases in Aqueous Solutions* (2nd ed., IUPAC Chemical Data Series No. 29, Pergamon Press: New York, 1982).
51. K. Nakamoto, *Infrared and Raman Spectra of Inorganic and Coordination Compounds* (John Wiley & Sons: New York, 1978), **3rd ed., p. 172.**
52. T. G. Belicheva, V. I. Ligus, Y. Failkov, Russ. *J.Inorg.Chem. (Eng. Transl.)* **18** (1973) 1701.
53. J.O'M. Bockris and D.A. Swinkels, *J.Electrochem.Soc.* **111** (1964) 736
54. E. Blomgren and J. O'M. Bockris, *J. Phys. Chem.* **63** (1959) 1475.
55. B. E. Conway, H. Angerstein-Kozlowska and H. P. Dhar, *Electrochim. Acta*, **19** (1974) 455.
56. M. Enayetullah, T. D. DeVilbiss and J. O'M. Bockris, *manuscript in preparation.*
57. R.W. Revie, B.G. Baker and J.O'M Bockris, *Surf. Sci.* **52** (1975) 664.
58. O. J. Murphy, T. E. Pou, J. O'M. Bockris, and L. L. Tongson, *J. Electrochem. Soc.* **131** (1984) 2785.
59. T. E. Pou, O. J. Murphy, V. Young, J. O'M. Bockris and L. L. Tongson, *J. Electrochem. Soc.* **131** (1984) 1243.
60. V. Jovancicevic, R. C. Kainthla, Z. Tang, B. Yang, and J. O'M. Bockris, *Langmuir* **3** (1987) 388.
61. N.C. Debnath and A.B. Anderson, *J.Electrochem Soc.* **129** (1982) 2169.
62. C. T. Chen and B. D. Cahan, *J.Electrochem.Soc.* **129** (1982) 17.
63. A. Schlegel, S. F. Alvarado and P. Wachter, *J. Phys. C: Solid State Phys.* **12** (1979) 1159.
64. K.W. Revie, B.G. Baker and J. O'M. Bockris, *J. Electrochem. Soc.* **122** (1975) 1460.
65. V. Jovancicevic, J. O'M. Bockris, J. L. Carbajal, P. Zelenay and T. Mizuno, *J.Electrochem.Soc.* **133** (1986) 2219.
66. B. Scharifker, M. A. Habib, J. L. Carbajal and J. O'M. Bockris, *Surf. Sci.* **173** (1986) 97.
67. J. R. Doring, W. H. Green and A. L. Martson, *J. Mol. Struct.* **2**

(1968) 209.
68. J. O'M. Bockris, B. Scharifker and J. L. Carbajal, *Electrochim. Acta* **32** (1987) 799.
69. *Tables of Interatomic Distances and Configuration in Molecules and Ions* (The Chemical Society, L.E. Sutton, Ed., Burlington House: London, 1958).
70. M.A.V. Devanathan, J O'M. Bockris and W. Mehl, *J. Electroanal. Chem.* **1** (1959) 143.
71. H. Flitt and J. O'M. Bockris, *Int. J. Hydrogen Energy* **7** (1982) 411.
72. D.Berclay, *J.Electroanal and Interfacial Electrochem.* **44** (1966)310.
73. H. Kalsz and R. Saillant, *Chem. Rev.* **72** (1932) 231.
74. A. Ginsberg in *Transition Metal Chemistry* (R. I. Carlin, Ed., Edward Arnold: London, 1965) **Vol. 1**.
75. J. de Boer, *Z.Physik.Chem. (Frankfurt am Main)* **B18** (1932) 49.
76. *Electron Energy Loss Spectroscopy and Surface Vibrations (H. Ibach and D. I. Mills, Academic Press: London, 1982) p. 277.*
77. A. M. Baro and W. Erley, *Surf.Sci.* **112** (1981) L759.
78. G. Socrates, *Infrared Characteristic Group Frequencies* (John Wiley and Sons, Inc.: New York, 1980).
79. *P. Atkins, Physical Chemistry* (W. H. Freeman Co.: San Francisco, 1978), p. 585.
80. *D. Steele, Theory of Vibrational Spectroscopy* (W. B. Saunders Co.: London, 1971), p. 127.
81. J. O'M. Bockris and H. Kita, *J. Electrochem. Soc.* **108** (1961) 679.
82. J. O'M. Bockris in *Modern Aspects of Electrochemistry* (Bockris-Conway, Ed., Plenum Press: New York, 1954) **Vol. 1**.
83. A. Yamaguchi, R. B. Penland, S. Mizushima, T. J. Laane, C. Curran, and J. V. Quagliano, *J. Am. Chem. Soc.* **80** (1958) 527.
84. H. Wroblowa and M. Green, *Electrochim. Acta* **8** (1965) 679.
85. V. Jovancicevic, B. Yang and J. O'M. Bockris, *J. Electrochem. Soc.* **135** (1988) 94.
86. V. Jovancicevic, B. Yang and J. O'M. Bockris, *Electrochimica Acta* **32** (1987) 1557.
87. M. P. Soriaga, J. H. White, D. Song and A. T. Hubbard, *J. Electroanal. Chem.* **171** (1984) 359.
88. K.C. Chandrasekaran and J.O'M. Bockris, *Surface Sci.* **175** (1986)623.
89. *J. O'M. Bockris, M. A. Habib and J. L. Carbajal, J.Electrochem. Soc.* **131** (1984) 3022.
90. E. Gileadi, L. Duic and J. O'M. Bockris, *Electrochim. Acta* **13** (1968) 1915.
91. C. D. Crauer, *Ed., Desk Book of IR Spectra* (The Cablentz Society, Inc., Kirkwood, 1980) 2d ed.
92. S. Kapusta and N. Hackerman, *J. Electrochem. Soc.* **13** (1984) 1511.
93. J. F. Durana and A. W. Mantz in *Fourier Transform IR Spectroscopy: Applications to Chemical Systems* (Eds. J. R. Ferraro and L. J. Basile, Academic Press: New York, 1979) **Vol. 2**.
94. K. O. Hartman and I. C. Hisatsune, *J. Chem. Phys.* **44** (1966) 1913.
95. B. Beden, A. Bewick, M. Gagaq and J. Weber, *J. Electroanal. Chem.* **139** (1982) 202.
96. O.A. Pettrii and G.P. Khomchenko, *J.Electroanal Chem* **106** (1980) 277
97. A.M. Baro and W. Erley, *Surf. Sci.* **112** (1981) L759.

# APPLICATION TO ELECTROCATALYSIS OF EMIRS (ELECTROCHEMICALLY MODULATED INFRARED REFLECTANCE SPECTROSCOPY) AND RELATED TECHNIQUES

Bernard BEDEN
*Laboratoire de Chimie n°1, UA. CNRS 350*
*Université de POITIERS*
*40, avenue du recteur Pineau,*
*86022, POITIERS (France)*

ABSTRACT. After a general survey of Electrocatalysis and of the techniques known as "in situ" infrared reflectance spectroscopies, the application of EMIRS (i.e. electrochemically modulated infrared reflectance spectroscopy) to Electrocatalysis is considered, in the light of recent results concerning the identification of adsorbed species produced in the chemisorption or oxidation of small organic molecules on noble metal electrodes. The role of poisons, which decrease the catalytic properties of electrode surfaces, is particularly stressed and it is shown how an extensive use of electrochemical techniques, like cyclic voltammetry or pulse potential voltammetry, is essential for a correct interpretation of the spectroscopic results. The adsorption of carbon monoxide on various substrates is studied quantitatively, in order to discuss the existence of interactions, either between different types of adsorbed species or with the substrate.

## INTRODUCTION

Infrared reflectance techniques have only recently been applied to the electrode-solution interface. This paper attempts to show how these techniques have, by their ability to provide "in situ" information on the structure of adsorbates, led to new ideas in the field of electrocatalysis. It can be said that these attractive techniques have done so much in the past five years for the comprehension of electrochemical mechanisms, that many groups of electrochemists are finding a renewed interest in electrocatalysis.

However, before surveying the infrared reflectance techniques and some of the most relevant results, it is necessary to define exactly what electrocatalysis is, and what are the traits that differentiate electrocatalytic reactions from the usual reactions encountered in electrochemistry.



*C. Gutiérrez and C. Melendres (eds.),*
*Spectroscopic and Diffraction Techniques in Interfacial Electrochemistry,* 103–131.

# 1. Electrocatalysis

## 1.1. INTRODUCTION

A commonly accepted definition of electrocatalysis is "the acceleration of a particular electrode reaction by an appropriate choice of electrode material" (1)(2)(3). However, to take into account some particular aspects developed in more recent work, one should add that electrocatalysis also aims to orientate electrode reactions towards a better selectivity. This, of course, implies a key role for adsorbed intermediates, and explains the interest for any technique able to give information on adsorbates. From this point of view, "in situ" infrared reflectance spectroscopy appears to be one of the most promising techniques.

Some of the pressing questions in electrocatalysis (2) are summarized in the following table:

TABLE I. Questions in electrocatalysis.

| |
|---|
| * To what extent can we understand the relation between the composition of catalysts, and their activity for given reactions ? |
| * What are the detailed mechanisms by which the electrocatalysts actually operate? |
| * Can we improve the existing electro-catalysts? (How can we design better electro-catalysts?) |
| * Where, and why, do we need electro-catalysts? |

| |
|---|
| "in situ" reflectance spectroscopic techniques may help to answer some of these questions |

## 1.2. FACTORS DETERMINING ELECTROCATALYTIC ACTIVITY

Many different parameters are known to affect the electrocatalytic behaviour of an electrode surface towards a given reaction. It is necessary to distinguish clearly between the activity of a catalytic surface (i.e. a characteristic of electrode materials), and the reactivity of a given molecule (i.e. a characteristic of electroactive species). All the systems listed in table II have been studied so far in our laboratory, and typical examples will be detailed in the following sections.

TABLE II
Factors determining electrochemical activity
(Survey of investigated systems)

| ELECTRODE MATERIALS |
| --- |
| -Polycrystalline metals<br>.noble, Pt, Rh, Pd, Au...<br>.non noble, Ni, Cd...<br>-Monocrystalline electrodes<br>.Pt(hkl), Ni(hkl)...<br>-Alloys<br>.Pt-Rh, Au-Pd, Pt-Pd, Pt-Au...<br>-Noble metals modified by u.p.d.<br>.adatoms: Bi, Pb, Cd, Cu, Tl, Re, S. |
| IMPORTANT ELECTROCATALYTIC PARAMETERS |
| -crystallographic orientation<br>-roughness factor<br>-pH<br>-temperature |
| ELECTROACTIVE SPECIES |
| -$C_1$ CO, HCOOH, $CH_3OH$, (HCHO)<br>-$C_2$ $C_2H_5OH$, $CH_2OH$-$CH_2OH$ (and related cpnds)<br>-$C_3$ glycerol, propanediols<br>-$C_4$ butanols<br>-$C_6$ glucose, fructose, sorbitol,... |

1.2.1.*Electrode materials*. The nature of the catalytic metal (Pt, Rh,...), and its crystallographic superficial structure (polycrystalline or monocrystalline), are perhaps the most important electrocatalytic parameters. However, the method of surface preparation is important as well. Especially the roughness of the surface, which varies according to the preparation technique (mechanical polishing, chemical polishing, electropolishing, thermal treatment, annealing,...), may lead to changes in the adsorbed species and therefore in the electrochemical mechanism. In these studies a most necessary condition is the cleanliness of the surface, which has to be as free of impurities as possible.

Another interesting way to change the surface properties is to use alloys. For instance, with two catalytic metals like platinum and rhodium, or gold and palladium, which are fully miscible over the whole domain of composition, the gradual investigation of the alloy electrocatalytic activity may reveal the existence of so-called "synergic effects", i.e. an enhanced activity of the alloy as compared with the simple weighted addition of those of the bare metals. However, the situation is generally much more complex (see Section 1.2.3.), due to preferential dissolution and surface segregation.

An alternative way to modify the surface activity is to deposit submonolayers of foreign metal atoms at potentials positive to the reversible potential of the Nernst equilibrium. The resulting underpotential deposited ("u.p.d.") atoms (also called adatoms), are known to enhance, in some cases greatly, the activity of noble metal electrodes for given electro-catalytic reactions. For instance the maximum current densities are multiplied by $\approx 60$ when the electrooxidation of formic acid is carried out on a platinum surface partially covered by Pb adatoms, as compared to bare platinum (4).

1.2.2. *Electroactive species*. The global reactivity of given organic molecules for electrocatalytic reactions depends on their structure (number of C atoms, shape of the skeleton, number and position of functional groups,...), but also on the way they adsorb on the surface. This means that superficial concentration and steric effects have also to be considered, and that the real boundary between "activity" and "reactivity" is not always easy to define, especially for rough catalytic materials.

Therefore, in order to study the reactivity of electroactive species, experiments have been conducted with series of compounds having : a) the same number of carbon atoms but different functional groups, to check the influence of the latter, b) the same functional group but variable number of carbons, to check the propagation of inductive effects, c) the same functional groups and the same number of carbons, but with different types of skeleton, to check isomeric effects.

Among the electrochemical techniques which can be used to evaluate reactivity, cyclic voltammetry (CV) is certainly very useful, providing easy qualitative information in a very didactic way. The following examples illustrate this point, although it has to be borne in mind that CV is only a first approach to the problem. Quantitative information has to be obtained by the use of criteria (5) or, better, by interpretation of Tafel slopes and calculation of exchange current densities (1).

## 1.3. APPLICATION OF CYCLIC VOLTAMMETRY TO ELECTROCATALYSIS

1.3.1. *Characterization of noble metal surfaces.* In cyclic voltammetry a repetitive triangular potential sweep, between two suitable limits $E_c$ (called "cathodic" or, preferably, lower potential limit), and $E_a$ ("anodic" or upper potential limit), is applied to the working electrode (Fig.1a). If the potential sweep rate, $v = dE/dt$, is constant, the response in current due to the potential perturbation has, for a platinum electrode immersed in a 0.5M $H_2SO_4$ solution, the shape displayed in Fig.1b. When folded along the $i$ axis, the resulting closed curve $i(E)$ is called a voltammogram. Classically three regions are distinguished, Faradaic processes being involved in

Figure 1: Cyclic voltammetry. a) Potential vs. time, E(t), and current vs. time, i(t), curves. b) Cyclic voltammogram, i(E), for Pt in sulfuric acid medium, 25°C, v = 50 mV $s^{-1}$. c) Voltammograms for several noble metals (from Ref. 6).

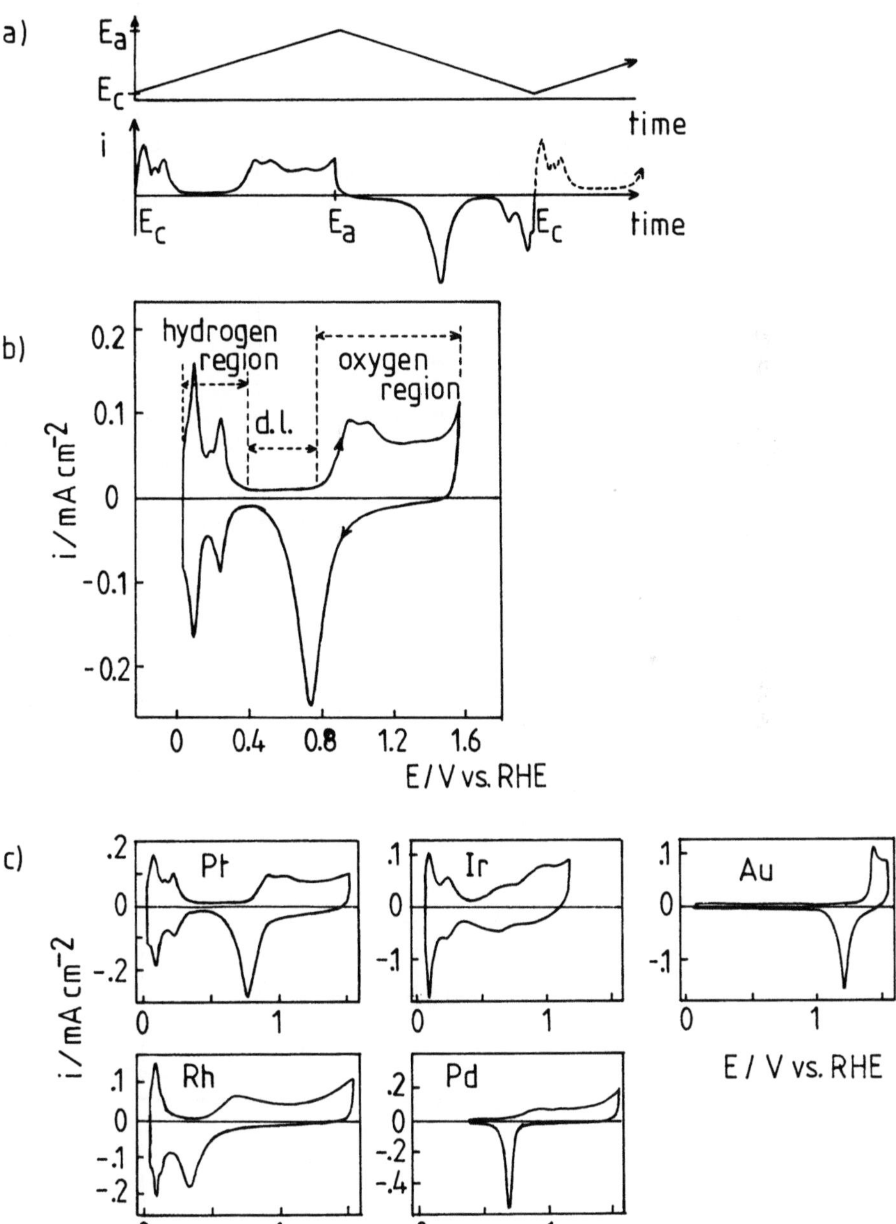

two of them. The Faradaic processes are quasi-reversible hydrogen adsorption-desorption processes in the range 0 to 0.25 V vs.RHE (i.e. with respect to a reversible hydrogen electrode in the same solution) and more irreversible oxygen adsorption-desorption processes in the range 0.8 to 1.6 V vs.RHE. In the intermediate region, between 0.25 and 0.8 V, only low capacitive current densities are observed, due to charge of the double layer.

Voltammograms recorded under given experimental conditions are specific of a metal surface (Fig.1c), and simple calculations can provide useful information on the stoichiometry of adsorbed layers, especially on noble metals (6). For instance, assuming that 1 adsorbed hydrogen (resulting from the reduction process $H^+ + e^- \rightarrow H_{ads}$ ) occupies one adsorption site on a smooth polycrystalline platinum surface with an average of $1.3 \times 10^{15}$ sites per $cm^2$, the charge associated with the formation of one monolayer should be

$$Q_H = 1.3 \times 10^{15}\ cm^{-2} \times 1.6 \times 10^{-19}\ C/e \times 10^6 = 208\ \mu C\ cm^{-2},$$

in excellent agreement with experimental values. This "reference charge" is used for further calculations, either of the roughness factor ( defined as the ratio of the real surface to the geometric surface) of an unknown Pt electrode, or of the coverage of the active

Figure 2. Voltammograms for the oxidation of 0.1M n-butanol in 0.1M NaOH, on noble metal electrodes.
a)Platinum. b)Gold. c)Rhodium. d)Palladium. 25°C, 50 mV $s^{-1}$

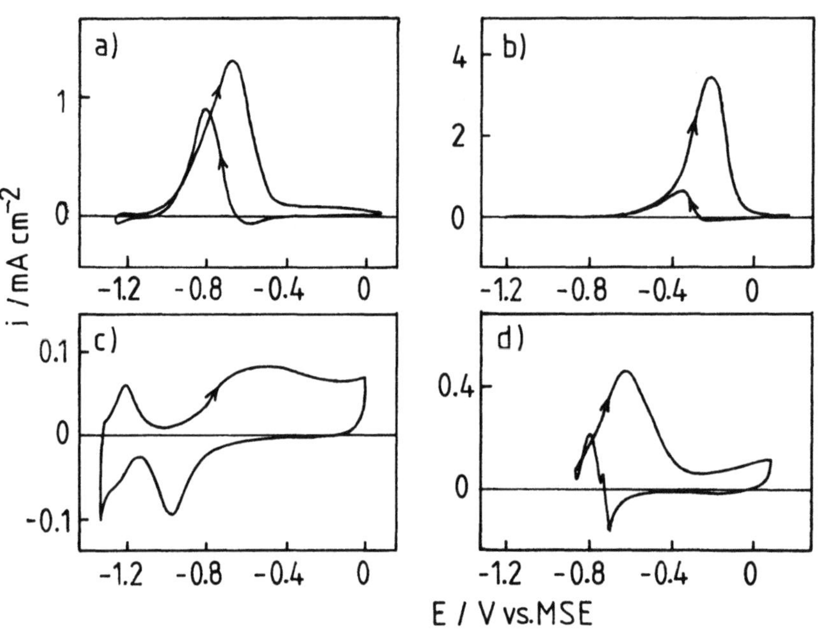

surface by other adsorbates capable of displacing $H_{ads}$. Similar reference values have been calculated for different faces of platinum single crystals (7).

1.3.2. *Activity of a metal surface.* Adding an electroactive species to the solution leads to an increase of the current densities, provided the experimental conditions (mainly the potential, but also time, sweep rate, pH...) are suitable for the oxidation or reduction of the species (see e.g. Fig.2a, where butanol is seen to be oxidized in both the positive and negative potential excursions). There are many examples in the literature which can illustrate the influence of the nature of the metal on a given electrocatalytic reaction (2)(3)(8). For instance, at the same concentration and same potential sweep rate, current densities of nearly 1.5 mA $cm^{-2}$ are obtained in acid medium for the oxidation of 0.1M HCOOH on platinum, whereas rhodium yields much lower current densities, and palladium and gold are almost inactive (9) (although gold becomes active at higher temperatures (10)). Contrarily, in alkaline medium gold appears to be a much better catalyst than platinum, especially for the oxidation of alcohols (11). Rhodium, although very similar to platinum from many points of view, is often a poor electrocatalyst as compared to platinum, whereas palladium has interesting but restricted properties, due partly to dissolution of hydrogen in the lattice. These general characteristics are illustrated in Fig.2 for the case of butanol oxidation in basic medium (12).

The interaction of the adsorbed species (especially the number of chemical bonds) with the surface will be mainly determined by the crystallographic orientation of the catalytic surface. It is therefore not surprising that the voltammograms for Pt single-crystal faces in 0.5M $H_2SO_4$ differ strongly (7) for the three low index planes, viz. (100), (110) and (111) (Fig.3), and that they exhibit a very different activity, not only with respect to maximum current densities, but also to the potential range over which electrooxidation occurs (see Ref.13 for oxidation of formic acid on Pt(hkl) in acid medium). Furthermore, the influence of anion adsorption may be very pronounced and structurally dependent, especially for methanol oxidation on Pt(111), whereas no similar effect is found with the (100) and (110) planes (14).

1.3.3. *Molecular structure and reactivity.* From Fig.4, where voltammograms for the oxidation of various alcohols are given, it can be concluded that an increase of the length of the carbon chain produces a decrease of the reactivity of the molecule. However, the maximum current densities are not very different, the main effect to consider being a positive shift of the oxidation peaks in the positive potential sweep. Poisoning phenomena, due to strong adsorption of side products but also to direct chemisorption residues, have certainly to be taken into account for a full explanation of this reactivity decrease.

Very interesting is the reactivity of molecules having several isomers, like butanols (12), or carbohydrates. Pronounced effects are

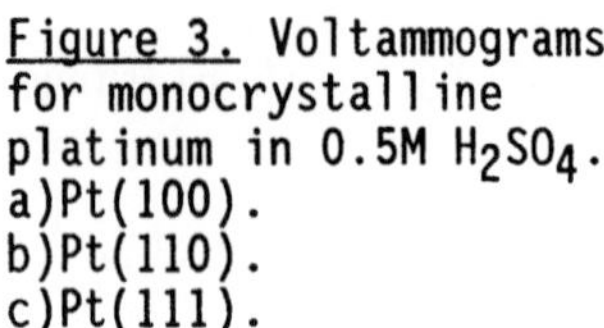

Figure 3. Voltammograms for monocrystalline platinum in 0.5M $H_2SO_4$.
a)Pt(100).
b)Pt(110).
c)Pt(111).

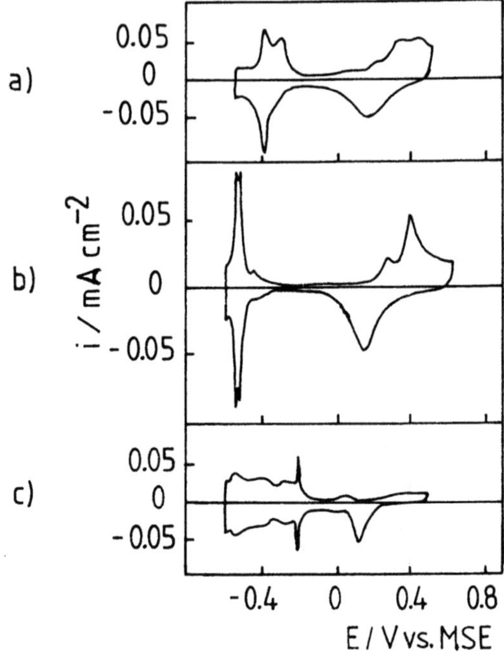


Figure 4. Voltammograms for the electrooxidation on Pt in 0.1M $HClO_4$, 25°C, 50 mV s$^{-1}$ , of different primary alcohols (0.1M). a)Methanol, b)Ethanol, c)n-butanol, d)Ethylene-glycol.

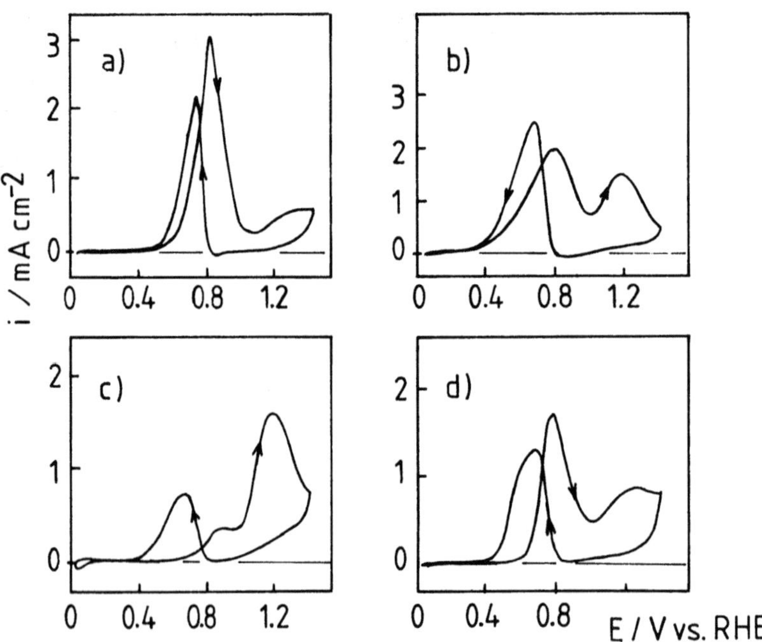

usually found for positional isomers (i.e. those in which the position of the functional group varies), leading for instance to high reactivity for n-butanol and almost no reactivity for tert-butanol. However, there seems to be no significant reactivity difference between conformational isomers, at least in the case of glucose (15), whose H atom in the asymmetric carbon can be equatorial (α-D-glucose) or axial (ß-D-glucose).

1.3.4. *Modification of surface composition.* Alloying two miscible metals leads to a voltammogram whose shape is intermediate between those of the two pure metals (6). By making the simple assumption that the potential of the oxide reduction peaks shifts linearly with the surface alloy composition, as originally noted by Rand and Woods (6), it is possible to measure the surface atom content of the alloy. Using this very simple technique, it was found that the surface composition can differ strongly from that of the bulk, by up to 30%. For instance, a Pt-Pd alloy having a superficial composition close to 33% platinum was found to be 10 times more active than Pt, and 100 times more active than Pd, for methanol oxidation (16). Other synergic effects were found with Pt-Rh or Au-Pd (17)(18).

Much more pronounced effects are found if the catalytic surface is modified by adatoms (see section 1.2.1.). Since the early work of Adzic et al. (19), many systems have been studied and reviewed (4)(20)(21). Table III gives the influence of different adatoms for the oxidation of ethylene-glycol on platinum in alkaline medium. The "catalytic" activity of the adatoms is estimated from the ratio of the peak current densities for the (Pt+adatom) electrode, to that for the bare Pt electrode. Very high enhancement factors are obtained with Pb, Bi and Tl. The effect of Pb adatoms, at the optimum concentration in solution ($10^{-4}$M), has also been studied on Pt-Au alloys. High current densities, above 100 mA $cm^{-2}$, were obtained with such ternary systems. Several theories have been advanced (20)(22), but there is no systematic rule which would predict the influence of a given adatom on a given electrode surface for a given reaction. This is probably due to the complexity of the problem : adatom superficial concentration and crystallographic structure of the catalytic substrate are important parameters, but bulk concentration of the electroactive species, potential, pH, adsorption time and, to a certain extent, temperature, are important as well. Competitive adsorption of adatoms, chemisorbed intermediates (including poisons), and final products cannot be ignored when modelling electrochemical mechanisms. This point will be addressed in the next sections, in the light of the results obtained by "in situ" infrared reflectance spectroscopy.

Table III. Characteristics of ethylene-glycol oxidation on a platinum electrode modified by u.p.d. of foreign metals.

| Adatom X | Optimum concn. mol $dm^{-3}$ | Peak characteristics during the positive potential sweep | | |
|---|---|---|---|---|
| | | $E_p$ / V/RHE | $i_p$ / mA $cm^{-2}$ | $i_p[(Pt+X)]$ / $i_p[Pt]$ |
| Bi | 2 $10^{-4}$ | 0.77 | 88 | 13.5 |
| Cd | 3 $10^{-5}$ | 0.60 | 5.5 | 0.8 |
| Cu | $10^{-5}$ | 0.74 | 9.5 | 1.4 |
| Pb | $10^{-4}$ | 0.75 | 96 | 14.5 |
| Re | 2 $10^{-4}$ | 0.72 | 7 | 1.1 |
| Ru | $10^{-5}$ | 0.76 | 9 | 1.4 |
| Tl | 2 $10^{-5}$ | 0.73 | 37 | 5.6 |

## 2. Infrared Reflectance Spectroscopy

Only recently has infrared spectroscopy been successfully applied to "in situ" studies of the electrode-solution interface (24). Two severe limitations had to be overcome : one was the considerable attenuation of the infrared beam by the solvent (especially in water), the other was the very high sensitivity required to detect submonolayers of adsorbed species, i.e. concentrations as low as $10^{14}$ to $10^{15}$ species per $cm^2$, undetectable by conventional spectrometers.

Several approaches have been followed to solve these problems. They are summarized in Table IV. They imply a choice of the type of <u>reflection mode</u> (internal or external) and of the type of spectrometer (dispersive or Fourier transform). The explicit names of the techniques are given in Table V, together with references to early works

### 2.1. COMPARISON OF THE TECHNIQUES

Each of the techniques listed in Tables IV and V has its own advantages and inconveniences, their relative merits being still a matter of discussion (25)(26)(27). A very high sensitivity (better than $10^{-5}$ absorbance unit) is achieved by EMIRS,i.e. by potential modulation, after only a few minutes to a few tens of minutes of successive wavelength scans, provided that the range investigated does not exceed a few hundreds of wavenumbers. Conversely, SNIFTIRS, due to the so-called multiplex advantage of FTIR, covers the whole range from 5500 to 400 $cm^{-1}$ in a single experiment, the difference spectrum being obtained after the Fourier transformation of two sets of accumulated interferograms. This may be considered as an advantage, although the sensitivity of SNIFTIRS is definitively slightly lower than that of EMIRS, even after long spectral accumulations (typically one to two hours).

TABLE IV.
Infrared reflectance spectroscopy at the electrode-solution interface.

| | |
|---|---|
| AIM: | to detect "in situ"<br>-adsorbed layers on metallic surfaces in contact with a solution<br>-species produced in the vicinity of the electrode. |
| AMOUNTS TO BE DETECTED: | -1 monolayer (or less) of adsorbed species (a few Angstroms thick)<br>-Superficial concentrations of $10^{14}$ to $10^{15}$ species per $cm^2$ |

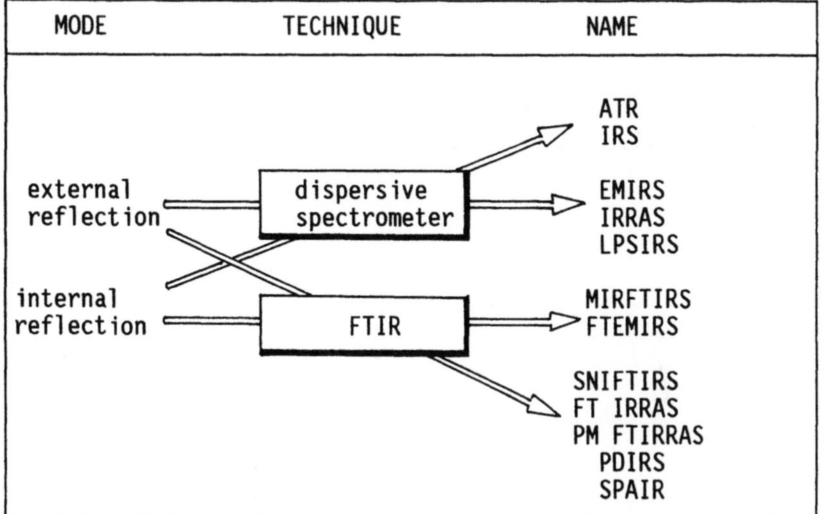


TECHNIQUES INVOLVED :

-Phase sensitive detection associated with a modulation, either of potential, or of polarization state of the beam

-Signal processing

-Spectral accumulation

-Averaging

-Spectral subtraction

TABLE V. Infrared reflectance techniques

| INTERNAL REFLECTION CELL | Ref. |
|---|---|
| Grating spectrometer | |
| *Internal reflection Spectroscopy | |
| IRS (Harrick,1967) | (31) |
| *Attenuated total reflectance | |
| ATR (Hansen,1970) | (32) |
| FTIR | |
| *Multiple internal reflection Fourier transform infrared spectroscopy | |
| MIRFTIRS (Brinda-Konopik,1981) | (33) |
| *Fourier transform electro-modulated infrared spectroscopy | |
| FTEMIRS (Chazalviel,1987) | (30) |

| EXTERNAL REFLECTION CELL | |
|---|---|
| Grating spectrometer | |
| *Electrochemically modulated infrared reflectance spectroscopy | |
| EMIRS (Bewick,1980) | (25)(34) |
| *Infrared reflection absorption spectroscopy | |
| IRRAS (Overend,Russell,1982) | (35) |
| *Linear potential sweep infrared reflectance spectroscopy | |
| LPSIRS (Kunimatsu,1982) | (36) |
| FTIR | |
| *Subtractively normalized infrared Fourier transform infrared reflectance spectroscopy | |
| SNIFTIRS (Pons,1981) | (25)(37) |
| *FT-IR (Bockris, 1984) | (38) |
| *Polarization modulated Fourier transform IR reflectance spectroscopy | |
| PMFTIRRAS (Kunimatsu,Golden,1984) | (28) |
| *Potential difference IR and *Single potential alteration IR | |
| PDIR, SPAIR (Weaver, 1986) | (39) |

Other techniques use polarization modulation, either with a dispersive instrument (IRRAS) or with a Fourier transform spectrometer (PM-FTIRRAS) (28)(29). Theoretically their main advantage is to allow a discrimination between surface and solution species. However, due to the strong absorption of the latter and the very weak absorption of the former, severe complications arise, especially as regards the determination of the base line (28). A very high sensitivity ( better than $10^{-6}$ a.u.) has been claimed recently by adding together the advantages of potential modulation and Fourier transform spectroscopy (30). However, the new technique, FTEMIRS, has not yet been applied to electrocatalysis.

## 2.2 TECHNICAL ASPECTS

2.2.1.*Internal or external reflection modes.* In the internal reflection mode, the infrared beam travels inside the IR transparent substrate (Ge, Si...) because the angle of incidence at the substrate-electrolyte interface is chosen to be higher than the critical angle. The number of reflections depends only on the angle of incidence and on the dimensions of the substrate. The evanescent wave originated at the interface penetrates a distance of the order of $\lambda/10$ into the electrolyte across a very thin metallic film (Pt, Au, Fe, Cd...) deposited on the substrate. Energy absorption by the solvent is therefore avoided, but the technique is restricted to the study of evaporated films, whose catalytic behaviour may be very different from that of smooth metals or single crystals.

In the external mode there is no restriction on electrode thickness and therefore any surface can be investigated (smooth, rough, polycrystalline or single crystal, modified by adatoms,...). The infrared beam crosses the infrared window and a very thin (a few microns) electrolyte layer before being reflected by the electrode surface and crossing back again through the solution. A limited number of materials are available for the window, $CaF_2$, Si and ZnSe being those most frequently used.

The flatness of the window and of the working electrode, which together determine the minimum thickness of the solution layer, and the incident angle of the infrared beam are very critical parameters (25)(27). Depending on the refractive indices of window materials and solutions, angles of incidence of 65° (for $CaF_2$) or 70° (for Si) are acceptable compromises between the high incidence angles necessary to maximize the resulting electric field vector at the interface, $\vec{E}_p$ (i.e. parallel to the incidence plane), and the low incidence angles necessary to avoid too much loss of energy by reflection on the outer surface of the flat window. As for the solution layer, an average thickness of a few micrometers (with aqueous solutions) allows to maintain electrochemical control of the interface without too much infrared radiation being absorbed. However, when the electroactive species is consumed by the reaction during spectral accumulation, severe mass transfer limitations can occur in the thin layer of electrolyte. This, of course, does not happen when the adsorption equilibrium is realized before spectral accumulation, and provided that the potential limits of the modulation signal are chosen well within the adsorption potential domain.

2.2.2. *Use of a grating spectrometer*. The most sensitive IR technique, originally developed by Bewick et al. (34) under the name of EMIRS, consists of modulating the potential of the electrode between two suitable limits, $E_c$ and $E_a$ (at which the reflectivities are $R_c$ and $R_a$ respectively), with a square pulse of 50 to 500 mV amplitude, at a frequency of 8 to 25 Hz. Synchronous detection of the infrared signal allows to reject the non-modulated information coming from the electrolytic solution to sort out the absorption bands due to vibrations of adsorbed species at the electrode surface from the background noise. The signal, $\Delta R_{(\bar{\nu})}$, is stored and averaged, either by means of a computer program (34) or by using a suitable digital oscilloscope. Then, it is divided by $R_{(\bar{\nu})}$ (i.e. the absolute reflectivity of the surface, previously recorded in a separate experiment), in order to obtain the normalized spectrum of the dimensionless differential reflectance ( $\Delta R/R$) vs. wavenumbers.

Using the same equipment it is possible to perform other interesting techniques, like LPSIRS (36). In this method, proposed by Kunimatsu, the wavelength is kept constant and the changing reflectivity during potential sweeps is recorded. Repetitive experiments (e.g. every 5 $cm^{-1}$ ), allow to build tridimensional diagrams [ $\Delta R/R$, E, $\bar{\nu}$ ], which contain the dependence of the spectra on the potential. However, the technique suffers from the very long spectral accumulation times necessary for improving the signal to noise ratio, and has to be restricted to investigations over very narrow ranges of wavenumbers, and also to very stable electrochemical systems.

By adding a polarization modulator to the same equipment as for EMIRS it becomes possible to modulate the polarization state of the incident beam between the s and p states, i.e. perpendicular and parallel to the incidence plane, respectively. Therefore, using the modulation signal (generally a few tens of kHz) as reference for the phase sensitive detection, the so-called surface selection rules allow us to obtain the IR spectra of adsorbed molecules whose dipole moment is normal to the surface. However, as underlined in Section 2.1., the sensitivity achieved with the present equipments is lower than that of EMIRS, and, in the absence of any improvement, the IRRAS technique has the only advantage of working at a fixed potential.

2.2.3. *Use of FTIR spectrometers*. By collecting series of interferograms at successive electrode potentials, it was shown by Brinda-Konopik et al. (33) that it was possible to take advantage of the high luminosity of FTIR spectrometers to follow "in situ" the formation and growth of multilayers at an electrode-aqueous solution interface. They used Multiple Internal Reflection (MIRFTIRS) on a very thin (tens of nm) metallic electrode deposited on an ATR (Attenuated Total Reflection) prism. However, the technique could not detect submonolayer levels, and the energy was not sufficient for an external reflection cell (much preferable as far as electrocatalysis is concerned). These limitations were soon overcome by Pons et al. (25), who proposed a new technique, called SNIFTIRS. Short series of interferograms are alternatively collected at two given electrode potentials, $E_c$ and $E_a$ , where the reflectivities are $R_c$ and $R_a$. After Fourier transformation, spectra of $[(R_a-R_c)/R]$ vs. $\bar{\nu}$ are obtained by a subsequent subtraction of the coadded series and ratioing to R (taken as $R_c$ or $R_a$).

SNIFTIRS differs from MIRFTIRS (33)(40) in that the latter is only a subtraction between accumulated series of spectra taken at various successive potentials (using one of them, generally the first, as a reference). SPAIR (39) is also a variant of SNIFTIRS, which uses subtractions between series of spectra accumulated at the same potential, but at various times, in order to follow the time dependence of species generated at the electrode surface or accumulated in its vicinity.

The advantage of SNIFTIRS over the other techniques is that it minimizes all the drifts, whatever their origin (chemical or electrical), thus allowing a higher sensitivity. For instance, a very good signal-to-noise ratio is currently obtained with the best commercial FTIR spectrometers after collection and processing of only several hundreds of 50-interferogram series (34), submonolayers of adsorbed species being easily detected under these conditions. Conversely, MIRFTIRS is more suited for the study of the growth of superficial layers (of oxides or polymers) on electrodes, and also for the identification of products accumulated in the vicinity of the electrode surface, since for these two applications less sensitivity is required.

Due to computer facilities and to the incredible advances in FTIRS, SNIFTIRS is relatively easier to use than EMIRS and has become a nearly routine technique for investigations at the electrode-solution interface. However, it must be said that the technique still suffers from the difficulty to separate the contribution of solution species, whose consumption (when non compensated because of mass transport limitations) leads to so strong "negative peaks" in the spectra that the weaker absorption bands of surface species can be masked. Such a difficulty is avoided in EMIRS, where potential modulation eliminates the non-modulated information associated with solution species.

In an attempt to solve this problem, Golden and Kunimatsu (28) applied polarization modulation to FTIR spectroscopy. The frequency (70 kHz) of the photoelastic modulator was chosen to be much higher than the frequency of the interferometer, which allowed the modulated interferograms to be demodulated before being Fourier transformed. Actually, in PM-FTIRS the spectra are the difference between s and p "polarized" spectra, and a further subtraction of spectra collected at two different potentials (as in SNIFTIRS) appears still to be necessary to minimize, if not to eliminate, the contribution of solution species. At the same signal-to-noise ratio, typical spectra accumulation times are notably shorter than in SNIFTIRS, but to the expense of simplicity.

A very attractive approach to solve the problem of the solution bands has been given recently by Chazalviel (30), who combined the advantages of both potential modulation and Fourier transform infrared spectroscopy. In this FTEMIRS technique the interferogram itself is electromodulated, which allows a better rejection of the solution bands, as compared with PM-FTIR. A sensitivity as high as $10^{-6}$ a.u. is obtained after one hour of spectral accumulation, using a specially designed Si internal reflection cell with 15 reflections. However, due to severe experimental constraints like the necessity to drive very smoothly the moving mirror at a very low speed (≈ 6 micrometers per second), which allows the electrode potential to be modulated at 100 Hz, the technique is not yet practicable with commercial FTIR instruments, even the most sophisticated ones.

## 3.Application of EMIRS to Electrocatalysis

Since 1980, i.e. the inception of "in situ" infrared reflectance spectroscopy, many papers have been published concerning EMIRS investigations on the solid-electrolyte interface (see e.g. the reviews in refs. 25 and 27). As said in Section 2.2.2., potential modulation makes EMIRS particularly appropriate for the study of adsorbed species. It is therefore a relevant technique for any problem connected with poisoning phenomena occurring on electrode surfaces during the course of electrocatalytic reactions. In particular, the unravelling of the formation of adsorbed carbon monoxide on electrodes during chemisorption processes has become a challenge since its first unexpected "in situ" identification as the major poisoning species on platinum when methanol (41), formic acid (42), or almost any small organic molecule, are electrooxidized in acid medium. Thus, adsorption of CO on noble metal electrodes has been extensively studied.

### 3.1.ADSORPTION OF CARBON MONOXIDE ON NOBLE METAL ELECTRODES

CO adsorption is commonly considered as a test reaction, both at the solid-gas and at the electrode-solution interfaces. Solid-gas studies initially used CO because it has very intense infrared absorption bands. It was soon discovered that it adsorbs under several forms, depending on the nature of the metal substrate and on the coverage (i.e. on the partial pressure of CO). From electrochemical measurements at metal electrodes, it is known that the coverage by $CO_{ads}$ species mainly depends on the electrode potential, but also on other parameters, including the pH of the solution and, possibly, anion adsorption. It is interesting that CO adsorption is structurally dependent (as shown by the dependence of the potentials of the voltammetric peaks on the crystallographic orientation of the surface), and that CO is obviously adsorbed under two or more different states. However, electrochemistry cannot yield more information on this point, even when performed quantitatively (43).

3.1.1.*EMIRS studies.* EMIRS has been able to provide a direct identification of the various $CO_{ads}$ species on noble metal electrodes (44). Fig.5 gives the spectra obtained under the same experimental conditions (13.5 Hz modulation signal of 400 mV amplitude, with a mid-pulse potential of 0.2 V vs. RHE), for Pt, Rh and Pd electrodes in 0.5M $HClO_4$ solutions saturated with gaseous CO. Striking differences occur in intensities and peak positions. Comparison with adsorption of CO at the solid-gas interface (45) allows to assign the bands (Table V)

Figure 5. EMIRS spectra of CO species adsorbed on noble metal electrodes. 0.5M $HClO_4$ solutions saturated with gaseous CO. $\Delta E$ = 400 mV, $\bar{E}$ = 0.2 V vs. RHE; f = 13.5 Hz a) Pt b)Rh c)Pd

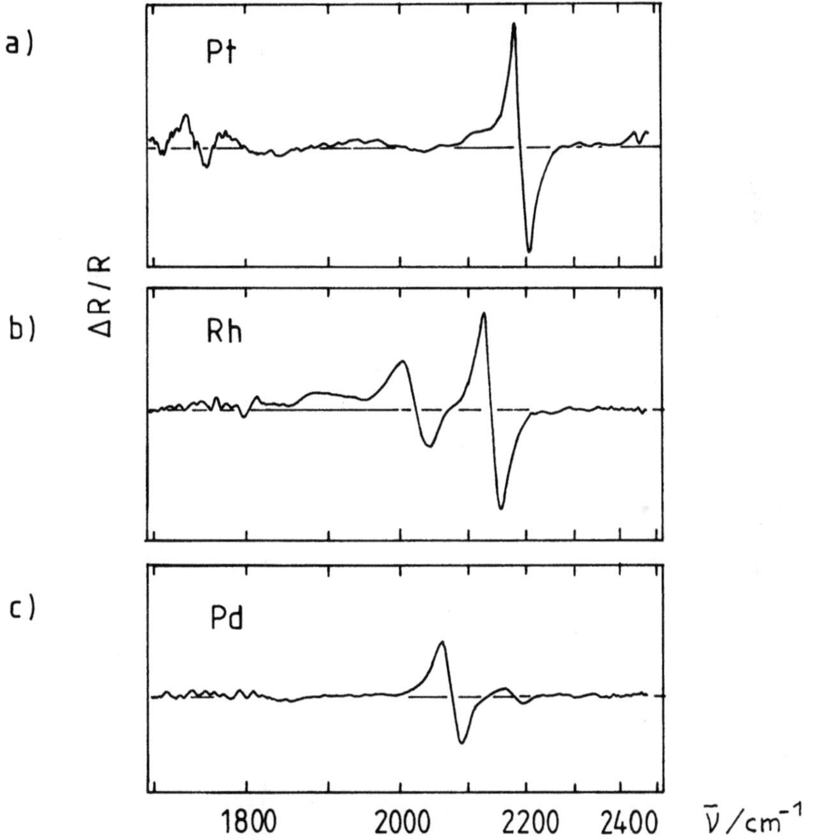


The pseudo-derivative shape of some of the EMIRS bands is related to the wavenumber shift with potential, at constant coverage of adsorbed species. So, for linearly bonded CO it has been measured that, on platinum , the shift of the band center vs. potential is ca. 30 $cm^{-1}$ $V^{-1}$ in aqueous medium, (46), and 20 $cm^{-1}$ $V^{-1}$ only in organic solvents (63). These values are interpreted in terms of the "Stark effect".

TABLE V. EMIRS bands for carbon monoxide adsorption on noble metal electrodes

| Metal | Band center $cm^{-1}$ | Shape | Intensity | Assignment |
|---|---|---|---|---|
| Pt | 2080 | deriv. | v.s. | $CO_L$ |
| | 1870 | s.s. | v.w. | $CO_m$ |
| Rh | 2040 | deriv. | v.s. | $CO_L$ |
| | 1920 | deriv. | s. | $CO_B$ |
| | 1820 | s.s. | v.w. | $CO_m$ |
| Pd | 2080 | deriv. | w. | $CO_L$ |
| | 1970 | deriv. | s. | $CO_B$ |

Shape: deriv., band with a derivative shape; s.s., single-sided band. (on the shape of EMIRS bands, see e.g. Ref.27).
Intensity: v.s., very strong; s., strong, w., weak, v.w., very weak.
CO band assignment: $CO_L$, linearly bonded; $CO_B$, bridge-bonded; $CO_m$, multibonded to the surface.

It is obvious from Fig.5 that $CO_L$ species are dominant on Pt surfaces, while $CO_B$ and $CO_L$ are both present at appreciable coverages on Rh, and that there is almost no $CO_L$, as compared to $CO_B$, on Pd surfaces. Furthermore, especially on Pt, other bands are consistently seen near 1640 and 1700 $cm^{-1}$, in spite of a higher noise level due to bulk water absorption in this range. The first band corresponds most probably to HOH deformation vibrations of adsorbed water molecules, in agreement with Kunimatsu (29). The second band at 1700 $cm^{-1}$ is not directly interpretable. Vibrations of carbonyls in a functional group (possibly .CHO or .COOH ...), are among the possibilities. Similar bands are found when $CH_3OH$ is chemisorbed on smooth (48)(49) or electrodispersed (50) platinum.

The dependence of the $CO_L$ band characteristics (i.e. band center, band width,...) upon potential has been studied in detail by PM-FTIRS, confirming the previous EMIRS conclusions (47).

3.2.2.*Surface structure effects.* CO adsorbs differently on the different faces of Pt single crystals. This was already known from electrochemistry (43), but is better seen (at least qualitatively) by EMIRS (Fig.6), even for CO produced by dissociative chemisorption of $CH_3OH$ . Three orientations were studied, viz. Pt(100), Pt(110) and Pt(111) (51)(52) (53). Roughly speaking, under similar experimental conditions it can be said that Pt(110) has the closest behaviour to that of a polycrystal. It is characterized by an intense, remarkably narrow $CO_L$ EMIRS band at high wavenumbers (ca. 2077 $cm^{-1}$) and a wide, weak, single-sided band at around 1795 $cm^{-1}$. On the contrary, Pt(100) has a wider $CO_L$ band centered at 2050 $cm^{-1}$, together with an intense nearly single-sided $CO_B$ band at 1870 $cm^{-1}$; the presence of a third, intense, band at 1700 $cm^{-1}$ is certainly related to the formation of methanol adsorbates. Pt(111) is different from the two other faces: the $CO_L$ band, at 2075 $cm^{-1}$, is weaker than for the two other orientations, and the $CO_B$ band seems to arise from at least two different contributions, a situation also recently found at the solid-gas interface (54).

All the bands are potential dependent, as also shown by Kunimatsu (55), and Kitamura (56), for $CO_{ads}$ on Pt(111). Furthermore, anion adsorption is likely to play a role, especially on the $CO_B$ bands, which are shifted to lower wavenumbers in sulfuric acid (see the work by Bewick and Sun, ref.53), as compared to perchloric acid (52)

Figure 6. EMIRS spectra of the adsorbed species resulting from the chemisorption of 0.1M $CH_3OH$ in 0.5M $HClO_4$ at Pt single electrodes. $\Delta E$ = 400 mV, $\bar{E}$ = 0.35 V vs. RHE, f = 13.5 Hz. Exposed faces : a)Pt(100). b)Pt(110). c)Pt(111).

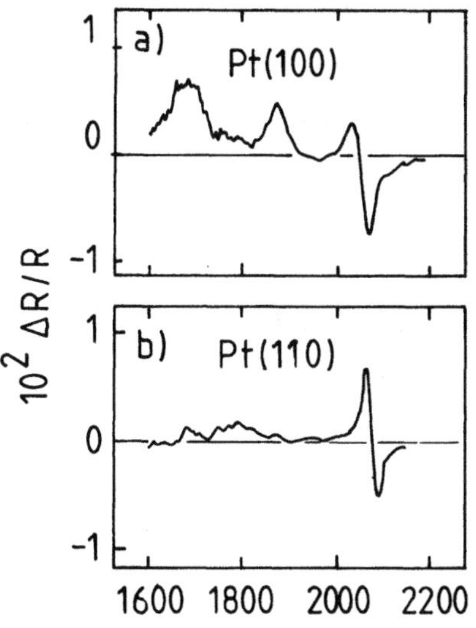


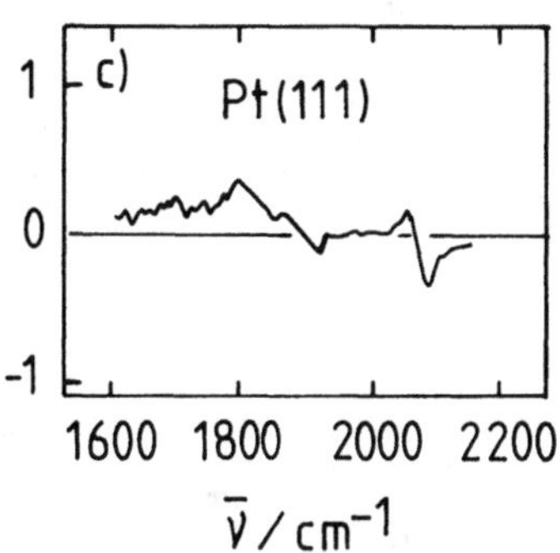


## 3.3.QUANTITATIVE DETERMINATIONS USING EMIRS.

Taking the chemisorption of formic acid on rhodium in acid medium as an example, an attempt has been made to obtain quantitative information from EMIRS measurements (57). Assuming that three types of $CO_{ads}$ species are simultaneously present on the surface (see Fig.5), then

$$\theta_{CO} = \theta_L + \theta_B + \theta_m = 1 - \theta_H - \theta_{org}$$

where $\theta_H$ and $\theta_{org}$ represent the respective coverages by $H_{ads}$ and organic adsorbates (if any). Following different assumptions (like the constancy of the product of the extinction coefficient, $\epsilon_i$, of each type of adsorbed CO, times the number of sites (1, 2 or 3.3) required by each type ($CO_L$, $CO_B$, $CO_m$, resp.) of $CO_{ads}$, and knowing the width at half-height of the absorption bands, it is possible to derive equations which allow the estimation of the relative fractions of the surface covered by $CO_L$, $CO_B$ and $CO_m$, and to follow the evolution of their distribution with respect to the concentration of formic acid in the bulk (Fig.7). At the electrode-solution interface, and at variance with the solid-gas interface, it is interesting to see that low concentrations favour the highly coordinated $CO_{ads}$ species, a difference which is certainly related to the presence of a solvent in the former.

Figure 7. Relative coverage of adsorbed CO species on a rhodium electrode, as a function of formic acid concentration in the bulk

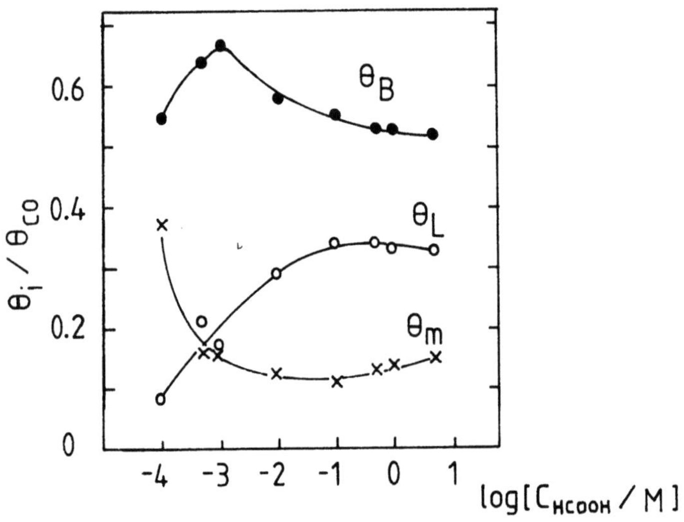


For 0.1M HCOOH in 0.5M $HClO_4$ solutions, about 33% of the sites of the rhodium surface covered by $CO_{ads}$ species are covered by $CO_L$, 55% by $CO_B$, and 12% by $CO_m$. In terms of number of species (i.e. taking into account that 1, 2 and 3.3 sites are necessary for $CO_L$, $CO_B$ and $CO_m$ to be formed), this means that the composition of the $CO_{ads}$ layer is about 53% of linearly bonded CO, 42% of bridge-bonded CO, and 5% of multibonded CO. Oxidation of such a layer would require an average of 1.51 electrons-per-site, a prediction which is in excellent agreement with the experimental electrochemical determination (57).

## 3.3. ADSORBED INTERMEDIATES AND POISONS

3.3.1. *"Methanol electrooxidation intermediates".* On the basis of electrochemical measurements, different mechanisms were proposed in the 70s to explain the oxidation of methanol on platinum electrodes in acid medium. They differed in the identity of the reactional intermediates (i.e. the "active adsorbed species"), which were either :CO (58), ≡C-OH (59), or a mixture of both. Surprisingly, none of them accounted for the formation of catalytic poisons, although their formation is most important for the use of methanol in fuel cells. Briefly, it can be said that the main controversy was whether on platinum the electro-oxidation mechanism proceeded according to scheme [1] or scheme [2] :

$$CH_3OH \longrightarrow \equiv C\text{-}OH + 3H^+ + 3e^-$$
$$\equiv C\text{-}OH + H_2O \longrightarrow CO_2 + 3H^+ + 2e^- \qquad [1]$$

$$CH_3OH \longrightarrow =C=O + 4H^+ + 4e^-$$
$$=C=O + H_2O \longrightarrow CO_2 + 2H^+ + 2e^- \qquad [2]$$

3.3.2. *Investigation by infrared reflectance spectroscopy.* The first work by EMIRS (41), in 1981, demonstrated unambiguously that the platinum electrode surface was covered by large amounts of adsorbed carbon monoxide as a result of methanol electrooxidation. It was recognized that CO was mainly linearly bonded to the surface, and the contribution of a minor species (of bridge-bonded type) was noted. These results were later confirmed by extensive studies using EMIRS (27), but also other infrared reflectance techniques, namely LPSIRS (36), SNIFTIRS (25) and PM-FTIRRAS (28). Actually, as said in Section 3.2.2., due to improvements of the technique, other bands are now currently found, especially when working at very low concentrations of methanol (48)(49), or when using a flow cell (58).

It is now understood that the composition (which determines its reactivity) of the adsorbed layer formed by methanol chemisorption, strongly depends on parameters like concentration in the bulk, adsorption time, crystallo-graphic orientation of the surface (52)(53), or roughness of the surface (50). It is now also understood that the EMIRS technique is more sensitive to the more strongly adsorbed species (this is why $CO_{ads}$ is so easily detected), and that several different adsorbates are formed during the early stages of the adsorption process (64). Thus, according to EMIRS results (49) .CHO or .COOH groups are likely to be initially present, but, being weakly adsorbed, are quickly removed from the surface by the so-called "poisons", i.e. the different $CO_{ads}$ species. For instance, on Fig.8a, which resulted from 25 averaged spectroscopic scans , the band at 1700 $cm^{-1}$ (attributed to .CHO or .COOH -type groups) is still there, but much less intense than at the first scan. Conversely, the $CO_L$ signal at ca. 2060 $cm^{-1}$, initially almost absent, has now an intensity comparable to that of the 1700 $cm^{-1}$ band. It continues to increase slightly with more spectral accumulations, until full coverage is reached. In the same spectrum (Fig.8a) weakly adsorbed $CO_2$ and C-H stretches are seen as well. Very similar conclusions are reached when ethanol is adsorbed instead of methanol (Fig.8b). It is therefore demonstrated that the C-C chain is easily broken during the chemisorption process of ethanol on platinum.

Figure 8. EMIRS spectra of the species resulting from the chemisorption of alcohols in 0.5M $HClO_4$ solutions ($\Delta E$ = 400 mV, $\bar{E}$ = 0.2 V vs. RHE, $f_s$ = 13.5 Hz, 25 scans), at a platinum electrode. a) $10^{-2}$ M $CH_3OH$. b) $10^{-3}$M $C_2H_5OH$.

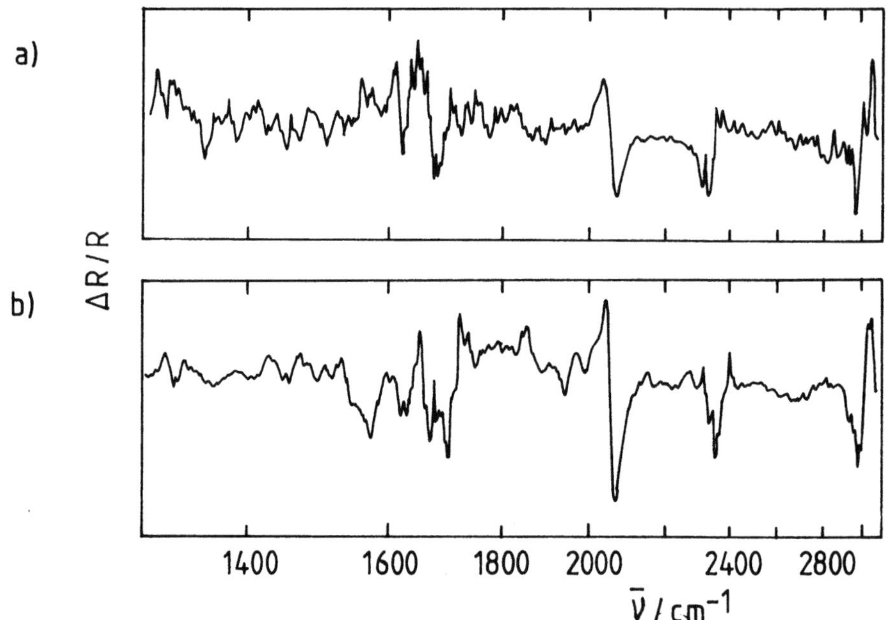


It is still an open question whether other species, like .CHOH or ≡C-OH, could be formed during methanol oxidation and could contribute, at least initially, to the methanol electrooxidation mechanism. According to Bewick et al. (58) some bands in the 1200 to 1350 $cm^{-1}$ region might be interpreted in this sense , especially if the spectroscopic investigation is carried out in a flow cell.

On the other hand, the "discrepancies" with SIMS, or DEMS (differential electrochemical mass spectroscopy) results (59)(60), which failed to detect CO, are now almost explained by differences in the surface state of the working electrode : to work with rough surfaces at $10^{-2}$M $CH_3OH$ in solution (like in DEMS), is not far from working with smooth electrodes in $10^{-4}$M solutions, i.e. experimental conditions under which EMIRS clearly shows a minimum of poisoning.

3.3.3.*"Poisoning and blocking"*. To say that CO is a poison is not enough to explain totally the characteristics of a CO covered surface. For instance it is known that during methanol oxidation on Pt(hkl) electrodes, the voltammograms for Pt(100) exhibit almost no current during the positive variation of potential up to ca. 0.75 V vs. RHE, potential at which there is a steep rise of current. Surprisingly, no similar current rise is found, neither during the reverse potential sweep on Pt(100), nor during both sweeps on Pt(110) or Pt(111). Taking into consideration the sign of the EMIRS single-sided $CO_B$ band at 1870 $cm^{-1}$ (Fig.6a), it is concluded that the $CO_B$ species are formed at the lower limit of potential. Thus, they coexist with $CO_L$ during the positive sweep, until their oxidation to $CO_2$ . During the reverse sweep, only $CO_L$ is formed until $E_c$ is reached. Therefore, lateral attractive interactions between $CO_B$ and $CO_L$ can be invoked during the positive sweep, but not during the negative one. This makes the difference between a <u>blocked surface</u> and a <u>poisoned surface</u>. Such a situation is not spectroscopically observed with the other orientations, in agreement with the fact that the i(E) curves for both the positive and negative sweeps are nearly coincident.

## 3.4. HOW DO ADATOMS WORK ?

Since the work by Adzic et al. (see ref.4 and Section 1.2.1.), it is known that underpotential deposited (upd) lead adatoms can enhance formic acid oxidation on rhodium electrodes, while cadmium adatoms do not (19). This is surprising since upd of both Pb and Cd occurs in almost the same potential range and, at the same bulk concentration of their precursor salt, both lead to a similar decrease of the hydrogen region. A first answer was given by UV- visible reflectance spectroscopy, which suggested that the difference in the adsorption energies of the two adatoms might explain their different behaviour.

EMIRS supplied a much more detailed information. The spectrum of the adsorbed layer (over the range 1600 to 2400 $cm^{-1}$), for concentrations of $Pb^{2+}$ in solution progressively increasing from 5 $10^{-7}$M to $10^{-3}$M is given in Fig.9. It is seen that only $CO_B$ species are affected (57). Therefore, it is concluded that i) the adatoms act by displacing the CO poisons of the surface, ii) Pb adatoms act selectively by removing the $CO_B$ species, i.e. Pb adatoms need two sites for their adsorption, in agreement with electrochemical measurements.

The effect of Cu (61) and Bi adatoms (62) was also studied by EMIRS. Bi is known to occupy three adsorption sites, and consequently the slight increase of the $CO_L$ band which accompanies the decrease of the $CO_B$ band was attributed to adsorption of Bi on two adjacent bridge sites, which releases one site favourable to $CO_L$ adsorption.

Cd adatoms behave differently from Pb, Cu and Bi, since no changes occur in the EMIRS spectra until very high concentrations of precursor salts in solution ( > $10^{-2}$ ) are reached. The observed catalytic effect may well be due to impurity adatoms (coming from the precursor salt), and it can be concluded that, being only weakly adsorbed, Cd adatoms cannot compete with CO adsorption, which explains their lack of activity. In order to summarize these results, Figs.10a and 10b give the dependence of the band.

Figure 9. Dependence of the EMIRS spectra of $CO_{ads}$ species on rhodium vs. the concentration of $Pb^{2+}$ in solution. Note the disappearance of the $CO_B$ band with increasing $Pb^{2+}$

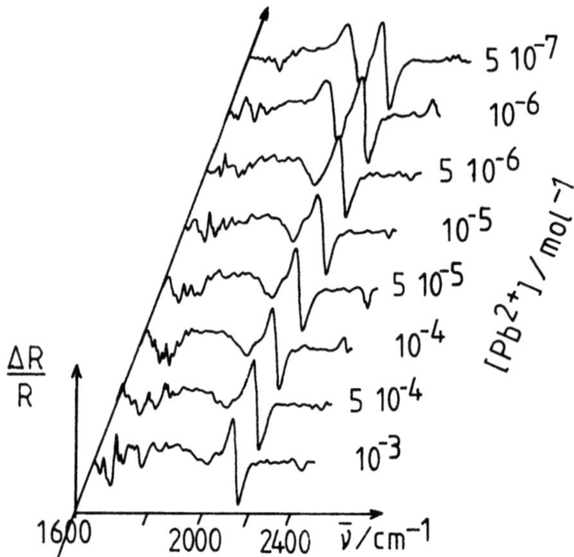


Figure 10. Dependence of the band intensities (of $CO_B$ and $CO_L$ on rhodium electrodes) vs. the concentration of precursor salt in solution. a) Effect of Pb adatoms. b) Effect of Cd adatoms. intensities of $CO_L$ and $CO_B$ , on the concentration of precursor salts in solution.

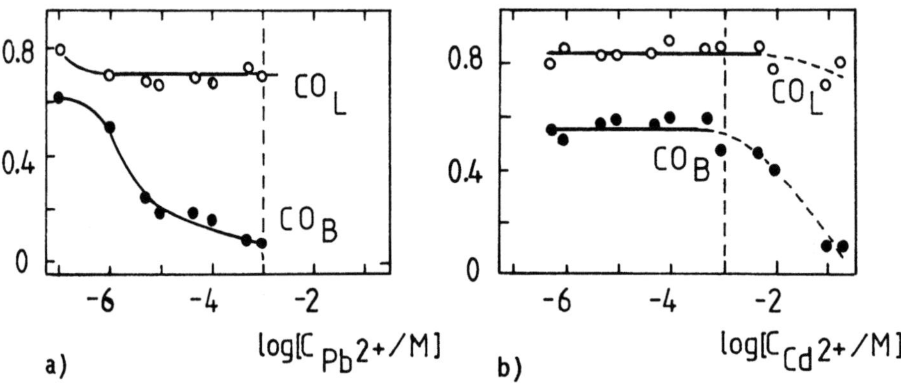

Are the same conclusions still valid for adatoms on platinum in acid medium ? An answer is more difficult to obtain, since one species, $CO_L$, is widely dominant on that surface. However, the same explanation (i.e. that Pb adatoms act by preventing the formation of multibonded carbon monoxide species) was given by Bewick et al. on the basis of preliminary experiments concerning the electro-oxidation of formaldehyde (25).

## Conclusions

The feasibility of applying infrared reflectance spectroscopy to investigations at the electrode-solution interface is now well demonstrated. Amongst the several variants which have been developed in the past ten years, EMIRS, i.e. electrochemically modulated infrared reflectance spectroscopy, due both to its high sensitivity and to its ability to sort out the vibrational information concerning adsorbed species at the electrode surface from the bands of the solution species, is particularly efficient for shedding some light on the nature of the intermediates involved in electrocatalytic reactions. However, the technique remains difficult as compared with SNIFTIRS (subtractively normalized interfacial Fourier transform infrared spectroscopy), and a good knowledge of the behaviour of the electrochemical system under investigation is absolutely necessary for a correct interpretation of the infrared spectra.

Application of EMIRS to the study of carbon monoxide adsorption on noble metal electrodes has led to a detailed interpretation involving different adsorption states (viz. linearly bonded, bridge-bonded, and multi-bonded to the surface), depending on the nature of the metal substrate. Furthermore, taking as example the chemisorption of formic acid on rhodium, it has been shown that quantitative information on the changes in the composition of the adsorbed layer that accompany the changes of concentration of formic acid in solution can be drawn from EMIRS spectra. It was thus shown that the lower concentrations favour the higher coordination states of $CO_{ads}$.

Due to its interest in fuel cell applications, the dissociative chemisorption of methanol at platinum electrodes (in aqueous solutions) was studied extensively, varying many experimental parameters. It was shown that various types of adsorbates are produced, whose surface distribution depends on the concentration of $CH_3OH$ in the bulk, the roughness and the crystallographic orientation of the surface, the pH of the solution, the adsorption potential and the adsorption time and, consequently, on the spectral accumulation time. This complexity has made clear that it is not possible to write "the" mechanism for methanol electrooxidation, since several different pathways are simultaneously possible, depending on the variable surface concentration of a few adsorbed species. $CO_{ads}$ species appear to be the more strongly adsorbed fragments and, therefore, are identified as the "catalytic poisons" of active surfaces. However, lateral interactions, particularly between linearly bonded and bridge-bonded CO, have also to be invoked for a full explanation of the electrochemical results.

The enhancement effects produced by adatoms on noble metal active surfaces are very important in electro-catalysis, although no industrial application has been reached as yet. By studying the perturbation induced by adatoms in the spectra of the $CO_{ads}$ layer formed by chemisorption of formic acid on rhodium it was shown that Pb adatoms act selectively by removing the bridge-bonded CO species, confirming that both Pb and $CO_B$ need two adsorption sites. Conversely, as no perturbation was produced by the presence of $Cd^{2+}$ in solution it has to be concluded that Cd adatoms are too weakly adsorbed to compete with $CO_{ads}$ formation, explaining why Cd adatoms do not lead to any activity enhancement.

These few examples demonstrate how powerful the EMIRS spectroscopic technique is. From the results obtained "in situ" at the electrode-solution interface, it is now evident that the composition of the adsorbed layer on a catalytic surface is more complex than expected, and that it changes with time due to competitive adsorption processes. Therefore, in the future a still higher sensitivity is needed, especially to investigate the early stages of adsorption processes, which require very short spectral accumulation times. Only then could one hope to answer definitively the very important question of the nature of the methanol reactional intermediates.

**Acknowledgments.** Most of the results presented here were obtained in the "Laboratoire de Chimie 1" at the University of Poitiers (France). The contribution of Prof. C. Lamy, Dr. J.M. Léger, F. Hahn, J.M. Perez, H. Huser, S. Juanto and M.Choy de Martinez is greatly acknowledged.

The author is also grateful to Dr. A. Bewick (University of Southampton, U.K.) for the opportunity he gave him, as a post-doc, to initiate in 1979 the applications of the EMIRS technique to electrocatalysis and for the collaboration ensuing, and to Prof S. Pons (University of Utah, U.S.A.) and Dr K. Kunimatsu (Hokkaido University, Japan) for their continuous interest and useful discussions.

**References**


(1) J.O'M.Bockris, B.E.Conway, E.Yeager and R.R.White, Comprehensive treatise of electrochemistry., Vol.**3**, *Electrochemical Energy conversion and storage*, Plenum Press, New York, (1981).
(2) D.Pletcher, J. Appl. Electrochem., **14** (1984) 403.
(3) C.Lamy, B.Beden and J.M.Léger, Bull. Soc. Chim., **3** (1985) 421; C.Lamy, Electrochim. Acta., **29** (1984) 1581.
(4) R.R.Adzic, Israel J. Chem., **18** (1979) 166.
(5) S.Srinivasan and E.Gileadi, Electrochim. Acta, **11** (1966) 321.
(6) R.Woods, in *Electroanalytical Chemistry*, A.J.Bard (ed), M.Dekker, New York, Vol.**9**, pp 1-162.
(7) J.Clavilier, R.Durand, G.Guinet and R.Faure, J. Electroanal. Chem., **127** (1981) 281.
(8) B.D.Mc Nicol, in Catalysis, **2** (1978) 243.
(9) A.Capon and R.Parsons, J. Electroanal. Chem., **44** (1973) 1; **44** (1973) 239; **45** (1973) 205; **65** (1975) 295.
(10) G.Crépy, C.Lamy and S.Maximovitch, J. Electroanal. Chem., **54** (1974) 161.
(11) B.Beden, I.Cetin, A.Kahyaoglu, D.Takky and C.Lamy, J.Catal., **104** (1987) 37.
(12) D.Takky, B.Beden, J.M.Léger and C.Lamy, J. Electroanal. Chem., **145** (1983) 461.
(13) J.Clavilier, R.Parsons, R.Durand C.Lamy and J.M.Léger, J. Electroanal. Chem., **124** (1981) 321.
(14) J.Clavilier, C.Lamy and J.M.Léger, J. Electroanal. Chem., **125** (1981) 249.
(15) L.H.Essis Yei, B.Beden and C.Lamy,J. Electroanal. Chem., **246** (1988) 349.
(16) F.Kadirgan, B.Beden, J.M.Léger and C. Lamy, J. Electroanal. Chem., **125** (1981) 89.
(17) N.R.De Tacconi, J.M.Léger, B.Beden and C.Lamy, J.Electroanal. Chem., **134** (1982) 117
(18) B.Beden, C.Lamy and J.M.Léger, Electrochim. Acta, **24** (1979) 1157.
(19) R.R.Adzic and A.V.Tripkovic,J. Electroanal. Chem., **99** (1979) 43.
(20) D.M.Kolb in *Advances in Electrochemistry and Electrochemical Enginering*, M.Gerischer and C.W.Tobias (Eds), **Vol 11**, J.Wiley, New-York (1978) 125.
(21) F.Kadirgan, B.Beden and J.M.Léger, J.Electroanal. Chem., **143** (1983) 135.
(22) N.Furuya and S.Motoo, J. Electroanal. Chem., **98** (1979) 189.
(23) B.Beden, F.Kadirgan and C.Lamy, 33rd ISE meeting, Lyon (France), 1982.
(24) A.Bewick, K.Kunimatsu and B.S.Pons, Electrochim. Acta, **25** (1980) 465.
(25) A.Bewick and S.Pons in R.J.H. Clark and M.E. Hester, (Eds), *Advances in Infrared and Raman Spectroscopy*, **Vol 12**, Heyden, London, (1985), Chap.1, p.1.

(26) J.K.Foley and S.Pons, Anal. Chem., **57**(8) (1985) 945A; J.K.Foley, C.Korzeniewski, J.L.Dashbach and S.Pons, in *Electroanalytical Chemistry*, **Vol 14**, A.J.Bard (Ed.), Marcel Dekker, New-York, (1986), p.309.
(27) B.Beden and C.Lamy in R.J.Gale (Ed.), *Spectroelectro-chemistry, Theory and Practice*, Plenum Press, New-York, (1988) Chap.**5**, p.189.
(28) W.G.Golden, K.Kunimatsu and H.Seki, J. Phys. Chem., **88** (1984) 1275.
(29) K.Kunimatsu and A.Bewick, Indian J. of Technology, **24** (1986) 407.
(30) F.Ozanam and J.N.Chazalviel, J. Electron. Spectr. relat. Phenom., **45** (1987) 323.
(31) N.J.Harrick, *Internal reflection spectroscopy*, Interscience Publ. inc., New York, (1967).
(32) W.N.Hansen, Symp. of the Faraday Soc., **4** (1970) 27.
(33) H. Neugebauer, G. Nauer, N. Brinda-Konopik and G.Gidaly, J. Electroanal. Chem., **122** (1981) 381.
(34) A.Bewick, K.Kunimatsu, B.S.Pons and J.W.Russell, J.Electroanal. Chem., **160** (1984) 381.
(35) J.W.Russell, J.Overend, K.Scanlon, M.W.Severson and A.Bewick, J. Phys. Chem., **86** (1982) 205.
(36) K.Kunimatsu, J. Electroanal. Chem., **140** (1982) 205.
(37) S.Pons, T.Davidson and A.Bewick, J. Electroanal. Chem., **160** (1984) 63. S.Pons, ibid., **160** (1984) 369.
(38) M.A.Habib and J.O'M Bockris, J. Electroanal. Chem., **180** (1984) 287.
(39) D.S.Corrigan and M.J.Weaver, J.Phys. Chem., **90** (1986) 5300. D.S.Corrigan, L.W.H.Leung and M.J.Weaver, Anal. Chem., **59** (1987) 2252.
(40) M.C.Pham, F.Adami, P.C.Lacaze, J.P.Doucet and J.E.Dubois, J. Electroanal. Chem., **201** (1986) 413.
(41) B.Beden, C.Lamy, A.Bewick and K.Kunimatsu, J. Electroanal. Chem., **121** (1981) 343.
(42) B.Beden, A.Bewick and C.Lamy, J. Electroanal. Chem., **148** (1983) 147; ibid. **150** (1983) 505.
(43) J.M.Léger, B.Beden, C.Lamy and S.Bilmes, J. Electroanal. Chem., **170** (1984) 305.
(44) B.Beden, A.Bewick, K.Kunimatsu and C.Lamy, J. Electroanal. Chem., **142** (1982) 345.
(45) N.Sheppard and T.T.Nguyen, in R.J.H. Clark and M.E.Hester, (Eds), *Advances in Infrared and Raman Spectroscopy*, Vol.15, Heyden, London, (1978), p.67.
(46) K.Kunimatsu, H.Seki, W.G.Golden, J.G.Gordon II and M.R.Philpott, Surface Sci., **158** (1985) 596.
(47) K.Kunimatsu, J. Electroanal. Chem., **213** (1986) 149.
(48) B.Beden, F.Hahn, S.Juanto, C.Lamy and J.M.Léger, J. Electroanal. Chem., **225** (1987) 215.
(49) B.Beden, F.Hahn, J.M.Léger, C.Lamy and M.I.Dos Santos Lopes, J. Electroanal. Chem., **258** (1989) 463.
(50) B.Beden, F.Hahn, C.Lamy, J.M.Léger, N.R. de Tacconi, R.O.Lezna and A.J.Arvia, J. Electroanal. Chem., **261** (1989) 401.

(51) S.Juanto, B.Beden, F.Hahn, J.M.Léger and C.Lamy, J. Electroanal. Chem., **237** (1987) 119.
(52) B.Beden, S.Juanto, J.M.Léger and C.Lamy, J. Electroanal. Chem., **238** (1987) 323.
(53) S.G.Sun, J.Clavilier and A.Bewick, J. Electroanal. Chem., **240** (1988) 147.
(54) M.Tushaus, E.Schweizer and A.M.Bradshaw, 5$^{th}$ International Conference on Vibrations at Surfaces, Garmisch Partenkirschen (FRG), (1987). Abst. n° P28. A.M.Bradshaw, private communication.
(55) N.Furuya, S.Motoo and K.Kunimatsu, J. Electroanal. Chem., **239** (1988) 347.
(56) F.Kitamura, M.Taiseda, M.Takahashi and M.Ito, Chem. Phys. Lett., **142** (1987) 318.
(57) M.Choy de Martinez, B.Beden, F.Hahn and C.Lamy, J. Electron Spectr. relat. Phenom., **45** (1987) 153. M.choy de Martinez, B.Beden, F.Hahn and C.Lamy, J. Electroanal. Chem., **249** (1988) 265.
(58) R.J.Nichols and A.Bewick, Electrochim. Acta, **33** (1988) 1691.
(59) H.W.Buschmann, S.Wilhelm and W.Vielstich, Electrochimica Acta, **31** (1986) 939.
(60) O.Wolter and J.Heitbaum, Electrochim. Acta, **31** (1986) 943; J.Willsau, O.Wolter and J. Heitbaum, J. Electroanal. Chem , **185** (1985) 163; J. Electrochem. Soc.,**132** (1985) 1635.
(61) K.Kunimatsu, R.O.Lezna and M.Enyo, J. Electroanal. Chem., **258** (1989) 115.
(62) B.Beden, M.Choy de Martinez and C.Lamy, VIII Reunion Latinoamericana de Electroquimica y Corrosion, Cordoba (Argentina) (1988), Ext.Abst. 1.14.
(63) M.R.Anderson, D.Blackwood and S.Pons, J. Electroanal. Chem., **256** (1988) 387. M.R.Anderson, D.Blackwood, T.G.Richmond and S.Pons, ibid., **256** (1988) 397.
(64) R.Parsons and T.VanderNoot, J. Electroanal. Chem., **257** (1988) 9.

## PHOTOACOUSTIC SPECTROSCOPY AND THE IN-SITU CHARACTERIZATION OF THE ELECTROCHEMICAL INTERFACE*

Catherine E. Vallet
Chemistry Division, Oak Ridge National Laboratory
P. O. Box 2008
Oak Ridge, Tennessee 37831-6197
USA

ABSTRACT. Photoacoustics is a new spectroscopic method which has been used for in situ characterization of the electrochemical interface during the past ten years. The basic principles of the photoacoustic effect and the principal results of the Rosencwaig-Gersho theory are discussed in light of the usefulness of the method in electrochemical studies. Different experimental arrangements suitable for in situ electrode studies are presented. A review of the use to date of photoacoustics in electrochemistry includes studies of electrochromic systems, semiconductor electrodes, passivation layers, and of electrocatalytic mixed oxides. These works demonstrated that, with relatively simple apparatus, it was possible to detect and to characterize very thin layers formed on electrodes. It is still not clear whether in most cases photoacoustic spectroscopy has an overwhelming advantage over well-established optical methods for absorption measurements; however, all the potentialities of the method have yet to be explored.

### 1. INTRODUCTION AND OVERVIEW

The photoacoustic effect was discovered in 1881 by Alexander Graham Bell.[1] Tyndall,[2] Rontgen,[3] Mercadier,[4] and Preece[5] also experimented on the effect. However, interest faded very rapidly because no clear analysis was developed at the time. Photoacoustics, called opto-acoustics, was applied to gas around 1940, but modern photoacoustics on solids started in the mid 70's with the development of the theory by Rosencwaig and Gersho.[6] Thus, although Sir Raman (1888-1970) was not born when the photoacoustic effect was first observed, photoacoustic spectroscopy is still in its infancy compared to the widely used Raman spectroscopy.

*Research sponsored by the Division of Materials Sciences, Office of Basic Energy Sciences, U. S. Department of Energy under contract DE-AC05-84OR21400 with the Martin Marietta Energy Systems, Inc.



*C. Gutiérrez and C. Melendres (eds.),*
*Spectroscopic and Diffraction Techniques in Interfacial Electrochemistry*, 133–153.

In an early experiment, Bell modulated with a voice-activated mirror a sunlight beam received on a selenium cell included in a conventional telephone circuit and was able to produce a telephone speech. During further experiments with what he called a photophone (or a spectrophone), shown in Figure 1, he could obtain a sound directly, i.e., without an electrical conversion. This was first done by focusing the chopped light on a thin diaphragm connected to a hearing tube. He

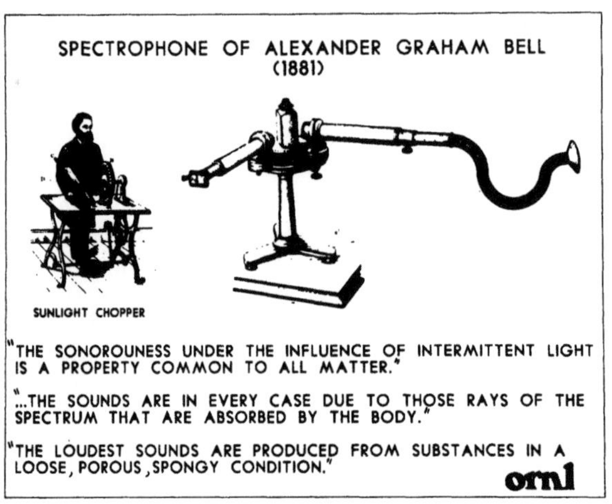


Figure 1. Spectrophone of Alexander Graham Bell (1881)

noticed also, that a sound was detectable from a solid enclosed in a glass tube attached to the hearing tube. Although Bell erroneously attributed the sound generation to adsorption/desorption of the surrounding gas during the cyclic heating of the sample, he recognized that the sound production 1) was a general phenomenon upon intermittent illumination of a solid sample, 2) was related to the optical absorption of the solid, and 3) that "loudest sounds are produced from substances in a loose, porous, spongy, condition." These few comments make photoacoustic detection attractive for obtaining optical absorption spectra of powdered, amorphous samples (e.g., hydrated passivation layers on electrodes).

A. Rosencwaig,[7-9] at Bell Laboratories, apparently was the first (in 1973) to design a photoacoustic cell for solids, and obtained absorption spectra of many different inorganic, organic, and biological solids. The photoacoustic spectrometer,[8,9] in Figure 2, included a high pressure Xe lamp, a monochromator, and a mechanical chopper. The signal from the microphone placed in an enclosed cell was detected through a phase-sensitive lock-in amplifier locked onto the chopping frequency. By this time, Rosencwaig had experimental evidence[10] that the sound was produced neither by adsorption/desorption of gas at the surface nor by gas compression resulting from the mechanical bending of the sample under cyclic heating, as had been proposed by Lord Rayleigh.[11] Rosencwaig[8-10,12] concluded that the light absorbed by the solid is converted completely (or partially) into heat by nonradiative

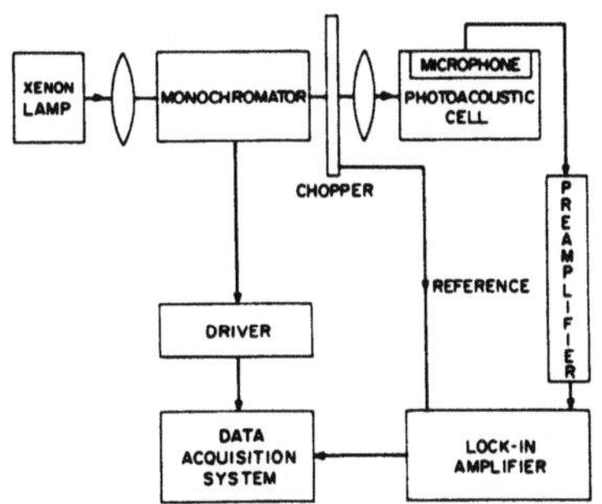


Figure 2. Typical single-beam photoacoustic spectrometer.[10]

processes. The periodic heat flow from the solid to the surrounding gas produces pressure fluctuations in the gas which are detected by the microphone. At the 1975 Pittsburgh Conference on Analytical Chemistry and Applied Spectroscopy, A. Rosencwaig summarized his photoacoustic experiments and had an enthusiastic reception.[12]

Figure 3 is an attempt to illustrate the volume of activity in photoacoustic spectroscopy over the years. The activity is represented by the number of citations obtained from the C A on line Data Base (American Chemical Society) for "Photoacoustic Spectroscopy." From 1975 to 1979, a remarkably fast increase in the number of photoacoustic works led to the First International Topical Meeting on Photoacoustic Spectroscopy. Approximately 75 papers were presented at the meeting. The meeting is presently held every two years and typically includes more than 100 posters in addition to plenary lectures and papers. Papers deal with the theory of the photoacoustic effect, with improvements in the design of photoacoustic cells, and with the introduction of new detection schemes, e.g., piezoelectric transducer,[13] or photothermal detection[14] (with the heat flow producing changes in the refraction index of the fluid near the sample surface). Various optical experiments have used the photoacoustic effect for monitoring radiationless deexcitation processes, e.g., Raman spectroscopy,[15] and interferometry[16] for the quantitative measurement of thin film thickness. Photoacoustics, which first has been used in the UV-visible region, found many applications in the IR wavelength region in studies of catalytic surfaces.[17,18] In the 1980's, a photoacoustic cell has become almost a standard option in commercial FTIR spectrometers. More recently, x-ray photoacoustic effect of solids has been reported[19] and may be useful in the study of x-ray absorption and the analysis of thermal properties of solids.

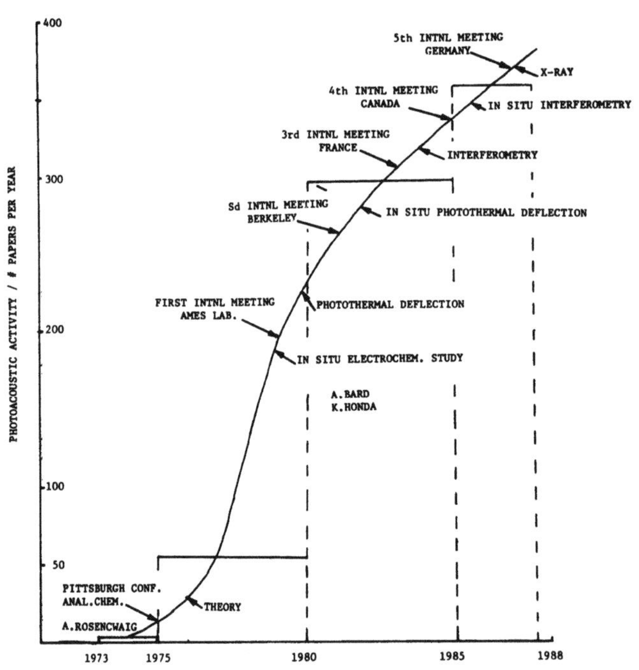


Figure 3. Volume of activity and focus of interest in photoacoustics with time.

A. Bard[20] may have been the first electrochemist to exploit the potentiality of photoacoustics for in situ electrochemical studies. He used[21] a gas-microphone detector to monitor the photo-induced gas evolution at an electrode/gas interface. To our knowledge, the first two in situ photoacoustic studies of the liquid/electrode interface were by R. E. Malpas and A. Bard[22] with a piezoelectric detector, and by A. Fujishima, H. Masuda, and K. Honda[23] with a gas-microphone assembly. At the Second International Topical Meeting there was one paper on an in situ electrochemical study,[24,25] but with the introduction of the photothermal deflection technique, an entire session was devoted to interface studies two years later. However, only 1.5% of the "photoacoustic spectroscopy" papers since 1979 relate to electrochemical studies. For comparison, the data base for 1981-1985 gives 1,480 citations for photoacoustic spectroscopy and 1,380 citations for Raman spectroscopy, but 3.8% of the Raman citations are related to electrochemistry. The scarcity of the photoacoustic/electrochemical studies may be due to the novelty of the method and to difficulties in obtaining quantitative results.[24] Although IR photoacoustic spectra of adsorbed surface species, present in submonolayer coverages, on silica and alumina powders have been reported,[25] IR reflection and Raman spectroscopies are more commonly used at the electrode/liquid interface. Photoacoustic spectroscopy may have its place in the development of an understanding of the electrochemical interface as a complementary method which provides insights into the energetics of the processes taking place.

## 2. THE ROSENCWAIG-GERSHO THEORY

The first attempt to interpret quantitatively the photoacoustic effect in a solid, transparent window in terms of the periodic heating of the gas near the absorbing window was made by Parker.[26] The general theory of the photoacoustic effect was developed primarily by Rosencwaig and Gersho[6,27] for a photoacoustic cell with a gas-microphone detector placed in front of the light absorbing sample. The calculations are based on the assumption that the intermittent light absorbed in the sample produces excitation of internal energy levels, and during the subsequent deexcitation, heat is produced within the sample. The resulting periodic temperature at the sample/gas interface generates a surface pressure disturbance which propagates through the gas column to the microphone. In the RG theory the exact expression of the temperature is calculated, but the transport of the perturbation through the gas is treated in approximate manner which is valid for nonresonant cells. Improvements made[28-30] in the treatment of the transport of the acoustic perturbation did not change the basic results except for conditions near the cell resonance. A further refinement to the theory included[31] the relatively small contributions to the signal from thermally induced vibrations in the sample.

### 2.1 Thermal Diffusion Equations

The one-dimensional analysis of the production of a photoacoustic signal was performed[6] on a simple cylindrical cell of diameter D (shown in Figure 4).

The total length of the cell, $L = \ell'' + \ell + \ell'$, is small compared to the wavelength of the acoustic signal. The microphone is placed for detecting the average pressure produced in the gas. The light-absorbing sample is a cylinder of diameter D and of thickness $\ell$. The length of the gas column is $\ell'$ and the backing material, with poor thermal conductivity, has a thickness of $\ell''$. In the following calculations, all of the sample parameters are unprimed symbols, the gas parameters are singly primed, and the backing material parameters are doubled primed symbols. The monochromatic light of wavelength $\lambda$,

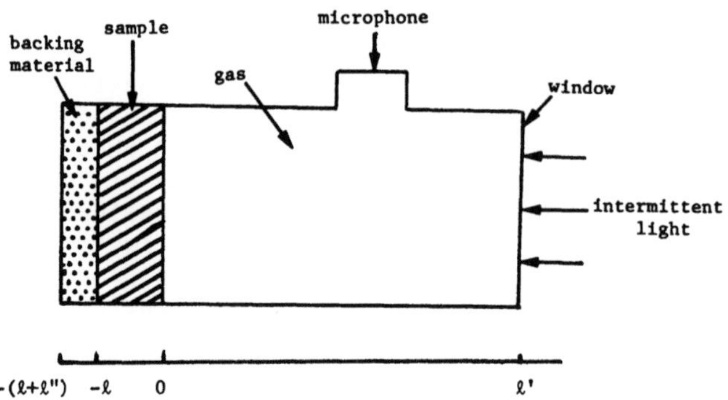


Figure 4. Cross-sectional view of gas-microphone photoacoustic cell.

sinusoidally chopped at a frequency $\omega$ (radians per second), impinges on the solid with intensity,

$$I = 1/2\ I_o\ (1 + \cos \omega t) \tag{1}$$

where $I_o$ is the monochromatic light flux (W $cm^{-2}$).

If $\beta$ is the optical absorption coefficient of the solid ($cm^{-1}$), the heat density produced at any point x within the solid from light absorption and subsequent radiationless deexcitation is

$$1/2\ \beta I_o\ e^{\beta x}\ (1 + \cos \omega t) \tag{2}$$

The thermal diffusion equation in the solid for the space-distributed heat source can be written as

$$\frac{\partial^2 \theta}{\partial x^2} = \frac{1}{\alpha}\frac{\partial \theta}{\partial t} - \frac{1}{2}\frac{\beta I_o}{\kappa} e^{\beta x}(1 + e^{i\omega t}) \tag{3}$$

where $\theta(x,t)$ is the temperature, $\alpha$ the thermal diffusivity ($cm^2\ sec^{-1}$) and $\kappa$ the thermal conductivity (J $cm^{-1}\ sec^{-1}\ °C^{-1}$).

For the backing material and for the gas, the thermal diffusion equations are, respectively

$$\frac{\partial^2 \theta}{\partial x^2} = \frac{1}{\alpha''}\frac{\partial \theta}{\partial t} \tag{4}$$

$$\frac{\partial^2 \theta}{\partial x^2} = \frac{1}{\alpha'}\frac{\partial \theta}{\partial t} \tag{5}$$

The equations are solved with the assumptions that the walls of the cell are at ambient temperature, i.e., the thermal diffusion lengths in the backing material and in the gas are smaller than $\ell''$ and $\ell'$, respectively. These conditions do not imply very large photoacoustic cells. For example, in an experiment with a light chopping frequency of 100 Hz, the thermal diffusion length ($\mu = (2\alpha/\omega)^{1/2}$ cm) of air is 2.5 x $10^{-2}$ cm and the thermal diffusion length of polyethylene backing material is 1.8 x $10^{-3}$ cm. The thermal diffusion length of water at the aforementioned frequency is 2.1 x $10^{-3}$ cm, i.e., on the same order of magnitude as for the propylene.

The complex amplitude of the periodic component of temperature at the sample/gas boundary (x = 0) is

$$\theta_o = \frac{\beta I_o}{2\kappa(\beta^2-\sigma^2)}\left[\frac{(r-1)(b+1)e^{\sigma\ell} - (r+1)(b-1)e^{-\sigma\ell} + 2(b-r)e^{-\beta\ell}}{(g+1)(b+1)e^{\sigma\ell} - (g-1)(b-1)e^{-\sigma\ell}}\right] \quad (6)$$

with $b = \frac{\kappa''a''}{\kappa a}$, $g = \frac{\kappa'a'}{\kappa a}$, $r = (1-i)\frac{\beta}{2a}$ and $\sigma = (1+i)a$.

One notes that b and g are dimensionless quantities characterizing the thermal heat flow through the backing material/sample and the gas/sample interfaces, respectively, while r is a complex quantity written only in terms of the optical absorption and the thermal diffusion coefficient of the sample.

## 2.2 Production of the Acoustic Signal

The acoustic signal in the RG theory is assumed to be produced only by the periodic heat flow from the solid to the gas. By solving (5) the AC temperature variation in the gas with time and with distance from the sample surface is written

$$T_{ac}(x,t) = e^{-a'x}[\theta_1\cos(\omega t-a'x) - \theta_2\sin(\omega t-a'x)] \quad (7)$$

where $\theta_1$ and $\theta_2$ are the real and imaginary parts of $\theta_o$ in (6). The variation with time of the temperature vanishes rapidly with increasing distance. At a distance of $2\pi/a' = 2\pi\mu'$ (e.g., 0.15 cm at 100 Hz in air), no periodic temperature variation in the gas subsists. Therefore, only the periodic compression/expansion of this thin gas layer is considered as acting as an acoustic piston on the rest of the gas column.

A spatially averaged temperature of the layer is calculated, and the displacement with time of the piston is calculated with the ideal gas law. One assumes that the rest of the gas responds adiabatically to the piston action. The complex envelope of the sinusoidal pressure variation is

$$Q = \frac{\gamma\ P_o\ \theta_o}{2\ \ell'a'T_o} \quad (8)$$

with $P_0$ the ambient pressure, $T_0$ the DC component of temperature at the sample surface and $\gamma$ and ratio of the gas specific heats.

The explicit form of Q can be obtained by combining (8) and (6). The expression is rather complicated and difficult to interpret. However, the discussion of special cases according to the relative values of the optical absorption length ($\ell_\beta = 1/\beta$), the thermal diffusion length ($\mu = \sqrt{2\alpha/\omega}$) and the sample thickness brings physical insights to the results. Since the thermal diffusion length varies with the chopping frequency (as $\omega^{-1/2}$), the relative order may be altered during one experiment by changing the chopping frequency. Figure 5 gives the simplified expressions of the pressure variation for an optically thin sample ($\ell_\beta > \ell$) and an optically thick sample ($\ell_\beta < \ell$) investigated over a wide range of chopping frequencies. At low chopping frequencies, at which the thermal diffusion length is larger than the thickness of the sample, (i.e., for a thermally thin sample), the pressure variation is proportional to $\beta\ell$ and depends on the thermal properties of the backing material. At higher chopping frequencies, with a thermal diffusion length smaller than the sample thickness, the pressure variation is proportional to $\beta\mu$ and depends only on the thermal properties of the sample. In this case, by increasing progressively the chopping frequency, measurements of absorption of increasingly thin layers can be carried out nondestructively. For an optically thick sample, as long as the thermal diffusion length $\mu$ is larger than the optical absorption length $\ell_\beta$, there is also photoacoustic opaqueness, i.e., the photoacoustic signal is independent of $\beta$. But, at higher chopping frequencies, when $\mu$ becomes smaller than $\ell_\beta$, the photoacoustic signal is proportional to $\beta\mu$ and again depth profiling measurements of absorption is possible on optically opaque samples.

## 2.3 Thermally Induced Vibrations of the Sample

The previous analysis of the production of sound by excitation of the surrounding gas considered only the heat energy transferred by means of heat conduction and diffusion. A more exact treatment should include the energy transferred through a coupling of the local heat energy to the vibrational modes of the material. The thermoelastic theory developed by R. M. White[32] computes the stresses produced by several transient heating conditions. The effect, when included into the calculation of the pressure variations produced in a gas-microphone assembly, is minor[33] except at high chopping frequencies and for weakly absorbing samples. However, the thermally induced vibrations of the sample are the physical reality measured by piezoelectric detection. The calculations of the piezoelectric signal are complex, with the results depending on whether all the sample surface is illuminated, on the position of the piezoelectric detector, and on the acoustic impedance of the bonding between piezoelectric device and sample. However, piezoelectric detection has some advantages over gas-microphone detection for measurements on weakly absorbing samples, for measurements in vacuum, and for measurements at high chopping frequency. The gas-microphone system is limited around 50kHz, but piezoelectric detectors may be used up to 1GHz.

OPTICALLY THIN SAMPLE ($\ell_\beta > \ell$):

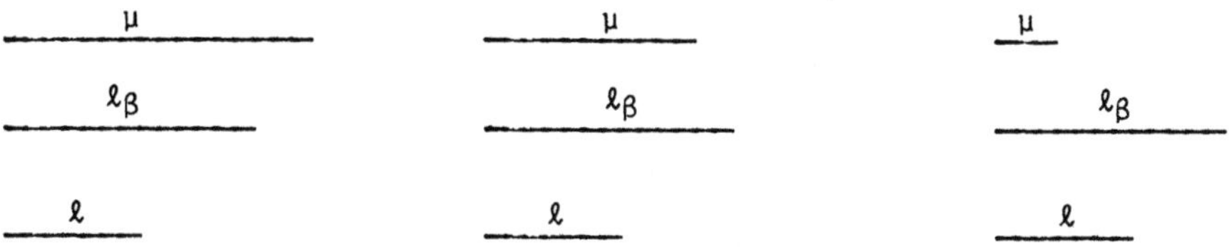

| | | |
|---|---|---|
| $\mu$ | $\mu$ | $\mu$ |
| $\ell_\beta$ | $\ell_\beta$ | $\ell_\beta$ |
| $\ell$ | $\ell$ | $\ell$ |
| $Q = Y \dfrac{(1-i)\ \beta\ell}{2a'} \left(\dfrac{\mu''}{\kappa''}\right)$ | $Q = Y \dfrac{(1-i)\ \beta\ell}{2a'} \left(\dfrac{\mu''}{\kappa''}\right)$ | $Q = -Yi \dfrac{\beta\mu}{2a'} \left(\dfrac{\mu}{\kappa}\right)$ |

OPTICALLY THICK SAMPLE ($\ell_\beta < \ell$):

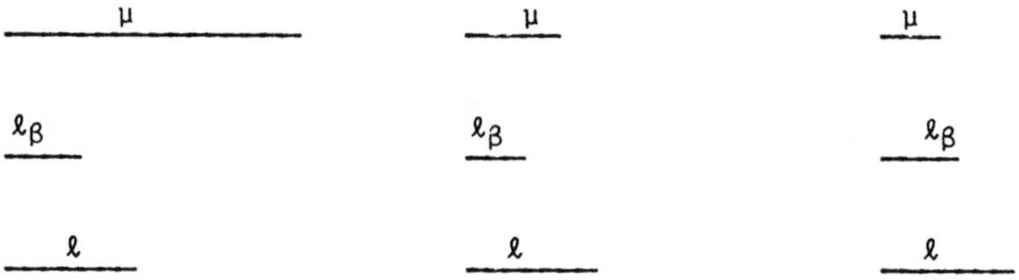

| | | |
|---|---|---|
| $\mu$ | $\mu$ | $\mu$ |
| $\ell_\beta$ | $\ell_\beta$ | $\ell_\beta$ |
| $\ell$ | $\ell$ | $\ell$ |
| $Q = Y \dfrac{(1-i)}{2a'} \left(\dfrac{\mu''}{\kappa''}\right)$ | $Q = Y \dfrac{(1-i)}{2a'} \left(\dfrac{\mu}{\kappa}\right)$ | $Q = -Yi \dfrac{\beta\mu}{2a'} \left(\dfrac{\mu}{\kappa}\right)$ |

Figure 5. Photoacoustic signal in simplified cases. Y is $\dfrac{\gamma P_o I_o}{2\sqrt{2}\ T_o \ell'}$ .

## 3. PHOTOACOUSTIC CELLS FOR ELECTROCHEMISTRY

### 3.1 Gas-microphone System

In a conventional photoacoustic cell with gas-microphone detection, the microphone senses the changes in pressure of a small volume of gas situated in front of the light-absorbing layer. The cell geometry had to be modified for in situ measurements at a solid electrode in solution. For such measurements, the electrode surface in contact with the solution absorbs the light and the thermal perturbation is detected by a microphone placed in an enclosed air volume at the back of the electrode. In order to have sizable temperature variations in the gas layer close to the back surface, the electrode through which the heat flow propagates must be thin, with a thickness smaller than the thermal diffusion length, $\mu$, (see equation 7). For instance, for a titanium electrode with thermal diffusion $\alpha = 6 \times 10^{-2}$ $cm^2$ $s^{-1}$, the thermal diffusion length at 26 Hz is about 0.3 mm. A. Fujishima et al.[23] were the first to our knowledge to report in situ measurements with a gas-microphone assembly. In a later paper,[34] they show experimental evidence that at 125 Hz with a gold plate of 0.05 mm thickness, the photoacoustic signals obtained with conventional detection and with back detection are equivalent. In addition, immersion of the plate in solution does not attenuate the photoacoustic signal by more than 10%. Figure 6 shows a typical photoacoustic cell used in ref. 35 with titanium electrodes 0.025 mm thick and optimized by a smaller gas volume than in the earlier version.[23]

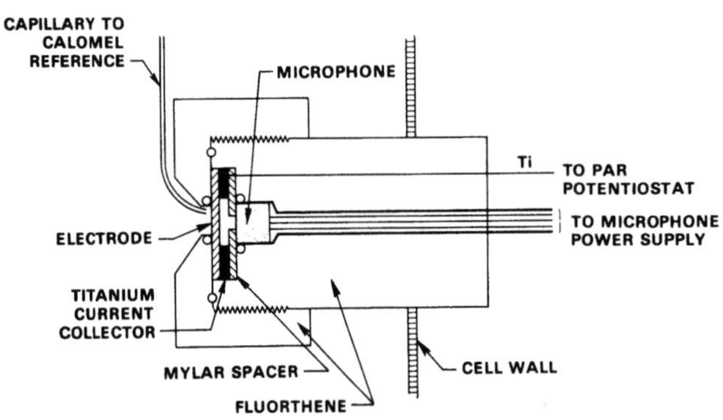


Figure 6. Electrode-microphone assembly for in situ photoacoustic measurements.

## 3.2 Piezoelectric Detection

Piezoelectric detection enables one to follow radiationless deexcitation processes by monitoring the thermal-induced stresses of the solid sample. The intrinsic advantages of this method include[36] the possibility of measurements on a broader frequency range and with better acoustic impedance matching than with a microphonic detection. Impulse responses, which are frequently less difficult to analyze, can be monitored. Malpas and Bard[22] attached a piezoelectric ceramic disk (Edo Western Corp., Salt Lake City) to either 0.05 mm Pt or 0.025 mm W foil with epoxy cement. In such arrangements, the electrode and the piezoelectric ceramic must be carefully electrically insulated, and the piezoelectric transducer must be insulated from the solution. Both insulations were provided[22] satisfactorily by the use of epoxy cement. A similar arrangement was used more recently[37] on iron electrodes (0.2 mm thick) with a TOKIN piezoelectric disk. In both applications, the chopping frequencies were less than 100 Hz. However, in the study of Malpas and Bard,[22] the fast response of the piezoelectric detection was used to study the coloration response time of electrochromic heptyl viologen ($HV^{2+}$) and anodic $WO_3$. Except for an initial transitory decrease of the signal when the potential was applied,[38] the piezoelectric signal followed closely the bleaching and coloration of $HV^{2+}$ on Pt and of anodic $WO_3$ on W with response times on the order of a few tenths of a second. Passivation of a cadmium electrode in KOH solution was studied[39] in situ with a piezoelectric detector. A pulsed tunable dye laser was used as the light source and the observed absorption signal was of few microseconds duration. The pulsed system detection limit was estimated to be on the order of 5A.

## 3.3 Mirage Effect Detection

Mirages stem[40] from the deflection of light beams passing through a region of air at higher temperature and, consequently, lower density and refraction index. The effect can be used as a means for detecting the thermal wave generated by the absorption of modulated light in a photoacoustic experiment. The theory and experimental verifications were developed by Jackson, Amer, Boccara, and Fournier.[41] In such an experiment, one laser beam (pump beam) is intensity-modulated and absorbed by the sample. A second laser beam (probe beam) travels through the region of perturbated refraction index, and the subsequent deflection is detected by lock-in amplification of signals given by position sensitive detectors (e.g., Silicon Detector Corp. Newbury Park, CA). Two geometrical arrangements can be used. In collinear photothermal deflection, the pump beam and the probe beam are nearly parallel-interacting over a small volume. This arrangement is the most sensitive ($\alpha \ell$ x pump power (W) $\sim 10^{-7}$-$10^{-8}$),[41] but can be applied only to optically transparent electrodes. The transverse photothermal deflection arrangement is suitable for photoacoustic spectroscopy of opaque electrodes since the probe beam travels near and parallel to the sample surface while the pump beam impinges normally to the surface. Mirage effect detection has been shown[42] to be two orders of magnitude more sensitive than microphonic or piezoelectric detection when used at a solid/liquid interface. Since the amplitude of the AC component of temperature falls sharply with distance from the electrode, the probe beam must travel at a distance smaller than the thermal diffusion length from the electrode surface. The probe beam, therefore, must be aligned close to the surface by focusing the laser beam. The signal is optimized for a flat sample with small lateral dimensions.[41] Another specific advantage of this detection mode appears to be the possibility of 3-D profiling of absorption by moving the region of beam overlap through the sample.

Roger, Fournier, and Boccara[43] demonstrated the use of in situ mirage detection at an electrochemical interface in the case of copper deposition on a platinum electrode (0.05 mm thick) using a (pump) Argon laser. The signal detected during electrochemical deposition reflected the changes in reflectivity,

$$[1 - (R + \Delta R)]\phi \tag{9}$$

where R is the reflectivity of the bare platinum electrode, $\Delta R$ the change in reflectivity from copper deposition, and $\phi$ the incident light flux. They pointed out that the deposition process is monitored on a weak signal ($\sim\Delta R$) superimposed on a large background. However, the deposition of 10 monolayers was well detected, with the underpotential deposition of the first monolayer barely observable. By using a polarization modulated pump beam, it was possible to suppress the large background from the bare electrode. The signal due to the bare electrode is

$$[(1-R_{//})\Phi_{//} - (1-R_{\perp})\,\Phi_{\perp}] \tag{10}$$

which can be made equal to zero by adjusting the $\Phi_{//}/\Phi_{\perp}$ ratio. With this arrangement, the deposition of the first monolayer gave rise to a large signal, and the possibility of detecting $10^{-3}$ monolayer was demonstrated. Royce, Voss, and Bocarsly[44] used mirage detection with a CW Xenon lamp as the pump source to study the photocorrosion of semiconducting ZnSe[45] and CdTe[46] electrodes. They were able to characterize the thin layer deposited at the electrode surface by photocorrosion process. Both groups[43,44] had been aware that the concentration gradient in the electrolyte may be probed simultaneously. In ref. 44, a potential step was used as the perturbation which was monitored by mirage detection. Measurement of the delay time of the deflection as a function of the distance from the electrode led to the diffusion coefficients of electrochemically produced species ($Fe(CN)_6{}^{-3/-4}$ in 0.5 $K_2SO_4$). More recently, Decker[47-49] treated the case of a mirage detection experiment at a solid electrode/liquid interface taking into account the contributions from the heat wave and from the concentration wave. The general expression[48] which relates the measured deflection, $\psi(x,t)$, of a probe beam traveling at distance x from the electrode surface to the time-dependent gradient of refraction index, n, is

$$\psi(x,t) = \frac{\ell}{n}\frac{\partial n(x,t)}{\partial x} = \frac{\ell}{n}\left[\frac{\partial n}{\partial T}\frac{\partial T(x,t)}{\partial x} + \frac{\partial n}{\partial C}\frac{\partial C(x,t)}{\partial x}\right] \tag{11}$$

where $\ell$ is the interaction length of the probe beam in the solution, and T(x,t) and C(x,t) are the time dependent temperature and concentration distributions, respectively. $\partial T(x,t)/\partial x$ and $\partial C(x,t)/\partial x$ can be written[6,47] in terms of the thermal and ionic diffusivities, $D_{th}$ and $D_{ion}$,

$$\frac{\partial T(x,t)}{\partial x} = C_1 \exp\left(-\sqrt{\frac{\omega}{2D_{th}}}\,x\right)e^{i\omega t}$$

$$\frac{\partial C(x,t)}{\partial x} = C_2 \exp\left(-\sqrt{\frac{\omega}{2D_{ion}}}\,x\right)e^{i\omega t} \tag{12}$$

where $\omega$ is the angular frequency of the chopped light. Thermal and ionic diffusivities are of different orders of magnitude (typically $D_{th} \sim 10^{-3}$ and $D_{ion} \sim 10^{-5}$ $cm^2$ $s^{-1}$). Therefore, the value of the slope of linear portions of the $\ln\psi$ vs x curves provides a test for discerning between experimental conditions where thermal diffusion or concentration gradient is predominant. This kind of experiment combined with photoacoustic gas-microphone detection has been used[49] to study dissipation mechanisms taking place within an illuminated semiconductor electrode.

## 4. EXAMPLES OF CONTRIBUTIONS OF THE PHOTOACOUSTIC TECHNIQUE TO ELECTROCHEMISTRY

### 4.1 Semiconductor Electrodes

Rosencwaig[50,51] showed that it was possible to measure accurately the band edges of direct-band and indirect-band semiconductors by photoacoustic spectroscopy on powdered samples. This result makes the method attractive for in situ studies of semiconductor electrodes where the surface of interest can not be polished as necessary for differential reflectance measurements, e.g., ZnO pressed powder[52] or platinized $TiO_2$ powder.[53]

Photoacoustic spectroscopy (with gas-microphone detector), coupled with voltammetry, was applied[34] to a CdS crystal electrode and to a ZnO polycrystalline electrode to monitor the potential dependent formation of a metallic layer. The complete stripping of the Cd layer during the anodic scan was confirmed by the return of the photoacoustic signal to its initial value. Formation of metallic layers on the order of a few Angstroms was detected easily from the increase in optical absorption outside the semiconductor absorption band.

A mirage detection photoacoustic spectrometer was used[45] to identify and to measure the thickness of the Se layer formed on semiconducting ZnSe by photocorrosion (at 460 nm) in electrolytic solutions of various controlled pH values.

Photothermal deflection spectroscopy, in conjunction with photocurrent spectroscopy, allowed[54] the monitoring of the nonradiative recombination channel in photoexcited CdS in a polysulfide electrolyte over a potential range including the flat band potential. As shown by a strong photothermal signal and a weak photocurrent, nonradiative recombinations are the dominant deexcitation pathway at potentials below the flat band potential.[55]

### 4.2 Modified Semiconductor Electrode

Semiconductors having a band gap energy in the solar spectrum degrade by anodic decomposition, while semiconductors having a larger band gap are more stable. A stable semiconductor can be sensitized to the solar spectrum via adsorption at the surface by a solar light absorbing substance. A photoacoustic study (with mirage detection) of sensitization by cobalttetraphenyl porphyrin on various oxides, i.e., $Al_2O_3$, $Ta_2O_5$,

$SiO_2$, $ZrO_2$, $NbO_5$, $TiO_2$, ZnO, and $SrTiO_3$, has been reported as a function of the acidity of the electrolyte.

### 4.3. Solution Species at the Electrochemical Interface

Mirage spectroscopy coupled with pulsed voltammetry showed[57] promise for real time spectroscopic detection of unstable species forming at the interface. The experiment was performed at a platinum electrode for the $I^-/I_2$ system. In order to discriminate between the signal from the concentration gradient arising from the establishment of the diffuse layer and the signal from the appearance of $I_2$ at the interface, the potential was pulsed at a frequency of 0.08 Hz and the pump light beam was modulated at 20 Hz. Absorption spectra characteristic of iodine were obtained by probing the electrode interface. Although iodine is not stable and transforms into $I_3^-$ in the bulk, the characteristic absorption band of $I_3^-$ at 350 nm was not present; therefore, the results demonstrated the capabilities of the method for probing interfacial phenomena.

### 4.4. Oxide Films

Very thin n-$TiO_2$ semiconducting films were characterized[35] for their structure by photoacoustic spectroscopy with gas-microphone detection. The films were formed by stepping the potential from 0.9 V to 5 V vs SCE at a polycrystalline titanium electrode (0.025 mm thick) in [0.5 M ($HSO_4^-$ + $SO_4^-$), pH = 1] or in [1M ($Cl^-$), pH = 1] solutions, and the estimated layer thicknesses were between 30 and 200 A. Diffuse absorption band edges were obtained which were interpreted in terms of a $TiO_2$ amorphous layer. Indirect band gap energies of 2.8 eV and 2.6 eV were measured[58] for layers grown between 3 V and 5 V vs SCE and at lower potentials, respectively. The direct band gap energy was found unambiguously for all the layers at 3.65 eV (350 nm), a value consistent with the band gap of crystalline anatase.[59] The large difference between indirect and direct transition points to an amorphous layer in which very little long range order remains. Photoacoustic spectroscopy appears to be, in this case, more sensitive than Raman spectroscopy[60] which was unable to detect the anatase layers formed at potentials lower than 4 V vs RHE.

Photoanodic oxygen evolution from water at a thin semiconducting $TiO_2$ passive film has been investigated[61] in situ by photoacoustic spectroscopy with gas-microphone detection. The study led to the in situ determination of the intrinsic quantum yield and of the Peltier heat for the photoelectrochemical process. Anatase n-$TiO_2$ layers of 0.3 μm were grown on Ti at 100 V. Photocurrent and photoacoustic signals were simultaneously measured with 330 nm light during a potential scan in (0.5 M $Na_2SO_4$, pH = 7). The potential dependence of the photoacoustic signal, P(U), normalized against the photoacoustic signal at the flatband potential, $P(U_{fb})$,

$$L = [P(U) - P(U_{fb})]/P(U_{fb}) \qquad (13)$$

was written in terms of the intrinsic quantum yield ($\eta_a$) of the photocurrent for the absorbed photon, the photon energy (E), the redox potential ($U_R$), and the Peltier heat ($T\Delta S_R$) as

$$L = \frac{\eta_a}{E} \left[ |e_o|(U-U_R) - T\Delta S_R \right] \tag{14}$$

$\eta_a$ was derived from the slope of the linear portion of L vs U curves in the saturation region of the photocurrent. Peltier heat and reflectivity, $R = (1-\eta/\eta_a)$, were also derived by combining photoacoustic and photocurrent results.

Passive films on iron in a weak alkaline buffer solution were investigated[37] in situ with piezoelectric detection. The results obtained by simultaneous voltammetry and photoacoustic spectroscopy (at 514.5 nm) led the authors to conclude that the passive film consists of two layers, $\gamma Fe_2O_3$ and $Fe_3O_4$. However, the results seem questionable since the photoacoustic experiment was executed only at one wavelength, and no attempt to change the light chopping frequency was made to confirm the hypothesis of a two-layer structure.

Another group[62-65] using a gas-microphone detection studied in situ oxide layers forming on copper, gold, and platinum electrodes by coupling photoacoustic spectroscopy with electrochemical measurements. In the more recent contributions,[64,65] the PA spectra were analyzed quantitatively. In ref. 64, intensity-calibrated PA spectra were shown to correspond to the complement $1-R(\lambda)$ of the reflectance spectrum $R(\lambda)$ of the electrode/solution interface. Mandelis[16] has shown the anticorrelation between the reflected light beam and the photoacoustic signal. He developed[16] a photoacoustic interferometric theory for the photoacoustic response from a transparent thin layer on an absorbing substrate. The theory indicates that the magnitude of the photoacoustic signal is related to the thickness of the thin layer. Experimental evidence was provided by the study of thin $SiO_2$ layers on Si. The photoacoustic method has a definite advantage[16] over reflectance measurements for nonreflecting substrates and light scattering layers. Photoacoustic interferometry was carried out in situ[65] during the deposition of a $PbO_2$ film on a platinum electrode in a $Pb(ClO_4)_2/HClO_4$ solution. The electrode was illuminated with a slightly converging beam at virtually normal incidence. By measuring the PA signals at constant wavelength as a function of time during constant-current deposition of $PbO_2$ a typical interference pattern was obtained. The reflectivity of the interface, $R_{1,2,3}$, was calculated[66] for a three-phase system, electrolyte (1)/$PbO_2$ layer(2)/Pt electrode (3), as a function of the index of refraction $n_2$ and of the absorption constant $k_2$ of the thin layer. The photoacoustic signal for the case of an optically thick and thermally thin platinum electrode is given as a function of the thin layer thickness, d, by

$$PAS(d) = P_{MAX}\,(1-R_{1,2,3}) \tag{15}$$

The starting values for $n_2$ and $k_2$ are taken from the position of the maxima and from the increase with distance of the average value between maxima and minima, respectively. A least-squares fit of the interference patterns, calculated at 50 nm intervals in the 300-750 nm range gave the first in situ determination of the optical constants of an electrochemically grown surface by a photoacoustic technique. Results were compared with existing data obtained by transmission[67] and in situ reflectance[68] on similar systems. Although a surprisingly good agreement was observed in all the $k_2$ values for $\lambda$ >550 nm, discrepancies attributable to differences in the structure and/or the composition of the layers were apparent at lower wavelengths. The key point in these measurements was to obtain an uniform growth of the layer.

Very thin passive layers (50 Å) electrochemically formed at implanted ruthenium-titanium alloys were investigated[35] by concurrent in situ photoacoustic spectroscopy (gas-microphone detection), photocurrent spectroscopy, and potentiodynamic measurements in the chlorine evolution potential region. The implanted alloy depth profile was measured by Rutherford backscattering before any electrochemical treatment, and the subsequent measurements monitored the changes in the passive layer properties versus the ruthenium content during the slow electrochemical depth profiling (2-3 Å $hour^{-1}$).[69] Absorption spectra obtained at 2 nm intervals over the 280-550 nm wavelength region were compared in terms of the dimensionless band gap absorption defined as

$$(PAS_{300nm} - PAS_{450nm}) \,/\, PAS_{300nm} \qquad (16)$$

The band gap absorption abruptly varied from a value measured for a semiconducting n-$TiO_2$ layer of similar thickness at low ruthenium contents to near zero values for increasing ruthenium contents. The sharp change in the band gap absorption correlated with the appearance of a 40 mV Tafel slope in the potentiodynamic scans, indicative of a metallic-like electrode, and with the quasi disappearance of the photocurrent. The critical ruthenium content was found to be of 7-10% for the transition from a semiconducting to a metallic electrode. The low percolation limit in the implanted samples may be indicative[70] of a different microstructure than that observed in mixed oxides conventionally prepared[71] by firing coatings from $RuCl_3$/$TiCl_4$ solutions.

## 5. CONCLUSION

The study of the electrochemical interface is a complex problem for which many questions remain unanswered. In situ methods, which have developed tremendously in recent years, should provide the most unambiguous answers about the nature of the interface during the electrochemical process. Because of the complexity of the system and the experimental constraints specific to each technique, results may be expected from the combination of an array of techniques rather than from one specific method.

The ability of photoacoustics to monitor optical absorption of absorbing and light scattering samples is an asset for in situ studies of "real" electrodes. The photoacoustic studies of electrochemical systems discussed above illustrate the versatility of the method to solve various aspects of the problem. Optical absorption of solid electrodes and of the solution in the diffuse layer were obtained on systems difficult to handle with a purely optical method. Since the photoacoustic signal is generated by radiationless deexcitation, the coupling of photoacoustics with a detection system sensitive to radiative deexcitation provides a means to test the various recombination pathways following optical excitation of semiconductor.

The most recent studies[65,72] indicate that photoacoustic spectroscopy enables quantitative measurements of optical constants and may be a useful complement to reflectance spectroscopy. The volume of accomplishments is quite impressive in view of the novelty of the method. In the future, more sophisticated variants, e.g., frequency domain photoacoustic spectroscopy,[73] should provide powerful tools for the nondestructive examination of multilayer systems.

## ACKNOWLEDGMENT

Prof. F. Decker is gratefully acknowledged for making available the preprint of his paper at the forthcoming International Conference on Chemistry and Physics of Electrified Interfaces, Bologna, Italy, August 30 - September 3, 1988.

## REFERENCES

1. A. G. Bell, Philos. Mag., **11**, 510 (1881).

2. J. Tyndall, Proc. R. Soc. London, **31**, 307 (1881).

3. W. C. Rontgen, Philos. Mag., **11**, 308 (1881).

4. M. E. Mercadier, C. R. Acad. Sci. Paris, **92**, 409 (1881).

5. W. H. Preece, Proc. R. Soc. London, **31**, 506 (1881).

6. A. Rosencwaig and A. Gersho, J. Appl. Phys., **47**, 64 (1976).

7. A. Rosencwaig, Bull. Amer. Phys. Soc., **18**, 357 (1973).

8. A. Rosencwaig, Opt. Commun., **7**, 305 (1973).

9. A. Rosencwaig, Science, **181**, 657 (1973).

10. A. Rosencwaig in "Optoacoustic Spectroscopy and Detection," edited by Yoh-Han Pao, Academic Press, New York, 1977, p. 209.

11. Rayleigh (Lord), Nature (London), **23**, 274 (1881).

12. T. H. Maugh II, Science, **188**, 38 (1975).

13. W. Lahmann, H. J. Ludewig and H. Welling, Anal. Chem., **49**, 549 (1977).

14. A. C. Boccara, D. Fournier and J. Badoz, J. Appl. Phys. Lett., 36, 130 (1980).

15. J. J. Barrett in "Chemical Aplication of Nonlinear Raman Spectroscopy," A. B. Harvey, Editor, Academic Press, New York, 1981, p. 89.

16. A. Mandelis, E. Siu, and S. Ho, Appl. Phys., A33, 153 (1984).

17. E. M. Eyring, S. M. Riseman and F. E. Massoth in "Catalytic Materials: Relationship between Structure and Reactivity," ACS Symposium Series 248, T. E. Whyte, Jr., R. A. Dalla Betta, E. G. Derouane, and R. T. K. Baker, editors, The American Chemical Society, Washington, DC (1984) p. 399.

18. B. S. H. Royce, S. McGovern, and J. B. Benziger, American Laboratory, March 1985, p. 16.

19. T. Masujima, H. Kawata, Y. Amemiya, N. Kamiya, T. Katsura, T. Iwamoto, H. Yoshida, H. Imai and M. Ando, Chem. Lett., 973 (1987).

20. A. J. Bard, Analusis, V.6, n°7, 277 (1978).

21. R. C. Gray and A. J. Bard, Anal. Chem. 50, 1262 (1978).

22. R. E. Malpas and A. J. Bard, Anal. Chem., 52, 109 (1980).

23. A. Fujishima, H. Masuda, and K. Honda, Chem. Lett., 1063 (1979).

24. "In Situ Characterization of Electrochemical Processes," NMAB 438-3, National Academy Press, Washington, DC (1987) p. 27.

25. M. J. D. Low and G. A. Parodi, Appl. Spectrosc., 34, 76 (1980); J. Mol. Struct., 61, 119 (1980).

26. J. G. Parker, Appl. Opt., 12, 2974 (1973).

27. A. Rosencwaig, "Photoacoustics and Photoacoustic Spectroscopy," John Wiley and Sons (1980).

28. H. S. Bennett and R. A. Forman, Appl. Opt., 15, 2405 (1976).

29. L. C. Aamodt, J. C. Murphy and J. G. Parker, J. Appl. Phys., 48, 927 (1977).

30. G. C. Wetsel, Jr. and F. A. McDonald, Appl. Phys. Lett., 30, 252 (1977).

31. G. C. Wetsel, Jr. and F. A. McDonald, Bull. Am. Phys. Soc., 22, 295 (1977).

32. R. M. White, J. Appl. Phys., 34, 3559 (1963).

33. F. A. McDonald and G. C. Wetsel, Jr., J. Appl. Phys., 49, 2313 (1978).

34. H. Masuda, A. Fujishima and K. Honda, Bull. Chem. Soc. Jpn., 53, 1542 (1980).

35. C. E. Vallet, S. E. Borns, J. S. Hendrickson, and C. W. White, J. Electrochem. Soc., 135, 387 (1988).

36. M. M. Farrow, R. K. Burnham, M. Auzanneau, S. L. Olsen, N. Purdie, and E. M. Eyring, Appl. Opt., 17, 1093 (1978).

37. K. Ogura, A. Fujishima, Y. Nagae, and K. Honda, J. Electroanal. Chem., 162, 241 (1984).

38. R. E. Malpas, R. A. Fredlein, and A. J. Bard, J. Electroanal. Chem., 98, 171, (1979).

39. M. Wun-Fogle, D. J. Milgaten, and W. C. Hwang, Appl. Opt., 21, 121 (1982).

40. D. C. Smith, IEEE J. Quantum Electron, QE-5, 600 (1969).

41. W. B. Jackson, N. M. Amer, A.C. Boccara, and D. Fournier, Appl. Opt., 20, 1333 (1981).

42. D. Fournier, A. C. Boccara, and J. Badoz, Appl. Opt., 21, 74 (1982).

43. J. P. Roger, D. Fournier, and A. C. Boccara, J. Phys. Colloq., C6, 313 (1983).

44. B. S. H. Royce, D. Voss, and A. Bocarsly, J. Phys. Colloq., C6, 325 (1983).

45. B. S. H. Royce, F. Sánchez-Sinencio, R. Goldstein, R. Muratore, R. Williams, and W. M. Yim, J. Electrochem. Soc., 129, 2393 (1982).

46. J. G. Mendoza-Alvarez, B. S. H. Royce, F. Sánchez-Sinencio, O. Zelaya-Angel, C. Menezes, and R. Triboulet, Thin Solid Films, 102, 259 (1983).

47. F. Decker, R. T. Neuenschwander, C. L. Cesar, and A. F. S. Penna, J. Electroanal. Chem., 228, 481 (1987).

48. F. Decker and M. Fracastoro-Decker, J. Electroanal. Chem., 243, 187 (1988).

49. F. Decker, M. Fracastoro-Decker, N. Cella and H. Vargas, International Conference on Chemistry and Physics of Electrified Interfaces, Bologna, Italy, Aug. 30 - Sept. 3 (1988).

50. A. Rosencwaig, Anal. Chem., **47**, 592A (1975).

51. A. Rosencwaig, Phys. Today, **28**, 23 (1975).

52. H. Masuda, S. Morishita, A. Fujishima, and K. Honda, J. Electroanal. Chem., 121, 363 (1981).

53. R. S. Davidson and C. L. Morrison, J. of Photochem., **24**, 323 (1984).

54. R. E. Wagner, V. K. T. Wong, and A. Mandelis, Proceedings 4th International Topical Meeting on Photoacoustic, Thermal and Related Sciences, Ville d'Esterel, Quebec, Canada (1985) p. MD8.

55. H. Gerisher, in Physical Chemistry: An Advanced Treatise, Vol. **9A**, H. Eyring, D. Henderson, and W. Jost, Eds., Acad. Press, New York, 1970, Ch. 5.

56. A-M. Merle, A. Cherqaoui, and C. Giannotti, J. Phys Colloq., **C6**, 291 (1983).

57. A-M. Dorthe-Merle, J-P. Morand, and E. Maurin, Proceedings 4th International Topical Meeting on Photoacoustic, Thermal and Related Sciences, Ville d'Esterel, Quebec, Canada (1985) p. MD-5.

58. J. Tauc in "Optical Properties of Solids," F. Abeles, Ed., p. 279, North Holland Publishing Company (1972); J. Tauc in "Optical Properties of Solids," S. Nudelman and S. S. Mitra, Eds. p. 123, Plenum Press, New York (1969).

59. N. C. Fernelius, Appl. Surf. Sci., **4**, 401 (1980).

60. T. Ohtsuka, J. Guo, and N. Sato, J. Electrochem. Soc., **133**, 2473 (1986).

61. J. Rappich and J. K. Dohrmann, Proceedings 5th International Topical Meeting on Photoacoustic and Photothermal Phenomena, Heidelberg, W. Germany, Springer-Verlag (1987) p. 204.

62. U. Sander, H-H. Strehblow, and J. K. Dohrmann, J. Phys. Chem., **85**, 447 (1981).

63. J. K. Dohrmann and U. Sander, J. Phys. Colloq., **C6**, 281 (1983).

64. J. K. Dohrmann, U. Sander, and H-H. Strehblow, Z. Phys. Chem. Neue Folge, **134**, 41 (1983).

65. J. K. Dohrmann and U. Sander, Ber. Bunsenges. Phys. Chem., **90**, 605 (1986).

66. O. S. Heavens, in "Optical Properties of Thin Solid Films," Butterworth, London, 1955.

67. W. Mindt, J. Electrochem. Soc., **116**, 1076 (1969).

68. K. D. Naegele and W. J. Plieth, Electrochim. Acta, **25**, 241 (1980).

69. E. J. Kelly, D. E. Heatherly, C. E. Vallet, and C. W. White, J. Electrochem. Soc., **134**, 1667 (1987).

70. G. E. Pike, in "Electrical Transport and Optical Properties of Inhomogeneous Media," J. C. Garland and D. B. Tanner, Eds., p. 366, American Institute of Physics, New York (1978).

71. W. A. Gerrard and B. C. H. Steele, J. Appl. Electrochem., **8**, 417 (1978).

72. J. T. Dodgson, A. Mandelis, and C. Andreetta, Can. J. Phys., **64**, 1074 (1986).

73. A. Mandelis and J. D. Lymer, Appl. Spectrosc., **39**, 473 (1985).

# Raman Spectroscopic Techniques in Interfacial Electrochemistry

R. K. CHANG
Section of Applied Physics and
Center for Laser Diagnostics
Yale University
Box 2157, Yale Station
New Haven, CT 06520
U.S.A.

ABSTRACT. Raman spectroscopy of molecular adsorbates at the electrode-electrolyte interface requires sensitive photon detectors or metal surfaces that exhibit surface enhanced Raman scattering (SERS). The following topics related to Raman spectroscopy and SERS are discussed: (1) The latest advances in Raman instrumentation with emphasis on the recently available two-dimensional detectors; (2) the electrochemical surface activation procedures necessary to achieve SERS on Ag and Au electrodes; (3) the possible role of the electrochemical oxidation-reduction cycle in producing and stabilizing metal adatoms and clusters as well as larger scale roughness; and (4) the connection between the fundamentals of the two accepted SERS models (electromagnetic and charge transfer) and the surface roughness.

## 1. Introduction

Surface enhanced Raman scattering (SERS) was first observed[1] in 1974 and identified[2] in 1977. The observed SERS was from pyridine adsorbed on Ag and Au electrodes which, under potential control, had undergone numerous oxidation and reduction cycles in an electrochemical cell. For the next ten years, vigorous research on SERS ensued in an attempt to fully understand the mechanisms leading to the observed, though not fully quantified, surface enhancement factor. Various forms of surfaces and microstructures of different types of metals were studied. Ag and Au, the two metals in which SERS was first observed, remain the favorite metals among the present SERS researchers. Several plausible mechanisms were proposed during the earlier years of SERS. The electromagnetic and chemical models are currently accepted as being of equal importance. Review articles on SERS provide a comprehensive background of the experimental and theoretical status of this field.[3-11]

This paper provides an introduction to SERS for scientists interested in molecular specific spectroscopic techniques at the electrode-electrolyte interface. The first section describes the basic Raman instrumentation used, with particular emphasis on the latest advances in optical multichannel analysis techniques. The second section summarizes the procedures for electrochemical surface activation to achieve large surface enhancement factors. Electrochemical procedures to further increase or decrease the enhancement factors are discussed, as such procedures provide additional clues to the origin of surface



*C. Gutiérrez and C. Melendres (eds.),*
*Spectroscopic and Diffraction Techniques in Interfacial Electrochemistry,* 155–180.

enhancement mechanisms. The third section reviews the theory essential to the electromagnetic and chemical enhancement models.

## 2. Instrumentation

The Raman signal strength from a monolayer of adsorbates on the surface of unenhanced electrodes is extremely weak—at most 1 photoelectron count per second (cps) with 200 mJ cw input laser energy. A long integration time is necessary so that the total number of photoelectron counts N will have a fluctuation $\sqrt{N}$ that is acceptable for a desired signal-to-noise ratio. For a monolayer of adsorbates on the surface of enhanced electrodes, the corresponding Raman signals can be in excess of $10^5$ cps. When such strong Raman signals are available, an experimentalist is often tempted to make more rapid spectral measurements as the electrode potential is swept, thereby again limiting the overall cps by $\sqrt{N}$ considerations. The typical experimental configuration and detection system are reviewed here with particular emphasis on recent detector developments that increase the performance of optical multichannel analyzers.

### 2.1. EXPERIMENTAL CONFIGURATION

Figure 1 shows the experimental configuration used for SERS measurements. The electrode is usually placed ≈1 mm from the electrochemical cell window, and the incident laser beam with wavelength $\lambda_i$ is tightly focused on the electrode by a spherical or cylindrical lens to a spot area with width w and height h. Should the laser beam "melt" the electrode surface, increasing h is wiser than decreasing the laser energy. The incident angle is usually larger than 45° relative to the electrode normal, and the axis of the collection optics is along the electrode normal. The collection lens is typically a high quality camera lens (with a low f-number ranging from 0.75 to 1.4) that converts the diverging Raman radiation with wavelength $\lambda_s$ to a parallel beam. A second lens (usually achromatic) aligned along the collection axis focuses this parallel beam onto the entrance slit width (≈500 μm) of a spectrometer or spectrograph determined by the desired spectral resolution. The f-number of this second lens must be higher than that of the dispersive instrument (typically 6) to avoid overfill of the collected radiation on the mirrors and gratings in the spectrometer/ spectrograph. Overfilling causes more of the intense elastically scattered radiation to spill into the $\lambda_s$ region. The magnification M of the illuminated spot on the electrode at the spectrograph slit is determined by the f-number ratio of the spectrograph and the first collection lens, e.g., M = 6/1.2. w of the illuminated spot on the electrode is set by the slit width, i.e., w = (slit width)/M. The spectrograph slit height is adjusted to be M x h because a higher slit would allow more of the unwanted elastic light to spill into the $\lambda_s$ region. The entire collection axis is carefully covered with black cloth to prevent stray light at $\lambda_i$ from entering the spectrometer/spectrograph slit.

### 2.2. SINGLE-CHANNEL DETECTION

The combination of a spectrometer and a photomultiplier (PMT) is referred to as an optical single-channel analyzer, since only one narrow wavelength width is detected at any given time interval and no Raman signals at wavelengths outside this width are detected as they are blocked by the prefilter stage and exit slits of the spectrometer. Tandem spectrometers, two or three identical spectrometers that can be scanned in unison, are needed to decrease

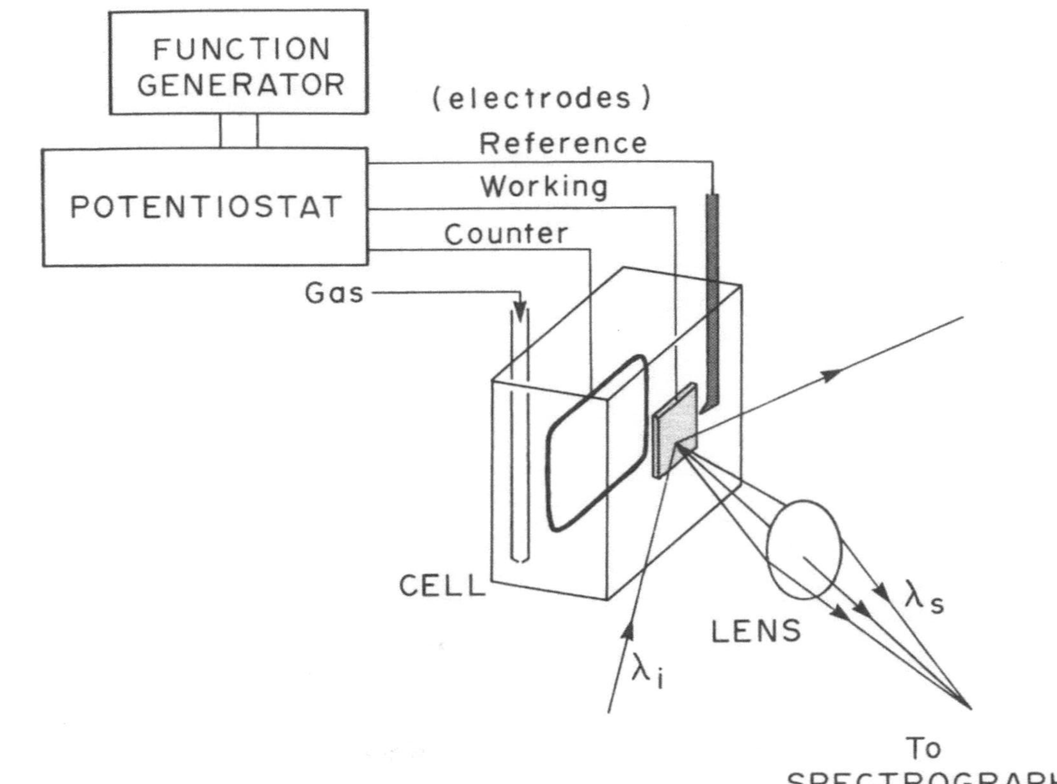


Fig. 1. Experimental configuration commonly used to investigate the temporal correlation between the SERS intensity and the cyclic voltammogram. To exclude oxygen from the electrolyte, $N_2$ gas is continuously bubbled through an electrochemical cell equipped with an optical window through which the incident laser beam enters and the SERS radiation exits. [Fig. from ref. 6]

the amount of stray light spilling from $\lambda_i$ into $\lambda_s$. Holographic gratings are generally selected for their low stray light characteristics. The mechanical linkage of the two or three gratings must track precisely as the gratings are rotated during the forward sweep with increasing $\lambda_s$. In the case of multiple spectrometer sweeps to increase N per wavelength width, Raman data are not recorded during the rapid backward sweep toward $\lambda_i$ or during the initial portion of the forward sweep that is necessary to remove the backlash in the grating mounts.

The spectral response of the PMT for Raman spectroscopy should be efficient in the $\lambda_s$ range and the photocathode area should be small in order to decrease the total dark count (typically <10 cps) even at a temperature cooled by thermoelectric coolers. Major advances have been made in III-V semiconductor photocathodes which have good quantum efficiency at wavelengths as long as 1 μm. The PMT also needs to be selected for good pulse height distribution for single photoelectron events, so that the discriminator of the photon-counting electronics can be set with its voltage window (between V and V + ΔV) to pass single photoelectron events. The discriminator can eliminate low voltage events caused by thermionic emission from the electron multipliers and high voltage events generated by cosmic rays. A computer is used to store N at each wavelength as the spectrometer is continuously swept with increasing $\lambda_s$ and to accumulate N for multiple sweeps. For single-channel analyzers, it is important to set the sweep rate or the step interval of the spectrometer advance in such a way that the spectral line shape is not distorted. Single-channel analyzers are currently being used less frequently in the field of Raman spectroscopy.

## 2.3. MULTICHANNEL DETECTION

The combination of a multistage spectrograph and a one- or two-dimensional detector is commonly referred to as an optical multichannel analyzer (OMA).[12] The spectrograph consists of a small filter stage and a large dispersive stage (see Fig. 2). The filter stage is a subtractive double monochromator containing two gratings. The first grating disperses the input radiation and the second, mechanically linked to the first, collapses the dispersion introduced by the first grating since the two grating angles are set opposite each other. The net result of these two gratings is zero dispersion, and most of the radiation at $\lambda_i$ is blocked by the adjustable bandpass slit.

The transmitted radiation, with a broad range of $\lambda_s$, is focused at the entrance slit of a larger dispersive single spectrograph. The dispersive stage contains one high resolution grating (mounted on a turret that can hold three gratings with different grooves/mm). A one- or two-dimensional detector is placed at the exit plane of the dispersive stage. A corrective lens is usually placed in the dispersive stage in order to have the spectral focus (along the horizontal axis) and the entrance slit image focus (along the vertical axis) coincide on the same focal plane. This corrective lens is essential if spectral information from many points along the slit height [$\lambda_s(z)$, where z is the distance along the slit to be simultaneously measured by a two-dimensional detector].

The advantages of the OMA have been reviewed.[12] Briefly, the entire Raman spectrum (within the broad bandpass of the spectrograph filter stage) is detected simultaneously. None of the Raman radiation with wavelengths larger than the spectral resolution is blocked by the slits, as in the case for single-channel analyzers. The detection time saved is equal to the spectral bandpass divided by the desired spectral resolution, e.g., 500 cm$^{-1}$ divided by 0.5 cm$^{-1}$ (assuming 1000 resolvable detector elements). The spectral bandpass of the OMA is determined by the product of the spectral resolution and the total number of detec-

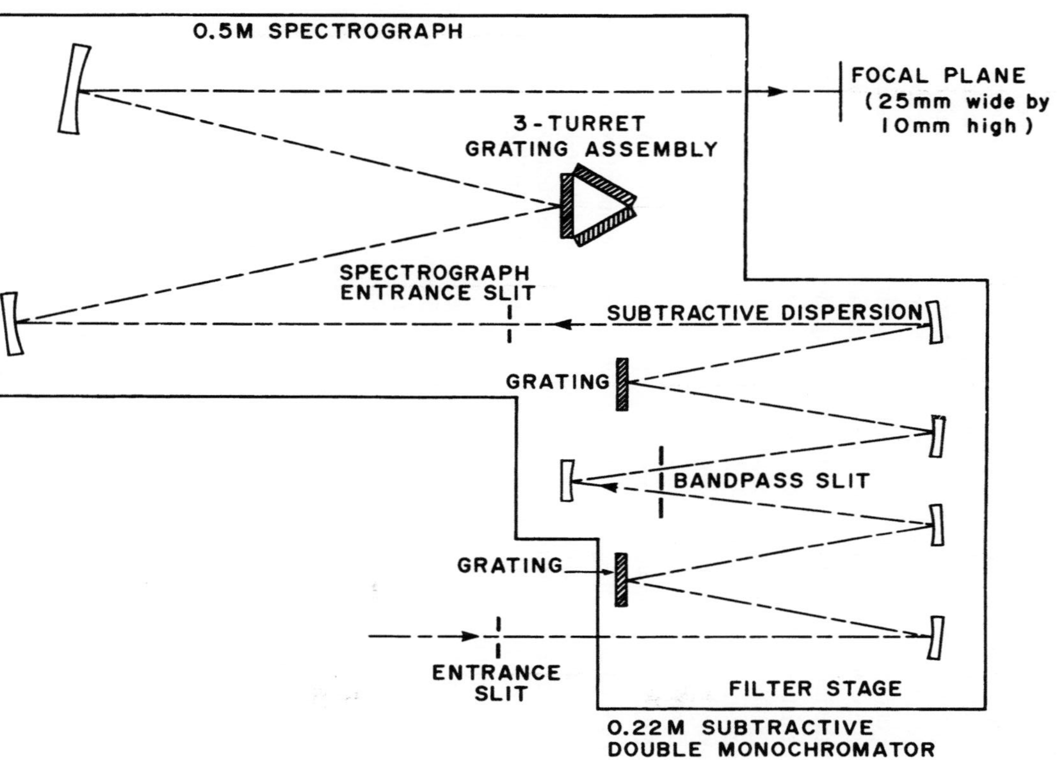


Fig. 2. Schematic diagram of the triple spectrograph. The filter stage consists of a subtractive double monochromator, and the dispersive stage consists of a single spectrograph. [Fig. from Spex Industries]

tor elements, including the "dead zone" between the elements. Furthermore, additional time is saved as the gratings do not need to be swept backwards in order to remove the grating backlash during multiple scans. Other advantages of the OMA are: (1) The input intensity need not be regulated since every part of the spectrum is equally affected by the intensity fluctuation; (2) temporal evolution of the entire Raman spectrum can be measured as the electrochemical potential is varied; and (3) since the Raman peaks and the unavoidable continuum background are detected simultaneously, data analysis of the peak line shape and height can be readily extracted from the continuum without concern for the unavoidable time changes in the collection optics and the tandem spectrometer.

The stray-light rejection from such triple-stage spectrographs is generally less than that from triple-stage spectrometers used in single-channel analyzers. Further reduction of the stray light at $\lambda_i$ can be achieved by placing a steeply rising bandpass interference filter between the two collections lenses, where the collected radiation is parallel. In general, the rising slope of the interference filter is not steep enough to block $\lambda_i$ and to transmit Raman radiation with shifts <200 $cm^{-1}$. For Raman shifts between 3 and 200 $cm^{-1}$, a heated cell containing $I_2$ crystals has been used to selectively absorb the stray light from the 514.5 nm output of a single-mode Ar-ion laser.[13] A temperature stabilized etalon is inserted within the Ar laser cavity to force lasing in a single mode within the Doppler-broadened gain profile of the 514.5 nm line. The single laser mode (with a linewidth ≈3 MHz) is purposely tuned to one of the many allowed $I_2$ vapor hyperfine transitions (v = 0 level of the $^1\Sigma_{g+}$ ground state to v = 43 of the $^3\Pi_{0u+}$ excited state) which has a temperature-broadened width of ≈750 MHz. The attenuation of $\lambda_i$ for a cell 11.4 cm long and heated to 90°C is 80 dB. The nearby $I_2$ hyperfine lines unavoidably contribute to the absorption of Raman radiation with shifts as far away as 3 $cm^{-1}$. Consequently, significant Raman line shape distortions can occur for Raman shifts <3 $cm^{-1}$.

Prior to 1980, the two detectors most frequently used with OMA's were one-dimensional intensified photodiode arrays (I-PDA) and two-dimensional vidicons with a silicon intensified target (SIT). Since 1980, two new two-dimensional detectors have become available to Raman spectroscopists, i.e., the CCD and the Mepsicron. Like the SIT vidicons, both these detectors offer the possibility of providing $\lambda_s(z)$ intensity information. Should the spatial information along the z-axis or slit be unnecessary, the counts along the z-axis of the two-dimensional detectors can be summed or binned to provide $\lambda_s$ intensity information.

A comparison of the CCD and Mepsicron with the I-PDA has recently been reported[14] using the very weak Raman scattering of the third-order phonon of Si as the test sample. Figure 3 shows the third-order Raman spectra for Si measured between 1200 and 1550 $cm^{-1}$ in the backscattered geometry.[14] The integration times are limited to about three hours for both the CCD and the Mepsicron, even though both detectors are capable of achieving a better signal-to-noise ratio with an even longer integration time. The I-PDA is allowed to integrate until the signal and the dark noise exceed the linear range, which occurs in less than one-half hour. Only the CCD and Mepsicron are able to detect the three peaks in the third-order Raman spectrum (labeled j, k, and $\ell$ in Fig. 3). Clearly, the low-light-level sensitivity of both detectors is at least an order of magnitude better than that of the I-PDA. The CCD and Mepsicron performed equivalently within the conditions of the Si experiment. Both have also been used to detect unenhanced Raman signals from submonolayers of adsorbates on metal surfaces.[15,16] The various sources of noise in the I-PDA, SIT, CCD and Mepsicron have been reported as well as the necessary nor-

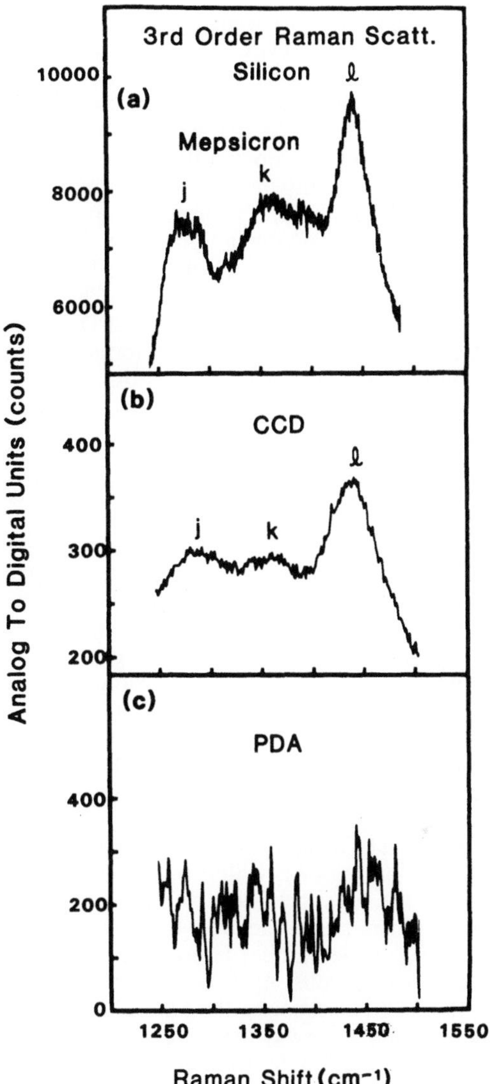


Fig. 3. Third-order phonon Raman spectra from Si detected with (a) a position sensitive resistive anode detector (Mepsicron), (b) a CCD, and (c) an intensified photodiode array (I-PDA). Appropriate normalization procedures were carried out for each detector. [Fig. from ref. 14]

malization and correction procedures for low-light-level signals.[12,14] A brief review of the features of the four detectors, the I-PDA, SIT, CCD, and Mepsicron, is presented below.

2.3.1. *I-PDA*. The I-PDA consists of an image intensifier, composed of a photocathode on one end followed by a multichannel plate (MCP), which multiplies the photoelectron distribution (with >2 x $10^4$ gain), and a phosphor on the other end. The phosphor output is optically coupled to the Si photodiode array through a fiber-optic plate. The array consists of 1024 Si photodiodes, each 2.5 mm high, 13 μm wide, and separated by 25 μm. The 2.5 mm diode height sets an upper limit on the height of the illumination spot on the electrode, i.e., $h < (1/M) \times 2.5$ mm. The main limitation of the I-PDA is the dark noise from the intensifier photocathode and from each Si photodiode. The dark noise limits the integration time to less than one-half hour (even when the I-PDA is cooled to -100°C). If longer integration time is used, the I-PDA output is no longer in the linear range. Other limitations are the low quantum efficiency of conventional intensifier photocathodes (usually 10%), the long lag time of the phosphor when the I-PDA is cooled, and the electronic readout noise associated with the preamps.[12]

2.3.2. *SIT*. The SIT vidicon consists of a photocathode and a target which has a dense matrix of p-type Si islands separated by 8 μm. An accelerated photoelectron (with energies between 3 and 9 keV), upon striking the diodes on the target, creates $\approx 10^3$ electron-hole pairs and intensification occurs on the target. The stored charge distribution is read by an electron beam. Several hours of integration time on the target is possible if the SIT vidicon is cooled to -40°C and the electron gun source is shut off to prevent its red glow from being channeled by the glass tube to excite photoelectrons at the photocathode. The main limitation of the SIT vidicon is the dark noise from the photocathode and from the target which that causes the output response to be nonlinear to the total number of photoelectrons bombarding the target. When the SIT vidicon is cooled to -40°C, many electron beam scans are needed to read the accumulated charge distribution on the target, thereby contributing significant preamp noise.[12]

2.3.3. *CCD*. The two-dimensional CCD array is a single-chip device with MOS capacitors as the active elements. Depletion regions, which result in potential wells when appropriately biased, are formed in the substrate beneath the electrodes. Incident photons of energy greater than the Si band gap create electron-hole pairs in the substrate. Minority carriers (electrons in our case), whose number is proportional to the number of incident photons are drawn into the potential wells of the depletion regions where they are stored until the readout process. During this process, the potential well depths are modified by clocking the electrode potentials in such a way that charges are transferred from well to well along the rows of MOS capacitors and finally reach an integrating capacitor and an on-chip preamplifier. The two-dimensional arrays are read out row by row.

Older CCD's required a two- or three-phase clocking sequence that needed overlapping gate architecture. This prevented some of the incident radiation from reaching the active region. To improve the quantum efficiency of these CCD's, several approaches have been devised to allow the incident photons to impinge directly on the active region with little or no obstruction by the gate electrodes. Two types of CCD's have recently achieved quantum efficiencies >60%: (a) a backilluminated, thinned, three-phase CCD; and (b) a frontilluminated virtual-phase CCD. The Si substrate in a backilluminated three-phase CCD is thinned (10 – 15 μm), thus allowing incident photons from the back to reach the active region relatively unattenuated. One disadvantage of the thinned CCD is that data correction is complicated by optical interference fringes introduced by the thinned Si substrate.[16] The virtual-phase CCD has only a single set of polysilicon electrodes covering half the front surface, leaving the remaining picture elements (pixels) bare except for a thin oxide overlayer. A stepped-potential well, created by ion implantation next to the single set of electrodes, enables the minority carriers to be clocked out by varying the voltage on the single set of electrodes between two values.

The CCD (Texas Instruments 4849) used in the backscattered Si experiment, an unintensified virtual-phase frontilluminated device with 584 x 390 elements, was part of an integrated Photometrics system consisting of a CCD chip in a liquid nitrogen cooling dewar, drive electronics, analog-to-digital converter, software control, and video display. With three hours integration time, the following noise-reduction procedures were carried out for the third-order phonon data shown in Fig. 3: (1) the vertical columns were binned since no spatial information was desired; (2) the readout-related bias was subtracted; (3) the dark current level (for equal integration time as the signal) was subtracted; (4) high-intensity spikes (in single picture elements or pixels) caused by cosmic ray events or defects in the CCD were removed; and (5) normalization was performed by the spatial response of the CCD (fixed pattern image) to uniform illumination of broadband radiation.[14]

2.3.4. *Mepsicron.* The Mepsicron, a photomultiplier with a position sensitive resistive anode detector (ITT Mepsicron F4146M), is shown schematically in Fig. 4. The bias voltages indicated were optimized by Surface Science Laboratories (SSL) for each detector. The multialkali photocathode (MA-2) has a broad spectral band ranging from 300 nm to nearly 900 nm and a quantum efficiency of 15% at 500 nm. A photoelectron, after being accelerated by $V_1$, enters a channel of the V-MCP stack (consisting of two MCP's). Each channel of the MCP plate has a length-to-diameter ratio of ≈40:1 and a bore diameter of 10 μm. To increase the overall electron multiplication, an additional Z-MCP stack (consisting of three MCP's) is used. The region between the V-MCP and Z-MCP stacks is reverse biased, allowing the electrons to spread laterally into an electron cloud and, thereby, make use of additional channels in the Z-MCP stack. The total amplification from the five MCP's is $\approx 5 \times 10^7$.

The electron cloud exiting the Z-MCP stack impinges on a resistive anode, causing current to flow through the four corner electrodes on the anode. The current at each of the four corner electrodes is sampled by charge-sensitive preamps, and the position of the electron-cloud centroid is determined by an analog-position computer (SSL 2401) which indicates that an event has occurred at location $x_1,y_1$. Even though the electron cloud spreads to a much larger area after traversing all the MCP's, the electron-cloud centroid can be determined with much better spatial resolution than the extent of the cloud.

The charge-sensitive preamps require ≈1 μsec to shape the current pulse into a higher level bipolar voltage pulse. Should two photoelectron events occur within the shaping time, a centroid calculation would produce erroneous values for $x_1$ and $y_1$. To indicate such an occurrence, a fifth preamp performs fast shaping on the sum of the four charge-sensitive preamps and provides the fast threshold logic circuit with a signal dependent on the total charge collected by the resistive anode. When two photoelectron events occur within 1 μsec, the signal from the fifth preamp will exceed the threshold for a single event, and both events will be ignored. The control logic also ignores events when the signal from the fifth preamp is below a low voltage threshold. The Mepsicron is, therefore, able to discriminate against noise based on the total charge imparted to the resistive anode, analogous to a discriminator with a voltage window ($V$ and $V + \Delta V$) in a photon counter for a PMT output.

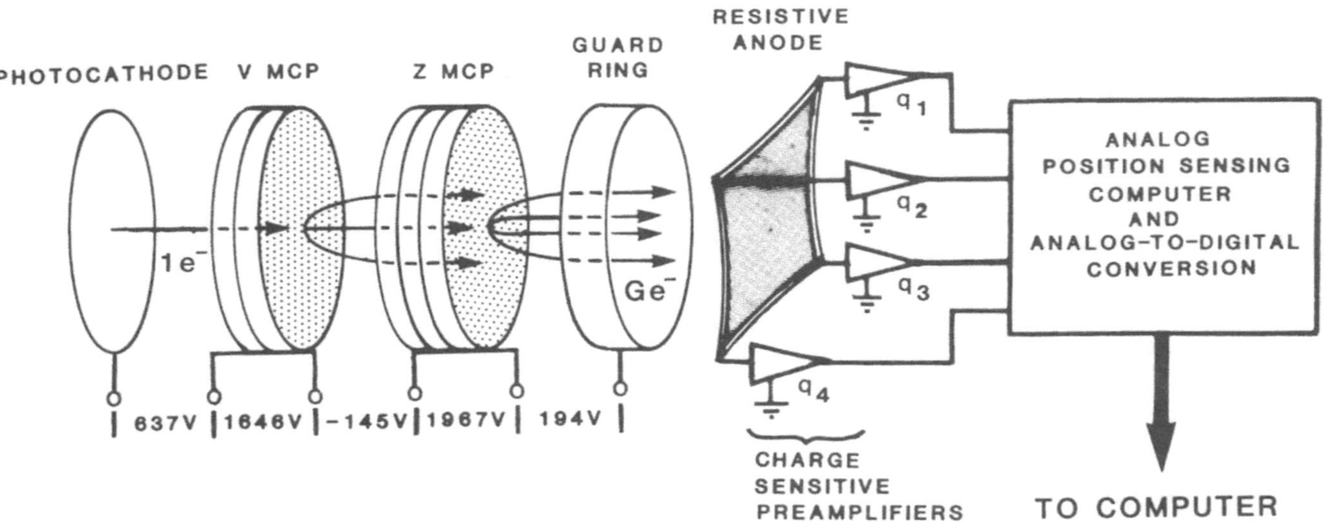


Fig. 4. Schematic diagram of the position sensitive resistive anode detector (Mepsicron). Typical MCP voltages are shown. [Fig. from ref. 14]

The low voltage threshold logic of the fifth preamp provides a major means of noise reduction. Thermionic emission from the MCP's, which is accumulated in the case of the intensified CCD and I-PDA, is ignored since these events do not produce a sufficient charge on the resistive anode. Only electrons emitted from the photocathode produce a sufficient charge on the resistive anode to be within the discriminator window of the threshold logic and thus be registered as events. When the photocathode is cooled to between -20°C and -30°C, the dark photoelectron rate is between 10 and 50 cps over the entire photocathode area. This corresponds to a dark count of $\approx 1 \times 10^{-5}$ cps/pixel with no effective readout noise.

The maximum count rate for the Mepsicron is $\approx 10^5$ cps. Thus, the Mepsicron is not so useful when the laser source has a low duty cycle, e.g., a Q-switched laser pulse with 20 ns duration and 10 Hz repetition rate. However, the Mepsicron is ideal for weak Raman signal detection with high repetition lasers, such as cw mode-locked lasers (e.g., at 80 MHz) or Q-switched mode-locked lasers (e.g., at 1 KHz). The threshold logic in the Mepsicron can also be used as an electronic gate. By externally introducing a threshold violation pulse into the control logic, the dark counts outside the gate window are ignored, resulting in an improvement in the signal-to-noise ratio. The spatial resolution of the Mepsicron is limited by two factors: (1) the lateral velocity of the photoelectrons; and (2) the thermal noise in the resistive anode. Using a line source input results in a Gaussian profile output with a FWHM of 61.7 µm.[14]

For the third-order Si phonon Raman data shown in Fig. 3, the integration time is also three hours, and no noise reduction procedures are necessary because of the lack of readout noise and the exceptionally low dark current ($\approx 10^{-5}$ cps/pixel) in the Mepsicron. If the entire area of the resistive anode is used, the dark current exhibits a spatial dependence because the photocathode and the MCP's are circular and effectively inscribe the resistive anode. However, in many spectroscopic applications, the detector is configured in such a way that only a horizontal strip near the center of the resistive anode is used. In general, spatial corrections of the data are not required if the strip is <25% of the detector diameter (40 mm).

## 3. Electrochemical Surface Activation

The most standard form of activating the electrode for SERS is by an electrochemical treatment in one of several specific electrolytes. The electrode must undergo one or several oxidation and reduction cycles (ORC's) with potential stepping or potential ramping. Since Ag and Au electrodes are most commonly used in SERS studies, the similarities and differences of these two coinage metal electrodes are briefly presented in order to understand the surface activation processes.[6]

### 3.1. Ag ELECTRODES

During the anodic step or ramping, dissolution of the Ag electrode occurs, i.e., $Ag^0 \rightarrow Ag^+ + e^-$, and the surface is roughened by this oxidation or anodization process. The $Ag^+$ combines with the anions in the electrolyte, and the Ag surface becomes partially covered by an adherent but porous layer of nearly insoluble Ag salt (such as AgCl) whose insulating nature limits the electrode current. In typical SERS experiments, about $10^2$ to $10^3$ layers of Ag atoms are oxidized. Should the Ag salt not adhere, it will fall from the electrode as a precipitate. Should the Ag salt be slightly soluble (e.g., $Ag_2SO_4$) or highly soluble (e.g., $AgClO_4$), the adherent layer is much thinner and some or all of the Ag salt dissolves into the electrolyte.

During the cathodic step or sweep and after the electrode current has changed sign, the Ag salt is reduced, i.e., $Ag^+ + e^- \rightarrow Ag^0$. For the adherent and insoluble Ag salt layer, all the $Ag^+$ ions originally oxidized can be reduced to $Ag^0$. During the reduction phase, reconstruction of the Ag surface occurs in a manner similar to that for crystal growth from vapor where deposition onto the surface is followed by surface diffusion and incorporation at the growth sites. However, there are three major differences between the electrolytic and vapor growth models: (1) the driving force for Ag growth is provided by the applied potential, not by the supersaturation of the vapor; (2) the rate of deposition in the electrolytic case is much slower that in the vapor case; and (3) the surface diffusion rate of the $Ag^0$ adatoms can be considerably slowed and even stopped by the presence of adsorbed anions such as $Cl^-$.

Figure 5 shows a simplified picture of the stabilization of the $Ag^0$ adatoms by adsorbates under potential sweep control in a halide electrolyte,[6] such as KC1. Along curve R of the voltammogram in this figure, the $Ag^+$ ions are still being reduced on the Ag surface and most of the electrode is covered with $Cl^-$ anions. These adsorbed anions surround and immobilize the $Ag^0$ adatoms, causing surface diffusion to be slowed or even stopped. At points B and C in Fig. 5, some the $Cl^-$ anions are desorbed and spatial rearrangement of some of the $Ag^0$ adatoms is possible, such as the formation of $Ag^0$ clusters. The remaining $Ag^0$ adatoms/clusters can be stabilized by the remaining adsorbed $Cl^-$. At point D, more $Cl^-$ anions are desorbed and, consequently, more of the $Ag^0$ adatoms/clusters diffuse and are incorporated at surface defect sites. At point E in Fig. 5, the potential is negative of the potential of zero charge (pzc) and some contact adsorbing cations with lower hydration energy (e.g., $K^+$, $Cs^+$ and $Rb^+$) can also stabilize the $Ag^0$ adatoms/clusters. However, not many $Ag^0$ adatoms and clusters remain during the time the potential is around pzc because only a few anions and cations are adsorbed on the surface. Evidence exists that, by themselves, adsorbed $H_2O$ molecules are not capable of stabilizing the $Ag^0$ adatoms/clusters. Reversing the potential sweep to potentials more positive of the pzc cannot liberate the $Ag^0$ adatoms/clusters from the surface defect sites. New adatoms/clusters can be created only during a new oxidation phase of the electrode (along curve O of Fig. 5) or during the reduction of the $Ag^+$ ions purposely added to the electrolyte in the form of a soluble salt (e.g., $AgNO_3$).

Anions with large hydration energy that do not contact adsorb (e.g., $ClO_4^-$ and $BF_4^-$) cannot be effective in immobilizing $Ag^0$ adatoms/clusters after $Ag^+$ has been reduced on the electrode surface. Nevertheless, the electrode surface does become roughened during the oxidation phase, whether or not the resultant Ag salt layer is adherent and soluble. However, SERS experiments indicate that such roughening alone does not cause giant surface enhancement factors. A close similarity exists in controlling the $Ag^0$ adatom/cluster diffusion rates with an applied potential of the electrode and with temperature of the metallic film. Applying a potential positive of the pzc is analogous to decreasing the temperature to slow the diffusion rates. Furthermore, cycling the potential positive of the pzc or cycling the temperature can induce surface rearrangement of both the adsorbates and the $Ag^0$ adatoms/clusters. However, once the temperature of the film has been raised or the potential has been swept positive of the pzc, cooling the film or applying the potential negative of the pzc will not restore the surface enhancement factor without re-evaporating more Ag on a cold substrate or re-oxidizing the electrode in a contact absorbing electrolyte.

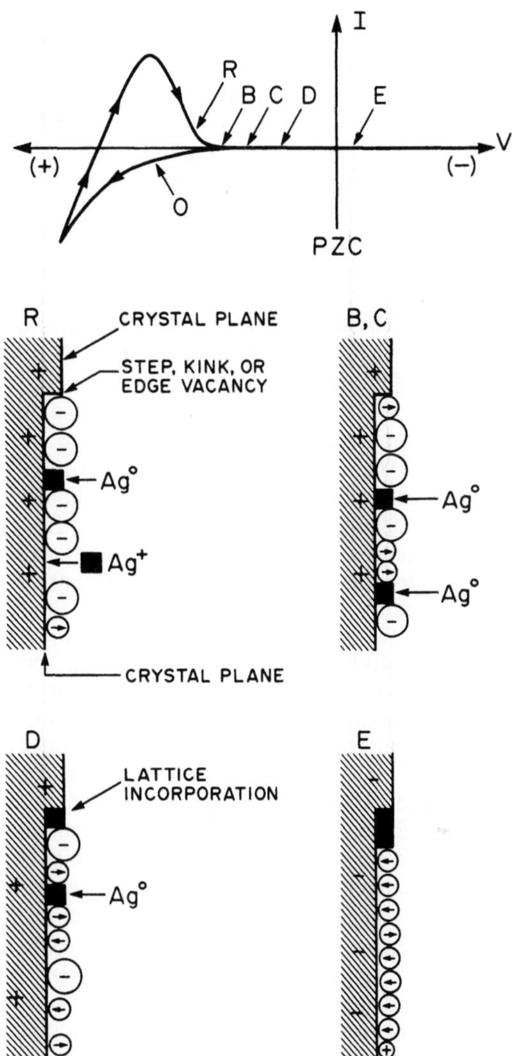


Fig. 5. The $Ag^0$ adatoms/clusters formed from the AgX layer and/or $Ag^+$ in the electrolyte are diffused along the surface and incorporated into surface defect sites, e.g., steps, kinks, edge vacancies, or holes. However, the potential-controlled absorption of anions, and possibly anion-molecule complexes can immobilize the adatoms and clusters (e.g., at potentials R, B, and C). When these adsorbates are only loosely held on the electrode (e.g., at potentials D and E), the surface diffusion of the $Ag^0$ adatoms/clusters can resume. [Fig. from ref. 6]

Several additional procedures can further increase the SERS intensity beyond the giant enhancement which occurs after ORC's. The following procedures can be combined or applied separately: (1) Keeping the laser beam on during the oxidation phase as the adherent Ag salt may be photoreduced and may nucleate morphologies more suitable for SERS than those created by electrochemical reduction in the dark;[17] (2) lowering the electrode and electrolyte temperature as the lower temperature decreases the surface diffusion of $Ag^0$ adatoms and clusters;[18] (3) increasing the electrolyte concentration, even for the same number of layers of oxidized Ag atoms, as more of the contact-adsorbing anions can stabilize the $Ag^0$ adatoms and clusters;[19] (4) waiting several minutes after the reduction current goes to zero while keeping the potential fixed as time appears to allow the optimum SERS morphology to evolve;[20] (5) vigorously generating $H_2$ gas just before the electrode activation cycle, as the clinging $H_2$ bubbles tend to clean the electrode surface and affect the surface morphology;[21] (6) adding molecules (such as pyridine or EDTA) to an electrolyte as such molecules tend to form complexes with $Ag^0$ adatoms/clusters and thereby stabilize their diffusion;[22] (7) optimizing the ORC with a number of potential steps rather than using a single ramping cycle causes the surface morphology to be more controllable and reproducible;[23] (8) adding $Ag^+$ to the electrolyte[24] as fresh $Ag^0$ sites are continuously produced by the electrochemical reduction of the diffusion controlled $Ag^+$; (9) adding an oxidizing salt to the electrolyte as the combined effect of chemical oxidation and electrochemical reduction generates a number of $Ag^0$ adatoms/clusters in steady state.[25]

There are several ways to decrease the surface enhancement on the Ag electrode. At first, it seems somewhat absurd to study ways of decreasing the SERS intensity. However, in the quest to understand the enhancement mechanism, the processes leading to the decrease of SERS are as informative as those which lead to the increased SERS intensity. The following procedures to decrease the SERS signal can also be combined or applied separately: (1) Applying a negative potential step or sweep beyond the pzc, which causes the contact-adsorbing anions to be desorbed and, therefore, no longer capable of stabilizing the $Ag^0$ adatoms/clusters;[26,27] (2) increasing the temperature of the electrode and electrolyte, which increases the surface diffusion rate;[28] (3) using the underpotential deposition (UPD) technique to deposit a non-SERS-active metal, such as Tl which appears to have affinity to SERS-active sites and to catalytically reduce $Ag^+$ complexes.[29] The UPD of Pb covers the Ag surface nonselectively and causes electromagnetic dampening of the metallic microstructure "antennas;"[30] (4) adding salts such as $S_2O_3^{2-}$ that form water soluble complexes with $Ag^+$ and thus cause less $Ag^+$ to be electrochemically reducible.[31]

The main disadvantage of the Ag electrodes is that the stability of the SERS-active sites and the potential-controlled coverage of the adsorbates can be decoupled. As a function of potential, the experimentalist is never certain if the decrease of the detected Raman intensity is associated with the decrease of the adsorbate coverage or with the decrease of the $Ag^0$ adatoms/clusters, which are also dependent on the adsorbate coverage for their immobilization.

## 3.2. Au ELECTRODES

From an electrochemical point of view, Au electrodes are more versatile than Ag electrodes since Au electrodes provide a wider polarizable potential window ($\approx 2$ V) even in aqueous media. Au electrodes were not used during the early phase of SERS development for two reasons: (1) After undergoing the usual electrochemical activation procedure described above for Ag electrodes, the SERS signal from Au electrodes is much weaker than that from Ag electrodes; and (2) Au electrodes require the use of red laser wavelengths (i.e., from a krypton ion laser) and red sensitive detectors rather than the more available blue and green laser wavelengths (i.e., from an argon-ion laser) and green sensitive detectors.

A straightforward electrochemical pretreatment procedure for Au electrodes has recently been reported.[32] This procedure yields extremely stable and intense SERS signals from a variety of adsorbates in a gas-phase, aqueous or nonaqueous media.[32] The fact that Au electrodes can be activated in one electrolyte and then be used in another electrolyte separates the issue of adsorbate-related morphology stabilization from the issue of adsorbate coverage with applied potential.

The pretreatment procedure for Au electrodes is similar to the ORC technique to activate Ag electrodes.[32,33] During the oxidation of Au electrodes in 0.1 M KCl (either by potential stepping or ramping to an anodic potential of ≈1 V), the dissolution of Au forms a predominantly soluble $AuCl_4^-$ complex. During the reduction phase, Au redeposition occurs via back diffusion of this soluble complex to the electrode surface. The optimum number of such ORC's depends on the relative cathodic charge passed per ORC. After this pretreatment, the Au electrode is rinsed with distilled water, transferred to another electrochemical cell of interest, and finally is potential controlled.

Had this been an Ag electrode, ending the ORC at a potential positive of the pzc or rinsing the electrode with distilled water, the SERS signal strength would have been greatly diminished as a result of $Ag^0$ adatom/cluster diffusion to defect sites. After the pretreatment of Au electrodes, noncontact-adsorbing anions (e.g., $ClO_4^-$) and cations (e.g., $H^+$) in the second electrochemical cell can be selected because these ions are not needed to stabilize the Au SERS-active sites. Consequently, the resultant SERS intensity from Au electrodes in a second electrochemical cell more closely reflects the potential controlled molecular or ionic coverage and there is no competition of the adsorbed species needed for active site stabilization.

It is not at all obvious why the Au electrode remains SERS active (after pretreatment), while the Ag electrode loses most of its surface enhancement. In scanning electron micrographs of pretreated Au electrodes and similarly electrochemically activated Ag electrodes, the metallic nodules appear to be similar in shape, size, and number density. Perhaps the use of a STEM can reveal the presence of small scale roughness in a pretreated Au electrode and the absence of small scale roughness in a similarly treated Ag electrode once the potential of both electrodes is positive of the pzc and is rinsed in distilled water.[32,33]

## 4. Theory

The classical description of the Raman scattering process for an isolated molecule involves the modulation of the electronic polarizability $\alpha_e$ by the interatomic distance R of the molecule which is vibrating with amplitude δR at frequency $\omega_v$, i.e., $\delta(\alpha_e)/\delta R \times \delta R$. The quantum description of the Raman Stokes scattering process involves a photon-induced virtual transition from the ground electronic level (with vibrational state at V = 0) to all the higher vibronic levels and photon-induced virtual transitions from these higher vibronic levels to the ground electronic level at V = 1. The initial state (thermally populated) of the Raman anti-Stokes scattering process is at V = 1 and the final state is at V = 0. Should the incident photon be less than any of the higher electronic levels (i.e., for the nonresonant Raman condition), the intensity ratio of the anti-Stokes and the Stokes is proportional to the $n_{BE}(\omega_v)/[1 + n_{BE}(\omega_v)]$, where $n_{BE}(\omega_v)$ is the Bose-Einstein function $n_{BE}(\omega_v) = [\exp(\hbar\omega_v)/kT$

- $1]^{-1}$. This intensity ratio is often used to deduce the temperature of the molecule, provided the nonresonant Raman condition is fulfilled.

The differential Raman scattering cross section per molecule ($\delta\sigma/\delta\Omega$) is extremely small (e.g., $10^{-29}/cm^2$). Raman scattering experiments on atomically smooth Ag faces in the UHV environment indicate that the effective $\delta\sigma/\delta\Omega$ for the adsorbates is not noticeably enhanced.[34] If low f-number collection optics, high throughput spectrographs, and long integration times are used with the I-PDA, the Raman spectra from submonolayers of adsorbates are just barely detectable.[34] This experimental result implies that the following enhancement mechanisms do not contribute in a major way to the giant enhancement factor (e.g., $10^4$ - $10^6$) that is normally observed from SERS-activated Ag and Au electrodes: (1) the image dipole mechanism involving the adsorbate and its image induced in the metal; (2) the modulated reflectivity mechanism associated with the vibration of the adsorbates; (3) the chemisorption mechanism involving the Ag surface and the adsorbate; (4) the resonance Raman enhancement mechanism involving the surface states of single crystal metal; and (5) the intercollation mechanism involving absorbates accumulated in surface crevices.

It is generally accepted that some form of surface roughness is necessary for giant enhancement in SERS, e.g., adatoms, clusters, and/or bumps. Recent experiments involving submonolayer UHV deposition of Ag on Rh (100) substrates show that Ag adatoms/clusters can enhance the Raman cross section by as much as 15-65 times.[35] However, without an in-situ technique that can measure the roughness (ranging from atomic- to micron-size roughness) and determine the number distribution as a function of roughness size, no quantitative SERS experiments on electrodes can verify the various existing roughness-based theoretical models which attempt to estimate the surface enhancement factor.

Most SERS investigators agree that there are two roughness-based models, the electromagnetic (EM) model and the charge transfer (CT) model. However, there is still disagreement over the relative and absolute enhancement factors for each. A summary of the EM and CT models follows.

## 4.1. EM MODEL

Since SERS is related to some form of surface roughness, the EM enhancement on a roughened Ag or Au electrode is simplified by considering the EM enhancement of a single isolated Ag or Au sphere or spheroid. In the EM model, an assumption is made that $\delta\sigma/\delta\Omega$ is unaffected by the proximity of the molecule to the microstructure. The enhancement factor results from the increase in the incident field intensity with wavelength $\lambda_i$ and, using a time reversal argument, from a similar increase in the Raman scattered field intensity with wavelength $\lambda_s$.

The input electric field enhancement is particularly easy to calculate in the Rayleigh limit, where only the dipole terms are important for a microstucture less than $\lambda_i$ or $\lambda_s$. The internal electric field is related to the applied electric field as follows:

$$E_{in} = (E_{applied})/[1 + \{[\varepsilon(\omega)/\varepsilon_0] - 1\}A] \qquad (1)$$

where A is the depolarization factor and is assumed to be size independent but shape dependent, i.e., A = 1/3 for a sphere and A = 0.174 for a prolate spheroid with axial ratio of 2. The dielectric constant of the electrolyte is $\varepsilon_0$ and that of the metallic microstructure is $\varepsilon(\omega)$. The external electric fields on the surface are related to $E_{in}$ by the usual boundary conditions for the normal and tangential components across the metal-electrolyte interface. Consequently, the external electric fields are much larger near the poles of the sphere or the tips of the spheroids (aligned parallel to the polarization of $E_{in}$) than at the equator. This enhancement at the poles or tips is often referred to as the "lightening-rod" effect.

$E_{in}$ becomes large at $\lambda_i$ when $Re[\varepsilon(\omega)] = (1 - 1/A)\varepsilon_0$ (commonly referred to as the surface plasmon condition) and when $Im[\varepsilon(\omega)]$ is small. Ag, Au and Cu are unique in SERS considerations for two reasons: (1) $Re[\varepsilon(\omega)] = (1 - 1/A)\varepsilon_0$ can be satisfied in the visible where tunable dye laser emission occurs and sensitive detectors exist; and (2) the accompanying $Im[\varepsilon(\omega)]$ is small. These two features of $\varepsilon(\omega)$ for the coinage metals can be readily understood by considering the two opposing contributions to $\varepsilon(\omega)$ from the nearly free electrons and the interband d-electrons.[36] The overall Raman enhancement factor is proportional to the product of the intensity enhancement factor at $\lambda_i$ and a similar enhancement factor at $\lambda_s$.

The relationship between $E_{in}$ and $E_{applied}$ becomes size dependent when the particle size increases to $0.1\lambda_i$ or larger. Retardation and multipole effects must be considered as the size increases beyond the Rayleigh limit. The dipolar contribution to $E_{in}$ greatly diminishes, shifts to a longer wavelength, and broadens. The quadrupolar contribution to $E_{in}$ appears at a wavelength on the blue side of the dipolar contribution with an amplitude much less than that estimated from Eq. (1). The intensity enhancement factor for larger size spheres and spheroids has been calculated as a function of $\lambda$ and size for a variety of transition and coinage metals.[37-42] Comparison has been made between the experimental and theoretical results[43] for the dipolar and quadrupolar contributions to the elastic scattering and extinction from a two-dimensional array of large Au microstructures (in the shape of a hemisphere, diameter $\approx 0.2$ µm, and center-to-center separation of 0.4 µm or 0.6 µm) immersed in air and various liquids in order to change $\varepsilon_0$.

Scanning electron micrographs of an electrochemically activated electrode reveal that the size of the metallic microstructures is beyond the Rayleigh limit and that they are definitely clustered. The validity of applying simple EM calculations of individual isolated metal particles to the case of actual surface roughness is undoubtedly questionable. Nevertheless, the EM model of individual particles provides a guide to selecting the incident wavelength to optimize SERS from electrodes of different metals. For closely spaced spheres, interactive effects must be considered. In fact, recent calculations indicate that interactive effects start to be important when individual microstructures are separated by less than two diameters apart.[44] In particular, the wavelengths of the dipolar and quadrupolar plasmon peaks for closely spaced spheres are shifted relative to those for isolated spheres.

In order to make the calculations more applicable to electrochemically activated Ag electrodes, other microstructure geometries have been considered, including the following: (1) Hemispheres with a Gaussian size distribution on a conducting surface;[45] (2) gratings with height comparable to the spacing rather than the usual approximation assuming that the height is much less than the spacing;[46] and (3) 2-, 4-, and 6- closely spaced sphere configurations that mimic cavities or pores which are believed to give rise to the EM enhancement factor.[47,48] For an actual electrode, it is hard to predict from the single isolated sphere calculation the optimal wavelength at which the EM enhancement will be the largest, much less knowing the exact enhancement factor for the electrode. The only certainty about the EM enhancement mechanism is that, when the wavelength is tuned to a region where the $\mathrm{Im}[\varepsilon(\omega)]$ is large (as a result of the interband transitions in the bulk metal),[36] the EM plasmons associated with the electrode microstructure are damped, and the expected enhancement factor diminishes.

## 4.2. CT MODEL

Non-EM related enhancement models are called the CT model, which is often referred to as the chemical effect, local effect, short-range effect, adatom-related or atomic-scale roughness effect, cluster-related or cluster-scale roughness effect, or electron transfer excitation. Surface enhancements from the EM and CT models are additive. A complete description of the CT model with ample experimental evidence in support of it has been reported.[7] Both the CT and EM enhancement mechanisms are currently accepted by SERS researchers even though some scientists continue to attempt to isolate the two mechanisms in order to quantify the surface enhancement factor associated with each model.

The distinguishing feature of the CT model is that the $\delta\sigma/\delta\Omega$ is enhanced. Whereas the EM model emphasizes the effective surface intensity, the CT model emphasizes the increase in the $\delta\sigma/\delta\Omega$ through resonant transitions to the excited states accessible by the incident laser photons. Thus, the CT model invokes a resonance Raman effect between the ground electronic state $|g\rangle$ and the excited electronic state $|j\rangle$ in this form:

$$\delta\sigma/\delta\Omega = (1/4\pi)\,(\hbar\omega_S)^4\,|\alpha_R|^2 \qquad (2)$$

$$\alpha_R = \Sigma \; \frac{\langle f\,|\,H'\,|\,j\rangle\langle\, j\,|\,H'\,|\,g\rangle}{(E_j - E_g) - \hbar\omega_i - i\Gamma_j}$$

where H' is the standard electron-photon Hamiltonian, $\Gamma_j$ is the linewidth of $|j\rangle$ and $\Sigma$ extends to all $|j\rangle$. In the CT model, the enhancement factor is determined by the matrix elements and by $\Gamma_j$, when the incident photons with energy $\hbar\omega_i = (E_j - E_g)$ can reach $|j\rangle$. The CT enhancement is large when the matrix elements are large and when $\Gamma_j$ is narrow. For the nonresonant Raman case, $\Sigma$ extends to all $|j\rangle$ levels that are not reached by $\hbar\omega_i$ [i.e., $(E_j - E_g) - \hbar\omega_i > 0$] and, thus, the linewidths of higher states $|j\rangle$ are not important in calculating the nonresonant contribution to $\delta\sigma/\delta\Omega$.

The main unknowns in the CT model are the appropriate $|g\rangle$ and $|j\rangle$ states and the $\Gamma_j$ values to be used in estimating the enhancement of $\delta\sigma/\delta\Omega$. For the case of an atomically smooth metal surface with a physisorbed molecule, $|g\rangle$ can be the metal Fermi level at $E_F$, rather than the ground state of the molecule. Further, $|j\rangle$ can be the molecular excited state which is located at $(E_j - E_F)$ above the Fermi level. Even though $\hbar\omega_i$ is not resonant with $|j\rangle$ starting in the molecular ground electronic state with energy at $E_g$, $\hbar\omega_i$ can be resonant with $(E_j - E_F)$, i.e., $\hbar\omega_i \approx (E_j - E_F) < (E_j - E_g)$. The $\Gamma_j$ for the molecules can be broadened even in the physisorbed case, since there are more decay channels available to shorten the lifetime of $|j\rangle$ (e.g., electron tunneling between the molecular orbitals $|j\rangle$ and the metal). The appropriate $|g\rangle$ can also be the ground state of the molecule and $|j\rangle$ can be states near the $E_F$ of the metal.[49] If the molecule is chemisorbed, some degree of hybridization can exist between the metal wavefunctions and the molecule orbitals, i.e., $|j\rangle$ is a hybridized state and, consequently, the matrix element in Eq. (3) can be increased by chemisorption.

However, chemisorption can further broaden $\Gamma_j$, which will decrease the total CT enhancement for $\hbar\omega_i \approx (E_j - E_F)$.

For both the physisorbed and chemisorbed cases with atomically smooth surfaces, the electrochemical potential can change the alignment between $E_j$ and $E_F$. The excitation spectrum for the SERS intensity can, therefore, shift with different potentials. Since experiments with atomically smooth Ag surfaces show that $\delta\sigma/\delta\Omega$ is not noticeably increased,[34] it is concluded that the resonant Raman enhancement involving such surfaces, including the existence of surface states that can be reached by the incident photon, is not of great importance.

Several variations of the CT model have been reported which explain the experimental results indicating that the SERS is large on roughened coinage metal electrode surfaces, on Ag deposited on Pt electrodes, and on submonolayers of Ag deposited on Rh substrates (Pt and Rh are known not to exhibit EM enhancement).[35,50,51] All these variations invoke some form of $Ag^0$ adatom[7] and $Ag_4^+$ clusters[52] which need to be stabilized on the electrode surface or substrate. The resonant Raman effect involves the $| g\rangle$ and $| j\rangle$ of such clusters and adsorbate complexes, since the electronic levels of these complexes (e.g., $Ag^0$ adatom, $Ag_4^+$ clusters, or multiatom metal structures covered by or embedded in $Cl^-$ ions, pyridine, or $H_2O$ molecules) can differ from the electronic levels of free molecules, metal surface, and bulk metal.[53,54] Since the vibrational frequencies of most molecules (over $10^2$ different types) observed in SERS experiments are quite similar to those of the free molecules, the implication is that only weak bonds are involved in such clusters and adsorbate complexes. Further, the implication is that it is unlikely that major modification of the molecular electronic states occurs on SERS-active electrodes.

For metal clusters consisting of tens of atoms, the associated electronic band structure approaches that of a bulk metal except that the k values for such clusters are not defined within a range of $\Delta k$, consistent with the Heisenberg uncertainty principle (i.e., $\Delta k \approx 2\pi/r$, where r is the cluster size).[11] The cluster can also be envisioned as localized roughness that supplies $\Delta k$ to the incident and Raman scattered radiation. This $\Delta k$ then allows intraband transitions within the nearly free s-band of the coinage metals that, for the smooth surface case, are normally not allowed due to momentum and energy conservation requirements. The CT process in this $\Delta k$-relaxed model involves the following steps: (1) the incident photon (with $\Delta k$ provided by the cluster) can now induce an intraband transition in the metal cluster; (2) the electron transfers from the metal cluster to the excited electronic level of the adsorbate ($| j\rangle$ with energy $E_j$) via tunneling for physisorbed adsorbates and via hybridization for chemisorbed adsorbates; (3) the electron transfers back from the excited adsorbate level to the metal cluster, leaving the adsorbate in the higher vibrational level of the molecular electronic ground state; and (4) photon emission with $\omega_s$ as the electron radiatively recombines with the hole originally created in step (1).

A similar process can take place involving the hole in the cluster. Once $\omega_i$ has created an electron-hole pair (e-h) in the cluster, an electron from the ground electronic level of the adsorbate ($| g\rangle$ with energy $E_g$) can charge transfer into this hole and then charge transfer back to $| g\rangle$, leaving the molecule in the excited vibrational state. The electron in the metal cluster then radiatively recombines with this hole which has lost exactly one vibrational state. The CT process is resonant when $\hbar\omega_i \approx (E_j - E_F)$ or $\hbar\omega_i \approx (E_F - E_g)$. In addition, the resonance condition is satisfied when $\hbar\omega_s$ is nearly equal to these energy differences. In the resonance Raman language, the former case is referred to as the "input resonance" and the latter case as the "output resonance."

Throughout the various steps in the photon-electron transition and the charge transfer, it is assumed that the hole on the metal side does not lose a continuous range of energy (e.g., via multiple hole-phonon interactions). Had such loss of hole energy occurred in the nearly free electron band of the metal, the radiative recombination between e-h would exhibit not only discrete peaks due to the vibrational excitation of the molecule but also a continuum (on the Stokes-shifted side) due to the continuous energy gain by the phonons in metal.[55,56] A continuum on the anti-Stokes side results if the hole gains energy from the phonons in the metal. The ratio of the anti-Stokes to Stokes intensity of the continuum at any inelastic frequency shift $\Delta\omega$ will be proportional to $n_{BE}(\Delta\omega)/[1 + n_{BE}(\omega)]$, where $n_{BE}(\Delta\omega)$ is the Bose-Einstein function $n_{BE}(\Delta\omega) = [\exp(\hbar\Delta\omega)/kT - 1]^{-1}$.

Experimental measurement of the continuum intensity of the anti-Stokes and Stokes (from as low as 3 $cm^{-1}$ to above 250 $cm^{-1}$ the incident frequency) has been correlated with the SERS intensity from the metal-halide stretching mode and the elastic scattering from the same electrode during the same ORC's.[57,58] Experimental results clearly indicate that the inelastic continuum due to continuous energy loss to, or gain from, the phonons on the metal side is another manifestation of SERS that should be considered along with the more commonly discussed discrete peaks due to the vibrational energy gain in the adsorbed molecule.[59,60]

The discrete vibrational mode need not be associated with the adsorbed molecule only. During the CT process, the electron can lose or gain energy by creating or absorbing one acoustic phonon on the metal side. The Raman shift must be commensurate with the metal cluster phonon dispersion curve, which can resemble that of the bulk metal with the k-vector defined only within a range of $\Delta k$. The observed low frequency at $\approx 8$ $cm^{-1}$ has been assigned to such Ag acoustic branch phonon loss or gain.[57-60] For atomic size clusters, $\Delta k$ can extend up to the Brillouin zone edge (located at $k = \pi/a$, where a is the Ag atomic spacing). The corresponding low frequency mode (due to acoustic phonons) in the SERS spectrum should then resemble the acoustic phonon density of states, since all modes within the Brillouin zone can be reached by the input and scattered photons.[61] For the case of $Ag_4^+$ clusters,[62] the electron during the CT process can lose or gain energy by creating or absorbing the vibrational modes of such a $Ag_4^+$ cluster at 73, 110, and 161 $cm^{-1}$. Should the adsorbate be an anion (e.g., a halide ion $X^-$), the discrete loss of energy during the CT process can also occur by creating or absorbing the stretching mode of the $Ag_4^+$ cluster and the $X^-$.

In conventional resonance Raman studies, many orders of overtones are readily observed (e.g., see the case for $I_2$ vapor).[63] The occurrence of multiorder Stokes shifts from isolated molecules is explained as follows: (1) for the resonance Raman scattering case, after the photon-induced transition to $|j\rangle$, the electron in $|j\rangle$ "lives" long enough for the molecule to undergo one or several vibrations, hence, the much larger $\delta\sigma/\delta\Omega$ for the vibrational mode and its overtones; and (2) for the nonresonant Raman scattering case, the photon-induced transition is virtual and thus "lives" in $|j\rangle$ for a time short compared to one vibrational period, hence, the much smaller $\delta\sigma/\delta\Omega$ for the vibrational mode and its overtones.

Even though the CT model invokes resonance Raman scattering with $|j\rangle$, the overtone SERS intensity is still very weak. One plausible argument for the weak overtone SERS intensity is that, after the CT process, the limiting lifetime is not how long the electrons live in the $|j\rangle$ of the adsorbate but how long the hole lives in the metal cluster. The hole-phonon interaction causes the hole in the cluster to continuously lose or gain energy and, thereby, to give rise to the SERS continuum rather than the discrete SERS overtone of the adsorbate vibrational mode.

Although the CT enhancement mechanism is by now accepted as readily as the EM enhancement mechanism, it will be some time before quantitative estimates can be made of the surface enhancement factor using current CT models. The initial appeal of the EM enhancement mechanism is in its simplicity to calculate the surface enhancement factor. However, EM calculations for the closely spaced and large size microstructures on SERS-activated electrodes will need to be developed further before quantitative estimates can be made of the surface enhancement factor.

## 5. Conclusion

While existing experimental results have explored different adsorbates, metal electrodes, and electrochemical parameters that give rise to SERS, no definitive experiment has yet been reported that provides direct correlation between the enhancement factor and the surface morphology (from angstroms to micrometers). Existing theoretical models have indicated the plausible mechanisms whereby an absorbate can exhibit increased surface electromagnetic intensity and an enhanced Raman scattering cross section. Validation of the theoretical predictions is not complete, however, and is usually only qualitative. Major advances in understanding the SERS field are, therefore, yet to come.

Raman instrumentation is now sufficiently improved that it is possible to detect the submonolayers of adsorbates on substrates that do not give rise to surface enhancement. Whenever present, surface enhancement always simplifies an experiment and provides the option of real-time measurements as specific parameters are temporally varied. Even at the present stage, the SERS technique can be utilized as a means to probe specific molecular and configuration details at the electrode-electrolyte interface under electrochemical potential control. Pretreatment of Au electrodes and recent advances in lithographically produced metal microstructures also provide a convenient, stable substrate for studying surface chemistry in a gaseous or vacuum environment. The frequent application of SERS as an interfacial probe is assured, and studies to understand and quantify surface enhancement mechanisms will continue to attract a new generation of scientists during the next decade, particularly when SERS-active electrodes can be examined by other diagnostic techniques such as STEM, X-ray diffraction, and IR absorption.

## 6. References


1. M. Fleischmann, P.J. Hendra, and A.J. McQuillan, "Raman Spectra of Pyridine Adsorbed at a Silver Electrode," *Chem. Phys. Lett.* **26**, 163 (1974).

2. D.L. Jeanmaire and R.P. Van Duyne, "Surface Raman Spectroelectrochemistry. I. Heterocyclic, Aromatic, and Aliphatic Amines Adsorbed on the Anodized Silver Electrode," *J. Electroanal. Chem.* **84**, 1 (1977).

3. R.P. Van Duyne, "Laser Excitation of Raman Scattering from Adsorbed Molecules on Electrode Surfaces," in *Chemical and Biochemical Applications of Lasers*, Vol. 4, C.B. Moore, ed. (Academic Press, New York, 1979), p. 101.

4. T.E. Furtak and J. Reyes, "A Critical Analysis of Theoretical Models for the Giant Raman Effect from Adsorbed Molecules," *Surf. Sci.* **93**, 351 (1980).

5. R.K. Chang and T.E. Furtak, eds., *Surface Enhanced Raman Scattering* (Plenum Press, New York, 1982).

6. R.K. Chang and B.L. Laube, "Surface Enhanced Raman Scattering and Nonlinear Optics Applied to Electrochemistry," *CRC Critical Rev. Solid State Materials Sci.* **12**, 1 (1984).

7. A. Otto, "Surface Enhanced Raman Scattering, 'Classical' and 'Chemical' Origins," in *Light Scattering in Solids IV, Topics in Applied Physics,* Vol. 54, M. Cardona and Güntherodt, eds. (Springer-Verlag, Berlin, 1984), p. 289.

8. M. Moskovits, "Surface-Enhanced Spectroscopy," *Rev. Mod. Phys.* **57**, 783 (1985).

9. D.A. Weitz, M. Moskovits, and J.A. Creighton, "Surface Enhanced Spectroscopy with Emphasis on the Liquid-Solid Interface," in *Chemistry and Structure at Interfaces - New Laser and Optical Techniques,* R.B. Hall and A.B. Ellis, eds. (VCH Publishers, Deerfield Beach, FL, 1986), p. 197.

10. S. Efrima, "Surface Enhanced Raman Scattering (SERS)," in *Modern Aspects on Electrochemistry,* Vol. 16, R.E. White, J. O'M. Bockris, and B.E. Conway, eds. (Plenum Press, New York, 1985), p. 253.

11. R.K. Chang, "Surface Enhanced Raman Scattering: A Status Report," *Ber. Bunsenges. Phys. Chem.* **91**, 296 (1987).

12. R.K. Chang and M.B. Long, "Optical Multichannel Detection," in *Light Scattering in Solids II, Topics in Applied Physics,* Vol. 50, M. Cardona and G. Güntherodt, eds. (Springer-Verlag, Berlin, 1982), p. 179.

13. K.F. Wall and R.K. Chang, "$I_2$-Vapor Notch Filter with Optical Multichannel Detection of Low-Frequency-Shift Inelastic Scattering from Surface-Enhanced Raman-Scattering Active Electrodes," *Opt. Lett.* **11**, 493 (1986).

14. W.P. Acker, B. Yip, D.H. Leach, and R.K. Chang, "The Use of a Charge-Coupled Device and Position Sensitive Resistive Anode Detector for Multiorder Spontaneous Raman Spectroscopy from Silicon," *J. Appl. Phys.,* in press.

15. J.C. Tsang, "Raman Spectroscopy of Thin Films on Semiconductors," in *Dynamics on Surfaces,* B. Pullman, J. Jortner, A. Nitzan, and B. Gerber, eds. (D. Reidel Publishing Co., Dordrecht, 1984), p. 379.

16. C.A. Murray and S.B. Dierker, "Use of an Unintensified Charge-Coupled Device Detector for Low-Light-Level Raman Spectroscopy," *J. Opt. Soc. Am. A* **3**, 2151 (1986).

17. F. Barz, J.G. Gordon II, M.R. Philpott, and M.J. Weaver, "Effect of Laser Illumination during Oxidation-Reduction Cycles upon Surface-Enhanced Raman Scattering from Silver Electrodes," *Chem. Phys. Lett.* **91**, 291 (1982).

18. B. Pettinger and L. Moerl, "Surface Enhanced Raman Scattering from $H_20$ and $D_20$ at Silver Electrodes. Influence of Temperature," *J. Electroanal. Chem.* **150**, 415 (1983).

19. J.F. Owen and R.K. Chang, "SERS of $H_20$ Adsorbed on a Ag Electrode: The Role of High Salt Concentration," *Chem. Phys. Lett.* **104**, 510 (1984).

20. A.M. Stacy and R.P. Van Duyne, "Time Resolved Studies of Electrochemical SERS: The Pyridine / $C1^-$ / Ag Model System," in *Time-Resolved Vibrational Spectroscopy,* G.H. Atkinson, ed. (Academic Press, New York 1983), p. 377.

21. F. Barz, J.G. Gordon II, M.R. Philpott, and M.J. Weaver, "Intense Electrochemical SERS Signal Following Hydrogen Evolution," *Chem. Phys. Lett.* **94**, 168 (1983).

22. J.F. Owen and R.K. Chang, "Increased Enhancement in Surface Enhanced Raman Scattering from Ag Electrodes with the Addition of Pyridine to the Electrolyte," *Chem. Phys. Lett.* **104,** 59 (1984).

23. D.D. Tuschel, J.E. Pemberton, and J.E. Cook, "SERS and SEM of Roughened Silver Electrode Surfaces Formed by Controlled Oxidation-Reduction in Aqueous Chloride Media," *Langmuir* **2**, 380 (1986).

24. H. Wetzel, H. Gerischer, and B. Pettinger, "Surface Enhanced Raman Scattering from Silver-Halide and Silver Pyridine Vibrations and the Role of Silver Ad-Atoms," *Chem. Phys. Lett.* **78**, 392 (1981).

25. P.B. Dorain, "The Observation with Surface-Enhanced Raman Scattering of the Sequential Electrochemical Formation of Adsorbed $O^{2-}$, $OH^-$, and $H_20$ on a Ag Electrode," *J. Phys. Chem.* **90**, 5808 (1986); "Molecular Precursors to the Formation of $H_2$ and the Reduction of $O_2$ at a Ag Cathode: A Surface-Enhanced Raman Scattering Experiment," *J. Phys. Chem.* **90**, 5812 (1986).

26. M.R. Philpott, F. Barz, J.G. Gordon II, and M.J. Weaver, "Time Dependence of Surface Enhanced Raman Scattering during and after Oxidation-Reduction Cycles," *J. Electroanal. Chem.* **150**, 399 (1983).

27. J.F.Owen, T.T. Chen, R.K. Chang, and B.L. Laube, "Irreversible Loss of Adatoms on Ag Electrodes during Potential Cycling Determined from Surface Enhanced Raman Intensities," *Surf. Sci.* **131**, 195 (1983).

28. S. H. Macomber and T.E. Furtak, "The Influence of Temperature on Surface Enhanced Raman Scattering in the Electrochemical Environment: Evidence for Adatoms," *Solid State Commun.* **45**, 267 (1983).

29. B. Pettinger and L. Moerl, "Influence of Foreign Metal Atoms Deposited at Electrodes on Local and Nonlocal Processes in Surface Enhanced Raman Scattering," *J. Electron Spectrosc. Related Phenomena* **29**, 383 (1983).

30. D.S. Kellogg and J.E. Pemberton, "Surface-Enhanced Raman Scattering of HCN at Pb-Modified Ag Surfaces," *J. Phys. Chem.* **91**, 1126 (1987).

31. T. Watanabe, O. Kawanami, K. Honda, and B. Pettinger, "Evidence for Surface $Ag^+$ Complexes as the SERS-Active Sites on Ag Electrodes," *Chem. Phys. Lett.* **102**, 565 (1983).

32. P. Gao, M.L. Patterson, M.A. Tadayyoni, and M.J. Weaver, "Gold as a Ubiquitous Substrate for Intense Surface-Enhanced Raman Scattering," *Langmuir* **1**, 173 (1985).

33. P. Gao, D. Gosztola, L.-W. H. Leung, and M.J. Weaver, "Surface-Enhanced Raman Scattering at Gold Electrodes: Dependence on Electrochemical Pretreatment Conditions and Comparisons with Silver," *J. Electroanal. Chem.* **233**, 211 (1987).

34. D.R. Mullins and A. Campion, "Unenhanced Raman Scattering from Pyridine Chemisorbed on a Stepped Silver Surface: Implications for Proposed SERS Mechanisms," *Chem. Phys. Lett.* **110**, 565 (1984).

35. X. Jiang and A. Campion, "Chemical Effects in Surface-Enhanced Raman Scattering: Pyridine Chemisorbed on Silver Adatoms on Rh (100)," *Chem. Phys. Lett.* **140**, 95 (1987).

36. H. Ehrenreich and H.R. Philipp, "Optical Properties of Ag and Cu," *Phys. Rev.* **128**, 1622 (1962).

37. B.J. Messinger, K.U. von Raben, R.K. Chang, and P.W. Barber, "Local Fields at the Surface of Noble Metal Microspheres," *Phys. Rev. B* **24**, 649 (1981).

38. P.W. Barber, R.K. Chang, and H. Massoudi, "Surface-Enhanced Electric Intensities on Large Silver Spheroids," *Phys. Rev. Lett.* **50**, 997 (1983).

39. P.W. Barber, R.K. Chang, and H. Massoudi, "Electrodynamic Calculations of the Surface-Enhanced Electric Intensities on Large Ag Spheroids," *Phys. Rev. B* **27**, 7251 (1983).

40. E.J. Zeman and G.C. Schatz, "An Accurate Electromagnetic Theory Study of Surface Enhancement Factors for Ag, Au, Cu, Li, Na, Al, Ga, In, Zn, and Cd," *J. Phys. Chem.* **91**, 634 (1987).

41. M.P. Cline, P.W. Barber, and R.K. Chang, "Surface Enhanced Electric Intensities on Transition and Noble Metal Spheriods," *J. Opt. Soc. Am. B* **3**, 15 (1986).

42. M. Meier and A. Wokaun, "Enhanced Fields on Large Metal Particles: Dynamic Depolarization," *Opt. Lett.* **8**, 581 (1983).

43. W.P. Acker, B. Schlicht, R.K. Chang, and P.W. Barber, "Quadrupole Mode in the Extinction and Elastic-Scattering Spectra of Gold Microstructures," *Opt. Lett.* **12**, 465 (1987).

44. M. Inoue and K. Ohtaka, "Enhanced Raman scattering by a two-dimensional array of dielectric spheres," *Phys. Rev. B* **26**, 3487 (1982).

45. E.J. Zeman and G.C. Schatz, "Electromagnetic Theory Calculations for Spheroids: An Accurate Study of the Particle Size Dependence of SERS and Hyper-Raman

Enhancements," in *Dynamics on Surfaces,* B. Pullman, J. Jortner, A. Nitzan, and B. Gerber, eds. (D. Reidel Publishing Co., Dordrecht, 1984), p. 413.

46. A. Wirgin and T. López-Ríos, "Can Surface-Enhanced Raman Scattering be Caused by Waveguide Resonances?," *Opt. Commun.* **48**, 416 (1984).

47. N. Liver, A. Nitzan, and J.I. Gersten, "Local Fields in Cavity Sites of Rough Dielectric Surfaces," *Chem. Phys. Lett.* **111**, 449 (1984).

48. E.V. Albano, S. Daiser, G. Ertl, R. Miranda, and K. Wandelt, "Nature of Surface-Enhanced-Raman-Scattering Active Sites on Coldly Condensed Ag Films," *Phys. Rev. Lett.* **51**, 2314 (1983).

49. B.N.J. Persson, "On the Theory of Surface-Enhanced Raman Scattering," *Chem. Phys. Lett.* **82**, 561 (1981).

50. J.E. Pemberton, "SERS from Thin Ag Films Electrochemically Deposited onto Pt Electrodes," *J. Electroanal. Chem.* **167**, 317 (1984).

51. J. Miragliotta and T.E. Furtak, "Enhanced Raman Scattering with One Monolayer of Silver," *Phys. Rev. B* **35**, 7382 (1987).

52. T.E. Furtak and D. Roy, "The Short-Range Mechanism of Surface Enhanced Raman Scattering," *Surf. Sci.* **158**, 126 (1985).

53. B. Pettinger, "Light Scattering by Adsorbates at Ag Particles: Quantum-Mechanical Approach for Energy Transfer Induced Interfacial Optical Processes Involving Surface Plasmons, Multipoles, and Electron-Hole Pairs," *J. Chem. Phys.* **85**, 7442 (1986).

54. B. Pettinger, "Quantum-Mechanical Model for Surface (Un-)Enhanced Resonance Raman Scattering of Adsorbates at Metals. Common Effects in SERRS and SUERS," *Ber. Bunsenges. Phys. Chem.* **91**, 308 (1987).

55. A. Otto, "Raman Spectra of $(CN)^-$ Adsorbed at a Silver Surface," *Surf. Sci.* **75**, L392 (1978).

56. V.V. Marinyuk, R.M. Lazorenko-Manevich, and Ya.M. Kolotyrkin, "The Background Signals in Raman Spectra of Silver-Electrode Surfaces," *Sov. Electrochem.* **17**, 527 (1981), translated by L.Ya. Karpov from *Electrokhimiya* **17**, 643 (1981).

57. K.F. Wall and R.K. Chang, "Separation of the Low-Frequency Mode from the Inelastic Scattering of a SERS Active Electrode," *Chem. Phys. Lett.* **129**, 144 (1986).

58. K.F. Wall, M.L.A. Temperini, and R.K. Chang, "Potential-Dependent Measurements of the Low-Frequency Mode, the Inelastic Continuum, Elastic Scattering, and SERS from a Ag Electrode," *Chem. Phys. Lett.* **129**, 253 (1986).

59. A.Z. Genack, D.A. Weitz, and T.J. Gramila, "Very Low Frequency Surface Enhanced Raman Scattering," *Surf. Sci.* **101**, 381 (1980).

60. D.A. Weitz, T.J. Gramila, A.Z. Genack, and J.I. Gersten, "Anomalous Low-Frequency Raman Scattering from Rough Metal Surfaces and the Origin of Surface-Enhanced Raman Scattering," *Phys. Rev. Lett.* **45**, 355 (1980).

61. I. Pockrand and A. Otto, "Surface Enhanced and Disorder Induced Raman Scattering from Silver Films," *Solid State Commun.* **37**, 109 (1981).

62. D. Roy and T.E. Furtak, "Vibrational Characteristics of Silver Clusters in Surface-Enhanced Raman Scattering," *Phys. Rev. B* **34**, 5111 (1986).

63. D.G. Fouche and R.K. Chang, "Observation of Resonance Raman Scattering below the Dissociation Limit in $I_2$ Vapor," *Phys. Rev. Lett.* **29**, 536 (1972).

# LASER RAMAN SPECTROSCOPY IN STUDIES OF CORROSION AND ELECTROCATALYSIS

C. A. Melendres
Materials Science and Chemical Technology Divisions
Argonne National Laboratory
Argonne, Illinois 60439
U.S.A.

**ABSTRACT.** Laser Raman Spectroscopy (LRS) has become an important tool for the in-situ structural study of electrochemical systems and processes in recent years. Following a brief introduction of the experimental techniques involved in applying LRS to electrochemical systems, we survey the literature for examples of studies in the inhibition of electrode reactions by surface films (e.g., corrosion and passivation phenomena) as well as the acceleration of reactions by electro-sorbates (electrocatalysis). We deal mostly with both normal and resonance Raman effects on fairly thick surface films in contrast to surface-enhanced Raman investigations of monolayer adsorbates, which is covered in another lecture.

Laser Raman spectroelectrochemical studies of corrosion and film formation on such metals as Pb, Ag, Fe, Ni, Co, Cr, Au, stainless steel, etc. in various solution conditions are discussed. Further extension of the technique to studies in high-temperature and high-pressure aqueous environments is demonstrated. Results of studies of the structure of corrosion inhibitors are also presented.

As applications of the LRS technique in the area of electrocatalysis, we cite studies of the structure of transition metal macrocyclic compounds, i.e., phthalocyanines and porphyrins, used for catalysis of the oxygen reduction reaction.

## 1. INTRODUCTION

Laser Raman Spectroscopy (LRS) has become an important tool for the study of electrochemical systems and processes because of the molecular specific information that it provides on the structure of the solid/solution interface in situ.[1,2a,b] The spectrum of the inelastically scattered light from a sample contains information related to the various normal modes of vibration of the system and constitutes a fingerprint by which the material can be identified. Such specific structural information nicely supplements the kinetic data contained in the conventional current-potential curve and is essential for arriving at a



*C. Gutiérrez and C. Melendres (eds.),*
*Spectroscopic and Diffraction Techniques in Interfacial Electrochemistry*, 181–222.

detailed interpretation of the mechanism of electron-transfer reactions. The technique also nicely complements vibrational information obtained by infrared spectroscopy. Numerous applications of LRS to electrochemistry have been made since Fleischmann and Hendra[2a] first reported use of the technique to study the composition of solid phases electrochemically formed on mercury. The study of anodically-formed surface phases and corrosion films continues to be an important application of the technique; work in this area since 1973 is surveyed here. We also cite applications of LRS to investigations of electrocatalysis. Techniques and instrumentation in Raman spectroscopy are well covered in another lecture; only a few important points are discussed here. We deal mostly with practical applications in order to illustrate the power of the technique. We also digress on the limitations and problems in its use to study interfacial electrochemical phenomena. Examples are taken from the literature and from the author's own work. The illustrations are not meant to be all inclusive, rather just of sufficient detail to be helpful to the beginner in the field.

## 2. THEORETICAL AND EXPERIMENTAL ASPECTS

When light of a single frequency $\nu_o$ strikes a sample (Fig. 1a), the scattered radiation is found to consist of a number of discrete frequencies which can be grouped into three categories, i.e. $\nu_o - \nu_{vib}$, $\nu_o$, and $\nu_o + \nu_{vib}$, where $\nu_{vib}$ is the frequency of a normal vibration[1,18] of the material. Raman spectroscopy consists in the measurement and analysis of the inelastically scattered light of frequency $\nu_o - \nu_{vib}$, $\nu_o + \nu_{vib}$, etc. The set of spectral lines of lower frequency than $\nu_o$, i.e., $\nu_o - \nu_{vib}$, are called the Stokes lines, while those with higher frequencies, $\nu_o + \nu_{vib}$ are referred to as anti-Stokes lines. The elastically scattered light of frequency $\nu_o$ is termed the Raleigh line. The change in energy states that accompany the scattering process is schematically illustrated in Fig. 1b, where a molecule is excited from the electronic state (with vibrational levels v = 0, 1) to a virtual higher electronic state. The corresponding Raman spectrum is shown schematically in Fig. 1c. From an analysis of such Raman spectra, it is possible to extract molecular electronic structural information from a calculation of the set of fundamental vibrational frequencies, $\nu_{vib}$.

The intensity of the scattered Raman light has been derived and shown to be[1b]

$$I_{scat} = k\,(\nu_o - \nu_{vib})^4\,(d\alpha_{zz}/dr)^2\,I_o \tag{1}$$

where $I_o$ is the incident light intensity, $(d\alpha_{zz}/dr)$ is the change of molecular polarizability $\alpha_{zz}$ with vibrational coordinate r, and k is a constant. It is important to note the dependence of the intensity on the fourth power of the frequency of the scattered light. Likewise, the essential requirement (selection rule) for observation of the Raman effect is that the polarizability of the molecule change with vibration, i.e., $d\alpha/dr \neq 0$.

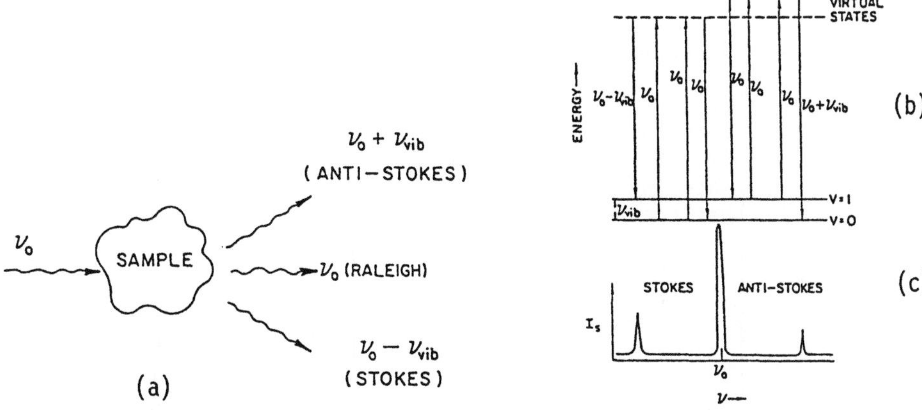


Figure 1. (a) Light scattering from a material; (b) Schematic energy level diagram illustrating transitions that give rise to Raman spectra; (c) Sample spectrum corresponding to (b).

The instrumentation required for LRS consists simply of a light source for exciting the sample, an analyzer to determine the wavelengths of the scattered light, and a detector to measure the intensity of the scattered radiation. The sample is contained in an electrochemical cell of suitable design. Commonly used source of high intensity and coherent light are lasers, e.g., $Ar^+$ or $Kr^+$ gas lasers, which provide discrete wavelengths in the visible region from about 450 to 680 nm. Dye lasers provide tunability over a wider range of wavelengths. Analysis of the frequency distribution of the scattered light is done with single-, double-, or triple-stage monochromators, which are outfitted with appropriate gratings and mirrors for dispersion of the incoming collected light beam. Choice of monochromators is often made in conjunction with the detector used. The latter can be a photomultiplier tube for single-channel photon counting or maybe one of a number of multichannel optical detectors, such as a diode array, silicon intensified target, or a silicon photon imaging detector. Further details on this subject are given in the paper by Chang in this proceedings volume. The ultimate aim of instrument design is to maximize the detection of the inelastically scattered light. The magnitude of this problem may be appreciated by realizing that this fraction is only about $10^{-6}$ that of the incident light intensity, $I_o$. The scattered light intensity, $I_{scat}$ from molecules on an electrode surface is given by,

$$I_{scat} = 2I_o \times N \times (\partial\sigma/\partial\Omega) \tag{2}$$

where N is the number density of molecules on the surface per $cm^2$ and $\partial\sigma/\partial\Omega$ is the differential cross section of the molecule for Raman scattering, usually in the range of $10^{-31}$ $cm^2$/steradian. The amount of light detected by a spectrometer system is

$$I_{det} = I_{scat} \times l \times t \times (\Omega/4\Pi) \times R \tag{3}$$

where l is the path length of the beam on the sample, t the time for data collection, $\Omega$ the solid angle subtended by the detector and R the instrument response. The latter is the amount of light detected divided by the amount entering the spectrometer. From equations (2) and (3), it follows that the detected light intensity is

$$I_{det} = k' \times N \times t \times R \times (\partial\sigma/\partial\Omega) \tag{4}$$

where the proportionality constants have been lumped into a single one, k'. Equation (4) shows how the Raman signal may be increased by increasing the density of molecules on the surface, the time for data accumulation, and the instrument response, R. The scattering cross section may be increased by making use of the resonance Raman effect.[1b] Veirs[2c] has compared the instrument response for a double monochromator fitted with a photomultiplier tube detector, a triple spectrometer equipped with an intensified photo-diode array, and a single monochromator with an imaging photomultiplier tube. He found R for these systems to be in the ratio of 1:13:300.

An experimental setup employed in the author's laboratory for in situ LRS of electrochemical systems is shown in Fig. 2. The three-compartment spectroelectrochemical cell[3] is fitted with a special high-purity quartz window inclined at an angle of 60° with respect to the horizontal. This is nearly the optimum angle for maximum scattering intensity for most metals as shown by Campion.[2b] The working electrode is brought to within 0.5 mm of this window to minimize the solution volume being traversed by the laser beam striking the electrode surface. The scattered radiation is collected at an angle of 90° with respect to the incident beam direction and then analyzed in the double monochromator. The light intensity is measured by a photomultiplier tube detector together with associated photon-counting electronics.

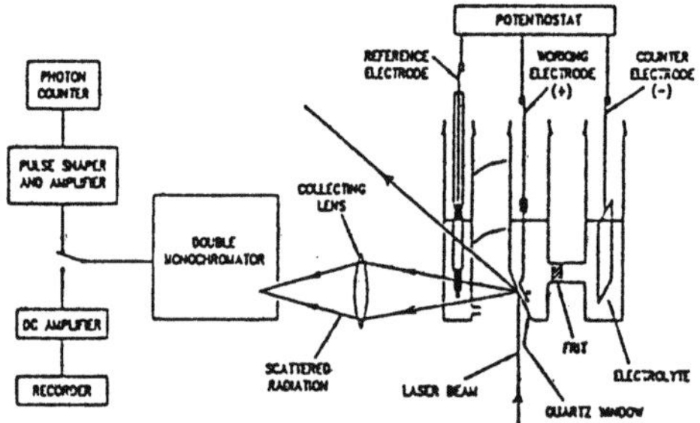


Figure 2. Experimental set-up for laser Raman spectroelectrochemical studies.

Of important consideration in the LRS experiments are the laser frequency for exciting the sample and the incident beam power. The latter is so chosen to be sufficiently intense to obtain a reasonably good signal without causing composition or structural changes in the sample. For a laser beam of 2 mm diameter incident on a sample, a power level of 50 to 150 mW is often adequate. The choice of excitation frequency is usually dictated by the laser sources available. Commonly used sources are $Ar^+$ and $Kr^+$ lasers with a number of usable wavelengths in the visible spectral range from about 460 nm to 650 nm. It is sometimes of great advantage to select a laser exciting frequency to be coincidental with the electronic absorption spectrum of the material, thus allowing for signal enhancement via the resonance Raman effect.

## 3. STUDIES OF CORROSION AND PASSIVATION FILMS ON METALS

The determination of the structure and composition of corrosion films on metals constitutes one of the most important practical application of laser Raman spectroelectrochemistry. An understanding of the relationships between composition and physico-electrochemical properties of passive films is essential to the technological development of methods for controlling the degradation of materials via corrosive mechanisms. Over a dozen metals have been studied by LRS under a variety of solution environments. Among these are Ag,[3] Au,[4] Pb,[5,6] Hg,[2] Cu,[7,8] Co,[9] Fe,[10-12] Ni,[13] and Fe-Cr alloys.[37] We review here results of studies on these systems and in the process illustrate the kind of information obtainable by the Raman technique.

### 3.1. Lead and Mercury

The corrosion films on lead and mercury were among the first to be studied by LRS for good reasons. High atomic number elements combine to form large molecules with highly polarizable electronic structures and hence have relatively large cross sections for Raman scattering. Raman spectral intensities are, therefore, high and detection sensitivities are better than for molecules composed of lower atomic number elements.

Fleischman and his group[2a] first studied the formation of solid phases on a Hg surface by LRS. By constant potential anodization in HCl and HBr solutions, they produced the surface phases $Hg_2Cl_2$ and $Hg_2Br_2$ respectively, which yielded characteristic Raman bands at 41, 168, and 273 $cm^{-1}$ for the chloride and 36, 135, and 218 $cm^{-1}$ for the bromide. HgO was prepared from NaOH solution and showed bands at 330 and 567 $cm^{-1}$.

Corrosion and film formation on Pb have been studied in aqueous chloride[5b] and sulfate[5a,6] solutions. Reid, Cooney, et al.[5b] investigated Pb in chloride media in order to verify the phases predicted by the Pourbaix diagram. They suspected the existence of the polymeric phase $[Pb(OH)]_n{\cdot}Cl_n$ which did not appear in the diagram. They studied films formed at different pH's as a function of applied potential in order to confirm their theory; they then proposed the inclusion of the phase $[Pb(OH)]_n \cdot Cl_n$ in the construction of the Pourbaix diagram of the $Pb/H_2O$ system. Investigations with a similar objective were undertaken by Thibeau et al.[5a,15] who found that the composition of surface films on Pb, which was corroded in 0.1 M sulfate, were not in complete agreement with the predictions of the calculated potential-pH diagram. The Pourbaix diagram[17] did not predict the formation of PbO under any condition; however, they obtained spectra of this phase at various potentials in acid and neutral solutions and at potentials above the immunity region in basic solutions. Pourbaix diagrams of Pb in the presence of $SO_4^{=}$ have been constructed by Thibeau et al.[15] and McMahon et al.[16] for sulfate activities of 0.1 and 6.4 X $10^{-4}$, respectively. The latter construction is presented in Fig. 3. As can be seen, both sulfate and

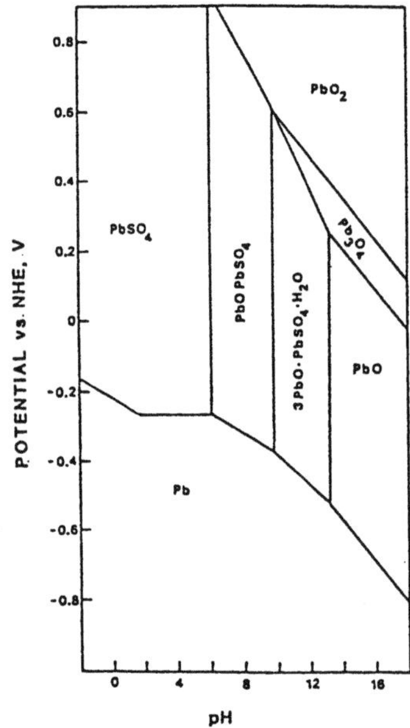


Figure 3. Pourbaix diagram of the Pb- $H_2O$ system at 25°C for a sulfate activity of 6.4 x $10^{-4}$ (from reference 16).

oxide surface phases are predicted to exist depending on the electrode potential and solution pH. We studied the corrosion of Pb in 0.05 M $H_2SO_4$ solution (pH=1.6) initially;[6] a cyclic voltammogram is shown in Fig. 4a. Typical active dissolution and passivation behavior is observed with the various regions indicated in the figure. The Pourbaix diagram predicts that controlled potential anodization at 0.16 V vs SHE (-0.45 V vs $Hg/Hg_2SO_4$) should lead to the formation of $PbSO_4$ on the surface. The Raman spectrum observed in-situ (Fig. 4b) shows the most intense bands of $PbSO_4$. Assignment of these bands was done by a comparison with the spectrum of a standard sample known to be $PbSO_4$ (Fig. 4c). Confirmation of this assignment was carried out by X-ray

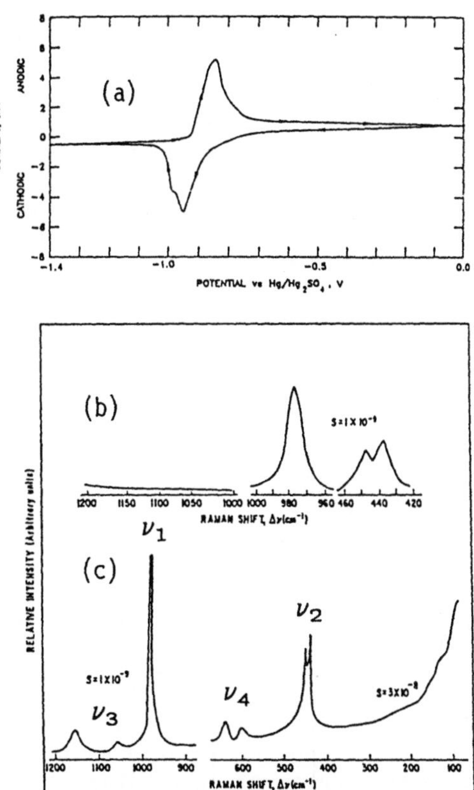


Figure 4. (a) Cyclic voltammogram of Pb in 0.05 M $H_2SO_4$ at 25°C (scan rate = 10 mV/sec); (b) In-situ Raman spectrum of anodic corrosion film on Pb at 0.16 V vs SHE; (c) Raman spectrum of standard $PbSO_4$; $Ar^+$ laser, 488 nm line, Power ≃ 280 mW, s = sensitivity scale.

diffractions analysis of the surface film. Moreover, from a theoretical analysis of the number and groups of vibrational frequencies observed, one can also surmise that the three bands observed correspond to two of the four modes of vibration of the $SO_4^=$ group that belong to different symmetry classifications[18]. The $SO_4^=$ anion is tetrahedral with $T_d$ symmetry. Theory predicts the appearance of four groups of bands in the Raman spectrum of $SO_4^=$ compounds. From Fig. 4c it is evident that four such groups of bands are indeed observed. The splitting of the degenerate bands $\nu_2$, $\nu_3$, and $\nu_4$ is apparent and is brought about by the internal electric fields in the crystal that the ion experiences.

Further anodic oxidation of the lead electrode at a higher potential should produce a new phase, e.g., $\beta$-$PbO_2$ in acid solutions.[20] On stepping the potential from 0.16 to 1.95 V vs SHE in 0.05 M $H_2SO_4$ in our experiment, we observed the white $PbSO_4$ layer to convert to a black coating which was identified by XRD as $\beta$-$PbO_2$. However, the Raman spectrum we observed was irreproducible and depended on the laser power incident on the sample. There seemed to be an induction time of a few minutes following laser irradiation before a spectrum could be observed. Similar observations were made on a standard $\beta$-$PbO_2$ sample. Rotation of the sample caused some of the peaks to disappear or to shift in frequency. On removal of the sample from the spectrometer, we found peach colored spots where the laser beam had impinged on the sample. It is evident that the laser beam had been changing the sample at the point of impact, causing some transformation of the $\beta$-$PbO_2$ to other forms of Pb compounds and leading to erroneous results. This is a good illustration of a problem associated with the study of dark and strongly light absorbing samples by the Raman technique. Sample heating by the laser beam can lead to changes in chemical composition or phase transformation of strongly light absorbing materials. It is, therefore, necessary to employ as low a laser power as possible consistent with obtaining a reasonable Raman signal. It is also important to check that no composition or phase change is occurring by obtaining spectra at different laser power levels.

## 3.2. Silver and Gold

Silver and gold have figured prominently in surface enhanced Raman spectroscopy (SERS) investigations where they are used as substrates for structural studies of absorbed molecules as well as for investigations into the mechanism of surface enhancement. Kötz et al.,[3] as well as Hamilton and co-workers,[8] investigated Ag in alkaline solutions while work in our laboratory was carried out in acid sulfate solutions.[7] In NaOH, the cyclic voltammogram of Ag exhibits two anodic waves at about 0.4 and 0.6 V vs Hg/HgO corresponding to the formation of surface layers of $Ag_2O$ and AgO, respectively.[3] AgO was identified by a strong band at 430 $cm^{-1}$ at a sensitivity of about 10 monolayers. On the other hand, $Ag_2O$ could not be detected because of laser photodecomposition which led to the formation of AgO.[3]

In 0.05 M $H_2SO_4$ solution,[7] anodic oxidation occurs at about 0.3 V vs $Hg/Hg_2SO_4$ resulting in the formation of a $Ag_2SO_4$ film on the electrode surface (Fig. 5). Raman spectra as a function of potential are shown in Fig. 6; the bands at 450, 608, 960 and 1130 $cm^{-1}$ are again characteristic of $SO_4^{=}$, while the low frequency band at ~229 $cm^{-1}$ is probably due to vibration of the whole $SO_4^{=}$ group with respect to the Ag surface or to $Ag^+$ adatoms on the surface. Conversion of the $Ag_2SO_4$ film to $Ag_7O_8SO_3$ (which is a AgO phase containing some $SO_4^{=}$ impurity) has been observed by X-ray diffraction analysis (XRD) following anodization at about 1.3 V.

Gao et al.[22] studied gold extensively as an ubiquitous substrate for obtaining intense SERS. They prescribed a series of oxidation-reduction cycles at well defined potential limits in 0.1 M KCl that can be used to generate optimum surface conditions for obtaining strong signals. Gold is, moreover, interesting because of the relatively wide electrochemical "window" available for investigating electron transfer

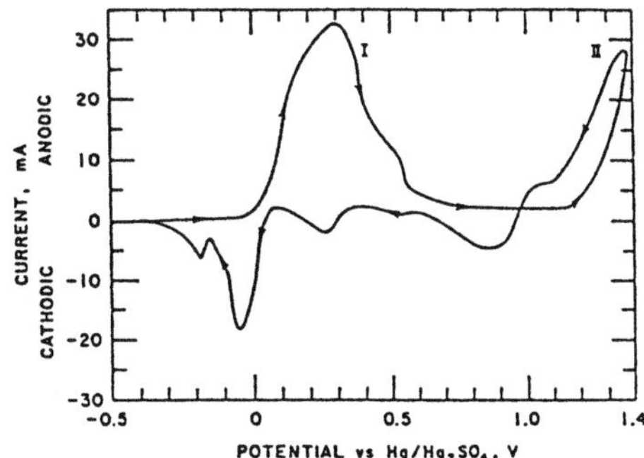


Figure 5. Cyclic voltammogram of Ag in 0.05 M $H_2SO_4$ at 25°C. Scan rate = 5 mV/sec, Electrode area = 0.9 $cm^2$.

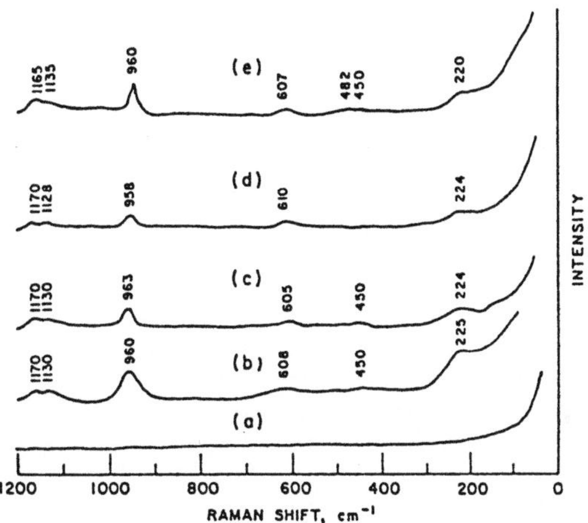


Figure 6. In-situ Raman spectra of Ag anodized at various potentials in 0.05 M $H_2SO_4$; (a) Open circuit (OCV) = -0.02 V vs $Hg/Hg_2SO_4$, (b) 0.18 V, (c) 0.55 V, (d) 0.75 V, (e) 1.30 V; $Ar^+$ laser, 514.5 nm line, Power = 80 mW.

reactions. The surface is nonetheless oxidized at fairly anodic potentials, and the structural changes have been studied by DeSilvestro and Weaver.[23] The latter work exploited the power of SERS to obtain information on the initial stages of film formation and growth in aqueous solutions. The authors found that during the anodic oxidation of Au in 1 M KOH, two Raman bands are observed; one ($\nu_1$) occurs at around 400-430 $cm^{-1}$ which appears at about -0.25 V vs SCE, and another band $\nu_2$ from 0.2 to 0.25 V (the so-called "pre-oxide" formation region of the current-potential, i-E, curve). Applying a more positive potential (~0.2 to 0.25 V) corresponding to the beginning of the voltammetric "anodic oxide" region resulted in the disappearance of $\nu_1$

and the appearance of a new band, $\nu_2$, at ~540 $cm^{-1}$. These bands were assigned to Au-OH vibrations based on their frequency shifts in similar experiments carried out using deuterated water as solvent. The model of anodic surface oxidation of noble metals has been proposed[24] to consist of two steps: 1) initial adsorption of O or OH, followed by 2) place exchange with the surface metal atoms along with further electron transfer to form the "oxide film." DeSilvestro and Weaver assigned the $\nu_1$ band to reversibly adsorbed $OH^-$ in step 1 and $\nu_2$ to the "oxide film" formed in step 2. The $\nu_2$ frequency of 540 to 580 $cm^{-1}$ however, appears to be lower than that due to surface $Au_2O_3$ (~650 $cm^{-1}$) as observed by Electron Energy Loss Spectroscopy (EELS) and for bulk $Au(OH)_3$ as obtained by Raman Spectroscopy (~635 $cm^{-1}$). The charge associated with the oxide film appears to be equivalent to or even less than one monolayer.

## 3.3. Copper and Cobalt

Copper too exhibits the SER effect; it is also of great interest industrially as a material for heat exchanger tubes and its corrosion behavior has been characterized electrochemically in a wide variety of aqueous solution environments. Raman spectroscopic studies had been carried out by Hamilton et al.[8] in alkaline solution; we have undertaken investigations in sulfate.[7] The voltammogram[8] of Cu in 0.1 M NaOH shows anodic waves at about -0.40 V and 0.0 V vs SCE with corresponding cathodic reduction waves at about -0.7 and -0.8 V, respectively. Using an optical multichannel detector, Hamilton et al. monitored the initial formation of $Cu_2O$ at 649 $cm^{-1}$ near 0.05 V with $Ar^+$ laser excitation using the 488 nm line. For potentials more anodic than about 600 mV a broad band appeared between 500 and 650 $cm^{-1}$. This band was assigned to $Cu(OH)_2$. The formation of $Cu_2O$ at -367 mV was visible in the spectra after passing 6.8 $mC/cm^2$ of charge leading the authors to conclude that they could detect a film ~40 Å thick. They also found the Raman scattering intensity to be proportional to the film thickness. The authors did not observe scattering from the $Cu_2O$ film using laser excitation with the 647.1 nm line. Since $Cu_2O$ has an optical absorption edge close to the 488 nm line of the $Ar^+$ laser being used they concluded that the detection of the $Cu_2O$ was by resonance Raman scattering.

The cyclic voltammogram of Cu in 1 M $Na_2SO_4$ is shown in Fig. 7a; Cu does not completely passivate in sulfate solution.[7] Film formation occurs, however, and the spectrum in-situ at 0.2 V is shown in Fig. 7b. The bands at 145, 214, and 644 $cm^{-1}$ for the red-orange film formed are assigned to $Cu_2O$, the presence of which is confirmed by XRD analysis. Interestingly, a $Cu_2SO_4$ film was not observed, this compound being rather soluble in the solution.

We have also examined the composition of surface films on Co in 0.05 M NaOH solution.[9] Previous spectroscopic investigations have consisted mainly of ellipsometry[25,26] and Mössbauer spectroscopy.[27,28] At pH 11.9 in 0.05 M NaOH, the Pourbaix diagram of the Co-$H_2O$ system predicts the formation of the phases $Co(OH)_2$, $Co_3O_4$, $Co(OH)_3$, and $CoO_2$ with increasing potential.[17] The current-potential curve shows an

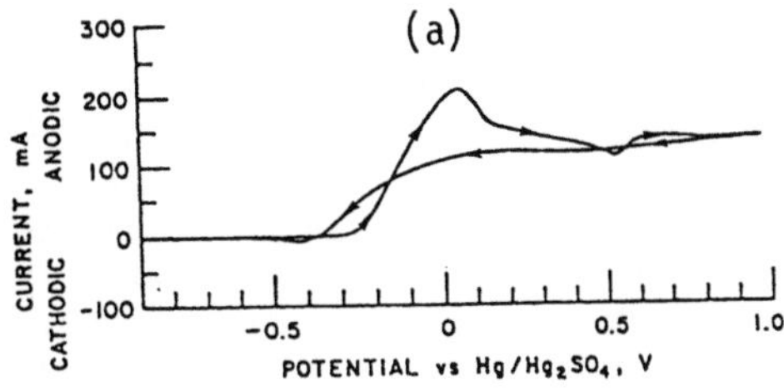


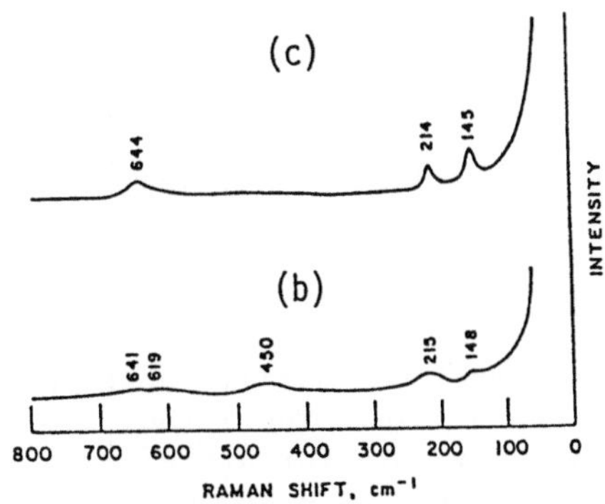


Figure 7. (a) Cyclic voltammogram of Cu in 1 M $Na_2SO_4$, Scan rate = 5 mV/sec, Electrode area = 1.3 $cm^2$; (b) In-situ Raman spectrum of anodic corrosion film on Cu at 0.2 V; (c) Ex-situ spectrum, $Ar^+$ laser, 488 nm line, Power = 40-100 mW.

anodic wave at about -0.5 V vs Hg/HgO assigned to the formation of CoO and two other waves between 0 and 0.5 V which had been assigned by Sato et al.[25,26] to the formation of $Co(OH)_3$ and $Co_3O_4$. Our laser Raman investigation of this system did not yield any spectrum in the primary passive region at about -0.2 V but showed a multiplet of bands in the secondary passive region at ~0.45 V. In-situ measurements showed weak bands at 475, 515, 587, and 690 $cm^{-1}$. Comparison of these lines with the spectra of standard samples of CoO, $Co_3O_4$ and $Co(OH)_2$ led us to conclude that the secondary passive film consists of a mixture of CoO and $Co_3O_4$. This is in agreement with Sato's finding that a bilayer film of CoO/$Co_3O_4$ oxide exists at the Co surface in the potential range that we have studied.

## 3.4. Titanium

Ohtsuka, Guo and Sato[48] studied the anodic oxide films on titanium at various potentials both in neutral phosphate and acidic sulfate solutions. Following anodic oxidation for one hour to form the film at constant potentials of 5.6 and 10.6 V vs RHE in neutral phosphate solution (0.1 M, pH 6.9), the electrodes were examined in-situ at 1.60 V to avoid oxygen evolution. Raman bands at 145, 400, 515 and 640 $cm^{-1}$ were observed which were identified with the anatase form of $TiO_2$. Similar bands were observed in 0.1 M $H_2SO_4$ solution after anodizing for one hour at 8.1 and 10.1 V then holding at 1.6 V vs RHE. They were unable to observe Raman bands below 4 V although the film thickness measured about 13 nm. They hypothesized that the film is still amorphous at this potential but transforms into a more crystalline structure at higher potentials. Thus they begin to see Raman bands at 5.6 V where they also observed the electrode surface to have a crater-like morphology. Strong bands due to the anatase modification were observed at 10 V where the oxide film appears to have a ripple-like morphology.

Hugot Le-Goff also reported work on thin $TiO_2$ films by interference enhanced Raman spectroscopy.[38a,b] She found that films grown by anodic oxidation of Ti consist of two different crystallographic forms that depended on the formation potential. Films prepared at potentials higher than 50 V were composed of anatase while those formed at lower potentials were microcrystalline, consisting of aggregates of 20 to 25 atoms each having the local order of anatase. The spectra of these two types were different; the anatase spectrum showed bands at 398, 520 and 625 $cm^{-1}$, while the microcrystals showed bands at 350 and 630 $cm^{-1}$. The author found that an electropolished Ti sample left in air for a long period of time developed a film whose spectrum was the same as the thin microcrystalline film. The author took advantage of enhancement of the Raman signal by interference effects in the film which varied as a function of thickness; a 10 to 20 times enhancement was obtained.

More recently Felske and Plieth[38c] obtained in-situ Raman spectra of oxide films electrochemically formed on Ti in 0.5 M $H_2SO_4$. They observed that below the formation potential of 10 V, the spectra were clearly due to rutile; at higher potentials, however, the spectra were neither characteristic of rutile nor anatase. Upon electrode emersion and drying of the film, spectra due to anatase were evident. An identical spectrum was obtained when the oxide film was formed in $HNO_3$. In all cases, the films appeared to be crystalline. The authors felt that differences in their results with those of others may be due to the film formation and ageing conditions.

## 3.5. Nickel and Iron

Nickel and iron have been most extensively studied electrochemically for their corrosion characteristics because of their industrial importance. Their electrochemical reactivities are also of great interest for the operation of Ni-Fe and other batteries. We studied the anodic corrosion of Ni in alkaline and near-neutral solutions[9] while Delichere, Hugot Le-Goff, and Yu[29] reported work in acid. DeSilvestro, Corrigan and Weaver[30] took advantage of the SERS effect on Au to study the oxidation and reduction of electrodeposited $Ni(OH)_2$ films on a Au substrate. The oxidation of Ni in near-neutral and alkaline solutions proceeds in two stages, the first leading to the formation of $Ni^{+2}$ and the second to the formation of a higher valency state commonly thought of as $Ni^{+3}$. The former leads to a very thin film of $Ni(OH)_2$ in base and most probably to NiO in acid. No Raman spectroscopy study has definitely identified the surface film on Ni in alkali in the passive region but UV-visible reflectance measurements by Hahn et al.[31] have been interpreted in terms of the species $Ni(OH)_2$. The film appears to be at best a few monolayers thick. DeSilvestro et al.,[30] obtained the Raman spectrum of five equivalent monolayers of $Ni(OH)_2$ cathodically deposited on a roughened Au surface by employing the SERS technique. Using the 647.1 nm line of a $Kr^+$ laser for excitation, they found a single band at 455 $cm^{-1}$ which they identified as the Ni-OH stretching mode. Oxidation of the $Ni(OH)_2$ gave rise to two bands at 480 and 560 $cm^{-1}$ which they identified as Ni-O (rather than Ni-OH) vibrations due to the absence of a deuterium isotope effect. The oxidized form could only be detected in thicker layers using 488 and 514.5 nm excitation leading the authors to believe that the enhancement was by the resonance Raman effect. The further oxidation of $Ni(OH)_2$ on Ni surfaces, in near-neutral and alkaline solutions, occurs in the transpassive potential region close to $O_2$ evolution. We have studied the structure of this higher oxide form by LRS and XRD techniques and have assigned the two bands at 480 and 560 $cm^{-1}$ to a structure consisting of layers of $NiO_2$ with intercalated $H_2O$ and ions such as $K^+$, $Na^+$, $OH^-$, $H^+$, etc. between the layers.[9,33] We found that the spectra of various preparations[37] of this higher oxide which have been referred to in the literature as NiOOH, $Ni_2O_3$, $Ni_3O_4$ hydrates, and $NiO_2$ yield the same vibrational spectrum (Fig. 8) and have very similar X-ray powder patterns

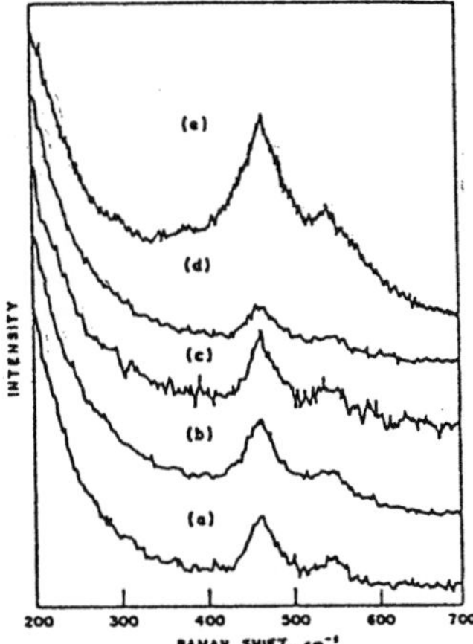


Figure 8. Raman spectra of various preparations of the higher oxide form of Ni (taken as powders pressed with KBr into a pellet), $Ar^+$ laser, 514.5 nm line, Power ~30-100 mW. (a) $\beta$-NiOOH (after Glemser)[37a]; (b) $\gamma$-NiOOH (after Glemser); (c) $Ni_2O_3 \cdot 2H_2O$ (after Cairns)[37b]; (d) $Ni_3O_4 \cdot xH_2O$ (after Cairns); (e) $NiO_2$ (purchased from Alfa).

(Fig. 9). We concluded that these various forms of higher nickel oxide have the same layered $NiO_2$ structure and differ only in the degree of structural disorder of the $NiO_2$ slabs, the nature and amount of intercalated ions and $H_2O$ molecules. Such a structure is depicted in Fig. 10. We propose that the higher oxide form of Ni be called "hydrous

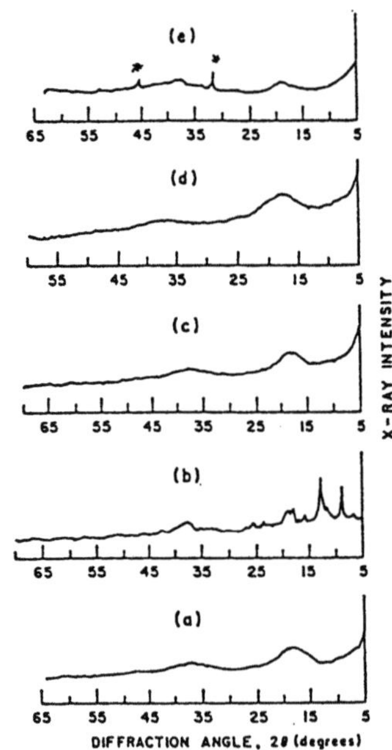


Figure 9. X-ray powder diffraction patterns of higher nickel oxide preparations. (a) β-NiOOH (prepared following Glemser)[37a]; (b) γ-NiOOH (prepared following Glemser); (c) $Ni_2O_3{\bullet}2H_2O$ (prepared following Cairns)[37b]; (d) $Ni_3O_4{\bullet}xH_2O$ (prepared following Cairns); (e) $Ni_2O_3{\bullet}xH_2O$ (Aldrich,* NaCl lines)

```
          O       O       O       O       O
          |       |       |       |       |
     -- Ni -- Ni  —  Ni -- Ni -- Ni --
          |               |       |       |
  ----O               O       O       ----
              HOH     H+     OH-     Na+
  -----          O       O               O---
              |       |       |       |       |
          -- N -- Ni -- Ni -- Ni -- Ni --
              |       |       |       |       |
              O       O       O               O
```

Figure 10. Proposed structure of "hydrous $NiO_2$" film.

$NiO_2$" instead of NiOOH as has been done for many years. The maximum valency of Ni in such a compound is +4, although the actual material prepared in the laboratory, most probably, consists of a mixture of $Ni^{+2}$ and $Ni^{+4}$. UV-visible spectroscopic evidence for the existence of $Ni^{+4}$ has been presented by Corrigan and Knight.[34]

The composition of corrosion films on Ni in acid solutions has been studied by Delichere, et al.[29] and also in our laboratory.[32,35] The former investigators identified the passive film on Ni in 1 M $H_2SO_4$ solution to be NiO; in the region of secondary passivity, they assumed that $Ni_2O_3$ is present. We have also obtained a spectrum of the surface film on Ni anodized in the transpassive region in 0.05 M $H_2SO_4$[32] but have not been able to assign it unequivocally to any oxide of Ni. However, in concentrated $H_2SO_4$ solution, our results[35] indicate that the corrosion film formed in the passive region is composed of $NiSO_4{\bullet}H_2O$. This assignment was confirmed by XRD studies.

A number of studies have dealt with the in-situ identification of the passive film on iron by LRS.[10,11,36-38] Thibeau et al.[39] obtained the spectrum of a number of iron oxides which they anticipated to be the corrosion products of iron. Their group also studied the oxidation of $Fe_3O_4$ on iron and steel surfaces.[40] Nazri[36] studied extensively the dissolution and film formation on iron in borate solutions using cyclic voltammetry (CV) and rotating ring disc electrode (RRDE) techniques. He used LRS in an attempt to identify the passive film on Fe. The film showed several bands at low frequency (200-800 $cm^{-1}$) which did not match those of the known oxides or oxyhydroxides of iron. Froelicher[11] et al. employed very fast oxidation-reduction cycles (scan rate = 6 v/sec) in order to increase the thickness of oxide films on iron in a number of solutions from 0.5 M $H_2SO_4$ to 1 M KOH. Thick films (~100 to 200 Å) were also produced by polarization maintained for a long time (~2 days). The films were thus artificially thicker than those of the passive films. For an iron sample polarized for 60 hours in 0.5 M $H_2SO_4$ solution, they obtained a Raman spectrum which corresponded to a mixture of $Fe_2O_3$ and $Fe_3O_4$. Dünnwald, Lossy, and Otto[10] appear to have been the first to make a definite in-situ LRS identification of the composition of the passive film on Fe in KOH solution. They observed bands at 543 and 674 $cm^{-1}$ in-situ but only after cycling the electrode 17 times from -1.3 to 0.5 V vs SCE at a scan rate of 1 mV/sec. These bands correspond to those of $Fe_3O_4$. Continued cycling of the electrode (over 20 cycles) resulted in the appearance of a reddish brown "rusty" coating which yielded bands at 251, 303, 390, 482, 552, 682, and 1003 $cm^{-1}$. This set of bands resembled closely those for $\alpha$-FeOOH. The coating could be washed off by a strong water jet, after which the sample appeared metallic again. The spectrum taken after washing corresponded to that of $Fe_3O_4$ once again. This is good evidence that the $\alpha$-FeOOH is an overlayer on top of the $Fe_3O_4$ which Dünnwald et al. believed to constitute the passive film. The existence of such an overlayer comes about due to the oxidation of $Fe^{+2}$ species in the solution and the resulting deposition on the electrode. Such a mechanism has been proposed by Nazri and Yeager.[41] It is also worthwhile to note here that Dünnwald et al. found that the same spectrum of $Fe_3O_4$ is obtained when the sample is allowed to stand in ambient air for one day. Thus, they concluded that exposure to gaseous $O_2$ is not critical in their experiments.

Hugot Le-Goff and Palotta[38a,b] reported results of in-situ LRS of Fe in 1 M $H_2SO_4$. Following 8 hours of anodic polarization, they found a slight increase in the reflectance of the electrode surface and were able to obtain a spectrum which corresponded to a mixture of $\alpha$-$Fe_2O_3$ and $Fe_3O_4$. The film thickness was measured by nuclear microanalysis to be about 10 nm. The authors equated the induction time needed to observe a spectrum to the time needed to age the film to the point where the "gel" structure developed into a more crystalline form.

Thanos[12] took advantage of the SER effect exhibited by small Ag particles to study the corrosion products of Fe and Fe-9% Cr alloy. He electrodeposited Ag on pre-passivated Fe-Cr alloy electrodes. He found that thicker corrosion products on Fe could be formed in $HNO_3$ by what he termed a reductive precipitation at -0.8 V vs SCE. $Fe_3O_4$ was identified at stationary electrodes while $Fe(OH)_2$ was found on rotated ones. He observed a broadening of the $Fe_3O_4$ band at 680 $cm^{-1}$ upon oxidation of $Fe(OH)_2$ which he attributed to the formation of higher defect oxides and hydroxides.

We have also studied the corrosion behavior of iron in phosphate solutions[42] from pH 1.6 to 11.9. Cyclic voltammetry measurements showed mostly active dissolution in acid solutions of 0.1 M $H_3PO_4$ and $NaH_2PO_4$; complete passivity is observed in alkaline 0.1 M $Na_2HPO_4$ and $Na_3PO_4$ (Fig. 11). The composition of the corrosion films formed in the

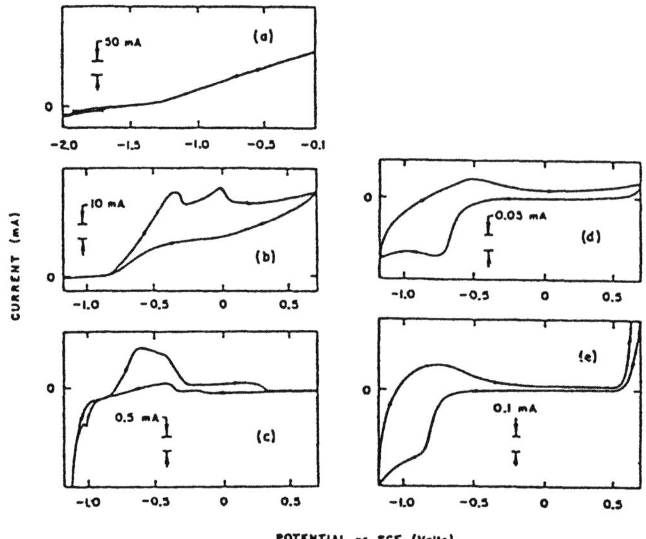


Figure 11. Cyclic voltammograms of iron in aerated phosphate solutions; (a) 0.1 M $H_3PO_4$ (pH 1.6); (b) 0.1 M $NaH_2PO_4$ (pH 4.4); (c) 1:1 volume mixture of 0.10 M $Na_2HPO_4$ and 0.16 M $NaH_2PO_4$ (pH 6.4); (d) 0.1 M $Na_2HPO_4$ (pH 8.7); (e) 0.1 M $Na_3PO_4$ (pH 11.9); All scan rate = 10 mV/sec.

latter solutions was determined by LRS to consist mainly of $Fe_3O_4$ (bands at 556 and 680 $cm^{-1}$) with $PO_4^{\equiv}$ (944 and 1068 $cm^{-1}$) incorporated into the lattice structure of the film (Fig. 12). In acidic 0.1 M $NaH_2PO_4$, multilayer film formation occurs. Open circuit corrosion

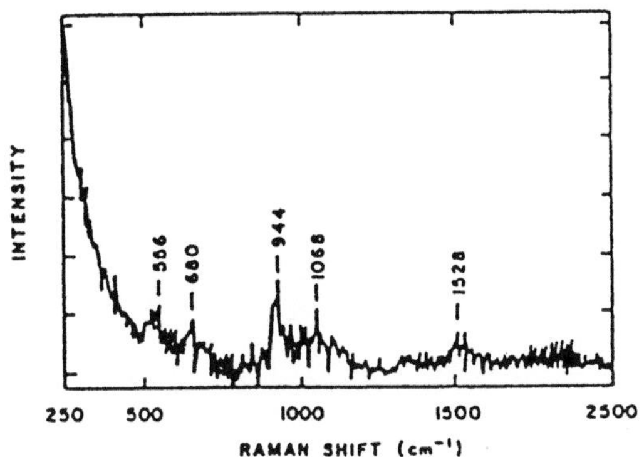


Figure 12. Raman spectrum of corrosion film on iron in 0.1 M $Na_3PO_4$ (pH 11.9) after potential hold at 0.4 V for 19 hours.

leads to the formation of a surface film whose in situ spectra are shown in Figs. 13a and 13b after two hours and two days of standing, respectively. The bands at 1020 and 1346 $cm^{-1}$ are due to material on the surface while those at 878 and 1082 $cm^{-1}$ are from $PO_4^{\equiv}$ in solution. It is interesting to note the change in spectra after two days immersion. The surface film was examined microscopically and found to consist of two layers: a very thin greyish-white outer layer and a bluish undercoating, which is adherent to the electrode surface. A composite Raman spectrum of the duplex film is shown in Fig. 13c while that of the bluish inner layer is shown in Fig. 13d. X-ray diffraction analysis of the latter shows the structure to be the same as that of the mineral vivianite or $Fe_3(PO_4)_2 \cdot 8H_2O$. The outer layer could not be analyzed by XRD, but we believe that it is the phase $Fe(H_2PO_4)_2$ in line with predictions of the Pourbaix diagram.

Finally, in a related study Keiser et al.[43,44] studied the ambient oxidation of $Fe_3O_4$ on iron and steel surfaces, as well as the electrochemical reduction of $\alpha$, $\gamma$, $\delta$, and amorphous FeOOH in aqueous sulfate solution. They found that $Fe_3O_4$ was oxidized to $\gamma$-FeOOH which converted to $\alpha$-FeOOH after several weeks. The electrochemical reduction of amorphous $\gamma$- and $\delta$-FeOOH on weathering steel substrates produced $Fe_3O_4$.

### 3.6. Metal Alloys

A number of papers have dealt with the LRS study of surface films produced by atmospheric oxidation of metal alloys[36,39,45,46]; very little work, however, has been done to identify the corrosion film formed in aqueous solution environments. Thanos[12] attempted to obtain in-situ spectrum of Fe-9% Cr alloy without much success. However, following

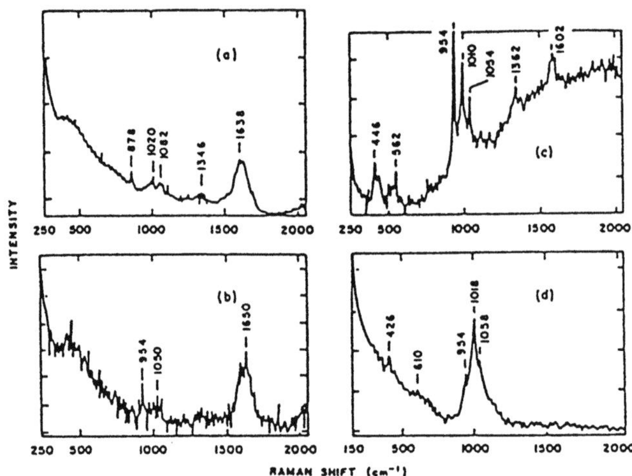


Figure 13. Raman spectra of corrosion films on iron in 0.1 M $NaH_2PO_4$, $Ar^+$ laser, 514.5 nm line, Power ≃50 to 150 mW.

electrodeposition of Ag on the prepassivated electrode (to take advantage of the SER effect), he was able to observe Raman active bands of thin films of oxides and hydroxides on the passivated alloy. A band due to $Fe_3O_4$ was observed as in the case of Fe which became broad at potentials greater than about 0 to 0.2 V vs SCE. Such broadening was explained by the author to be due to lattice disorder because of random distribution of $Fe^{+2}$ and $Fe^{+3}$ ions.

Thierry et al.[47] also studied the anodic corrosion films formed on single crystals of Fe-Cr and Fe-Cr-Mo alloys using LRS in combination with X-ray Photoelectron Spectroscopy (XPS) and nuclear microanalysis. Films were formed in the passive and transpassive regions in 1 M KOH. The former had a thickness in the range of 5 to 6 nm and was determined to most likely be a mixed $\alpha$-$(Fe,Cr)_2O_3$ phase. The transpassive film was found to be 5 to 25 nm thick and was formed in various stages. The latter stages were identified to include the formation of 5 nm of $\alpha$-$(Fe,Cr)_2O_3$, followed by 10 nm of Fe-Cr spinel, and a top layer of 10 nm $\gamma$-FeOOH. They found no significant differences in spectra from corrosion films formed on Fe-Cr and Fe-Cr-Mo single crystals.

A study of Ni-10% Mo alloy has been reported by Delichere et al.[29] Polarization of the alloy in 1 M $H_2SO_4$ solution showed the existence of a NiO film over a more extended range than was the case of pure Ni in the same solution.

### 3.7. Raman Spectroscopy Studies of Corrosion Inhibitors

The inhibition of corrosion through the addition of small quantities of organic and inorganic substances to the solution is of great practical importance. Considerable research activity goes on in this field[49] but few have taken advantage of the potential of LRS in elucidating the mechanism of corrosion inhibition. A number of theories have been proposed to explain the action of inorganic inhibitors, and these theories most probably apply to organic ones also. Among these are the chemical reaction theory, electrochemical polarization theory, adsorption theory, and electrostatic polarization theory.[50] Whereas one mode of action may be dominant in a certain case, it is not always a straightforward process to delineate this and it is likely that quite often several of the mechanisms are operative to various extents.

Kester, Furtak, and Bevolo[51a] first utilized the SER effect to study the structure of benzotriazole (BTA) on Cu. They obtained spectra of BTA at different potentials but were unable to assign the observed frequencies to the various normal modes of vibration of the molecules. They established that the observed signals were due to SER scattering. They also observed a hysteresis in the Raman intensity as a function of applied potential which they attributed to a bonding rearrangement of the molecules in the protective surface layer. They further noted that the structure of the surface complex of BTA with Cu was altered by laser illumination during anodization which resulted in better protective ability of the coating.

Thierry and Leygraf[51b] studied the adsorption of 1, 2, 4 triazole and imidazole on microlithographically prepared Cu surfaces at the corrosion potential. They compared their spectra with those on Cu surfaces subjected to oxidation-reduction cycles (ORCs) and found the spectra to be very similar, suggesting that the ORC had no influence on the orientation or possible dissociation of the molecules on Cu. The spectra were also similar to those in ethanol, indicating that the molecules did not undergo any dissociation upon adsorption. They were able to obtain an intense spectrum of the 1, 2, 4 triazole on microlithographed Cu (with particle size of 2000 to 3000 Å and periodicity of 5500 Å) at the corrosion potential of -150 mV vs SCE in 0.1 M NaCl; they did not, however, observe a spectrum on a smooth Cu surface which was polished with diamond paste. This demonstrated that an ORC is not necessary to obtain SERS but microscopic surface roughness appears to be essential.

Loo et al.[52] also studied the adsorption of some imidazoles on Cu and Ag electrodes in 0.1 M KCl (pH 7.7). They found that whereas the Raman spectrum of free imidazole exhibits complex spectral features in the region between 2500 and 3100 $cm^{-1}$ due to intermolecular hydrogen bonding, the SER spectrum does not. Thus, they concluded that the imidazole is adsorbed on Cu and Ag surfaces through its nitrogen atom with the five membered ring perpendicular to the substrate; otherwise, hydrogen bonding would occur for flatly oriented molecules giving rise to a more complex spectrum. Loo[53] also studied the adsorption of thiourea on Cu and Ag electrodes in 0.1 M KCl solutions. The solution spectrum of thiourea, $SC(NH_2)_2$ shows bands at 409, 479, 733, 1096, and 1404 $cm^{-1}$. Upon adsorption of the molecule on a Ag or Cu surface, the 733 $cm^{-1}$ band showed a large red shift of ~30 $cm^{-1}$ and exhibited a large reduction in relative intensity. This band was assigned to a C-S stretching frequency; since it was the most strongly affected band, Loo concluded that the thiourea is bonded to the Cu or Ag surface via the S atom. X-ray, IR, and Raman studies suggest that the electronic structure of the thiourea may be represented by equal contributions from the resonance hybrid structures, a, b, and c, shown below.

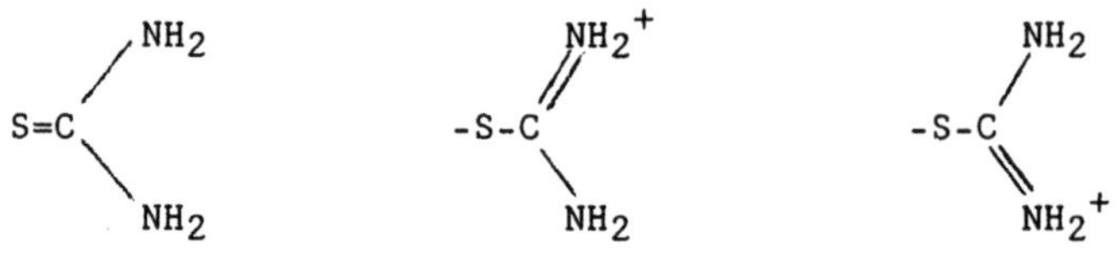

$$S{=}C(NH_2)_2 \qquad -S{-}C(NH_2){=}NH_2^+ \qquad -S{-}C(NH_2){=}NH_2^+$$

(a) (b) (c)

Loo also found a low frequency vibration at 221 $cm^{-1}$ for thiourea adsorbed on Ag and one at 228 $cm^{-1}$ for adsorption on Cu; these he attributed to the metal adatom-thiourea complex vibration rather than to Ag-Cl or Cu-Cl complex, respectively. The metal thiourea complex was proposed to be $AgSC(NH_2)_2{}^+Cl^-$ and $CuSC(NH_2)_2{}^+Cl^-$.

Rubin, Gutz, and co-workers[54] also studied the adsorption of benzotriazole (BTA) on a Cu electrode as a function of potential in different chloride solutions at various pHs. They concluded that the protective film on Cu in the presence of BTA is copper (I) benzotriazolate or $CuC_6H_4N_3$ [Cu(I)BTA]. The material is polymeric with the structure,

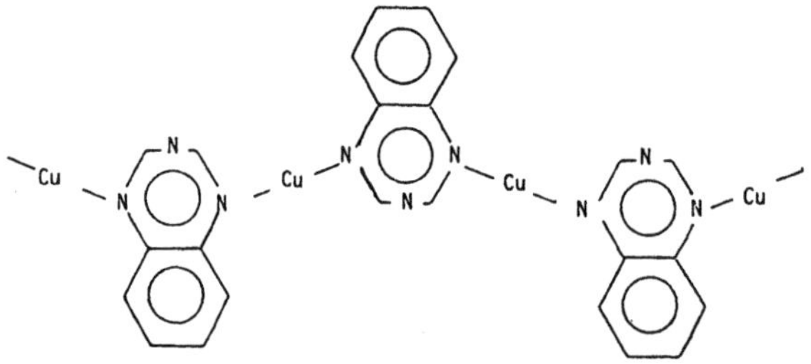

The CuBTA polymer film grows from a chemisorbed monolayer to form a protective layer which is 3000-4000 Å thick. The growth of such thick films involves transport of Cu ions through the layer. The Cu ions are then thought to react with benzotriazolate anions at the solid/liquid interface to produce more polymer.

Thierry and Leygraf[55] compared the inhibitor properties of five structurally related N-heterocyclic molecules towards the corrosion of Cu in NaCl solutions. Polarization resistance measurements in 0.01 and 0.1 M NaCl solutions showed that the percentage inhibition efficiency increases in the order mercapto-benzimidazole (MCBIA)> benzimidazole (BIA)> methylbenzimidazole (MBIA)> hydroxybenzimidazole (HBIA). Raman spectral measurements showed all the molecules to be adsorbed on the Cu surface without dissociation. For BTA and BIA, they found significantly larger differences in frequency shifts for vibrational modes originating from the triazole or imidazole part of the molecule than for modes originating from the benzene part. They interpreted this to mean that adsorption of these molecules occurs via the triazole and imidazole part and most likely through interaction of all three nitrogen or carbon atoms with the Cu substrate. MCBIA, HBIA, and MBIA, on the other hand, showed no preferential enhancement in intensity or frequency shift originating from the imidazole part; therefore, they are oriented differently than BTA or BIA. AC impedance measurements were used to obtain double layer capacitances of Cu in 0.1 M NaCl with added 5 mM inhibitors. The measured capacitances decreased in the order BIA< BTA< MCBIA; the values also decreased with increasing anodic potential (in the range of -700 to -100 mV vs SCE). Thus, the more efficient inhibitor exhibited lower double layer capacitance which the authors attributed to a thicker polymer layer. The difference in inhibition efficiency of the five molecules investigated was explained by the number and type of functional groups or atoms in the triazole or imidazole ring that interact with the Cu. This interaction was thought to involve N-N-N and N-C-N in the case of BTA and BIA, N and S from the N-H and S-H groups in the case of MCBIA, and only N from the N-H group in the case of HBIA and MBIA. From the behavior of the molecules studied, the inhibiting efficiency of various atoms in functional groups follows the order S > N > O.

Ohsawa and Süetaka[56] studied the intensity enhancement and spectral change in the Raman scattering of mercaptobenzotriazole (MCBTA) adsorbed on a Ag electrode from 0.1 M $Na_2SO_4$ solution. They concluded that MCBTA is adsorbed on Ag in the form of ionized thiol. They observed dramatic changes in Raman intensity in changing the concentration of MCBTA at $1 \times 10^{-8}$ M and $1 \times 10^{-6}$ M. They ascribed the former to a change in orientation of adsorbed molecules and the latter to the formation of a multilayered film. They found the largest intensity enhancement to be for the species adsorbed in normal orientation with respect to the surface. The addition of $I^-$, $Br^-$, or $Cl^-$ resulted in the conversion of the thiol species to the thione form. They believe that this conversion and the occupation of adsorbed sites by halide ions is related to the effect of halide in impeding the protective action of MCBA.

### 3.8. High Temperature Raman Spectroelectrochemistry

The extension of LRS techniques for in-situ measurements in high-temperature/high-pressure (HTHP) aqueous solution environments is of considerable practical interest. Such conditions are encountered in a number of applications, such as, studies on the stress corrosion cracking of metals used in nuclear reactor coolant systems, thermodynamic processes in geothermal environments, general and localized corrosion in industrial boiler water and condenser systems, etc. These facilities normally operate at temperatures above 200°C and pressures over 1000 psi. LRS measurements in-situ on corrosion films produced under such environmental conditions require the construction of special apparatus;[57,58] a spectro-electrochemical cell for HTHP application is shown in Fig. 14. The cell body (A) is made of titanium for corrosion resistance and has dimensions of 5.7 cm diameter by 7.6 cm long. It is

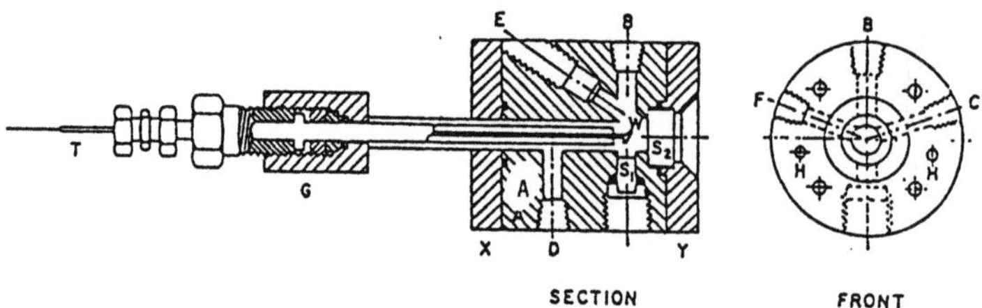


A - Titanium Cell Body
B - Counter Electrode Port
C - Reference Electrode and Electrolyte Exit Port
D - Electrolyte Inlet
E - Light Exit or Thermocouple Port
F - Viewing Port
G - Shaft Seal

H - Heaters
$S_1$ - Sapphire Window for Laser Beam Inlet
$S_2$ - Sapphire Window for Scattered Raman Light
T - Teflon Sheath on Electrode Lead
W - Working Electrode

Figure 14. Cell for in-situ laser Raman spectroscopy and electrochemical studies of high-temperature/high-pressure aqueous solutions.

bolted to the end flanges "X" and "Y" with Teflon O-ring seals between the parts. Provision is made to admit the laser beam through a port fitted with a 0.95 cm dia. by 1.11 cm thick sapphire window ($S_1$); the inelastically scattered light is collected through a second sapphire window ($S_2$) which is 2.22 cm dia. by 0.95 cm thick. The corrosion sample is normally prepared in the form of a 0.47 cm dia. disc and is used as the working electrode (W). In one study,[16,57] for example, lead was deposited on a nickel disc by splat cooling from a molten lead bath. The electrode connection is made through a 0.16 cm nickel rod sheathed in heat-shrinkable Teflon (T); it is held inside a 0.635 cm cylindrical shaft which can be moved back and forth or rotated through the use of a especially designed shaft seal (G). These movements are necessary for proper alignment of the electrode surface with respect to the laser beam. The counder electrode is introduced through either port B or E; we have used a loop consisting of several turns of Pt. wire. Either one of ports B or E is fitted with a thermocouple for temperature measurement. A silver-silver sulfate ($Ag/Ag_2SO_4$) reference electrode is introduced through port C. This electrode is constructed in a manner similar to the Ag/AgCl quasi-reference described by Danielson.[59] It has a potential of 0.041 V with respect to $Hg/Hg_2SO_4$ at 25°C. We have calibrated it by comparison with a Ag/AgCl reference up to a temperature of 290°C.

The cell is heated using two "Calrod" heater cartridges (H) which are imbedded in the cell body. Two thermocouples are similarly mounted. All seals (black shading in Fig. 14) are made of Teflon; metal-reinforced Teflon rings, obtained commercially (Bal-Seal Engineering Co., CA), are used for the shaft-seal and to seal the sapphire windows.

A viewing port (F) fitted with a sapphire window is also provided to facilitate positioning of the working electrode and the cell inside the illumination chamber of the spectrometer.

The electrolyte solution is refreshed in order to prevent buildup of corrosion products inside the cell; it flows through the cell, entering at port D and exiting at C. The recirculation system is shown in Fig. 15. The electrolyte solution is prepared in a stainless steel reservoir (R) and deoxygenated by sparging with dry nitrogen. The

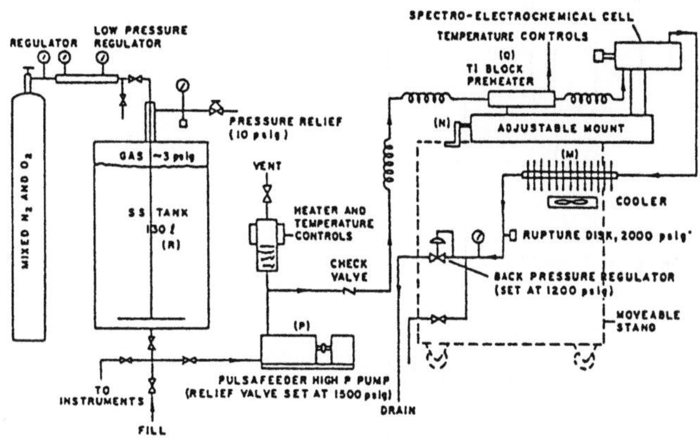


Figure 15. High-temperature/high-pressure spectroelectrochemical facility.

solution is then pumped into the cell using a high-pressure pump (P) after passing through a Ti block preheater (Q). The effluent from the cell is cooled by passing it through an air-cooled heat exchanger (M) and then discarded into the drain. The spectro-electrochemical cell and Ti block preheater are mounted on a movable cart fitted with a screw drive system (N) which allows positioning of the cell in the illumination chamber of the Raman spectrometer. The electrochemical equipment and Raman instrument used have been described previously[9] and are shown schematically in Fig. 2.

Initial testing of the HTHP facility was undertaken through a study of the anodic corrosion of lead in $7 \times 10^{-4}$ M $Na_2SO_4$ (i.e., 100 ppm $Na_2SO_4$ solution, pH = 6.1). The choice of this system was based on our previous studies[(6)] which showed that the corrosion of lead results in the formation of a surface layer of $PbSO_4$ that has a large cross section for Raman scattering. Other studies[60] also indicated that the $PbSO_4$ film is stable at high temperatures. The low concentration (100 ppm) of $Na_2SO_4$ was designed to simulate the level of impurities often present in reactor coolant water. Fig. 16 shows cyclic voltammograms of Pb in $7 \times 10^{-4}$ M $Na_2SO_4$ ($O_2$-free) at 25°C and 280°C. It is evident that typical active dissolution-passivation behavior is observed at both temperatures; the voltammetric behavior is basically the same at intermediate temperatures. Reversibility is apparent from the presence of the corresponding cathodic wave. Fig. 17 shows Raman spectra

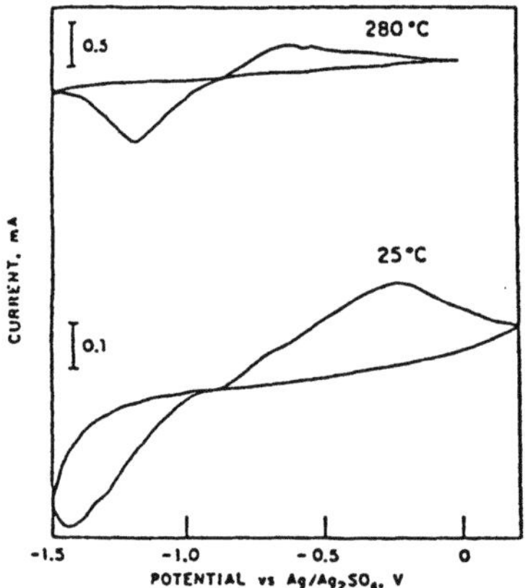


Figure 16. Cyclic voltammogram of Pb in 7 x $10^{-4}$ M $Na_2SO_4$ at 25° and 280°C. Scan rate = 1 mV/sec.

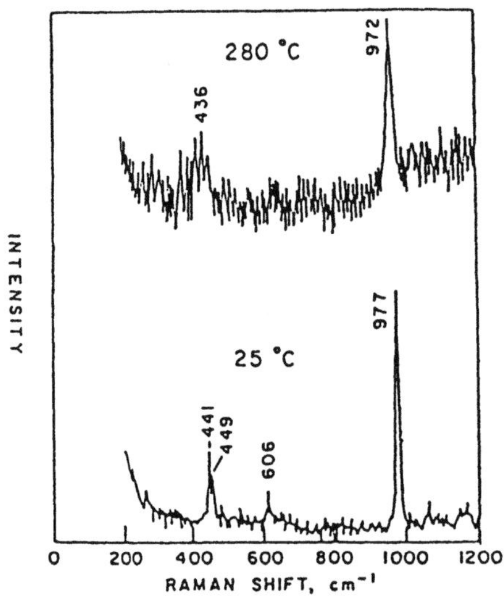


Figure 17. In-situ Raman spectra of passive film on Pb anodized at -0.1 V vs $Ag/Ag_2SO_4$.

obtained in-situ at 25 and 280°C for the passive film formed by constant potential anodization at -0.10 V vs $Ag/Ag_2SO_4$. Assuming that all the charge under the cathodic wave in Fig. 16 is due to the reduction of the passive film, we estimated the film thickness to be on the order of 1-3 μm. Based on the observed vibrational spectrum, the anodic corrosion film is identified as $PbSO_4$, consistent with the spectrum of this material observed in our previous work.[6] The same spectra are obtained upon anodization at intermediate temperatures. We observed no significant shift in band frequencies but found a steady decrease in signal-to-noise ratio as the temperature was raised from 20 to 280°C.

We have also observed the formation of a surface film on corrosion at the open-circuit voltage (OCV) of about -0.96 V (Fig. 18). The

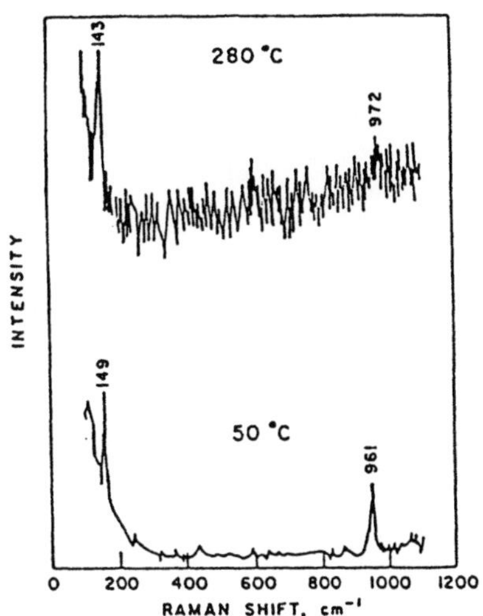


Figure 18. In-situ Raman spectra of corrosion films on Pb formed at open circuit.

Raman spectrum showed bands due to $SO_4^{=}$ vibrations, as well as one at 147 $cm^{-1}$, which we attributed to Pb-O stretching. X-ray diffraction study of corrosion films scraped off the electrode surface indicated that the film consisted of the phase $3PbO \bullet PbSO_4 \bullet H_2O$ (or $Pb_4O_3SO_4 \bullet H_2O$). This film persisted up to about 110°C. Above this temperature we found the phase $PbO \bullet PbSO_4$ (or $Pb_2OSO_4$) to predominate as exemplified by the bands at 143 and 972 $cm^{-1}$ (Fig. 18).

Following successful demonstration of the operation of the HTHP facility, a study was subsequently conducted on the corrosion and passivation behavior of Ni, an important component of stainless steel and other alloys used in reactor coolant systems. Fig. 19 shows cyclic voltammograms if Ni in 7 X $10^{-4}$ M $Na_2SO_4$ solution as a function of temperature. Both primary (A) and secondary (B) passivation peaks are clearly evident at temperatures below 120°C. Passivity breaks down above this temperature and the metal actively corrodes. The hysteresis in the current-potential curve above 200°C is interpreted to indicate the occurrence of pitting corrosion. Fig. 20 shows in-situ Raman spectra obtained by potentiostating the specimen in the secondary passivation region (about 0.55 V vs $Ag/Ag_2SO_4$). The spectra observed correspond to that of "hydrous $NiO_2$" as we have shown in previous works.[13,33]

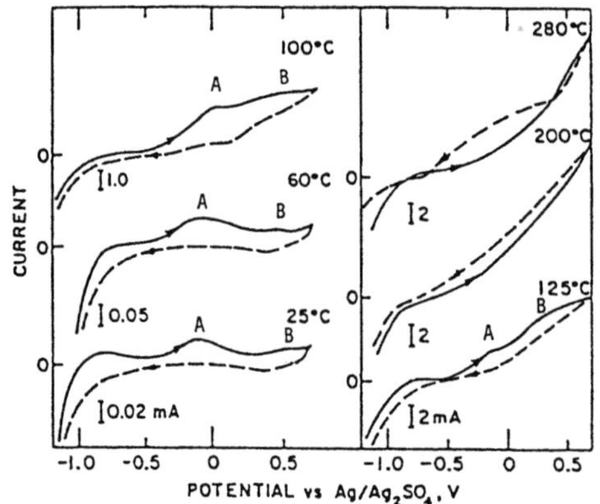


Figure 19. Cyclic voltammograms of Ni in 7 x $10^{-4}$ M $Na_2SO_4$ as a function of temperature. Scan rate = 10 mV/sec.

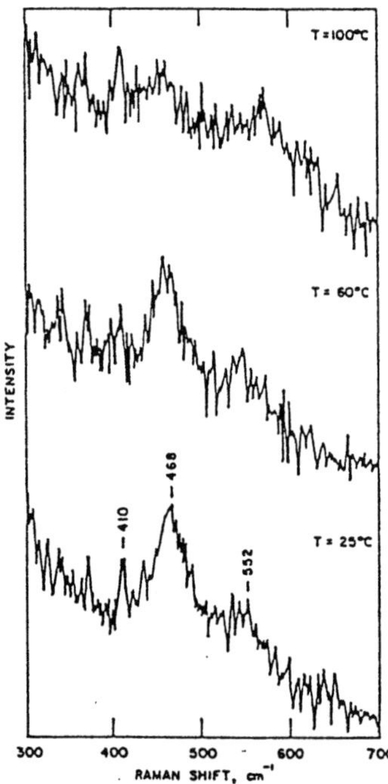


Figure 20. In-situ Raman spectra of secondary passive film on Ni in 7 x $10^{-4}$ M $Na_2SO_4$ at various temperatures.

The film thickness is estimated to be about 100 to 200 Å. Attempts to obtain a spectrum of the film in the primary passivation region were not successful, probably because of the thinness of the film and its low Raman scattering cross-section.

We have demonstrated that it is possible to fruitfully undertake in-situ laser Raman spectro-electrochemical investigations of corrosion and passivation phenomena on metals in aqueous environments at high temperature and pressure. It is possible to extend such measurements to the critical point of water (374°C) or even to supercritical conditions by using appropriate seals. The cell could be modified slightly for UV-visible reflection spectroscopy. Moreover, by replacing the sapphire window ($S_2$) with a metal disc counter electrode, a more uniform current distribution could be achieved in purely electrode-kinetic measurements using fast transient techniques. The working electrode must also be modified accordingly for such studies.

## 4. APPLICATIONS TO STUDIES IN ELECTROCATALYSIS

Several works have appeared in the literature on LRS investigations to elucidate the processes involved in electrocatalysis. Of great interest, in particular, is the use of transition metal macrocyclic compounds such as metallated phthalocyanines and porphyrins for the catalyzed electrochemical reduction of oxygen,[65,66,67] hydrogen,[68] and carbon dioxide,[69,70] and of nitrite (to ammonia).[71] In view of space limitation, we will confine our discussions to the first application. This lecture will also serve as a cursory review of recent literature on the subject.

We digress briefly on the electrochemistry of oxygen reduction. The literature on this subject is very extensive;[65] the understanding of this reaction is of great importance in such applications as fuel cells and corrosion, because it is often the rate controlling process in such systems. The reaction under consideration is thought to proceed in two parallel paths,[65] i.e., for example in acid solutions:

1) The "peroxide" mechanism

$$O_{2(g)} \longrightarrow O_{2(ads)} \xrightarrow[2H^+]{2e^-} H_2O_2 \xrightarrow[2H^+]{2e^-} 2\,H_2O \tag{5}$$

$$H_2O_2 \xrightarrow[\text{catalyst}]{} H_2O + 1/2\,O_2$$

2) Direct 4-electron mechanism

$$O_{2\,(g)} \longrightarrow O_{2(ads)} \xrightarrow[4H^+]{4e^-} 2\,H_2O \tag{6}$$

The first mechanism involves the production of peroxide which can be reduced further to $H_2O$ or can be decomposed catalytically to form $O_2$ and $H_2O$. This pathway appears to predominate on such electrode surfaces as graphite, gold, transition metal oxides, and some metal macrocyclic compounds like cobalt tetrasulfonated phthalocyanine. Direct 4-electron reduction appears to proceed on electrodes like platinum, palladium, silver, tetrasulfonated iron phthalocyanine and a cobalt face-to-face porphyrin.[66]

It is evident that establishment of the proper mechanism for oxygen reduction depends critically on the detection and identification of the intermediates formed; structural transformations that the catalysts undergo are also of utmost importance.

### 4.1. Electrocatalysis by Transition Metal Macrocyclic Compounds

The $O_2$ reduction reactions (5) and (6) proceed at very slow rates under ordinary conditions so that it is necessary to employ a catalyst in practical applications. Transition metal macrocyclic compounds appear to be the most promising at present and have been studied extensively since they were first discovered by Jasinski.[74] Different structural modifications, e.g., phthalocyanines, porphyrins, tetraazaannulenes, cyclic amines, etc. have been tested for catalytic activity.

The literature on electrochemical studies using macrocyclics is rather extensive and no attempt will be made here to review it completely. The interested reader can refer to an earlier review by Jahnke, Schönborn, and Zimmerman.[75] We confine ourselves to studies using metalphthalocyanines and porphyrins. Yeager and his group[62,63] at Case Western Reserve University in particular have made significant contributions to the advancement of the field. They have characterized the electrochemical and spectroscopic properties of phthalocyanines and porphyrins both as thick films and adsorbed monolayers on carbon and metal substrates. They established that cobalt tetrasulfonated phthalocyanine adsorbed on graphite surfaces promoted $O_2$ reduction via a 2-electron process with the formation of peroxide; iron tetrasulfonated phthalocyanine allowed a direct 4-electron transfer to $H_2O$. Van Veen[72] in the Netherlands and Savy et al.[73] in France have likewise conducted detailed studies of phthalocyanine and porphyrin films supported on carbon. They investigated the effect of the central metal atom, electrolyte, and catalyst pretreatment on activity and sought correlations between the structure of the macrocycle and catalytic activity, as well as stability. Van Veen[72] established the order of activity among the phthalocyanines in both acidic and alkaline electrolyte to be Fe> Co≃ Ru> Pd ≃ Pt> Zn. He also confirmed that heat treatment of the macrocycles in $N_2$ or Ar atmosphere at temperatures up to 800°C enhanced the stability and activity of carbon-supported cobalt phthalocyanine (CoPc) and iron phthalocyanine (FePc). This effect was discovered earlier by Bagotzky[76] and his co-workers in the USSR.

Savy, et al.[73] found that polymeric iron phthalocyanines were more active catalytically than the monomeric form. They pointed out the importance of the electronic state of the central metal atom in determining activity and concluded from x-ray, optical, and ESCA measurements, that the active form of the iron ion in FePc has a high spin, $Fe^{+3}$ configuration. More recently, Savy et al.[77] carried out UV-visible reflectance and electroreflectance measurements in conjunction with voltammetric and capacitance studies on vacuum deposited and adsorbed metal phthalocyanines on glassy carbon and Au electrodes. They found that the catalytic activity of FePc is related to the occurrence of a two-metal-centered redox process in the potential range between oxidation and reduction of the phthalocyanine ligand. They claimed that the formation of $\mu$-oxo and $\mu$-peroxo bridges between two iron centers is essential for catalytic activity.

Van der Putten et al.[67] likewise studied the electrochemical behavior of FePc, CoPc, and cobalt tetraazaannulene (CoTAA) irreversibly adsorbed on pyrolitic graphite as a function of pH using cyclic voltammetry and rotating disc electrode techniques. They measured the redox potentials of the macrocycles as a function of pH and studied their behavior towards $O_2$ reduction. They found a clear correlation between the redox potential of the central metal atom and the $O_2$ reduction behavior at low pH. Thus for CoTAA with a redox potential of 470 mV more positive than CoPc, the half wave potential, $E_{1/2}$, for $O_2$ reduction was about 400 mV more positive than $E_{1/2}$ for CoPc, i.e., CoTAA was more catalytically active than CoPc.

Van der Brink et al.[80] studied extensively the kinetics of $O_2$ reduction on thick films of cobalt and iron phthalocyanine. They employed such techniques as cyclic voltammetry and rotating disc (with ring) to determine the rate constants for $O_2$ reduction; they further used AC impedance and ellipsometry to characterize the state of the electrocatalytic surface. They proposed a redox mechanism for the electrocatalysis of the $O_2$ reduction reaction. The reduction kinetics on CoPc was found to be different from that of FePc. The stable reaction product at low cathodic overpotentials on CoPc was $H_2O_2$; only at more negative potentials does further reduction to $H_2O$ occur. No direct reduction of $O_2$ to $H_2O_2$ was found over the entire potential range investigated. No O-O bond rupture occurred in the formation of $H_2O_2$. In the case of FePc, reduction of $O_2$ in alkaline media proceeded directly to $OH^-$, with $HO^{-2}$ formation occurring as well only in a very narrow potential region. $HO^{-2}$ was either reduced to $OH^-$ or decomposed chemically on FePc electrodes. On the scheme of redox catalysis, $O_2$ is adsorbed on the CoPc electrocatalyst at the site of the central metal atom which results in a partial electron transfer to the $O_2$. Electron transfer to the CoPc-$O_2$ complex leads to the production of $HO_2^-$. In the case of FePc, the Fe(III)Pc-$O_2^-$ is reduced in a series of fast electron transfers to produce $OH^-$ and $HO_2^-$.

A number of electrochemical studies have also been carried out by Kuwana and others[78,79] on water soluble porphyrins and their derivatives in solution or adsorbed on carbon electrodes. It appears that these compounds catalyze the reduction of $O_2$ only to the peroxide stage via an electrochemical-chemical (EC) mechanism. As an example, for iron tetra-4-N-methylpyridyl-porphyrin pentachloride, i.e. $Fe(III)TMP^{5+}5Cl^-$, we have the reactions

$$Fe(III)TMP^{+5} + e^- \rightarrow Fe(II)TMP^{+4} \tag{7}$$

$$Fe(II)TMP^{+4} + 1/2\ O_2 + H^+ \xrightarrow{k} Fe(III)TMP^{+5} + 1/2\ H_2O_2. \tag{8}$$

Anson and his group[81,82] have also undertaken extensive investigations on both dissolved and adsorbed metal porphyrins as $O_2$ reduction electrocatalysts. They studied $O_2$ reduction at graphite electrodes coated with thick layers of iron (III) protoporphyrin IX (FePP IX), iron (III) meso-tetraphenylporphine (FeTPP), and iron (III) meso-tetra (3-pyridyl) porphine (FeTPyP). Using rotating disk and ring-disk techniques, they showed that the first two compounds were effective catalysts for the reduction of $O_2$ and $H_2O_2$ to $H_2O$ over a wide pH range. They also proposed a redox-type mechanism which involved an initial reaction between the reduced form of the porphyrin and the substrate. The product of the $O_2$ reduction undergoes further rapid reaction at the electrode. Only the outer portions of the catalyst coating participated in the catalysis.

Collman and his group together with Anson,[66,82] synthesized dicobalt porphyrin dimers (so called face-to-face porphyrins) and found them to have high catalytic activity towards the reduction of oxygen to water without producing significant amounts of hydrogen peroxide. This suggests that $O_2$ bridging of two-cobalt centers facilitates cleavage of the O-O bond and thus direct 4e-reduction to $H_2O$. The catalytic activity of the porphyrin dimers investigated by Collman et al.[66,82] was found to be a sensitive function of the geometry, the type of metal centers and their distance of separation. The mechanism of reduction is again believed to be initially a $2e^-$ transfer resulting in the reduction of two Co(II), followed by reoxidation of these two centers and then bridging by $O_2$; finally, a further $2e^-$ reduction and protonation occurs with the formation of water molecules as shown below.

$$Co^{III}\cdots Co^{III} \xrightarrow{+2e^-} Co^{II}\cdots Co^{II} \xrightarrow{+O_2} Co^{III}\text{-}O\text{-}O\text{-}Co^{III} \qquad (9)$$

$$Co^{III}\text{-}O\text{-}O\text{-}Co^{III} \xrightarrow{+H^+} \left[Co^{III}\text{-}O\text{-}O(H)\text{-}Co^{III}\right]^+ \xrightarrow[3H^+]{+2e^-} Co^{III}\cdots Co^{III} + 2H_2O \qquad (10)$$

We have also studied the reduction of $O_2$ on thick FePc films using cyclic voltammetry and rotating disk (with ring) techniques.[83] We found that the loss of catalytic activity of the film on cycling and with time is due to the production of $H_2O_2$ which results in oxidative dissolution of the catalyst. Heat treatment of carbon-supported phthalocyanines at temperatures up to 900°C as was found by Bagotzky[76] apparently increases the stability of the catalyst most probably by producing modified structures which are more resistant to oxidative dissolution.

There are many other electrochemical studies worthy of note, but this short lecture will not allow reviewing all of them; the author apologizes to those whose work has been inadvertently omitted. This brief survey is meant only to be an overview of the field and the interested reader can refer further to the many other original references cited in the papers quoted herein.

## 4.2. Raman Spectroscopy of Adsorbed Macrocycles

Two types of surface films of metal phthalocyanines and porphyrines have been studied: 1) adsorbed macrocycles on electrodes at monolayer or near-monolayer thickness,[84,85] and 2) films from 100 Å to a few microns thick.[86-88] We consider first some of the investigations on adsorbed layers.[90]

Kötz and Yeager[83b] first obtained the in situ Raman spectra of cobalt tetrasulfonated phthalocyanine (CoTSPc) adsorbed on a silver electrode in 0.05 M $H_2SO_4$ solution. They found that the intensity of the Raman bands increased linearly with the amount of charge transferred during the electrochemical activation of the Ag. The spectra of the adsorbed CoTSPc was very much like that of dissolved CoTSPc, indicating that the molecule interacted only weakly with the surface. They measured the variation in intensity of a number of bands with electrode potential, and found that the intensity of four Raman bands was modified by the presence of $O_2$ in the solution; i.e., bands at 1609 and 1597 $cm^{-1}$ are enhanced significantly, a band at 694 $cm^{-1}$ is diminished in intensity while a new band appears at 1153 $cm^{-1}$ when oxygen is present. They found a surprising similarity between changes in the Raman spectra of the solution species due to $O_2$ saturation and changes in the surface vibrational spectra due to the application of negative potentials. They also observed significant change in spectra at potentials in the hydrogen evolution region which indicate a major alteration in the structure of the adsorbed species.

Simic-Glavaski et al.[84] have characterized extensively the spectroscopic behavior of the tetrasulfonated forms of metal free phthalocyanine ($H_2$TSPc), iron phthalocyanine (FeTSPc), and cobalt phthalocyanine (CoTSPc). All molecules were adsorbed on Ag electrodes and were studied in acidic, basic, and neutral solutions. They also studied the solution phase species as well as the effect of $O_2$ on the system using UV-visible and Raman spectroscopy. The UV-visible absorption spectra of $H_2$TSPc and CoTSPc showed bands at about 335 nm (Soret, $\gamma$ or B(0,0)), 630 nm ($\beta$ or Q(1,0)) and 660 nm ($\alpha$ or Q(0,0)). These bands are thought to be due to $\pi \rightarrow \pi^*$ transitions within the delocalized phthalocyanine ring system. Marked changes in the absorption spectra for $H_2$TSPc were observed when the pH was varied. The Q band splitting observed in neutral and alkaline solutions disappeared in acid solution due to protonation which changed the molecular symmetry from $D_{2h}$ to $D_{4h}$. The effect of pH on the absorption spectra of CoTSPc was not as marked, however, variation in concentration of the macrocycle resulted in higher intensity of the Q(0,0) band and a shift toward longer wavelength by as much as 20 nm as the CoTSPc concentration was lowered. A shift in the Q (0,0) band was also found upon $O_2$ saturation of acid and alkaline solutions. This shift is believed to be due to the formation of an $O_2$ adduct with the macrocycle.

The Raman spectrum of adsorbed phthalocyanine was found to be very much like that in the solid and in solution; this suggests a rather weak interaction between the macrocycle and the Ag electrode, i.e., physisorption rather than chemisorption. The cyclic voltammetry curves showed two peaks in the anodic and cathodic cycles for $H_2$TSPc, CoTSPc and FeTSPc which led the authors to conclude that these correspond to the oxidation and reduction of the TSPc ligand rather than the central metal atom. Raman spectra were recorded over the entire pH range of 1-13. The bands did not show any shift in frequency as a function of pH but the intensities showed significant change when the electrode potential was altered. The 612 $cm^{-1}$ band for FeTSPc, 763 $cm^{-1}$ for CoTSPc, and 506 $cm^{-1}$ for $H_2$TSPc were found to shift in frequency as a function of potential. The 612 $cm^{-1}$ band for FeTSPc showed two characteristic inflection points in the plot of frequency shift vs potential which appeared to coincide with the two oxidation peaks in the cyclic voltammogram. The authors followed the change in intensity of selected bands as the voltage was scanned and then constructed what they called Voltage-Raman (VR) plots. The curves often exhibited a maximum at certain voltages which shift as the pH of the solution was varied. The most noticeable change in the Raman band intensities occurred in potential regions of the oxidation-reduction peaks; the intensity of a Raman band appeared to be a function of the oxidation state of the adsorbed phthalocyanine.

Substantial changes were also observed in the Raman intensities of the adsorbed macrocycles upon oxygenation of the solution. The maxima in the VR curves were found to shift to more negative potentials in the oxygenated solution compared to the deoxygenated ones. This potential shift paralleled a similar shift of the VR maxima with pH which was explained on the basis of a shift in the redox potential of the $O_2$ adduct with pH.

We have also obtained the Raman spectra of iron phthalocyanine (FePc) adsorbed on Cu and Au electrodes[85a] in 0.05 M $H_2SO_4$. Adsorption was originally carried out by immersing the metal electrode in a dilute solution of the macrocycle in pyridine as solvent. Transfer of the electrode to the aqueous solution resulted in the appearance of a band at 1529 $cm^{-1}$ in situ which was interpreted as being due to interaction of the FePc with water molecules at the bridging nitrogen positions. A decrease in intensity and broadening of some vibrational bands was observed on increasing cathodic polarization; these were attributed to a lifting of the degeneracy of some vibrational modes due to a change in symmetry of the adsorbed molecules brought about by polarization induced by the double-layer field.

We have likewise studied the surface Raman scattering and electrochemistry of iron protoporphyrin IX chloride, Fe(III)PPCl, adsorbed on a silver electrode in sodium borate solution.[85b] Modest catalysis of the $O_2$ reduction reaction was found in the presence of Fe(III)PPCl. It was possible to discriminate between the reduction of Fe(III) occurring in the solution and that on the electrode surface. The Raman data indicated that an intermediate-spin, 4-coordinate iron (III) porphyrin (Fe(III)PP) was adsorbed on the electrode and a high-spin, more highly coordinated iron (II) species (Fe(II)PP) existed in the solution.

## 4.3. Investigations of Thick Films

Electrochemical and spectroscopic characterization of fairly thick phthalocyanine films had been carried out in a number of laboratories.[86-95]

Aroca and his co-workers[96,97] studied the Raman spectra of a variety of phthalocyanine films supported on glass substrates. They gave a description of the various modes of vibration of these macrocycles and showed some correlations in the vibrational properties of the series Mg, Cu and Zn phthalocyanines. They also used polarization Raman techniques (i.e., a study of the scattering of polarized light) to probe the molecular orientation of submicron films of phthalocyanines. They found a splitting of certain vibrational modes of the molecules which is due to the crystalline field of the film. This correlation field or "Davydov splitting" could provide unique and useful information about the crystalline nature of organic thin films.

Kahl et al.[88] studied the reversible oxidation and reduction of magnesium phthalocyanine films (1000-2000 Å thick) supported on a Au substrate in aqueous solutions of NaCl, $NaNO_3$ and potassium hydrogen phthalate (KHP). They observed changes in the Raman spectra of MgPc during oxidation and reduction which they associated with the vibrational modes of the Pc rings; these led them to the conclusion that the oxidation occured at the Pc ligand instead of the central metal atom. They also demonstrated that the anions from the solution moved into and out of the organic film during the oxidation and reduction processes as would be necessary for proper charge compensation. Their results also showed that anion size determined the rate of oxidation, i.e., smaller anions ($Cl^-$, $ClO_4^-$) allowed rapid rates while larger ones, e.g. the biphthalate anion, permitted only limited or "surface" oxidation. $SO_4^{=}$, $HPO_4^{=}$ and $SiF_6^{=}$ did not allow oxidation of the MgPc at potentials up to 0.9 V vs SCE.

We have also studied the electrochemical and laser Raman spectroscopic behavior of thick iron phthalocyanine films supported on carbon.[90] We found that different methods of preparing the film resulted in different polymorphic forms of FePC on the surface, which in turn yielded varying catalytic activity for $O_2$ reduction. The films were found to deactivate during voltammetric cycling; we subsequently showed this to be due to the anodic dissolution of the film. LRS indicated coordination of the FePc with $H_2O$ molecules in the solution. The vibrational spectra, however, did not show any significant changes in the potential range from 0.4 to -0.9V vs $Hg/Hg_2SO_4$.

### 4.4. Theoretical Analysis of Vibrational Spectra

The successful interpretation of Raman spectral data to obtain information on molecular structure depends acutely on a theoretical model of the atomic assembly comprising the system and the force of the interactions between atoms. The normal modes of vibration of a molecular system consisting of N atoms can be determined by a procedure developed by Wilson and co-workers that has come to be known as the FG Matrix method[98] of normal coordinate analysis. Details of this theoretical method appear in a number of publications[18,98] and need not be elaborated here.

Normal coordinate treatments of the vibrational modes of metalloporphyrins have been performed by a number of workers.[99-101] Jennings et al.[102] have made assignments of the vibrational modes of Mg, Cu and Zn phthalocyanines. Using normal coordinate analysis, we have assigned the planar modes of vibration of iron phthalocyanine.[103] Good agreement was obtained between calculated and observed frequencies using an adjusted valence force field (AVFF) and incorporating empirical correlations for related types of force constants. We found the assignments of the normal modes of vibration to be consistent with other available experimental and theoretical information for phthalocyanine complexes and related metalloporphines. However, the extensive mixing of the internal coordinates within the calculated normal modes showed clearly that assignments of specific vibrational frequencies to individual internal coordinates is generally not justified. The vibrational kinematics appear to involve concerted motions of the various interior and exterior ring systems.

## 5. LIMITATIONS OF THE RAMAN SPECTROSCOPIC TECHNIQUE

While we have highlighted the advantages of the LRS technique for studies in corrosion and electrocatalysis, it is also important to point out the limitations. There is plenty of room for improvement in detection sensitivity to enable the study of corrosion and passivation films at near monolayer levels. This is inherently difficult for oxide or hydroxide films because the corrosion process itself causes roughening of the metal surface and thus increases the background of diffuse scattered light. The low frequency of the metal-oxygen vibration (generally less than 1000 $cm^{-1}$ and hence closer to the Raleigh line) compounds the problem of picking up a weak signal from a high background. Metal oxides have, morever, also relatively low cross sections for Raman scattering.

Black and dark colored samples generally have a high absorption coefficient for visible light and hence exhibit weak Raman scattering. Sample heating by the incident laser can be minimized by spinning the sample or through the use of other related techniques. It is also very important experimentally to check that the Raman spectra obtained are not due to the decomposition products of photoreactive samples. One way of doing this is to obtain spectra at different laser powers and to use the lowest power possible on the sample.

The laser beam used to probe the sample can cause other structural and compositional changes on the sample, e.g. the drying of normally hydrous surface films, sample decomposition, etc. The need to determine film compositions in-situ cannot be overemphasized and spectra obtained ex-situ should always be interpreted with caution.[38c,42]

## 6. FUTURE OUTLOOK

We have given here a survey of the applications of Raman spectroscopy in conjunction with electrochemical studies in aqueous corrosion and electrocatalysis. The survey is by no means complete and will undoubtedly be rather quickly outdated. The author anticipates more and more use of the technique in electrochemistry. New developments in detector technology and advances in other aspects of Raman instrumentation, such as better monochromators and filters, will enhance sensitivity to the region of monolayer and submonolayer quantities of electrocatalysts and corrosion films. Such capability has already been demonstrated for cyanopyridine on Rh.[104] The phenomenon of SERS will continue to be applied to appropriate systems using Au, Ag, and Cu electrodes. The development of Fourier transform Raman spectrometers should bring about significant reduction in the cost of Raman facilities and hence encourage wider use of the technique. The combination of FTIR and Raman spectroscopy in a single instrument is becoming popular and would represent a powerful capability for electrochemical and surface science laboratories.

## 7. ACKNOWLEDGEMENT

The author acknowledges support of his research by the U.S. Department of Energy, Office of Basic Energy Sciences, Division of Materials Science, under contract W-31-109-Eng-38. He also thanks his colleagues and students, among them W. Ruther, J. J. McMahon, S. Xu, X. Feng, W. Paden, C. Rios, W. Walczak, T. Tipton, N. Camillone III, L. Pena-Cordova, R. Grubbs, and J. Solis-Fernandez, for their help in the conduct of laboratory experiments. The continued encouragement of V. A. Maroni and D. M. Gruen at Argonne National Laboratory is gratefully acknowledged.

## 8. REFERENCES


1a. M. C. Tobin, Laser Raman Spectroscopy, Wiley Interscience, New York (1971).

b. P. R. Carey, Biochemical Applications of Raman and Resonance Raman Spectroscopies, Academic Press, NY (1982).

2a. M. Fleischmann, P. J. Hendra and A. J. McQuillan, J. Chem. Soc. Chem. Communications (1973), 80.

b. A. Campion, Vibrational Spectroscopy of Molecules on Surfaces, J. T. Yates Jr. and T. E. Madey, editors, Plenum Press, NY, 1987. p. 345.

c. D. K. Veirs, V. K. F. Chia, and G. M. Rosenblatt, Applied Optics, **26**, 3530 (1987).

3. R. Kötz and E. Yeager, J. Electroanal. Chem. **111**, 105 (1980).

4. J. DeSilvestro and M. J. Weaver, J. Electroanal. Chem. **209**, 377 (1986).

5a. R. J. Thibeau, C. W. Brown, A. Z. Goldfarb and R. H Heidersbach, J. Electrochem. Soc. **127**, 37 (1980).

b. E. S. Reid, R. P Cooney, P. J. Hendra and M. Fleischman, J. Electroanal. Chem. **80**, 405 (1977).

6. R. Varma, C. A. Melendres and N. P. Yao, J. Electrochem. Soc. **207**, 1416 (1980).

7. C. A. Melendres, S. Xu and B. Tani, J. Electroanal. Chem. **162**, 343 (1984).

8. J. C. Hamilton, J. C. Farmer, and R. J. Anderson, J. Electrochem. Soc. **133**, 739 (1986).

9. C. A. Melendres and S. Xu, J. Electrochem. Soc. **131**, 2239 (1984).

10. J. Dünnwald, R. Lossy and A. Otto in Passivity of Metals and Semiconductors, M. Froment (editor), Elsevier Science Publishers, Amsterdam (1983) p. 107.

11. M. Froelicher, A. Hugot Le-Goff, C. Pallotta, et al. - ibid. p. 101.

12. I.C.G. Thanos, Electrochimica Acta **31**, 811 (1986).

13. K. Carr, S. Baer and C. A. Melendres, Corrosion, **42**, 307 (1986).

14. J. DeSilvestro, D. A. Corrigan and M. J. Weaver, J. Electrochem. Soc. **35**, 885 (1988).

15. R. J. Thibeau, C. W. Brown, et. al., J. Electrochem. Soc. **127**, 1913 (1980).

16. J. J. McMahon, W. Ruther, and C. A. Melendres, J. Electrochem. Soc. **135**, 557 (1988).

17. M. Pourbaix "Atlas D'equilibres Electrochimiques", Gauthiers-Villars, Paris (1963) p. 489.

18. K. Nakamoto, Infrared and Raman Spectra of Inorganic and Coordination Compounds, Third Edition, John Wiley, New York (1978).

19. W. G. Fateley, F. R. Dollish, et.al Infrared and Raman Selection Rules for Molecular and Lattice Vibrations: The Correlation Method, Wiley Interscience, New York (1972).

20. J. Burbank, A. C. Simon and E. Willihnganz in Advances in Electrochem. and Electrochem. Enginnering, edited by C. W. Tobias and P. Delahay, vol. 8, (1971) p. 157.

21. R. K. Chang and T. E. Furtak, editors, Surface Enhanced Raman Scattering, Plenum Press, NY (1982).

22. P. Gao, M. L. Patterson, et al., Langmuir, **1**, 173 (1985).

23. J. DeSilvestro and M. J. Weaver, J. Electroanal Chem. **209**, 377 (1986).

24. B. E. Conway and H. Angerstein-Kozlowska, NBS Publication **455**, 107 (1976).

25. N. Sato and T. Ohtsuka, J. Electrochem. Soc. **125**, 1735 (1978).

26. K. Kudo, N. Sato and T. Ohtsuka in "Passivity of Metals", R. P. Frankenthal and J. Kruger, editors, the Electrochem. Soc. monograph series, Princeton, New Jersey (1978) p. 918.

27. H. Leidheiser Jr, G. W. Simmons and E. Kellerman, J. Electrochem. Soc. **120**, 1516 (1973).

28. G. W. Simmons, E. Kellerman and H. Leidheiser Jr, J. Electrochem. Soc. **123**, 1276 (1976).

29. P. Delichere, A. Hugot Le-Goff and N. Yu "Proceedings of the Tenth International Conference on Raman Spectroscopy, W. L. Peticolas and B. Hudson, editors, University of Oregon Press, Eugene, Oregon (1986) p. 5-45.

30. J. DeSilvestro, D. A. Corrigan and M. J. Weaver, J. Phys. Chem. **90**, 6408 (1986).

31. F. Hahn, B. Beden, J. J. Croissant and C. Lamy, Electrochimica Acta **31**, 335 (1986).

32. K. Carr, S. Baer and C. A. Melendres, Corrosion **42**, 307 (1986).

33. C. A. Melendres, W. Paden, B. Tani and W. Walczak, J. Electrochem. Soc. **134**, 762 (1987).

34. D. A. Corrigan and S. L. Knight, J. Electrochem. Soc. **136**, 613 (1989).

35. C. A. Melendres and B. S. Tani, J. Electrochem. Soc. **133**, 1059 (1986).

36. G. Nazri, Ph.D. Thesis, Case Western Reserve University, 1981. (University Microfilms #3267, Ann Arbor, Michigan).

37a. O. Glemser and J. Einerhand, Z. Anorg. Chem. **261**, 26 (1950; Z. Elektrochemie **54**, 302 (1954).

b. R. W. Cairns and E. Ott, J. Amer. Chem. Soc. **55**, 527 (1933); **55**, 534 (1973).

c. P. Bro and D. Cogley, J. Electrochem. Soc. **113**, 521 (1966).

d. K. Nakagawa, R. Konaka, and T. Nakata, J. Amer. Chem. Soc. **27**, 1597 (1962).

38a. A. Hugot Le-Goff and C. Pallotta in Proceedings of the Tenth International Conference on Raman Spectroscopy, W. L. Peticolas and B. Hudson, editors, University of Oregon Press, Eugene, Oregon (1986) p. 6-16.

b. A. Hugot Le-Goff and C. Pallotta, J. Electrochem. Soc. **132**, 2805 (1985).

c. A. Felske and W. J. Plieth, Electrochimica Acta, **34**, 75 (1989).

39. R. J. Thibeau, C. W. Brown and R. H. Heidersbach, Applied Spectroscopy **32**, 532 (1978).

40. J. T. Keiser, C. W. Brown and R. H. Heidersbach, Corrosion, **38**, 352 (1982).

41. G. Nazri and E. Yeager, Extended Abstracts, Electrochem. Soc. Meeting, San Francisco, CA., May 8-13, 1983; Abstract No. 54.

42. C. A. Melendres, N. Camillone III and T. Tipton, Electrochimica Acta., **34**, 281 (1989).

43. J. T. Keiser, C. W. Brown and R. H. Heidersbach, Corrosion, **38**, 357 (1982).

44. J. T. Keiser, C. W. Brown and R. H. Heidersbach, Proceedings of the Second International Conference on Environmental Degradation of Materials, Virginia Polytechnic Institute and State University, Blacksburg, VA, Sept. 22-23, 1981, p. 43.

45. D. J. Gardiner, M. Bowden and P. R. Graves, Phil. Trans. Roy. Soc. Lond. **A320**, 295 (2986).

46. R. L. Farrow, R. E. Benner, A. S. Nagelberg and P. L. Mattern, Thin Solid Films, **73**, 353 (1980).

47. D. Thierry, D. Persson, C. Leygraf, et. al., J. Electrochem. Soc. **132**, 305 (1985).

48. T. Ohtsuka, J. Guo and N. Sato, J. Electrochem. Soc. **133**, 2473 (1986).

49. See for example, Proceedings of the Sixth European Symposium on Corrosion Inhibitors, Ferrara, Italy, Sept. 16-20, 1985, vols. I & II.

50. G. H. Cartledge, Corrosion **18**, 316t (1962).

51a. J. J. Kester, T. E. Furtak, and A. J. Bevolo, J. Electrochem. Soc., **129**, 1716 (1982).

b. D. Thierry and C. Leygraf, J. Electrochem. Soc. **133**, 2236 (1986).

52. B. H. Loo, Y. G. Lee and A. Etlage in Proceedings of the Ninth International Conference on Raman Spectroscopy, Tokyo, Japan, Aug. 27-Sept. 1, 1984, p. 538.

53. B. H. Loo, Chem. Phys. Letters **89**, 346 (1982).

54. J. Rubin, I. G. R. Gutz, O. Sala and W. J. Orville-Thomas, J. of Molecular Structure, **100**, 571 (1983).

55. D. Thierry and C. Leygraf, J. Electrochem. Soc. **132**, 1009 (1985).

56. M. Ohsawa and W. Süetaka, J. Electron Spec. and Related Phenomena **30**, 221 (1983).

57. C. A. Melendres, J. J. McMahon and W. Ruther, J. Electroanal. Chem. **208**, 175 (1986).

58. W. Ruther, J. J. McMahon and C. A. Melendres, in Surfaces, Inhibition and Passivation - Proceedings of an International Symposium, E. McCafferty and R. J. Brodd, editors, vol. 86-7, The Electrochem. Soc. Inc., Princeton, New Jersey (1986).

59. M. J. Danielson, Corrossion **35**, 200 (1979).

60. M. Lepeintre, C. Mahieu and J. Monjou, C. R. Acad. Sci. Paris **261**, 3389 (1965).

61. R. Heidersbach, C. Brown, R. Thibeau and A. Goldfarb, Technical Report No. 4, Contract No. N00014-76-C-0889, Office of Naval Research, Metallurgy Program - Code 471, Aug. 1978.

62. E. Yeager, in Proceedings of the Symposium on Electrode Materials and Processes for Energy Conversion and Storage, edited by J. D. McIntyre, S. Srinivasan and F. G. Will, The Electrochem. Soc. Inc., Princeton, NJ (1977), p. 149.

63. J. H. Zagal-Moya, Ph.D. Thesis, Case Western Reserve University (1978), University Microfilms, No. 7816494, Ann Arbor, MI, 48106.

64. B. Z. Nikolic, R. R. Adzic and E. Yeager, J. Electroanal. Chem. **103**, 281 (1979).

65. M. Tarasevich, A. Sadkowski and E. B. Yeager in Comprehensive Treatise in Electrochemistry, B. E. Conway, J. O'M. Bockris, et al., editors, Plenum Press, NY (1983); vol. VII, p. 301.

66. J. P. Collman, P. Denisevich, P. Konai, M. Marrocco, C. Koval and F. C. Anson, J. Amer. Chem. Soc. **102**, 6027 (1980).

67. A. Vander Putten, A. Elzing, W. Visscher and E. Barendrecht, J. Electroanal. Chem. **221**, 95 (1987).

68. R. M. Kellett and T. G. Spiro, Inorg. Chem. **24**, 2378 (1985).

69. S. Kapusta and N. Hackerman, J. Electrochem. Soc. **131**, 1511 (1984).

70. D. Masheder and K. P. J. Williams, J. Raman Spec. **18**, 387 (1987).

71. M. H. Barley, K. Takeuchi, W. R. Murphy Jr. and T. J. Meyer, J. Chem. Soc. Chem. Commun. **1985**, 507.

72. J. A. R. Van Veen and C. Visser, Electrochimica Acta **24**, 921 (1979).

73. M. Savy, C. Bernard and C. Magner, Electrochimica Acta **20**, 383 (1975).

74. R. Jasinski, J. Electrochem. Soc. **112**, 526 (1965).

75. H. Jahnke, M. Schönborn and G. Zimmerman, Topics in Current Chemistry, **61**, 133 (1976).

76. V. S. Bagotzky, M. R. Tarasevich, K. A. Radyushkina, et. al., J. Power Sources **2**, 233 (1977/78).

77. D. Van den Ham, C. Hinnen, G. Magner and M. Savy, J. Phys. Chem. **91**, 4743 (1987).

78. T. Kuwana, M. Fujihira, K. Sunakawa and T. Osa, J. Electroanal. Chem. **88**, 299 (1978).

79. N. Kobayashi, M. Fujihira, et al., J. Electroanal. Chem. **101**, 269 (1979); **103**, 427 (1979).

80. F. Van der Brink, W. Visscher and E. Barendrecht, J. Electroanal. Chem. **157**, 283 (1983); **157**, 305 (1983); **172**, 301 (1984); **175**, 279 (1984).

81. R. Durand Jr. and F. C. Anson, J. Electroanal. Chem. **143**, 273 (1982).

82. J. P. Collman, M. Marrocco, P. Denisevich, C. Koval and F. C. Anson, J. Electroanal. Chem. **101**, 117 (1979).

83a. C. A. Melendres and X. Feng. J. Electrochem. Soc. **130**, 811 (1983).

b. R. Kötz and E. Yeager, J. Electroanal. Chem. **113**, 113 (1980).

84. B. Simic-Glawaski, S. Zecevic and E. Yeager, J. Electroanal. Chem. **150**, 469 (1983); **194**, 155 (1985); J. Raman Spectroscopy **14**, 338 (1983); J. Physical Chem. **87**, 4555 (1983); J. Amer. Chem. Soc., **107**, 5625 (1985).

85a. C. A. Melendres, C. B. Rios, X. Feng and R. McMasters, J. of Physical Chem. **87**, 3526 (1983).

b. J. J. McMahon, S. Baer, and C. A. Melendres, J. of Physical Chem., **90**, 1572 (1986).

86. F. R. Fan and L. R. Faulkner, J. Amer. Chem. Soc. **101**, 4779 (1979); H. Tachikawa and L. R. Faulkner, J. Amer. Chem. Soc. **100**, 4379 (1978).

87. J. M. Green and L. R. Faulkner, J. Amer. Chem. Soc. **105**, 2950 (1983).

88. J. L. Kahl, L. R. Faulkner, K. Dwarakanath and H. Tachikawa, J. Amer. Chem. Soc. **108**, 5434 (1986).

89. C. A. Melendres and X. Feng, J. Electrochem. Soc. **130**, 811 (1983).

90. C. A. Melendres and F. A. Cafasso, J. Electrochem. Soc. **128**, 755 (1981).

91. A. J. Appleby and M. Savy, in Proceedings of the Symposium on Electrode Materials and Processes for Energy Conversion and Storage, edited by J. D. E. McIntyre and S. Srinivasan, The Electrochem. Soc., Princeton, New Jersey, 77-6, 321 (1977).

92. M. Savy, P. Andro, C. Bernard and G. Magner, Electrochem. Acta. **18**, 191 (1973).

93. C. Paliteiro, A. Hamnett, and J. B. Goodenough, J. of Electroanal. Chem., **249**, 167 (1988).

94. D. Scherson, S. B. Yao, E. B. Yeager, et al., Applications of Surface Science, **10**, 325 (1982).

95. D. A. Scherson, S. B. Yao, E. B. Yeager, et al., J. Electroanal. Chem. **150**, 535 (1983).

96. R. Aroca, C. Jennings, R. O. Loutfy and A. Hor, J. Phys. Chem. **90**, 5255 (1986).

97. C. Jennings, R. Aroca, A. Hor and R. O. Loutfy, J. Raman Spec. **15**, 34 (1984).

98. E. B. Wilson Jr., J. C. Decius and P. C. Cross, Molecular Vibrations, McGraw Hill, NY (1955).

99. S. Sunder and H. J. Bernstein, J. Raman Spectroscopy, **5**, 351 (1976).

100. H. Ogoshi, Y. Saito and K. Nakamoto, J. Chem. Phys. **57**, 4194 (1972).

101. M. Abe, T. Kitagawa and Y. Kyogoku, J. Chem. Phys. **16**, 4526 (1978).

102. C. Jennings, R. Aroca, A. M. Hor and R. O Loutfy, J. Raman Spectroscopy **15**, 35 (1984).

103. C. A. Melendres and V. A. Maroni., J. Raman Spectroscopy **15**, 319 (1984).

104. C. Shannon and A. Campion. J. Phys. Chem. **92**, 1385 (1988).

# UV-VISIBLE REFLECTANCE SPECTROSCOPY IN ELECTROCHEMISTRY

W. Plieth
Institute of Physical Chemistry
Free University of Berlin
Takustr. 3, D-1000 Berlin 33

ABSTRACT. A description of the reflection of UV-visible radiation in two-phase and three-phase systems is followed by a discussion of the special problems of reflection measurements in an electrochemical system. While former reviews have concentrated on the influence of the metallic phase (metal electroreflectance), here the influence of the electrolyte side of the double layer is also considered. Examples for the application of reflectance measurements to the adsorption of anions and organic adsorbates are given. Some applications of UV-visible reflectance measurements to quasi monomolecular oxide layers are discussed.

## 1. INTRODUCTION

The development of in-situ UV-visible reflectance spectroscopies for examining electrode-electrolyte interfaces began approximately 20 years ago /1,2/. An alternative method, ellipsometry, had been applied earlier but at a single wavelength /3/. After an initial enthusiastic development of optical spectroscopies, the discovery of surface enhanced Raman spectroscopy /4/ shifted the focus of interest towards vibrational spectroscopy. This process continued with the introduction of modulated IR /5/ and Fourier transform IR reflectance spectroscopies /6,7/. However, there has been a revival of interest in UV-visible reflectance spectroscopies. Here we will summarize their theoretical background and their main application areas. The results will be demonstrated with the help of some selected experimental samples.

## 2. LIGHT AS AN ELECTROMAGNETIC VIBRATION

Maxwell's theory /8/ describes light as an alternating oscillation of an electric and a magnetic vector, E and H,



*C. Gutiérrez and C. Melendres (eds.),*
*Spectroscopic and Diffraction Techniques in Interfacial Electrochemistry*, 223–260.

respectively. For a plane wave travelling in a homogeneous, non-absorbing medium the electric and magnetic vectors propagate along a direction z:

$$E = E_0' \cos\left[ 2\pi\nu \left(t - \frac{z}{v}\right) + \delta \right] \tag{1}$$

$$H = H_0' \cos\left[ 2\pi\nu \left(t - \frac{z}{v}\right) + \delta \right] \tag{2}$$

where $v = c/\sqrt{\varepsilon\mu}$ is the velocity of light in the medium of dielectric constant $\varepsilon$ and magnetic permeability $\mu$, c the velocity of light in vacuum, $\delta$ the phase displacement, $\nu$ the frequency of the electromagnetic vibration, t the time, and $E_0'$ and $H_0'$ the amplitudes of the electric and magnetic vectors, respectively. With the definition of the refraction index $n = c/v$ one obtains

$$n = \sqrt{\varepsilon\mu} \tag{3}$$

It is often convenient to use the complex exponential notation:

$$\hat{E} = E_0' \ \exp i\left[ 2\pi\nu \left(t - \frac{n}{c} z\right) + \delta \right] \tag{4}$$

$$\hat{H} = H_0' \ \exp i\left[ 2\pi\nu \left(t - \frac{n}{c} z\right) + \delta \right] \tag{5}$$

In order to include absorbing media in the described formalism the exponential decay of the electrical amplitude $E_0'$ with distance z has to be considered:

$$E_0' = E_0 \exp\left(- \frac{2\pi k}{\lambda} z\right) = E_0 \exp\left(- \frac{2\pi\nu k}{c} z\right). \tag{6}$$

where k is the extinction coefficient. By inserting eq. (6) into eq. (4) one obtains

$$\hat{E} = E_0 \exp i\left[ 2\pi\nu\left(t - \frac{n - ik}{c} z\right) + \delta \right] \tag{7}$$

An absorbing medium follows the Maxwell theory if the refraction index n is substituted by its complex counterpart $\hat{n}$ defined as

$$\hat{n} = n - ik \quad (8)$$

Alternatively, the complex refraction index, $\hat{n}$ , can be substituted by the complex dielectric constant, $\hat{\varepsilon}$,

$$\hat{\varepsilon} = \varepsilon_{re} - i\ \varepsilon_{im} \quad (9)$$

which is related to $\hat{n}$ by the equations

$$\varepsilon_{re} = n^2 - k^2 \quad (10)$$

$$\varepsilon_{im} = 2\ nk. \quad (11)$$

provided the magnetic permeability $\mu$ is unity.

An alternative concept to describe absorbing media is the following one. In the case of metals, whose optical properties largely depend on the presence of free electronic charges, Maxwell's equation includes a term containing the conductivity $\sigma$:

$$\hat{\varepsilon} = \varepsilon_{re} - i\ \frac{2\sigma}{\nu} \quad (11)$$

From eqs. (11) and (12) one obtains the following relation between the absorption coefficient and the surface conductivity:

$$nk = \frac{\sigma}{\nu} \quad (13)$$

## 3. REFLECTION AT AN ELECTRODE/ELECTROLYTE INTERFACE

### 3.1 Reflection of Light

If the electromagnetic wave encounters a phase boundary (between phase 1 and 3) (Fig. 1), the light is partially transmitted and partially reflected. The reflected electromagnetic wave is described by the Fresnel reflection coefficient, which is the ratio of the amplitudes of the reflected (r) and incident (i) light beams,

$$\hat{r} = E_r/E_i. \quad (14)$$

The Fresnel reflection coefficient depends on the polarisation of light, i.e., on the orientation (perpendicular ($\perp$) or parallel ($\parallel$)) of the electric vector with respect to the plane of incidence, which is that defined by the incident and reflected beams (Fig. 1). Assuming that the magnetic permeability is unity, it is:

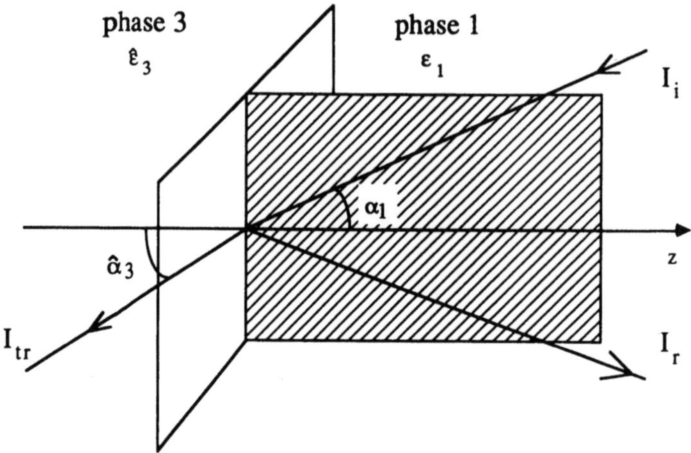


Fig. 1 Reflection at the phase boundary between phase 1 (electrolyte) and phase 3 (electrode). $I_i$, $I_r$, $I_{tr}$ are the incident , reflected, and transmitted intensities, respectively. $\varepsilon_1$ and $\alpha_1$ are the dielectric constant and the angle of incidence in the electrolyte. $\hat{\varepsilon}_3$ and $\hat{\alpha}_3$ are the complex dielectric constant and the complex angle of incidence in the electrode. z is the direction normal to the surface.

$$\hat{r}_{\perp} = \frac{n_1 \cos \alpha_1 - \hat{n}_3 \cos \hat{\alpha}_3}{n_1 \cos \alpha_1 + \hat{n}_3 \cos \hat{\alpha}_3} \tag{15a}$$

$$\hat{r}_{\parallel} = \frac{\hat{n}_3 \cos \alpha_1 - n_1 \cos \hat{\alpha}_3}{\hat{n}_3 \cos \alpha_1 + n_1 \cos \hat{\alpha}_3} \tag{15b}$$

where $\alpha_1$ and the complex quantity $\hat{\alpha}_3$ are the angles of incidence and refraction, respectively, measured with respect to the interface normal (the angle $\hat{\alpha}_3$ is complex because phase 3 is absorbing) (Fig. 1). $\hat{n}_3$ is the complex index of refraction of the solid phase (phase 3). It is assumed that phase 1 is a transparent dielectric for which $k_1 = 0$.

## 3.2 Reflection in Presence of a Surface Film

Obviously, the presence of a surface film (phase 2 in Fig. 2) will change the reflectivity. The new reflection coefficient is given by /9/:

$$\hat{r}_{123} = \frac{\hat{r}_{12} + \hat{r}_{23} \exp(-i\, 4\pi d/\lambda\, \hat{n}_2 \cos\hat{\alpha}_2)}{1 + \hat{r}_{12}\, \hat{r}_{23} \exp(-i\, 4\pi d/\lambda\, \hat{n}_2 \cos\hat{\alpha}_2)} \tag{16}$$

where the index 2 refers to the intermediate layer and d is the layer thickness. For many electrochemical systems it is possible to linearize eq. (16), since for them $d/\lambda \ll 1$, in which case it is /9/:

$$\hat{r}_{\perp,123} = \hat{r}_{\perp,13} \left[ 1 - 4\pi \frac{d}{\lambda} \frac{\sqrt{\varepsilon_1} \cos\alpha_1}{(\varepsilon_1 - \hat{\varepsilon}_3)} (\hat{\varepsilon}_2 - \hat{\varepsilon}_3) \right] \tag{17a}$$

$$\hat{r}_{\parallel,123} = \hat{r}_{\parallel,13} \left[ 1 - 4\pi \frac{d}{\lambda} \frac{\sqrt{\varepsilon_1} \cos\alpha_1}{(\varepsilon_1 \cos^2\hat{\alpha}_3 - \hat{\varepsilon}_3 \cos^2\alpha_1)} \cdot \right.$$

$$\left. \cdot \left[ \cos^2\hat{\alpha}_3 (\hat{\varepsilon}_2 - \hat{\varepsilon}_3) + \hat{\varepsilon}_3^2 \sin^2\hat{\alpha}_3 \left(\frac{1}{\hat{\varepsilon}_2} - \frac{1}{\hat{\varepsilon}_3}\right)\right] \right] \tag{17b}$$

where the film-free surface constitutes the reference value, $\hat{r}_{13}$, for the film-covered case. These equations can easily

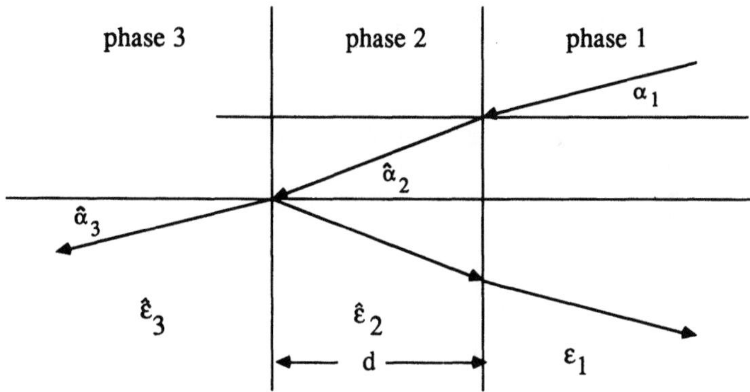


Fig. 2 Reflection at a system with an intermediate surface layer (phase 2) of thickness d and complex dielectric constant $\hat{\varepsilon}_2$ (three-layer model).

be extended to the case of a film with uniaxial anisotropy whose dielectric constant is represented by a tensor with two components, $\hat{\varepsilon}_t$ and $\hat{\varepsilon}_n$, tangential and normal to the surface, respectively. The following changes then have to be made in eq. 17:

$$(\hat{\varepsilon}_2 - \hat{\varepsilon}_3) \rightarrow (\hat{\varepsilon}_{2,t} - \hat{\varepsilon}_3) \tag{18a}$$

$$\left(\frac{1}{\hat{\varepsilon}_2} - \frac{1}{\hat{\varepsilon}_3}\right) \rightarrow \left(\frac{1}{\hat{\varepsilon}_{2,n}} - \frac{1}{\hat{\varepsilon}_3}\right) \tag{18b}$$

Unfortunately, the Fresnel reflection coefficient, $\hat{r}$ ,is not directly measurable. The magnitude that can be measured is the reflectivity, R, which is the product of $\hat{r}$ by its complex conjugate $\hat{r}^*$.

$$R = \frac{I_r}{I_i} = \hat{r}\,\hat{r}^* \tag{19}$$

Provided $d/\lambda \ll 1$, as is the case in many electrochemical systems, from formula 17 one can obtain an equation for the differential reflectivity $(R_{123} - R_{13})/R_{13} = \Delta R/R$ which can be simplified to:

$$\left(\frac{\Delta R}{R}\right)_{\perp} = 8\,\frac{d}{\lambda}\sqrt{\varepsilon_1}\cos\alpha_1\,\mathrm{Im}\,\frac{\hat{\varepsilon}_2 - \hat{\varepsilon}_3}{\varepsilon_1 - \hat{\varepsilon}_3} \tag{20a}$$

$$\left(\frac{\Delta R}{R}\right)_{\parallel} = 8\,\frac{d}{\lambda}\sqrt{\varepsilon_1}\cos\alpha_1 \cdot$$

$$\cdot\,\mathrm{Im}\,\frac{\cos^2\hat{\alpha}_3\,(\hat{\varepsilon}_2 - \hat{\varepsilon}_3) + \hat{\varepsilon}_3^2\sin^2\hat{\alpha}_3\left(\frac{1}{\hat{\varepsilon}_2} - \frac{1}{\hat{\varepsilon}_3}\right)}{\varepsilon_1\cos^2\hat{\alpha}_3 - \hat{\varepsilon}_3\cos^2\alpha_1} \tag{20b}$$

Again, uniaxial anisotropy can be taken into account by effecting the substitutions indicated in eqs. (18a) and (18b).

### 3.3. Transition-Layer Model of the Double Layer

In order to apply the reflection theory to the electrode/electrolyte interface an optical model is required. In the simple three-layer model of McIntyre and Aspnes /10/ the double layer and possible surface films are approximated by an isotropic intermediate phase with fixed phase boundaries, and then eqs. (17a, 17b) and (20a, 20b) can directly be used. This model, although often applied to describe both the double layer and film-covered surfaces, does not take into account that the double layer has no fixed phase boundaries since the variation, real and imaginary part, of the complex dielectric constant, $\hat{\varepsilon}$ , across the electrode/electrolyte interface is continuous. Obviously, the same applies for the variation of the dielectric loss function, $1/\hat{\varepsilon}$. This is schematically shown in Fig. 3 for two different applied potentials. Various approaches have been developed to take into account this problem. For instance, Dignam et al. /67/ used a multilayer model, and obtained an effective dielectric constant by averaging over the different layers. Abeles /68/ used the concept of uniaxial anisotropy surface conductance.

A Fresnel reflection coefficient for the transition-layer model with a gradual change of $\hat{\varepsilon}$ and $1/\hat{\varepsilon}$ along the z direction have been derived by Plieth and Naegele /11/. The differences $(\hat{\varepsilon}_2-\hat{\varepsilon}_3)$ and $(1/\hat{\varepsilon}_2-1/\hat{\varepsilon}_3)$ have been replaced by integrals over dz. The moduli of $\hat{\varepsilon}(z)$ and $1/\hat{\varepsilon}(z)$ are represented in Fig. 3. The curves for the potential of zero charge (curves 2) have been taken as reference. The integration

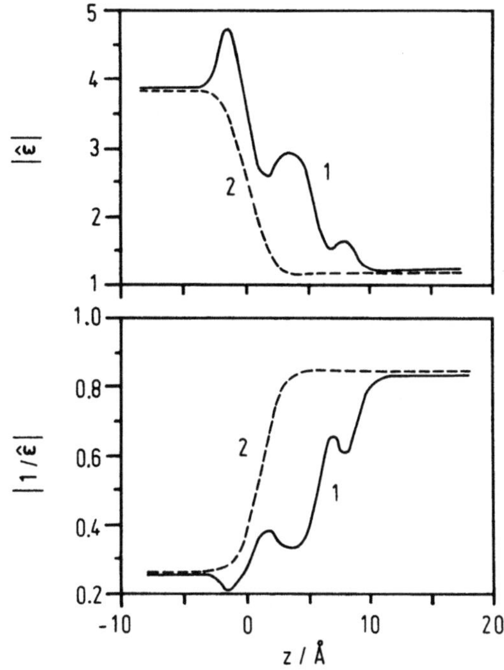


Fig. 3 Transition-layer model of the double layer: Schematic plots of the moduli of $\hat{\varepsilon}$ (the complex dielectric constant $\hat{\varepsilon}$) and of $\hat{\varepsilon}^{-1}$ (the dielectric loss function) as a function of z are shown for two different applied potentials, 1) for a charged surface, 2) at the potential of zero charge (pzc). The areas between curves 1 and 2 determine the change of the optical properties of the interface (here called 'the dielectric surface excess') produced by a potential step.

has its natural limits in phase 1 and phase 3 where curves 1 and 2 coincide. It is noteworthy that the separation into two integrals takes into account a possible uniaxial anisotropy. Then the substitutions indicated in eqs. (18a) and (18b) are used for the integrals.

While a three layer model might be isotropic or anisotropic depending on the properties of the intermediate film, a continuous variation of $\hat{\varepsilon}$ with z brings about that $\int \hat{\varepsilon} dz$ and $\int (1/\hat{\varepsilon}) dz$ will always be different (inherent anisotropy). Therefore, the validity of the simple three layer model for the description of the double layer is the exception and will mostly lead to quantitatively wrong results.

The integrals have similarities with the thermodynamic surface excess known from the thermodynamic theory of the double layer. Therefore, it was suggested to designate each integral as a 'dielectric surface excess' /12/.

## 3.4 Determination of the Phase Difference

For a complete description of the optical properties of an interface on the basis of the classical theory four quantities are necessary, the real and imaginary parts of the surface excess of the dielectric function and of the surface excess of the dielectric loss function. This does not yet take into account the possible presence of longitudinal plasma waves at frequencies near the bulk metal plasma frequency (section 4.5).

The determination of the four parameters is possible if the complete Fresnel reflection coefficient for two different polarizations of the light (eq. (22)) can be measured. A calculation method have been given in /16/. It has been applied to thin layers of platinum oxide on platinum /17/.

It must be stressed that the calculation of n and k for the film still requires to know the film thickness d and the complex dielectric constant $\hat{\varepsilon}_3$ of the metal, and that the calculated n and k values depend very strongly on the values taken for d and $\hat{\varepsilon}_3$.

As an alternative to ellipsometry, interferometry has been suggested for determining the phase shift in reflectance /18/. However, this absolute method had the disadvantage that mechanical effects, i.e. electrostriction, could not be separated from electronic effects /19/.

## 3.5 Experimental Set-up

### 3.5.1 Differential Reflectance

#### 3.5.1.1 Double Beam Reflectometer

The usual spectroscopic equipment is a double beam spectrometer. A similar set-up was developed for electrochemical applications (Fig. 4). The problem with a double beam spectrometer is the demand for identical properties of the two electrochemical systems. The optical components can be made identical within a tolerance of ±0.01% allowing ΔR/R measurements down to $10^{-3}$ ± 10%. An uncertainty remains in the reflectivity of the two working electrodes. After several electrochemical cycles, identical reflectivity is guaranteed only in special cases. Few electrode materials might fulfill such demands, e.g. platinum.

### 3.5.1.2 Single Beam Reflectometers

In principle, electrochemistry offers the possibility to simulate the double beam spectrometer by switching the potential between two values, one of which acts as reference (potential step method). The principle of such measurement is shown in Fig. 5. While the demands on the optical components are drastically reduced, the problem with the electrode surface stability still remains. Only metals like platinum can be used. Applying signal averaging techniques, the difference in the reflectance between two potentials can be measured down to ΔR/R being approx. $10^{-4}$ ± 10%.

The problem of the time stability of electrochemical systems can be overcome with rapid scanning spectrometers; also their sensitivity is less than $10^{-3}$ ΔR/R.

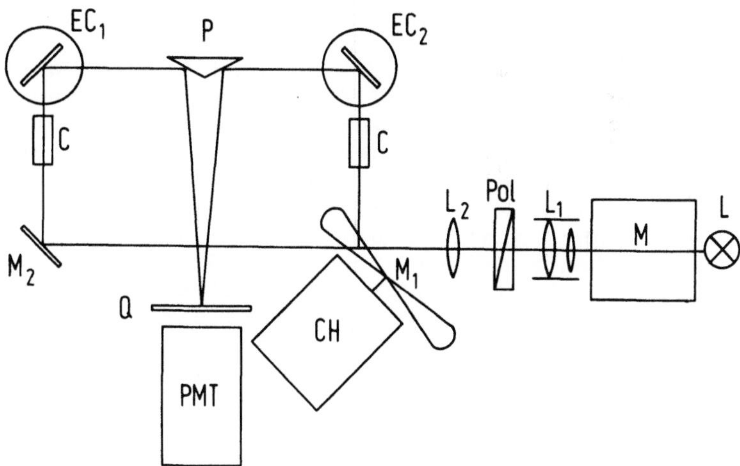


Fig. 4 Double beam reflectance spectrometer: $EC_1$, $EC_2$: electrochemical cells, one serving as reference cell, L: light source, M: monochromator, $L_1$, $L_2$: lenses, POL: polarizer, CH: light chopper (combined with mirror $M_1$), $M_2$: second mirror, C: compensator cells to correct for intensity differences, P: prism, Q: diffuser plate, PMT: photomultiplier; Ref. /65/.

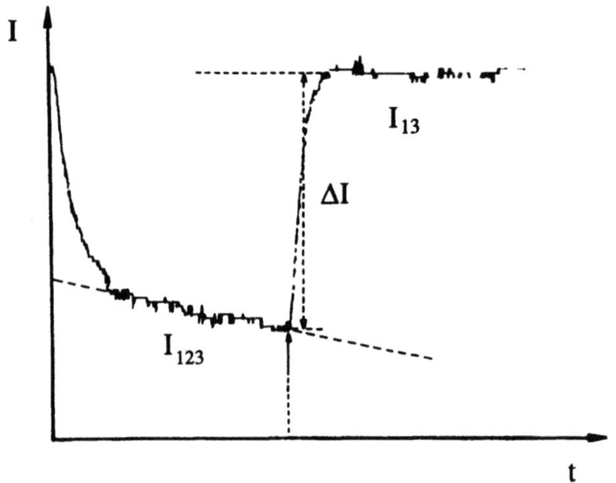


Fig. 5 Schematized description of the potential step method. Oxide layer formation on Pt in 0.5 M $Na_2SO_4$/ 0.005 M $H_2SO_4$. ΔI: change in the reflected intensity produced by a potential step from 0.5 V (SHE) into the oxide layer region; $I_{13}$: intensity of light reflected from the bare surface; $I_{123}$: intensity of light reflected from the oxide-covered surface.

### 3.5.2 Modulated Reflectance

#### 3.5.2.1 Potential Modulated Reflectance Spectroscopy

An increase in sensitivity is achieved by the superposition of a sine wave, e.g. ±100 mV, 33 Hz, to a constant or slowly varying potential. The expression "electroreflectance" is best reserved for systems where only physical changes are produced by the potential modulation. For systems where chemical changes are expected the expressions "potential modulated reflectance spectroscopy" (PMRS) or "electrochemically modulated reflectance spectroscopy" (EMRS) are in use in the literature. A schematic description of an electromodulation spectrometer is given in Fig. 6. More experimental details are given in the chapter by Gutierrez.

There is a pronounced difference between differential and modulated reflectance spectroscopy in the sense that in the former method one measures the reflectance difference between a film-covered and a film-free surface. Modulation spectroscopy, on the other hand, measures the potential-induced reflectance changes of a film-covered surface. Three examples, band reduction due to electrochemical oxidation or reduction, band splitting due to broadening, and appearance of a bipolar band due to band shift, are shown in Fig. 7.

The dependence of $\Delta R/R$ on the modulation frequency can

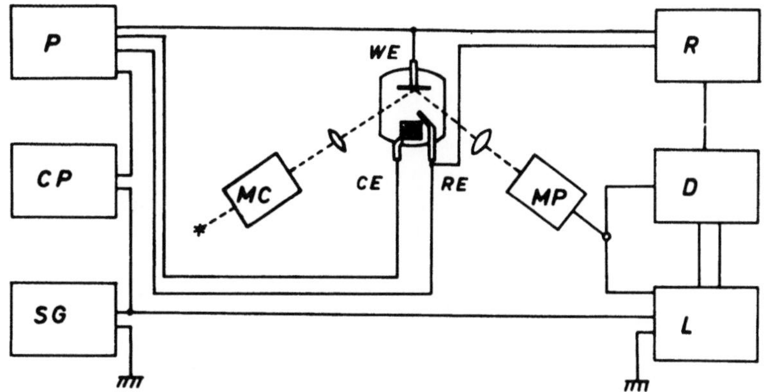


Fig. 6 Schematic description of a spectrometer for potential modulated reflectance spectroscopy. P: potentiostat, CP: bias potential, SG: signal generator for modulation potential, MC: monochromator, MP: multiplier, L: lock-in amplifier measuring the reflectivity modulation ΔR, D: divider measuring the reflectivity R and forming the ratio ΔR/R, R recorder, WE, CE, RE: working, counter, and reference electrode, respectively.

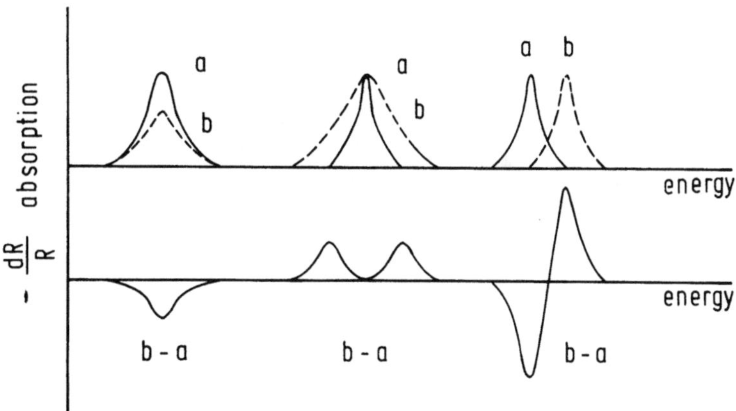


Fig. 7 Different shapes of modulation spectra. Above: absorption spectra at two different potentials, a ——— and b -----. Below: modulated reflectance (ΔR/R) spectra.

indicate the origin (physical or chemical) of the optical signal. If the modulation is mainly affecting the electronic properties, no relaxation between the modulation and the optical response is expected. On the contrary, for redox reactions, adsorption/desorption reactions or other modulations of chemical properties the response signal depends on the modulation frequency. The method can then be used for kinetic studies.

#### 3.5.2.2 Laser-modulated Reflectance Spectroscopy

Other methods of varying the optical properties can be used to modulate the optical properties of the interface region. One possibility is the illumination of the surface by a laser beam. This method was used for spectroscopy at semiconductors /49,50/ and recently applied to electrochemical systems /51/. The experimental set-up is shown in Fig. 8. The laser beam can heat the surface (thermomodulation) or modulate the optical properties by excitation of electron/hole pairs. Their separation in the space charge region will also lead to changes in the potential drop (photopotentials). Thus the field strength in the surface region can be changed by optical means (photomodulation).

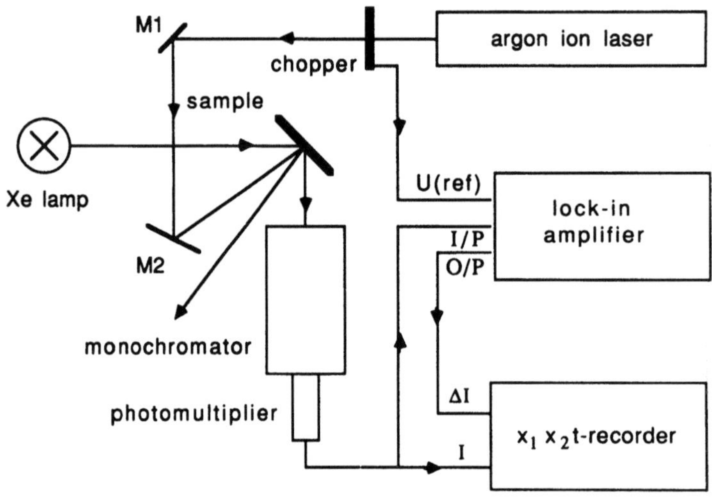


Fig. 8 Experimental set-up for the measurement of Laser-modulated reflectance spectra; the modulation is achieved by chopped illumination of the surface with a CW laser beam.

## 4. METALLIC ELECTROREFLECTANCE

### 4.1 The Free Electron Model, Plasma Frequency

We will briefly discuss the general optical properties of metals as described by the Free Electron Model /20/, which leads to a simple equation for the dielectric constant $\hat{\varepsilon}$.

$$\hat{\varepsilon} = 1 - \frac{\omega_p^2}{\omega^2 - i\omega/\tau} \qquad (21)$$

where $\omega_p$ is called the bulk plasma frequency, and $i\,\omega/\tau$ is a damping contribution due to collisions and other sources of energy loss. Collective longitudinal oscillations of an electron gas are possible at frequencies $\omega > \omega_p$. These plasma oscillations are quantized in plasmons. The bulk plasma frequency is given by

$$\omega_p^2 = \frac{4\pi N e^2}{m^*} \qquad (22)$$

where N is the number of free electrons per $cm^3$ in the bulk metal, m* the effective mass of the electrons. A good example of a plasma metal in which modulation of the potential E leads to a direct modulation of the electron density N (within the very thin (0.1 nm) penetration depth of electric fields in metals) are the coinage metals (Prostak and Hansen /21/):

$$\Delta N = \mathrm{const} \cdot \Delta E \qquad (23)$$

Consequently, the plasma frequency can be modulated by the potential,

$$\frac{d\,\omega_p}{dE} = \frac{1}{2}\,\frac{4\pi e^2}{m^*}\,\frac{dN}{dE} \qquad (24)$$

this being the reason for the strong electroreflectance of Cu, Ag and Au, first observed by Feinleib /1/ (Fig. 9). However, the free electron model does not explain the fine structure of the metallic electroreflectance (ER) spectra. It was therefore suggested that the dielectric function had a contribution from the bound electrons /21,22/:

$$\hat{\varepsilon} = \hat{\varepsilon}_f + \hat{\varepsilon}_b \qquad (25)$$

The experiments with single crystals of Cu, Ag and Au described in the experimental part support, in principle, this concept.

### 4.2 In-Plane Polarisation Anisotropy of The ER spectra in Single Crystals

The reflection from an electrode surface depends, among others, on the structure of the first atomic layers of the metal. Obviously, single crystals must be used for detect-

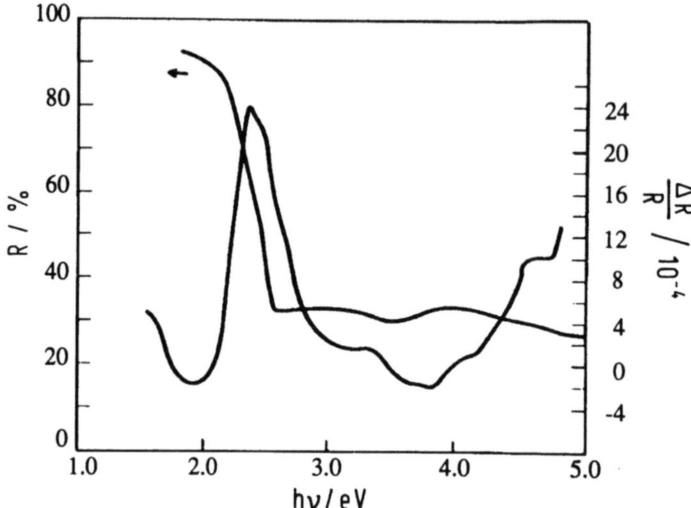


Fig. 9 ER spectrum of gold in a KCl electrolyte. Modulation: 2 V at 35 Hz (after Feinleib /1/).

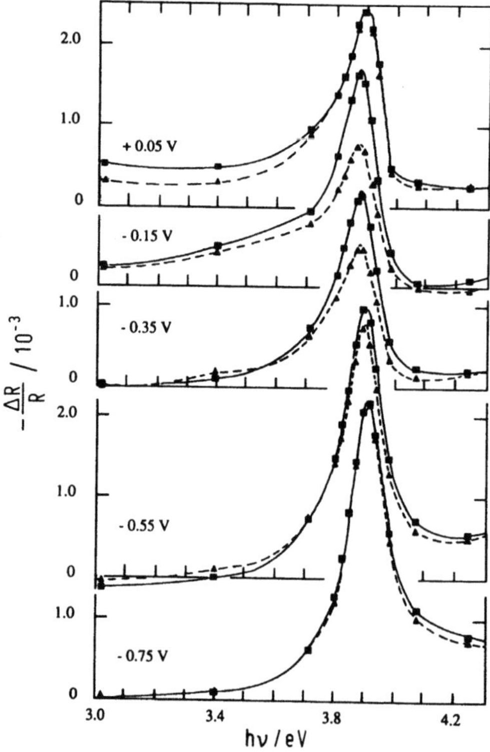


Fig. 10 ER spectra of the (110) surface of single crystals of silver in 1 M KOH at quasinormal light incidence with light polarized parallel to the $[1\bar{1}0]$ direction (▲) or parallel to the $[00\bar{1}]$ direction (■). The potential vs. SHE is shown on each curve. Modulation was 0.1 V at 1 kHz (after Furtak and Lynch /25/).

ing these effects, such as the in-plane (i.e. at normal incidence) polarization anisotropy of the electroreflectance. It was first observed by Furtak and Lynch /24,25/ who found that the magnitude of the ER signal from the (110) surface of an Ag single crystal at quasi-normal light incidence depended on the orientation of the polarisation plane of the light, parallel to the [00$\bar{1}$] or [1$\bar{1}$0] direction, respectively (Fig. 10). The effect occurs near the bulk plasma frequency, 3.88 eV. According to Ho, Harmon und Liu /28/, polarization anisotropy originates from optical transitions into empty surface states which are allowed only for light parallel to the [001] direction. A linear increase of the difference between the ER maxima for the two light polarizations (Fig. 10) with increasing potential was found exclusively at potentials positive of the pzc (Fig. 11).The disappearance of the effect above 0 V is caused by oxidation of the metal.

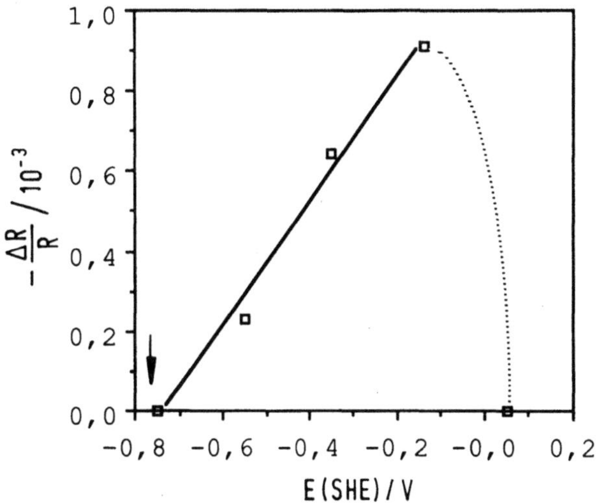


Fig. 11 Potential dependence of the surface anisotropy of the maxima of the ER spectra of Fig. 6. The potential of zero charge (-0.77 V, /61/) is indicated by an arrow.

## 4.3 Detection of Surface States in Single Crystal Faces by ER spectra

ER spectra of gold (100) and (111) faces by Kofman, Garrigos and Cheyssac /26,27/ are shown in Fig. 12. While in the first papers dealing with the electroreflectance of gold /1,21/ only the free and bound electron concept was used for interpretation of the results, it is evident from Fig. 12, where large differences between the ER spectra of two different faces of a gold single crystal are clearly apparent, that more sophisticated models have to be taken into account. The first model given by the authors in Ref. /26/ was based on a non-local analysis of the dielectric constant of the metal. Lateron, first-principle self-consistent electronic structure calculations of surface states by Ho, Harmon, and Liu /28/ were in perfect agreement with the ER results. Besides, while the ER from the (111) face, and especially the 2.5 eV band, was rather insensitive to electrode polarization, the ER signal from the (100) face varied dramatically with potential. Similar observations were reported by Nguyen Van Huong, Hinnen et al./29,30/. Surface states were also detected in ER spectra of silver electrodes (Boeck and Kolb /31/). The experimental results again agree with calculations by Ho et al. /32/.

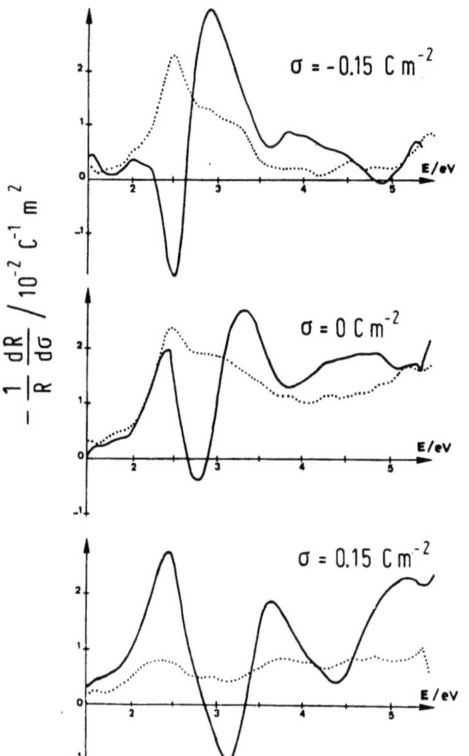


Fig. 12 ER spectra of the Au(100) (———) and Au(111) (.....) faces at negative, zero, and positive surface charge; after Kofmann et al. /27/.

### 4.4 Dependence on Potential of the Surface Plasmon Excitation Energy

In the case of Ag, excitation of surface plasmons (SP) overlaps the spectral region where absorption by surface states is found. Similar to bulk plasmons, surface plasmons are collective oscillations of electrons but localised in the surface region. The electromagnetic wave of a SP travels parallel to the surface, while perpendicular to it the fields decay exponentially. Within the framework of the Free Electron Theory, a simple relationship exists between the bulk and surface plasmon frequency /63/

$$\omega_p = \frac{\omega_s}{\sqrt{2}}. \tag{26}$$

Although surface plasmons were first observed in rough surfaces, they can be detected even in smooth single crystals faces by two different techniques described by Kretschmann /33/ and Otto /34/, respectively. Surface plasmons on single-crystal Ag electrodes were measured by Tadjeddine and Kolb /35/. A linear dependence of the SP excitation energy on the potential (Fig. 13) was observed at potentials positive of the potential of zero charge (pzc), while at poten-

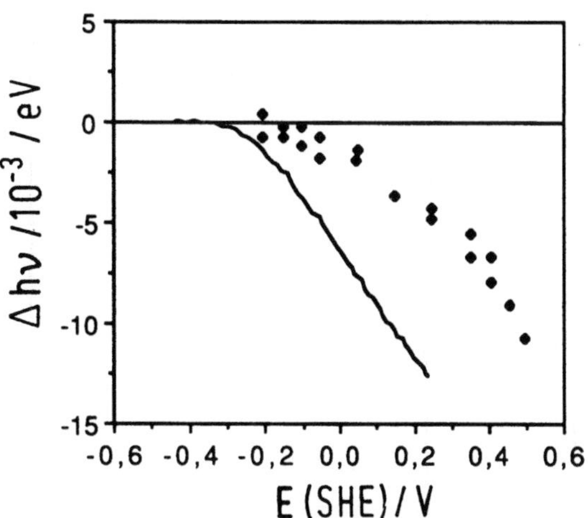


Fig. 13 Potential dependence of the Surface plasmon excitation energy of Ag(111) single crystals (♦) and polycrystalline silver (——). Electrolyte: 0.5 M $NaClO_4$. $\lambda$ = 475 nm and 400 nm, respectively. After Tadjeddine and Kolb /35/.

tials negative of the pzc the SP excitation energy was independent of the potential. This linear potential dependence is similar to that found for the anisotropy of the electroreflectance peak of the (110) surface of silver electrodes (Fig. 11). One possible origin of this dependence is that the chemisorption of anions changes the optical properties of the metallic surface near the plasma edge while cations mostly do not absorb. An alternative explanation is that the electron density at a metal surface can be decreased by positive polarisation, but it cannot be increased above the bulk level by negative polarisation. This linear dependence is obviously not a unique optical property of silver; it was also recently observed for the ellipsometric parameters $\Delta$ and $\psi$ of platinum /36/.

### 4.5 Non-Local Metal Optics Near the Bulk Plasma Frequency: Longitudinal Plasma Waves

Near the bulk plasma frequency, $\omega_p$, optics in metals differ from the simple picture of the Maxwell theory since longitudinal plasma waves (LPW, /14/), in which the direction of the oscillation is parallel to the direction of propagation, can be excited by parallel-polarized light at non-normal incidence. Then the dependence of the dielectric function on the k-vector (spatial dispersion) must be considered, and the equation for the longitudinal dielectric function, $\varepsilon_{long}(k,\omega)$ is, with good approximation,

$$\hat{\varepsilon}_{long}(k, \omega) = 1 - \frac{\omega_p^2}{\omega^2 - \beta k^2} \tag{27}$$

with

$$\beta = \frac{3}{5} \frac{\hbar^2 (3 \pi^2 N)^{2/3}}{m^2} \tag{28}$$

For the plasma condition $\varepsilon_{long}(k, \omega) = 0$ one obtains the relation governing the longitudinal plasma wave dispersion

$$\omega^2 = \omega_p^2 + \beta k^2. \tag{29}$$

The inclusion of longitudinal plasma waves eliminates the discrepancy between experimental results and classical Maxwell calculations for the electroreflection of p-polarised light at non-normal incidence on bulk silver and on thin silver films on copper near the bulk plasma edge as shown in Fig. 14 (Forstmann, Kempa and Kolb /23/).

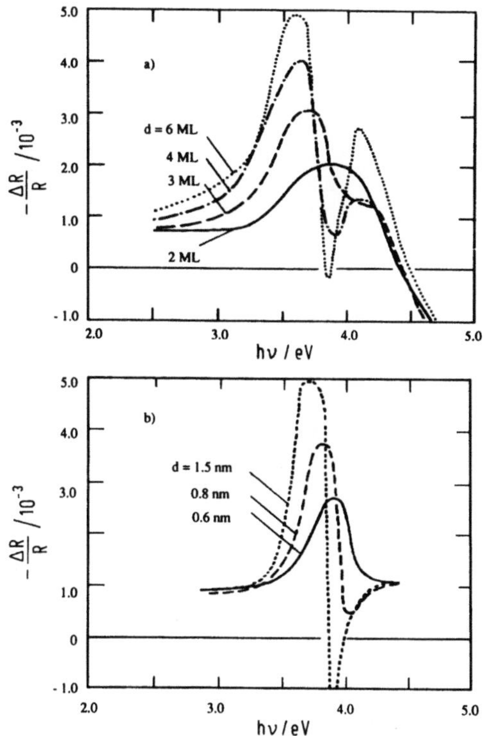


Fig. 14 a) Differential reflectance spectra of thin silver films of different thickness d (expressed as monolayers, ML) on copper (111) single crystals near the bulk plasma frequency of silver (3.9 eV) for p-polarized light; $\alpha_1 = 45°$; potential step from -0.45 to -0.15 V (SCE). b) theoretical calculations including longitudinal plasma waves, i.e. spatial dispersion or non-local optics (after Forstmann, Kempa and Kolb /23/).

## 4.6 EFFECT OF SURFACE ROUGHNESS ON ER AND ELLIPSOMETRY

While the three effects of the metal structure on ER and SP discussed in paragraphs 4.2-4.4 have only to be taken into account for those metals with a plasma frequency in the easily accessible region ( $\lambda \geq 200$ nm), i.e. Cu, Ag and Au, the optical properties of all metals depend on surface roughness the influence of which on optical measurements can be described by the effective medium theory introduced by Maxwell-Garnett /37/. An alternative treatment of this theory was given by Bruggemann /38/. Further developments relevant to electrochemical systems were introduced for example by Lukes et al. /39/ and by Hunderi /40/.

A schematic description of a rough surface is given in Fig. 15. The surface roughness can be treated as an intermediate layer of a thickness equal to the maximum amplitude of the roughness and can be described as an effective medium by the mixing of two components with dielectric constants $\hat{\varepsilon}_3$ (metal) and $\varepsilon_1$ (electrolyte). Then the effective dielectric constant of the intermediate layer $\hat{\varepsilon}_{eff}$ is

$$f \frac{\varepsilon_1 - \hat{\varepsilon}_{eff}}{\varepsilon_1 + 2\hat{\varepsilon}_{eff}} + (1-f) \frac{\hat{\varepsilon}_3 - \hat{\varepsilon}_{eff}}{\hat{\varepsilon}_3 + 2\hat{\varepsilon}_{eff}} = 0 \tag{30}$$

where f is the volume fraction of phase 1 of the intermediate layer. In particular, ellipsometry should be suitable for the measurement of the thickness and the optical constants of the intermediate layers, as shown by Hunderi /40/. An interesting experiment, off-specular ellipsometry, was carried out by Williams and Aspnes /41/.

metal

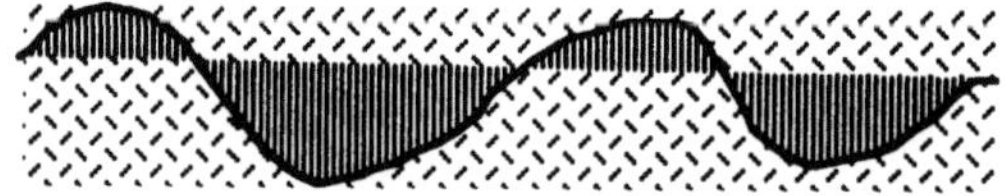

electrolyte

Fig. 15 Model of a rough surface. The intermediate layer is treated by the Bruggemann effective medium theory /38/.

## 5. OXIDE FILMS

### 5.1 Study of Oxide Films on Platinum by Differential Reflectance

Oxide film formation is a case of special interest in spectroelectrochemistry since the strong light absorption of some oxides originates large changes in the optical properties of the electrode surface. Oxides were among the first electrochemical systems studied by ellipsometry /3/ and differential reflectance spectroscopy /46/. The latter results were analyzed by application of the simple three layer model /46/ which, especially for thin oxide layers such as those formed on noble metals like platinum and gold, might be questionable. For platinum in dilute sulfuric acid the more elaborate transition layer model (Fig. 3) was applied by Naegele and Plieth /17a/ who measured the differential reflectance spectra of the oxide layer by the potential step method already schematized in Fig. 5. The potential was stepped between 0.5 V vs. SHE (in the double layer region) and different potentials in the oxide layer region. The differential reflectance spectra at near-normal incidence are shown in Fig. 16. Application of the Kramers-Kronig (KK) relation yielded the spectra of the phase differences which allowed the calculation of $\Delta\varepsilon_{re}$ and of $\Delta\varepsilon_{im}$. With thickness values taken from Kim, Paik and Bockris (62), the mean values of n and k could be calculated exclusively from the near-normal incidence measurements; in the case of anisotropic layers they represent the components of n and k tangential to the surface. The calculated values of n and k vs. $h\nu$ are shown in Figs. 17 and 18, respectively. They depended strongly on the values of the film thickness and of the complex dielectric constant of bulk Pt. The measurements were repeated with light polarized parallel and perpendicular to the plane of incidence; a peak at 4.8 eV was obtained from the KK analysis and was interpreted as a transition perpendicular to the surface plane /17b/.

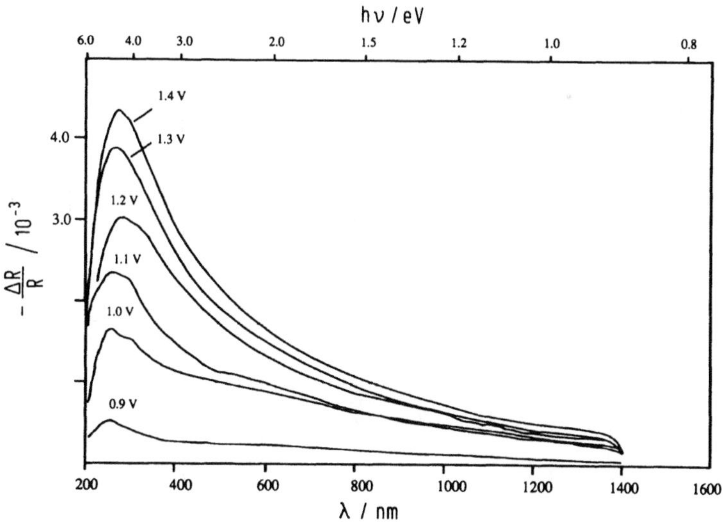


Fig. 16 Normalized relative change $\Delta R/R_{13}$ at near-normal incidence as a function of the wavelength $\lambda$, for a platinum electrode in 0.005 M $H_2SO_4$ + 0.5 M $Na_2SO_4$. The potential was switched from 0.5 V (SHE) to different potentials in the oxide layer region.

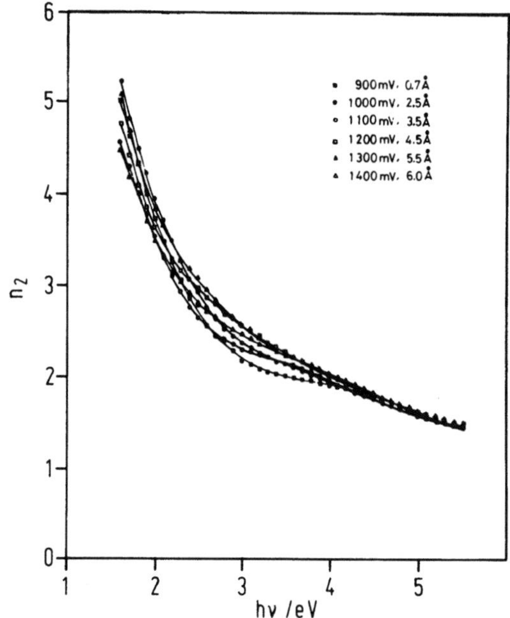


Fig. 17 Dispersion curves of the calculated refraction index $n_2$ of oxide layers on Pt at different potentials.

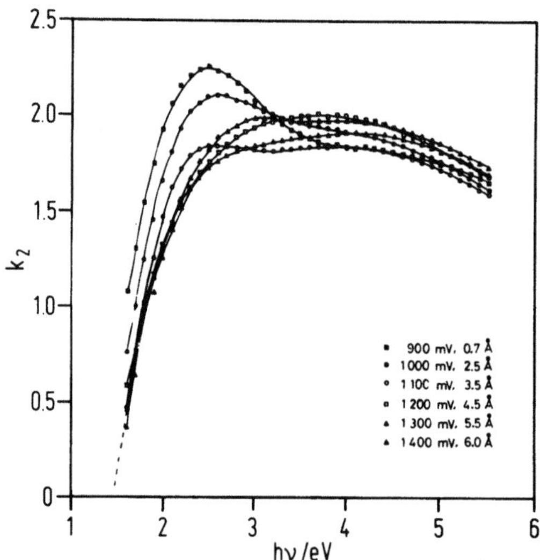


Fig. 18 Dispersion curves of the extinction coefficient $k_2$ of oxide layers on Pt at different potentials.

### 5.2 Study of Oxide Films by Modulated Reflectance Spectroscopy

#### 5.2.1 Potential Modulated Reflectance Spectroscopy

The classical example of a passive layer is the oxide film on iron. Reflectance spectra cannot be measured by a differential method, in this case due to the lack of a reference spectrum, since the clean oxide-free metal surface undergoes corrosion. The potential modulation technique provides an alternative for such systems. Examples are given Gutierrez (ref. /48/ or his chapter).

#### 5.2.2 Laser-modulated Reflectance Spectroscopy

An alternative source for modulating the optical properties of a surface film is a periodic illumination of the sample with chopped light from a laser. This photoreflectance (PR) technique has been applied to semiconductor/air surfaces /49,50/ and, recently, also to semiconductor/electrolyte interfaces /51/. The experimental set-up was shown in Fig. 8.The PR structure results from a modulation of the temperature or of the electric field in the Schottky surface barrier by the photoexcited charge carriers /49/.

For anodically grown titanium dioxide films the illumination resulted in a thermomodulation and results for titanium dioxide/ air interfaces are shown in Fig. 19. The peak-to-peak height decreases with increasing formation potential of the $TiO_2$ film until eventually the photoreflectance changes sign for formation potentials above 10 V. This is comparable with the results obtained by EMRS of $TiO_2$-layers, e.g. by Paatsch /52/ and Blondeau et al. /64/. The results can be described quantitatively by assuming a shift of the band gap energy of 10 meV/°C /66/.

## 6. STUDY OF ION ADSORPTION BY DIFFERENTIAL REFLECTANCE

The adsorption of ions changes the optical properties of interfaces as was detected in a number of earlier studies, e.g. with chloride ions /42,43/. These changes are probably due to bond formation between the anion and the metal, which will change the electronic states at the metal surface and, consequently, its reflectivity even if the adsorption layer itself is nonabsorbing. The dependence of the change in reflectivity on the electrode potential (i.e. on $Br^-$ surface coverage $\Theta$) is shown in Fig. 20 /44/. The approximately linear increase of $\Delta R/R$ with increasing potential, up to 0.6 V, is due to Temkin adsorption in which the coverage increases linearly with potential. The electrosorption va-

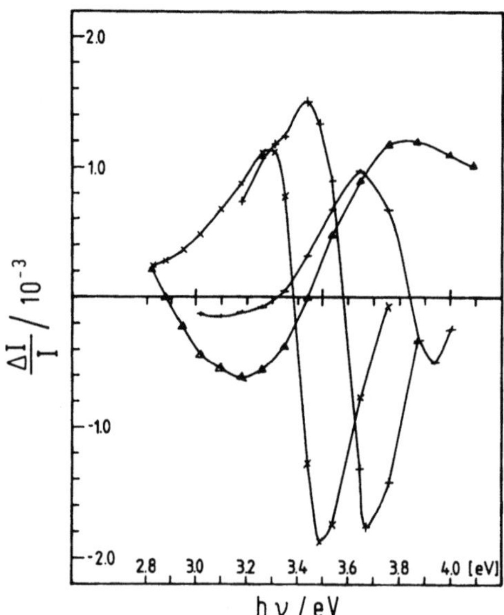


Fig. 19 Laser-modulated reflectance spectra of oxide layers of $TiO_2$; formation of the oxide in 0.5 M $H_2SO_4$ up to +4 V (8 nm thickness, +), +6 V (12 nm, x), +12 V (24 nm, -●-), +14 V (28 nm, ▲). Chopper frequency 20 Hz, laser power 2.5 W.

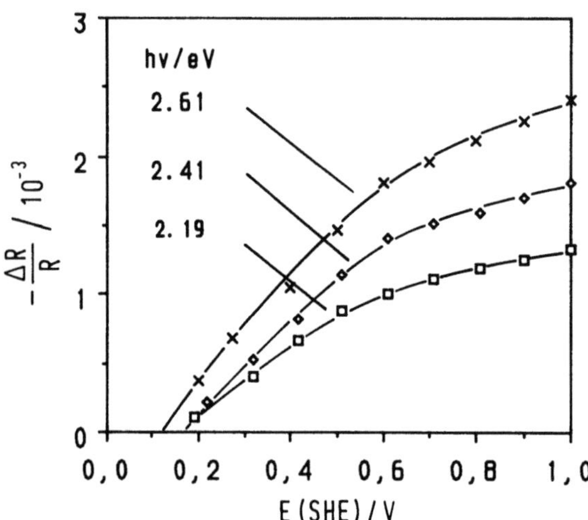


Fig. 20 Adsorption of bromide on platinum in a 0.5M$H_2SO_4$/ $0.625\cdot10^{-3}$MKBr solution. Dependence of the relative change of reflectivity ΔR/R on the electrode potential E for three photon energies. Parallel polarisation, 45° angle of incidence.

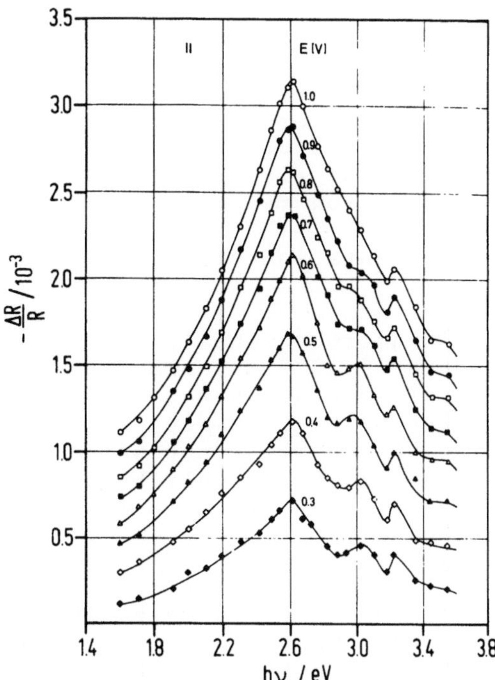


Fig. 21 Differential reflectance spectra of bromide adsorbed on Pt at different potentials (vs. SHE) shown against each curve. The electrolyte was 0.5 M $H_2SO_4$ with $0.64 \times 10^{-3}$ M KBr. Parallel polarization, 45° angle of incidence. The scale of the y-axis is valid for E = 0.3 V; the spectra at higher potentials are shifted upwards by 0.1 units each.

lency (net electronic charge transferred across the interface per adsorbed atom) was determined to be 0.8 (i.e., the mean formal charge of each bromide anion was -0.2). The differential reflectance spectra (Fig. 21) show a potential dependent peak at 3.1 eV which is not present in the electroreflectance spectra in a bromide-free electrolyte. A similar study has been carried out by Adzic, Yeager and Cahan /45/.

## 7. STUDY OF ORGANIC ADSORPTION LAYERS BY POTENTIAL-MODULATED REFLECTANCE

Organic adsorption layers play an enormous role in the inhibition of metal deposition (plating) and metal dissolution (corrosion). They may also appear as intermediates in organic electrochemical redox processes. Besides, organic adsorbates can both decrease (poison) or increase (catalyze) the rate of electrocatalytic reactions. In all these cases it is interesting to know the structure of these films which can be studied by UV/VIS modulation reflection spectroscopy /53/. We will discuss two applications of UV/VIS modulated reflectance spectroscopy: the determination of the orientation of adsorbate molecules and of changes in its orientation, and the influence of the electric field on the equilibrium constant of a protonation reaction.

## 7.1 Determination of the Orientation of the Adsorbate by PMRS

The PMR spectrum of p-Aminonitrobenzene (p-ANB) adsorbed on Pt in 0.5 M sodium sulphate ($5\cdot10^{-5}$ M p-ANB) is shown in Fig. 22 /54/. The potential modulation was within the double layer region of platinum at pH 6, extending from +0.2 up to +0.6 V vs. SHE. Oxidation of the $NH_2$ group was not observed, and reduction of the $NO_2$ group began only below $< -0.2$ V. The spectrum is strongly anisotropic with respect to the polarization of the incident light.

The transition moment (and also the dipole moment) of p-ANB is parallel to the long axis of the molecule, and therefore at least some of the adsorbed molecules must be oriented with this long axis perpendicular to the surface since only parallel-polarized light has a component of the electrical field vector in this direction and can be absorbed by molecules with their transition moment in the same direction.

A second condition is necessary for a potential modulation of the reflectance to be observed: the dipole moment itself or, at least, a component of it must be parallel to the electric field vector. Then a Stark interaction can occur which obviously is the predominant origin of the reflectance modulation. If the dipole and the transition moments are parallel to the surface, molecules lying flat on the surface cannot be observed. Therefore, at least part of the molecules have to be in an upright position.

The height of the peaks increases with concentration up to $10^{-5}$ M where saturation is reached (Fig. 23). At a concentration of $5\times10^{-5}$ M p-ANB the height of the peaks is independent of the potential in the range 0.3 and 0.6 V vs. SHE (Fig. 24). Both facts lead us to believe that for the concentration used in the experiments all molecules are oriented perpendicular to the surface. As a matter of fact, the interaction of the dipole moment (which is parallel to the long axis of the molecule) with the electric field across the interface contributes to the adsorption energy. An oblique orientation can be excluded due to the absence of a PMR-spectrum for perpendicularly polarized light.

The PMR peak at 436 nm (peak 1) was attributed to the $n-\pi^*$ transition observed at 380 nm in the absorption spec-

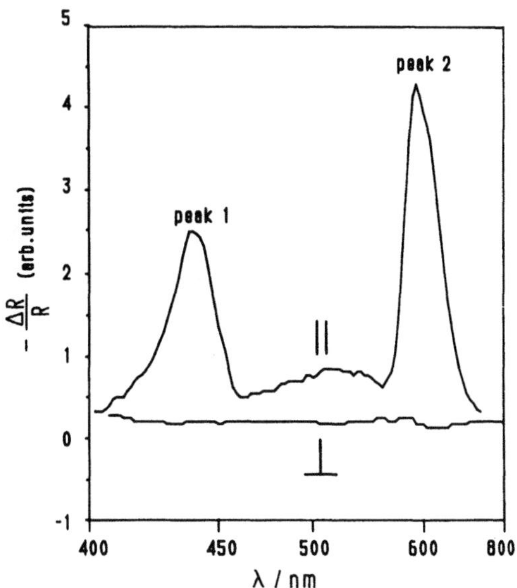


Fig. 22 Potential Modulated Reflectance spectrum of a smooth platinum electrode in 0.5 M $Na_2SO_4/5x10^{-5}$ M p-ANB solution at 440 mV vs. SHE. Modulation amplitude: ±50 mV, modulation frequency: 33 Hz; angle of incidence: 65°; parallel and perpendicular polarized light. The height of the peaks in the normalized PMR spectrum is of the order of $10^{-5}$.

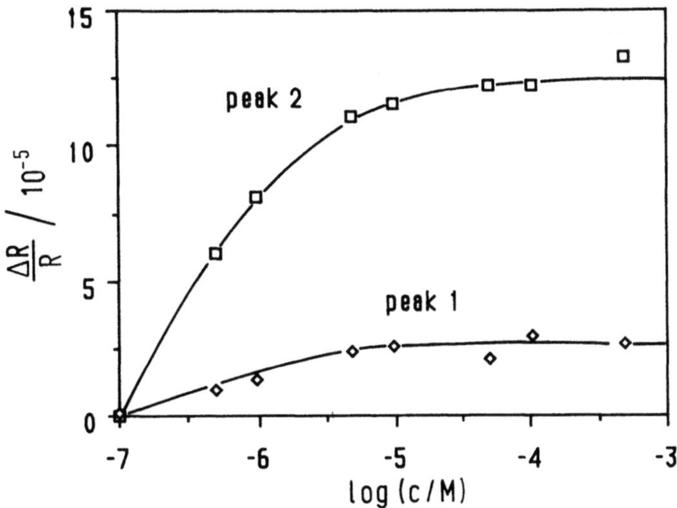


Fig. 23 Dependence of the PMR peak height on the concentration of p-ANB for the two PMR peaks of Fig. 22.

trum of p-ANB in 0.5 M $Na_2SO_4$. A second peak at 620 nm appeared in the PMR spectrum. Since a shoulder at 600 nm is observed in the diffuse reflectance spectrum of p-ANB powder which was attributed to aggregation /55/, it is possible that the second peak is the result of a strong interaction between the adsorbed, perpendicularly oriented molecules.

The potential modulation is described by the following equation taken from Ref. /54/ (k is used instead of $\Delta k$, assuming that at the interface only the adsorbed p-ANB molecules absorb light):

$$\frac{1}{R}\left(\frac{dR}{dE}\right)_{\parallel} = B_{\parallel}\left[\Theta(E)\frac{\partial k_n}{\partial E} + k_n(E)\frac{\partial \Theta}{\partial E}\right] \quad (31)$$

The independence of the PMR peak height from the stationary electrode potential (Fig. 24) means that the Θ-modulation term in Eq. (40) is negligible (i.e. the coverage Θ does not change with potential) and the entire optical effect is due to the Stark effect. Thus, the following simplified version of Eq. (39) can be used:

$$\frac{1}{R}\left(\frac{dR}{dE}\right)_{\parallel} = B_{\parallel}\,\Theta(E)\frac{dk_n}{dE} \quad (32)$$

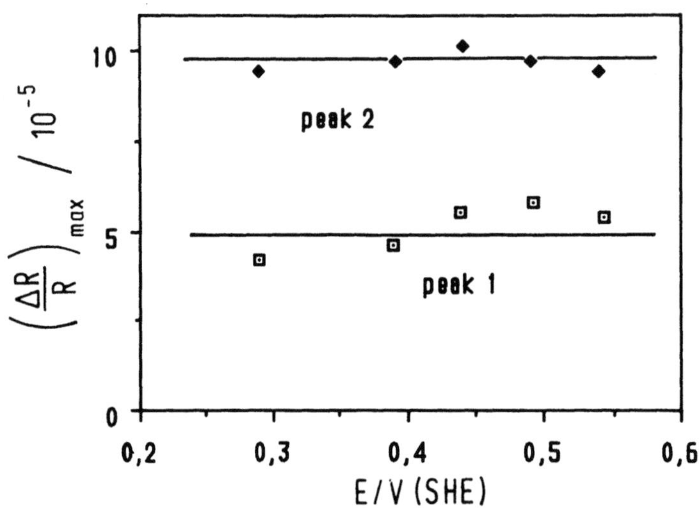


Fig. 24 Dependence of the peak height (for the PMR peaks of Fig. 22) on the electrode potential. Modulation amplitude: ±100 mV.

In the case of a pure dipole-field interaction dk/dE is given by

$$\frac{dk}{dE} = -\left(\frac{\partial k}{\partial \omega}\right)\left(\frac{\partial \omega}{\partial E}\right) = \frac{b}{h} |\mu_f - \mu_i| \cos\Theta \frac{\partial k}{\partial \omega} \tag{33}$$

where $\omega = 2\pi\nu$ is the angular frequency of the light, $|\mu_f - \mu_i|$ is the absolute value of the dipole moment difference between the initial and final states, cos Θ is the angle of the dipole with respect to the direction of the field, and b = $X/(E-E_o)$ is the proportionality coefficient between field strength X and potential E; $E_o$ is the potential of zero charge). The product b cos Θ can be determined from the frequency difference observed between the PMR maximum of adsorbed molecules and the absorption maximum of molecules in solution $(\omega - \omega_o) \simeq -0.19$ eV ($\lambda$ = 380 nm, $\lambda_o$ = 405 nm) at E = 440 mV, since the dk terms in both sides of eq. (33) cancel and therefore

$$b \cos\Theta = -\frac{h\ (\omega - \omega_o)}{|\mu_f - \mu_i|\ |E - E_o|} \tag{34}$$

With the values of $E_{pzc}$ = +0.18 V /56/ and $|\mu_f - \mu_i|$ = (14 - 6.7) D = 7.7 D /57/, one obtains:

$$b \cos\Theta = -\ 5.6\text{x}10^7\ \text{cm}^{-1}. \tag{35}$$

Assuming cos Θ = -1 (perpendicular orientation, $NO_2$-group next to the metal surface), a value of b = $5.6\text{x}10^7$ $\text{cm}^{-1}$ for the field strength/potential relation is obtained.

A second example is given by the molecule p-dimethylamino-nitrostilbene (p-DMANSt) which is insoluble in neutral electrolytes but can be dissolved in 2 M $H_2SO_4$ where it is converted to its protonated form. Its PMR spectrum has a peak structure similar to that found for p-aminonitrobenzene /58/. The dependence of the peak heights on the electrode potential E is shown in Fig. 25 where it can be seen that a PMR spectrum is observed only at potentials above 550 mV vs. SHE. The sudden appearance of the peaks at this potential, as well as their linear increase in height with positive potential, were interpreted as being due to reorientation at 550 mV to an upright position (of the trans-form) and to a steady increase with further potential increase of the surface coverage of molecules with this perpendicular orientation. It has to be noted that the onset potential coincides with the zero charge potential of platinum in acid electrolytes as given by Gileadi, Argade and Bockris /59/.

## 7.2 Study by PMRS of the Field Dependence of Chemical Equilibria: The Dissociation-Field Effect

We have studied by PMRS the protonation equilibrium of 4-NN-dimethylamino-4'-sulfo-azobenzene /60/, usually known as methyl orange or helianthin. Its sodium salt has the structure

$$(Me)_2N-C_6H_4-N{=}N-C_6H_4-SO_3^-\ Na^+$$

The dissociation/protonation equilibrium is

$$MH \rightleftharpoons M^- + H^+. \tag{36}$$

$$\frac{[M^-]\ [H^+]}{[MH]} = K_a \qquad pK_a = 3.5 \tag{37}$$

A Pt electrode and a 0.5 M $Na_2SO_4$ electrolyte buffered with 0.04 M $CH_3COOH/CH_3COONa$ were used. PMR spectra at a constant potential vs. SHE, but at several pH values, are shown in Fig. 26. The spectra show anisotropy with respect to light

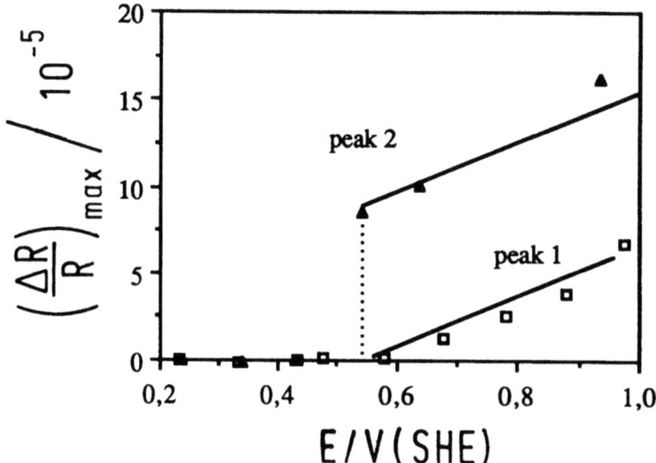


Fig. 25 Dependence on electrode potential of the peak height in PMR spectra of $10^{-5}$ M p-dimethyl-aminonitrostilbene adsorbed on platinum in 2 M $H_2SO_4$. Modulation amplitude: ±50 mV; parallel polarization. Stationary electrode potential +540 mV vs. SHE.

polarization as was the case with p-ANB, and therefore the molecules detected in the spectra are also oriented perpendicular to the surface. Both the zero point crossing of the PMR spectra, and the absorption maximum in solution, depend on pH (Fig. 27). The parallelism between the two dependencies confirms that the PMR spectra originate from a field modulation of the absorption band. A red shift of 0.3 eV is observed which means that the energies of the initial and the final states of the electronic states are affected by adsorption to a different extent.

The dependence of the peak-to-peak height of the PMR spectra on the pH at +440 mV vs. SHE is shown in Fig. 28. The peak-to-peak height shows two maxima which can be correlated with the protonation equilibria of the dye and of the buffer, respectively. The modulation of the reflectance is due to the modulation of the degree of dissociation of the dye ($\alpha_I$) and of the buffer system ($\alpha_{II}$):

$$\frac{1}{R}\left(\frac{dR}{dE}\right) = \frac{1}{R}\left(\frac{\partial R}{\partial \alpha_I}\right)\left(\frac{\partial \alpha_I}{\partial E}\right) + \frac{1}{R}\left(\frac{\partial R}{\partial \alpha_{II}}\right)\left(\frac{\partial \alpha_{II}}{\partial E}\right) \tag{38}$$

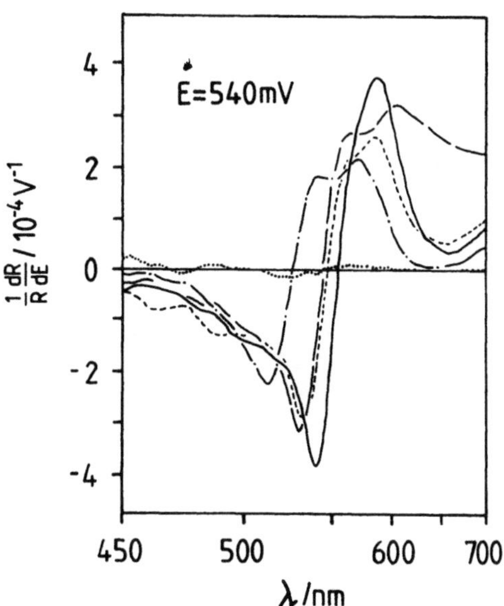


Fig. 26 PMR spectra of methyl orange adsorbed on Pt at different pH values: 3.27 (.....), 3.66 (-----), 3.84 (———), 4.14 (- - - -), 4.44 (—·—·—·—).

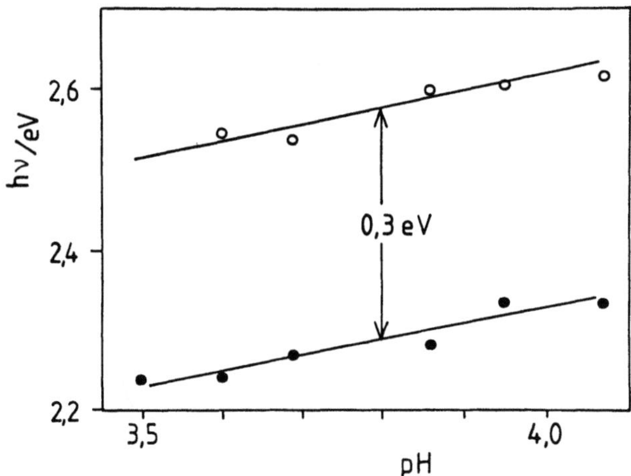


Fig. 27 Dependence on pH of the absorption maxima of the transmission spectra of methyl orange in the homogeneous solution (o) and of the zero-point crossing of PMR spectra of methyl orange adsorbed on Pt (o).

Besides,

$$\frac{d\alpha}{dE} = \frac{d\alpha}{d(pK - pH)} \frac{dpK}{dE} \tag{39}$$

and

$$\frac{d\alpha}{d(pK - pH)} = - \frac{1}{4} \frac{1}{\cosh^2 \left[ 1/2 \ (pK - pH) \right]} \tag{40}$$

Since $d\alpha/d(pk-ph)$ has a maximum at pH = pK, one can obtain the pK values of the dye and the buffer in the electric field of the double layer from Fig. 28. pH-dependencies were measured for different stationary electrode potentials E, and the $pH_{max}$-values (equal to the pK-values) were plotted versus E (Fig. 29). The result is a direct determination of the dissociation field effect in the double layer which is

$$\frac{dpK_I}{dE} = + 1.0 \ /V^{-1} \quad \text{for the dye} \tag{41a}$$

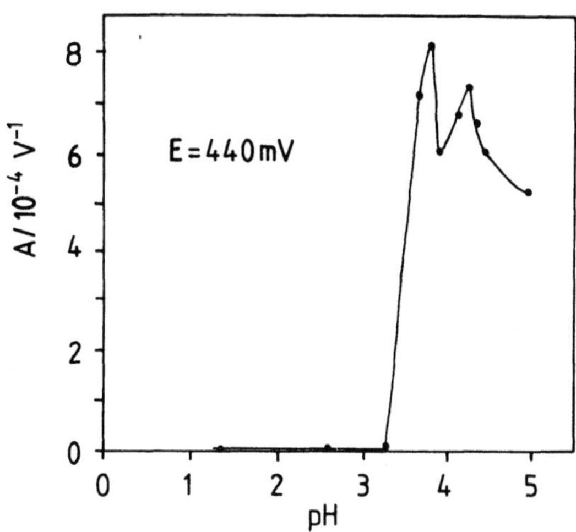


Fig. 28 Dependence of the peak-to-peak height of PMR spectra (similar to those in Fig. 25) on the pH value.

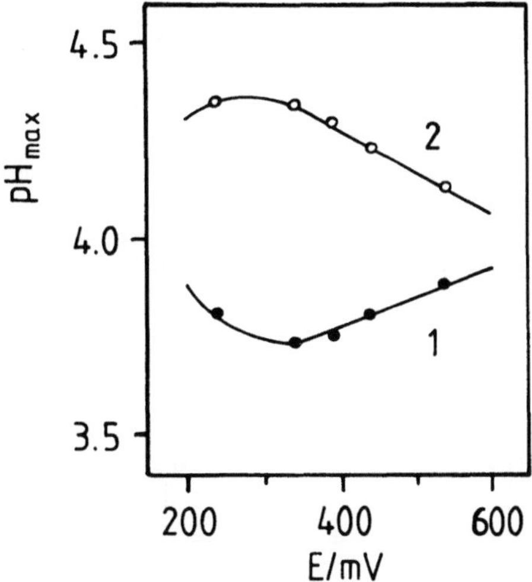


Fig. 29 Dependence of the $pH_{max}$ values of the maxima in the plot of Fig. 28 on the electrode potential.

and

$$\frac{dpK_{II}}{dE} = -0.8 \;/V^{-1} \qquad \text{for the buffer.} \qquad (41b)$$

The two dissociation equilibria are not independent of each other since they are coupled through reaction (42):

$$M^- + HAc \rightleftharpoons M + Ac^-. \qquad (42)$$

Since

$$pK_{III} = pK_I + pK_{II} \qquad (43)$$

the dissociation field effect of reaction (42) is

$$\frac{dpk_{III}}{dE} = -0.2\ V^{-1} \qquad (44)$$

## 8. CONCLUSION

The results have shown that UV-VIS spectroscopy, due to its intense development during the last two decades, can now be applied to all types of electrochemical interfaces. Monomolecular, and even submonomolecular, coverages can be studied. The quality of the spectra has reached the standard of spectroscopists. This has made possible to achieve new insights into the in-situ structure of the metal surface and also into the arrangement of molecules and their chemistry inside the electrochemical double layer. Recent papers have already gone one step further studying the time dependence of the optical phenomena. Therefore, optical spectroscopy, one of the oldest techniques in spectroelectrochemistry, has still a wide potential in research and development. However, a combination of spectroscopic techniques is required in order to achieve a microscopic picture of the double layer.

## References


/1/ J. Feinleib, Phys. Rev. Lett. 16 (1966) 1200.

/2/ D.F.A. Koch, D.E. Scaife, J. Electrochem. Soc. 113 (1966) 302.

/3/ A.B. Winterbottom, Trans. Faraday Soc. 42 (1946) 487.

/4/ M. Fleischmann, P.J. Hendra, A.J. McQuillan, Chem. Phys. Lett. 26 (1974) 163.

/5/ A. Bewick, K. Kunimatsu, B.S. Pons, Electrochim. Acta 25 (1980) 465.

/6/ H. Neugebauer, G. Nauer, N. Brinda-Konopik, G. Gidaly,

J. Electroanal. Chem. 122 (1981) 381; H. Neugebauer, G. Nauer, N. Brinda-Konopik, K. Kellner, Fresenius Z. Z. Anal. Chem. 314 (1983) 266.

/7/ T. Davidson, B.S. Pons, A. Bewick, P.P. Schmidt, J. Electroanal. Chem. 125 (1981) 277.

/8/ M. Born, E. Wolf, 'Principles of Optics', 4th ed., Pergamon Press, New York 1970.

/9/ H. Meyer, 'Physik dünner Schichten', Bd. 1, Wissensch. Verlagsgesellschaft, Stuttgart 1950, p. 144.

/10/ J.D.E. McIntyre, D.E. Aspnes, Surf. Sci. 24 (1971) 417.

/11/ W.J. Plieth, K. Naegele, Surf. Sci. 64 (1977) 484.

/12/ W.J. Plieth, a) J. de Physique, 38 (1977), C5-215; b) Israel J. Chem. 18 (1979) 105.

/13/ F. Forstmann, Z. Physik 32 (1979) 385.

/14/ F. Forstmann, R.R. Gerhardt, in: 'Festkörperprobleme, Advances in Solid State Physics', J. Treusch (Ed.), Vol. 22, Vieweg, Braunschweig 1982, p. 291.

/15/ J.S.Toll, Phys. Rev. 104 (19569 1760.

/16/ W.J. Plieth, K. Naegele, Surf. Sci.50 (1975) 53.

/17/ a) K. Naegele, W.J. Plieth, Surf. Sci. 50 (1975) 64; b) ibid. 61 (1976) 504.

/18/ W.J. Plieth, H. Bruckner, Surf. Sci. 66 (1977) 357.

/19/ H. Bruckner, Ph.D. Thesis, Freie Universität Berlin, 1980.

/20/ R.E. Hummel, 'Optische Eigenschaften von Metallen und Legierungen', Springer Verlag, Berlin-Heidelberg 1971.

/21/ W.N. Hansen, A. Prostak, Phys. Rev. 174 (1968) 500.

/22/ J.D.E. McIntyre, Surf. Sci. 37 (1973) 658.

/23/ F. Forstmann, K. Kempa, D.M. Kolb, J. Electroanal. Chem. 150 (1983) 241.

/24/ T.E. Furtak, D.W. Lynch, Phys. Rev. Lett. 35 (1975) 960.

/25/ T.E. Furtak, D.W. Lynch, J. Electroanal. Chem. 79 (1977) 1.

/26/ R. Kofmann, R. Garrigos, P. Cheyssac, Thin Solid Films 82 (1981) 73; Surf. Sci. 101 (1980) 231.

/27/ R. Kofmann, R. Garrigos, L. Duteil, P. Cheyssac, J. Electroanal. Chem. 150 (1983) 253.

/28/ K.M. Ho, B.N. Harmon, S.H. Liu, Phys. Rev. Lett. 44 (1980) 1531.

/29/ C. Nguyen van Huong, C. Hinnen, J. Lecoeur, R. Parsons, J. Electroanal. Chem. 92 (1978) 239.

/30/ S.H. Liu, C. Hinnen, C. Nguyen van Huong, N.R. de Tacconi, K.M. Ho, J. Electroanal. Chem. 176 (1984) 325.

/31/ W. Boeck, D.M. Kolb, Surf. Sci. 118 (1982) 613.

/32/ K.M. Ho, C.L. Fu, S.H. Liu, D.M. Kolb, G. Piazza, J. Electroanal. Chem. 150 (1983) 235.

/33/ E. Kretschmann, Z. Physik 241 (1971) 313.

/34/ A. Otto, Z. Physik 216 (1968) 398.

/35/ A. Tadjeddine, D.M. Kolb, 4th Int. Conf. Solid Surfaces, Cannes 1980, Vol.1, p.615; A. Tadjeddine, J. Electroanal. Chem. 169 (1984) 129.

/36/ M. Chao, M. Costa, Intern. Conf. 'Chemistry and Physics of Electrified Interfaces', Bologna, August 1988.

/37/ J.C. Maxwell-Garnett, Philos. Trans. Royal Soc. 203 (1904) 385; 205 (1906) 247.

/38/ D.A.G. Bruggemann, Ann. Phys. Leipzig 24 (1935) 636.

/39/ I. Ohlidal, K. Navratil, F. Lukes, FOLIA Facultatis Scientiarium Naturalium Universitatis Purkynianae Brunensis, XV, Physica 16,4 (1974) 1.

/40/ O. Hunderi, Corrosion Sci. 19 (1979) 621; Surf. Sci. 96 (1960) 1.

/41/ M. Williams, D. Aspnes, Phys. Rev. Lett. 41 (1978 )667.

/42/ R.S. Sirohi, M.A. Genshaw, J. Electrochem. Soc. 116 (1969) 910.

/43/ T. Takamura, K. Takamura, E. Yeager, J. Electroanal.

Chem. 29 (1971) 279.

/44/ H.D. Sedlmaier, W.J. Plieth, J. Electroanal. Chem. 180 (1984) 219.

/45/ R. Adzic, E. Yeager, B.D. Cahan, J. Electroanal. Chem. 85 (1977) 267.

/46/ J.D.E. McIntyre, D.M.Kolb, Symp. Faraday Soc. 4 (1970) 99.

/47/ W.J. Plieth, W. Höpfner, Werkstoffe und Korrosion 36 (1985) 373.

/48/ C. Gutierrez, M.A. Martinez, J. Electrochem. Soc. 133 (1986) 1873.

/49/ J.L. Shay, Phys. Rev. B 2 (1970) 803.

/50/ E.Y. Wang, W.A. Albers,Jr., Phys. Lett. 27A (1968) 347.

/51/ W.J. Plieth, F. Niebuhr, W. Badawy, BESSY-Jahresbericht 1984, p. 408.

/52/ W. Paatsch, Ber. Bunsenges. Phys. Chem. 79 (1975) 922.

/53/ W.J. Plieth, P. Gruschinske, H.-J. Hensel, Ber. Bunsenges. Phys. Chem. 82 (1978) 615.

/54/ P.H. Schmidt, W.J. Plieth, J. Electroanal. Chem. 201 (1986) 163.

/55/ G. Scheibe, 'Theoretische Organische Chemie' (Fiat-Ber.), Vol. 35, Part II, 1953, p. 51.

/56/ O.A. Petri, A.N. Frumkin, Yu.G. Kotlov, Sov. Electrochem. 5 (1969) 435.

/57/ J. Czekalla, G. Wick, Ber. Bunsenges. Phys. Chem. 65 (1961) 727.

/58/ P. Schmidt, W.J. Plieth, J. de Physique, Colloque C10, 44 (1983) C10-175.

/59/ E. Gileadi, S.D. Argade, J.O'M. Bockris, J. Phys. Chem. 70 (1966) 2044.

/60/ P. Schmidt, W.J. Plieth, Ber. Bunsenges. Phys. Chem. 91 (1987) 323.

/61/ A. Frumkin, B. Damaskin, N. Grigoryev, I. Bagotskaya, Electrochim. Acta 19 (1974) 69.

/62/ S.H. Kim, W.K. Paik, J.O'M. Bockris, Surf. Sci. 33 (1972) 617.

/63/ R.H. Ritchie, Phys. Rev. 106 (1957) 874.

/64/ G. Blondeau, M. Froelicher, M. Froment, A. Hugot-LeGoff, Thin Solid Films 38 (1976) 261; 42 (1977) 147; G. Blondeau, M. Froelicher, V. Jovancicevic, A. Hugot-LeGoff, ibid. 80 (1979) 151.

/65/ G. Schäfer, Ph.D. Thesis, Freie Universität Berlin 1973.

/66/ F. Niebuhr, W. Plieth, unpublished.

# STUDY OF ANODIC OXIDES BY UV-VISIBLE POTENTIAL-MODULATED REFLECTANCE SPECTROSCOPY

C. GUTIERREZ
Departamento de Electroquímica
Instituto de Química Física"Rocasolano", C.S.I.C.
C. Serrano, 119, 28006-Madrid
Spain

ABSTRACT. Details are given about the experimental measurement of UV-Visible Potential-Modulated Reflectance (PMR) spectra. Two different possible origins of a PMR signal are discussed, viz. physical and redox. It is postulated that the PMR transitions observed with most anodic films correspond to the absorption edge of oxides present in the film. Results are presented for the anodic oxides on Fe, Ni and Co, and it is shown how in favourable cases the anodic oxides can be identified in situ by PMR spectroscopy. Finally, interference effects in the thick electrochromic oxide film on Ir are presented and analysed.

## 1. INTRODUCTION

Potential-Modulated Reflectance Spectroscopy in the UV-VIS range (in the following designated as PMRS) has been applied since 1970 (1) to the study of anodic oxides on metals. However, although PMRS combines the advantages of very high sensitivity and moderate cost, it has been scarcely used for corrosion studies, due to the difficulty of interpreting PMR spectra. In this chapter we will try to show that, at least in favourable cases, the compounds present in passivating layers can be identified by PMRS.

### 1.1. The PMRS technique. Experimental setup.

In the PMRS technique the magnitude measured is the (usually very small) change, $\Delta R$, produced by a change of the electrode potential in the intensity, R, of the light reflected from an electrode. The heart of the technique is the lock-in amplifier (LIA), a device which first became commercially available in the mid 60s. A LIA is able to rescue from a noisy signal only that portion of it which is in phase with (or at a chosen phase angle from) a reference signal, in this case taken from the same oscillator used to superimpose a (usually sinusoidal) modulation on the electrode potential. With a LIA a reflectance change, $\Delta R/R$, as low as $10^{-6}$ can be easily rescued from a much higher noise level.

A schematic diagram of a PMRS setup is given in Fig. 1. A beam of monochromatic light is focused onto the electrode, the reflected light being focused in turn on the cathode of a photo-multiplier tube (PMT). An electronic servo device is used to keep constant the anode current (which is proportional to R) of the PMT, changing as required the PMT high voltage obtained from a programmable power supply. In this way the output, R, of the LIA directly yields the PMR spectrum.



*C. Gutiérrez and C. Melendres (eds.),*
*Spectroscopic and Diffraction Techniques in Interfacial Electrochemistry*, 261–279.

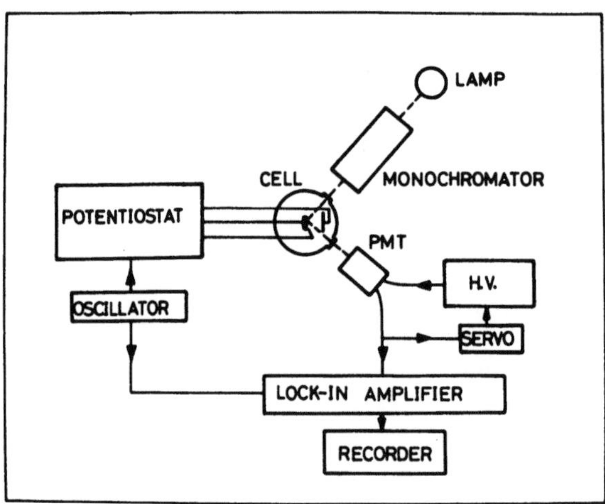


**Fig.1.** Schematic diagram of setup for Potential-Modulated Reflectance Spectroscopy (PMRS).

## 1.2. Experimental details.

When using PMRS it is very important to take a few simple precautions, required because no commercial equipments are available: Since the setups must be built in the laboratory, they are rather artefact-prone.

1.2.1. Lamps. We prefer to use a deuterium lamp for the UV range and a halogen lamp for the visible range, instead of a Xe arc lamp for the whole VIS-UV range. The use of a deuterium lamp, with practically no output in the visible, greatly reduces stray ligh in the UV, since a Xe lamp emits most of its radiant energy in the visible. Stray light can cause the appearance of false PMRS maxima in the UV range, as clearly seen in Fig. 2a, where both spectra were taken with the same (Instrumatic B102) monochromator, but with different (deuterium and Xe) lamps.

Another inconvenience of Xe lamps is that for ignition they require a 20 kV discharge, which in many occasions damaged the operational amplifiers of our potentiostat, although it was unplugged from the mains before starting the Xe lamp. On the contrary, deuterium lamps are started with a few hundred volts, and therefore their ignition cannot damage electronic components.

1.2.2 Monochromators. Ideally a monochromator with two gratings should be used, one of which should be blazed at 200 nm in order to increase the light throughput in the UV range. It must be remembered that the reflectivity of metals in the UV is lower than in the visible range, not only because of a greater sensitivity to surface roughness, but also because most metals become more transparent (i.e. less reflecting) at shorter wavelengths (see e.g. (2)).

Diffraction gratings may have "ghosts" at one or more wavelengths. They are easily detected by putting the PMT, supplied with a constant, suitable high voltage, at the exit slit of the monochromator. At each wavelength the anode current will be the product of [light irradiance of the lamp] x [monochromator efficiency] x [PMT responsivity], and, since both the irradiance of the (halogen or deuterium) lamp and the PMT responsivity change smoothly with wavelength, the "ghosts" of the diffraction grating will appear as humps in the [PMT anode current] vs [wavelength] curve.

In our case, the diffraction grating blazed at 300 nm had a "ghost" at 500 nm, which caused the appearance of false maxima or humps in PMR spectra (Fig. 2b).

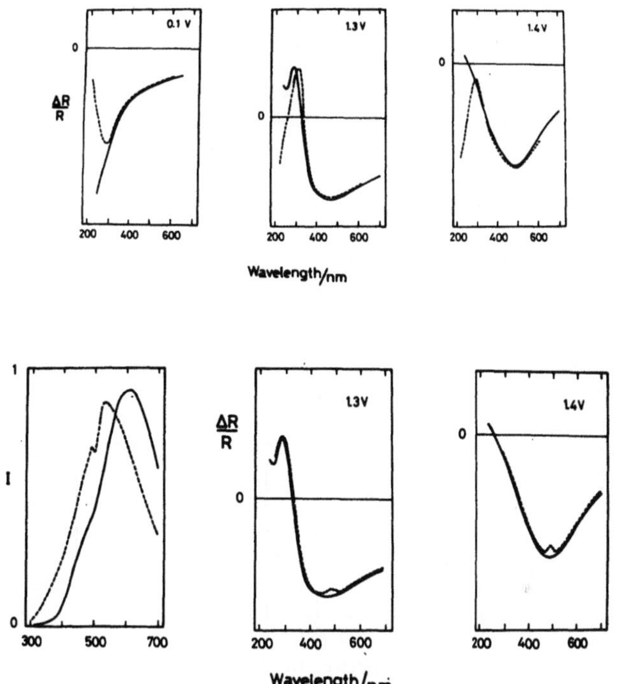


**Fig.2.** False PMR maxima due to: **a)** stray light in the UV region. Solid line: negligible stray light. Dashed line: appreciable stray light up to 300 nm; **b)** presence of a "ghost" in the diffraction grating. Solid line: ghost-free diffraction grating. Dashed line: diffraction grating with a ghost at 490 nm. (Reprinted by permission of The Electrochemical Society, Inc.)

1.2.3. Detectors. The use of a 13-stage PMT is not advisable, since the amount of light reaching the photocathode is very high, and consequently the total voltage applied to the dynode chain must be very low, in order not to exceed the maximum allowed steady-state anode current of the PMT. It is preferable to use a 9-stage side-window PMT, and even so the maximum steady-state anode current will be reached with a few hundred volts.

The use of a PMT together with a programmable power supply and a servo device allows the direct measurement of $\Delta R/R$, as above explained. This system is much better than the alternative one of chopping the light beam at a frequency much higher (e.g. 15 times) than the potential modulation frequency, measuring the reflected light (R) with a second LIA, and electronically dividing ΔR by R. Since the high voltage of the PMT is kept constant, the anode current of the PMT becomes very low over given wavelength ranges, which degrades R and, especially, ΔR. So, while only a very broad maximum had been detected with this system in the passivating layer on iron in 0.1M NaOH at V *vs* RHE (3), using a servo device we could clearly distinguish two maxima (4).

## 2. ORIGIN OF MODULATED REFLECTANCE SIGNALS

The first question that comes to mind when analyzing a PMR signal is that of its origin. Since the technique is so sensitive, many different phenomena can yield a PMR signal. All these phenomena can be classified as physical (without electron transfer) or chemical (better designated as redox processes, since in this case the chemical change is due to electron transfer).

### 2.1. PMR signals of physical origin.

All phenomena giving origin to PMR maxima can be classified into physical or chemical, the designation Electroreflectance (ER) being recommended in the first case. Due to its simplicity, the Electrolyte Electroreflectance (EER) technique, discovered in 1965 by Cardona et al. (5), is that most used for measuring electroreflectance. EER is applied mostly to semiconductors; the basic theory was advanced by Aspnes in the early 70s (6), and is still being developed (7).

As is well known, in the absence of faradaic reactions most of the potential changes applied at the semiconductor–electrolyte interface drop in the space–charge region of the semiconductor, provided it is not highly doped, since its capacity is much smaller than that of the Helmholtz layer. Therefore, a modulation of the electrode potential will modulate the charge of the semiconductor space–charge region, and consequently its dielectric constant, thereby modulating the light reflected from the electrode. The EER spectrum will be dominated by the band–gap transition, at which wavelength a mono– or bipolar structure (depending on the state of the surface) appears. Fitting of this structure to Aspnes' equation (7) yields the band–gap energy.

Since negligible faradaic reactions occur in the passive range, and passivating compounds are known to be oxides or oxyhydroxides, it could be thought that a quantitative interpretation of the PMR spectra of passive layers with the help of Aspnes' equations would be a fairly easy task. This is not so, since:

1. Passive layers are only 10–50 Å thick, i. e. much thinner than the penetration depth (≈μm) of both light and electric fields in a bulk semiconductor.
2. The structure of passive layers is thought to be very different from that of bulk crystalline semiconductors, being probably amorphous.
3. It is necessary to know the equivalent electric circuit of the metal/oxide/electrolyte interface in order to calculate the fraction of the applied modulation actually dropping in the oxide film.

In spite of the above, it is reasonable to assume that the PMR spectrum of a passive layer will be dominated by a structure at the bandgap energy of the constituting oxide or oxides. A good example is Ni, whose passive layer can be clearly identified as NiO by a PMR maximum at the bandgap energy, as will be shown below.

We will not consider here other possible physical origins of PMRS signals, since they are not relevant to corrosion studies, and, besides, are of a much smaller magnitude. We will show below that the PMR spectrum of the ferrous metals in the passive range is mostly due to EER of the passive film.

## 2.2. PMR signals of redox origin.

It is obvious that in the cases where modulation of the electrode potential produces a modulated faradaic current, there will appear a PMR signal if either the oxidized or reduced species, or both, absorb light in the wavelength range being used. The simplest case would be that of solution species, since their kinetic response to simple potential perturbations is well known, and the optical behaviour of solution species in reflectance spectroscopy is simply that of conventional transmission spectroscopy. We are more interested in the redox reactions occurring in electroadsorbed films, especially in the thick (often electrochromic) films that can be grown on many metals (e.g. Ir, Rh and Fe) by repetitive potential cycling. Provided a redox process involving an absorbing species can follow the potential modulation, a PMR signal must appear. With the usual sign convention in modulated reflectance (ΔR taken as positive if the reflectivity increases, i.e. light absorption decreases, at the more positive potential), the PMR spectrum will show positive and negative maxima of the reduced and oxidized species, respectively.

As shown below, thick electrochromic layers on Ir yield PMR spectra of redox origin, and the same can be said of the transpassive electrochromic processes on Fe and Ni.

In the following we will describe the results obtained by PMRS with a few metals

## 3. FERROUS METALS

Iron, cobalt and nickel are closely related metals of obvious industrial importance. We have carried out a PMRS study of the passive layers on these metals, hoping to obtain a better characterization of these layers.

### 3.1. Iron.

The passive film on iron is perhaps the one yielding more structures in PMR spectra, and more variation with potential of these structures. This metal was studied in 1979 by Hara and Sugimoto (3), in 1985 by Plieth et al. (8), and more recently in our laboratory (4)(9).

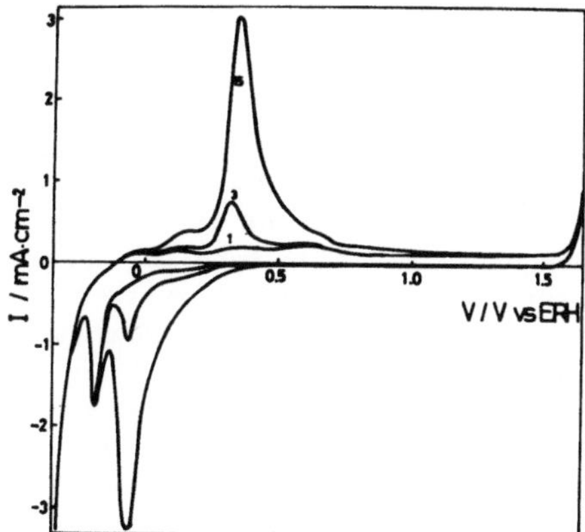


**Fig.3.** Successive cyclic voltammograms of Fe in 0.1M NaOH between −0.35 and 1.65 V *vs* RHE. (With permission of The Electrochemical Society).

Obviously the changes in PMR spectra with potential are best seen at alkaline pH, where Fe does not corrode and PMR spectra can be obtained over a very wide potential range. The voltammogram is given in Fig. 3, where it is seen that an electrochromic peak (anodic peak $A_3$) can be grown by potential cycling. The colour changes from transparent at low potentials to brown at high potentials, and the process is thought to involve an Fe(II) and an Fe(III) oxycompound, respectively.

The PMR spectra of an uncycled Fe electrode in 0.1M NaOH at seven increasingly positive potentials are given in Fig. 4. The position of the PMRS maxima under these conditions, as well as after 3 and 15 previous potentials cycles at 40 mV/s from −0.35 to 1.65 V *vs* RHE, is given in Table 1. The electronic transitions of the iron oxycompounds possibly present in the passive layer are given in Fig. 5. Assuming, as said above, that PMR maxima correspond to absorption edges of compounds present in the passive layer, we have made the following assignments:

- At ≈0 V (first anodic peak) the only maximum, at 260−270 nm, corresponds to FeO, whose absorption edge lies at ≈280 nm.
- Between 0.15 and 0.35 V (second and third anodic peaks, respectively) the wavelength of the PMR maximum increases from 320 to 370 nm with increasing potential. We think that over this region there is a progressive oxidation of an Fe (II) compound to an Fe(III) compound, probably an oxyhydroxide, since the absorption edge of $\alpha$−, ß− and $\gamma$−FeOOH lie at 380−420 nm. This is in agreement with the brown colour of the cycled electrode at potentials more positive than the electrochromic peak (peak $A_3$), since compounds that absorb in the blue are brown. The progressive transformation of an Fe(II) compound into an Fe(III) compound with increasing positive potential is in agreement with the XPS results of Haupt and Strehblow (10). At 0.35 V there remains a shoulder at 315 nm, which we assign to an Fe (II) oxycompound.
- In the potential range between 0.35 and 1.0 V (fourth anodic peak $A_4$, and beginning of passivation) the main feature in the spectrum is the maximum at 370−400 nm (again, its wavelength increases with increasing potential), which we again assign to one or more Fe(III) oxyhydroxides. The shoulder at 320−330 nm could be due to an Fe(II) compound, perhaps $Fe(OH)_2$; although we have found no spectroscopic data for this compound, it is known that it is white when unoxidized, and therefore its absorption edge must lie in the UV. If this assignment could be confirmed, it would yield direct spectroscopic evidence of the "sandwich" structure of the passive layer, proposed (already 30 years ago) to be $Fe/Fe_3O_4/Fe_2O_3$ by Vetter (11) and Nagayama and Cohen (12).
- The passive region, between 1.0 and 1.5 V, shows a wealth of structures in the PMR spectra. The Fe(II) and Fe(III) compounds are still present, and at 1.0 V two new PMR maxima appear at 470 and 520 nm. The evolution with increasing potential of the maxima at 400−420 nm and at 520 nm is most informative: while the intensity of the 400−420 nm maximum (due to FeOOH) decrease, that of the 520 nm maximum increases and its position shifts to 550−570 nm, in good agreement with the band−gap of $\alpha$−$Fe_2O_3$. This can be simply explained as a progressive dehydration of the Fe(III) oxyhydroxide into $\alpha$−$Fe_2O_3$. Actually dehydration might have begun at 0.5 V, where a very weak shoulder at ≈500 nm appeared. As for the shoulder at 470 nm that first appears at 1.0 V and is present up to 1.6 V, we think it corresponds to $\gamma$−$Fe_2O_3$, whose absorption edge lies at 490 nm. It would be formed by dehydration of FeOOH, as we have also assumed for $\alpha$−$Fe_2O_3$.

Therefore, the composition of the passive layer seems to be even more complex than the "sandwich" structure proposed by Vetter (11) and Nagayama and Cohen (12). We have reproducibly found four structures in the PMR spectrum at 1.0 V, and think that each one of them corresponds to a different compound. The PMRS technique does not allow us to conclude if these compounds are arranged in parallel layers, e.g.

$$Fe/Fe(OH)_2/FeOOH/\alpha\text{−}Fe_2O_3/\gamma\text{−}Fe_2O_3$$

**Table 1. Wavelength of maxima of PMR spectra of Fe / nm**

| Vp/V | −0.05 | 0.15 | 0.35 | 0.5 | 1.0 | 1.5 | 1.6 | 1.8 |
|---|---|---|---|---|---|---|---|---|
| **1M NaOH** | | | | | | | | |
| no cycling | 265 | 320 | $330^{sh}$ | $330^{sh}$ | $330^{sh}$ | | 300 | |
| | | | 370 | 380 | $\underline{400}$ | 430 | | |
| | | | | (500) | (450) | 460 | 460 | |
| | | | | | 540 | $\underline{560}$ | $\underline{515}$ | |
| 3 cycles | 270 | 320 | | (330) | $330^{sh}$ | | | |
| | | | 375 | 385 | $\underline{405}$ | 430 | | |
| | | | | | (455) | 470 | 470 | |
| | | | | (490) | 530 | $\underline{560}$ | 515 | |
| 15 cycles | 270 | 320 | | (330) | $310^{sh}$ | | | |
| | | | 390 | 400 | $\underline{410}$ | 430 | | |
| | | | | (490) | (450) | 470 | 470 | |
| | | | | | 525 | $\underline{560}$ | $\underline{525}$ | |
| **0.1M NaOH** | | | | | | | | |
| no cycling | 270 | 300 | $315^{sh}$ | (310) | (310) | | 300 | |
| | | | 370 | 385 | $\underline{400}$ | 420 | | |
| | | | | | (470) | 470 | 470 | |
| | | | | (510) | 520 | $\underline{550}$ | $\underline{525}$ | |
| 3 cycles | 260 | 270 | | (310) | (310) | | 300 | |
| | | | 365 | 390 | $\underline{400}$ | 425 | | |
| | | | | | (465) | 465 | $\underline{465}$ | |
| | | | | (500) | 520 | $\underline{560}$ | 510 | |
| 15 cycles | 260 | 270 | 370 | 380 | $\underline{390}$ | $\underline{400}$ | 400 | |
| | | | | | (460) | (450) | (430) | |
| | | | | | $510^{sh}$ | 550 | (520) | |
| **Borax, pH 9.2** | | | | | | | | |
| no cycling | 265 | 305 | 320 | (320) | $320^{sh}$ | $320^{sh}$ | 320 | 320 |
| | | | | 380 | $\underline{400}$ | 410 | 430 | |
| | | | | | (455) | 475 | 470 | 480 |
| | | | | (500) | 515 | 570 | $\underline{565}$ | $\underline{530}$ |

**Sulphuric or Nitric acid, pH 1–3**

| Vp/V | 0.75 | 1.0 | 1.5 | 1.6 | 1.8 |
|---|---|---|---|---|---|
| | $320^{sh}$ | $320^{sh}$ | $320^{sh}$ | 320 | 320 |
| | 400 | $\underline{405}$ | 415 | 430 | |
| | | (455) | 480 | 480 | 485 |
| | 510 | 530 | $\underline{565}$ | $\underline{565}$ | $\underline{565}$ |

sh: shoulder     (): very weak shoulder     _: main maximum

or if they occur in patches over the iron surface. This problem could perhaps be tackled by effecting the PMRS measurements with a thin enough (≈ μm) beam of light, and scanning the electrode surface.
- At the onset of $O_2$ evolution at 1.6 V, FeOOH and $\alpha$–$Fe_2O_3$ (maxima at 430 and 560 nm, respectively) disappear, while the Fe(II) compound and $\gamma$–$Fe_2O_3$ (maxima at 300 and 470 nm, respectively) still remain. A new maximum appears at 520 nm; it could perhaps correspond to the formation of soluble $FeO_4^{2-}$ which has an absorption band at 510 nm.

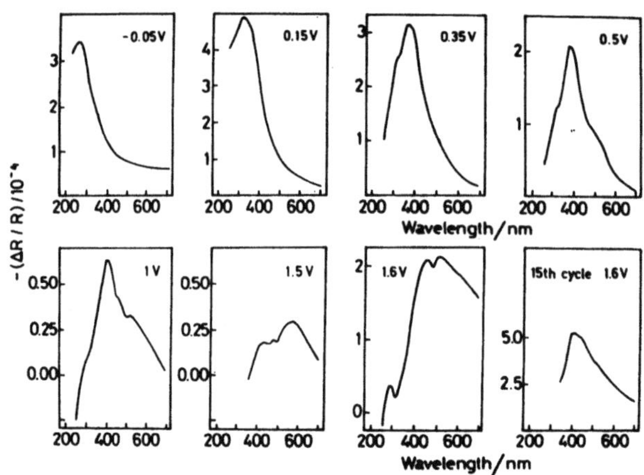


**Fig.4.** PMR spectra of Fe in 0.1M NaOH at seven increasingly positive potentials. The spectrum at 1.6 V *vs* RHE after 15 previous cycles as in Fig. 3 is also included. (Reprinted by permission of The Electrochemical Society).

3.1.1. Influence of previous cycling. In 0.1M NaOH the ohmic–drop–corrected impedances remained nearly the same after subjecting the Fe electrode to up to 15 potential cycles. In agreement with this behaviour, the intensities of PMR maxima in the passive range did not change with cycling. However, in the 0.15–0.35 V range (second and third anodic peaks) the PMR intensities increased by a factor of 2–3 in 1M NaOH, and by a factor of 5–15 in 0.1 M NaOH.

It is obvious from the above findings that the presence of the thick electrochromic layer neither changes the composition of the passive layer, nor originates PMR structures in the passive range. This agrees with the fact that the electrochromic layer is so weakly attached to the electrode that it can be easily wiped off it. The minor changes observed at 1.6 V in 0.1M NaOH are discussed elsewhere (9).

### 3.2. Nickel.

The electrochemical behaviour of Ni has been intensively studied, due to its importance in industry. Several PMRS studies have appeared (13–15). Successive cyclic voltammograms at pH 7, 9.2 and 13 are given in Fig. 6, where it can be seen that at pH ≥ 9.2 a reversible transpassivation peak appears at ≈ 1.4 V *vs* RHE. The transpassivation peak corresponds to an electrochromic process, the electrode colour changing from metallic at lower potentials to black at higher potentials. The position of the maxima in the PMR spectra of Ni in the pH range 1–14 and over the potential range where measurement was possible is given in Table 2.

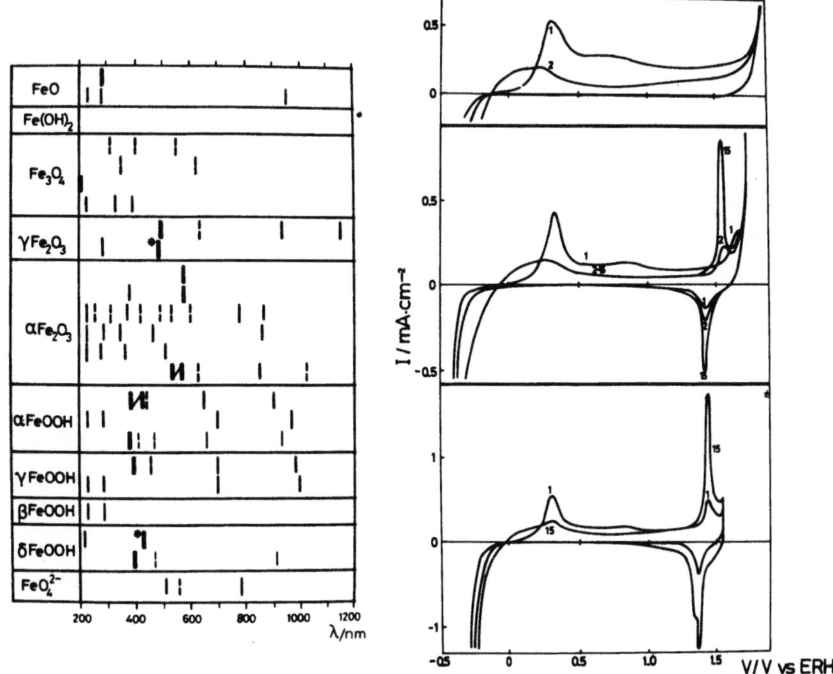


**Fig.5.** Electronic transitions of selected Fe oxygenated compounds. **Thick segment**: absorption edge; **sh**: shoulder; **w**: weak maximum; *: estimated from reported spectrum; | / | : variable range, depending on particle size; | = | : flat maximum. (Reprinted by permission of The Electrochemical Society).

**Fig.6.** Successive cyclic voltammograms of Ni electrode from −0.5 to 1.75 V *vs* RHE (to 1.55 V in NaOH) at 40 mV/s: **a)** Phosphate buffer, pH 7. **b)** Borax, pH 9.2. **c)** 0.1M NaOH. Cycle number is indicated against each curve. (Reprinted by permission of The Electrochemical Society, Inc.)

The PMR spectra of Ni in 0.1M NaOH at six increasingly positive polarization potentials are given in Fig. 7. The presence of NiO in the passive region is clearly evidenced by the maximum at 290 nm, which is the bandgap of NiO; it appears over the whole investigated pH range, 1–14. Therefore, NiO is present not only at acid pH, as universally accepted, but at alkaline pH as well, in agreement with *in-situ* and *ex-situ* findings using other techniques (16–18). The maximum at 290 nm still appears at the most positive potentials where measurements could be made, i.e. above the transpassive peak in alkaline media. This means that not all of the NiO is oxidized to a higher (Ni(III) and/or Ni(IV)) oxyhydroxide, a (probably inner) layer of NiO still remaining. This finding is in agreement with the XPS results of Strehblow (16) and Sherwood (19).

In acid media we have observed for the first time a positive PMR maximum at 400 nm, which appears at the lower potentials in the passive range (Table 2). Since many Ni(II) compounds present a transition at this wavelength, we attribute the PMR maximum to a Ni(II) species involved in the transition from corrosion to passivation, as proposed by Sato and Okamoto (20) and Arvia (21).

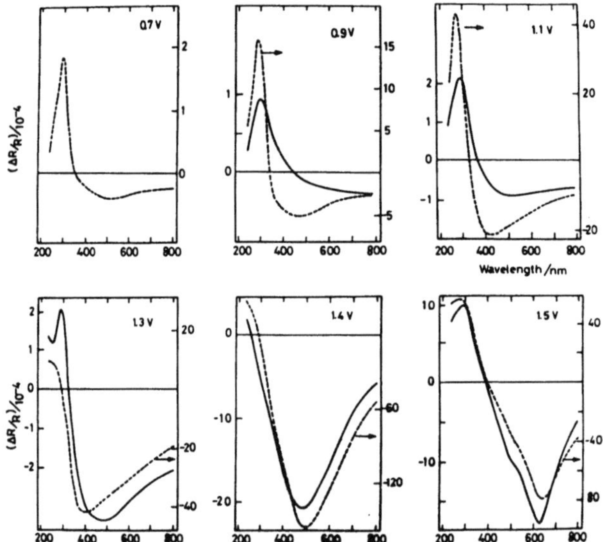


**Fig.7.** PMR spectra of Ni in 0.1M NaOH at different potentials. The potential is indicated against each curve. (Reprinted by permission of The Electrochemical Society, Inc.).

In the potential region of the transpassive peak, which, as said above, only appears at pH ≥ 9.2, there is a negative PMR maximum at ≈ 630 nm (Table 2). It must be produced by the black oxycompound (usually postulated to be NiOOH) responsible for the electrochromic effect, since both the intensity of the PMR maximum and the thickness of the electrochromic compounds increased with previous potential cycling. An absorption maximum at 605 (22) or 630 nm (17) has been reported in the differential reflectance spectrum of the electrochromic oxide, which is in agreement with our finding.

At pH ≥ 7 a negative PMR structure at ≈ 500 nm appears at high positive potentials. It is a maximum at pH 7, and only a shoulder at pH ≥ 9.2. Actually, even at alkaline pH it appears as a maximum in a very narrow potential range around the transpassive peak, just before the 630 nm maximum appears. Therefore, it could correspond to a Ni species with an oxidation state intermediate between those of NiO and the electrochromic oxide.

Summing up, NiO and the same two higher Ni oxides would be present in the transpassive region over all the alkaline range, pH 9.2–14.

### 3.3. Cobalt.

The corrosion and passivation of Co have been less studied than those of Fe and Ni. The first three cyclic voltammograms of a Co electrode in 0.1M borax, pH 9.2 are shown in Fig. 8a. It can be seen that passivation is not perfect, and that a transpassive peak is seen.

Before recording the PMR spectra, the electrode was held 15–30 min. at −0.5 V, after which the potential was stepped to 1.5–1.7 V in order to produce a passive layer without appreciable electrode dissolution, and then PMR spectra were recorded at several, increasingly cathodic potentials. The wavelengths of the PMR maxima at several pH values are reported in Table 3. In general, the results obtained at increasingly anodic potentials were the same, although some times shifted by ≈0.1 V due to irreversibility of the electrochemical processes involved.

**Table 2. Wavelength of maxima of PMR spectra of Ni / nm**

**$H_2SO_4$, pH 1–3**

| Vp | 0.9 | 1.1 | 1.3 | 1.5 | 1.6 | 1.7 |
|---|---|---|---|---|---|---|
| | 300 | 290 | 280 | 270 | – | 280 |

**Phosphate buffer, pH 7**

| Vp | 1.1 | 1.3 | 1.5 | 1.7 | 1.8 | 1.9 |
|---|---|---|---|---|---|---|
| | 290 | 280 | 270 | 270 | 260<br>510 | 270<br>520 |

**Borax, pH 9.2**

| Vp | 1.1 | 1.3 | 1.5 | 1.6 | 1.65 |
|---|---|---|---|---|---|
| uncycled | 290 | 290 | 270 | 300<br>$520^{sh}$<br>660 | 310<br>$520^{sh}$<br>670 |
| three cycles | 290 | 280 | 260<br>500 | 270<br>$530^{sh}$<br>640 | 280<br>$530^{sh}$<br>660 |
| fifteen cycles | 290 | – | $260^{sh}$<br>520 | 300<br>$520^{sh}$<br>640 | 300<br>$540^{sh}$<br>660 |

**0.1M NaOH**

| Vp | 0.7 | 0.9 | 1.1 | 1.3 | 1.4 | 1.5 |
|---|---|---|---|---|---|---|
| uncycled | | 290 | 290 | 290 | –<br>490 | 290<br>$520^{sh}$<br>630 |
| three cycles | 300 | 290 | 280 | 280 | – | 280 |
| fifteen cycles | 290 | 280 | 270<br>$410^{a}$ | –<br>$390^{a}$ | –<br>500 | 260<br>$520^{sh}$<br>640 |

**1M NaOH**

| Vp | 0.7 | 0.9 | 1.1 | 1.3 | 1.35 | 1.5 |
|---|---|---|---|---|---|---|
| uncycled | | 300 | 290 | 290 | –<br>490 | 300<br>$510^{sh}$<br>630 |
| three cycles | 300 | 290 | 280 | 260 | –<br>$430^{a}$ | 270<br>630 |
| fifteen cycles | 290 | 280 | 270<br>$400^{a}$ | –<br>$360^{a}$ | –<br>$420^{a}$ | 290<br>650 |

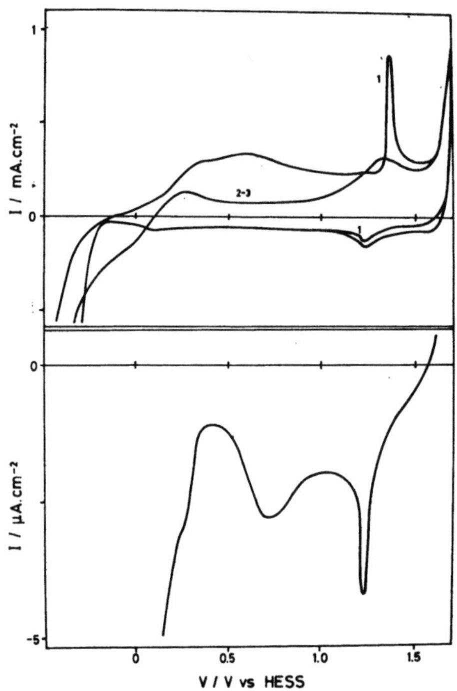


**Fig.8. Top**: First three cyclic voltammograms of Co in 0.1M borax, pH 9.2, from −0.5 to 1.7 V *vs* RHE at 50 mV/s. **Bottom**: Cathodic sweep from 1.7 to −0.5 V at 1 mV/s.

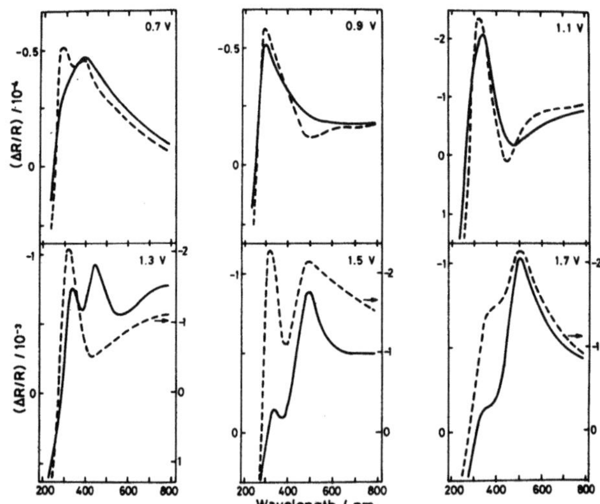


**Fig.9**. PMR spectra of Co in 0.1M borax at different potentials (V *vs* RHE). (—) uncycled electrode. (−−) electrode previously subjected to three potential cycles.

**Table 3. Wavelength of maxima of PMR spectra of Co / nm**

**Phosphate buffers, pH 4 and 7**

| Vp/V | 1.5 | 1.6 | 1.7 |
|---|---|---|---|
| | 320 | 320 | 340 |
| | | 550 | 510 |

**Borax, pH 9.2**

| Vp/V | 0.7 | 0.9 | 1.1 | 1.3 | 1.5 | 1.6 |
|---|---|---|---|---|---|---|
| uncycled | 290$^{sh}$ | 310 | 340 | 340 | 340 | 360$^{sh}$ |
| | 390 | | | 450 | 490 | 500 |
| | | | ≥ 800 | ≥ 800 | | |
| three cycles | 290 | 300 | 320 | 330 | 330 | 380$^{sh}$ |
| | 390 | | | | 510 | 510 |
| | | | ≥ 800 | ≥ 800 | | |

**0.1M NaOH**

| Vp/V | 0.7 | 0.9 | 1.1 | 1.3 | 1.5 | 1.6 |
|---|---|---|---|---|---|---|
| uncycled | 290 | 320 | 340 | 340 | 360$^{sh}$ | 360$^{sh}$ |
| | 400 | | | 450 | 490 | 490 |
| | | | ≥ 800 | ≥ 800 | | |
| three cycles | 290 | 290 | 320 | 320 | 320 | 340 |
| | 440 | 400 | | | | 510 |

**1M NaOH**

| Vp/V | 0.7 | 0.9 | 1.1 | 1.3 | 1.5 |
|---|---|---|---|---|---|
| uncycled | 290 | 320 | 340 | 340 | 360$^{sh}$ |
| | 460 | | | 460 | 480 |
| | | ≥ 800 | ≥ 800 | 720 | |

sh: shoulder

Over the pH range 4–14 there appeared a shoulder or maximum whose wavelength increased from 290 to 360 nm as the potential increased from 0.7 V (passive region) to 1.6 V (transpassive region and beginning of $O_2$ evolution). No absorption edges in this range have been found in the spectra of Co oxycompounds. We assume that the maximum corresponds to CoO, which is olive–green when stoichiometric, but is easily oxidized, its colour changing to brown and eventually black as the Co (III) content increases. Thus, the increase with potential of the wavelength maximum would be due to an increasing nonstoichiometry of the CoO, as proposed in (23).

A maximum at 400 nm appeared at 0.7 V (passive region) in alkaline media, pH 9.2–14. Again, no absorption edges in this range were found in the spectra of Co oxycompounds. The maximum cannot be due to the well–known pink $Co(OH)_2$, whose absorption edge lies at 250 nm (our estimate). It could perhaps correspond to another form of $Co(OH)_2$. A blue variety, $\alpha$ $Co(OH)_2$, with water molecules between the O–Co–O layers, can be obtained electrochemically (24). It can be noted in Fig. 8a. that there is no change in current at 0.7 V in the cathodic sweep at 50 mV/s, although the PMRS spectrum shows a maximum at 400 nm that was not present at higher potentials. Therefore, after stepping the electrode potential to 1.7 V a cathodic sweep at a much lower speed, 1mV/s, was carried out (Fig. 8b), and, effectively, now a broad cathodic peak appeared at 0.7 V. This shows how the presence of an otherwise unobserved electrochemical process can be inferred from PMRS results.

A broad maximum at ≥800 nm appeared at 1.0–1.3 V (transpassive peak) at pH 9.2–14. It must correspond to $Co_3O_4$, whose absorption edge is ≈820 nm (our estimate), in agreement with its black colour. At 1.5 V the maximum broadened into a plateau, probably because of overlap with another maximum at 490–500 nm (see below). Therefore, it cannot be decided if at potentials ≥1.5 V $Co_3O_4$ is oxidized or not.

A maximum at 500 nm appeared at pH 9.2–14 in the transpassive peak, at potentials ≥1.3 V when the PMR spectra were recorded at increasingly more positive potentials, and at ≥1.2 V when they were recorded at increasingly more negative potentials. It must correspond to the oxidized species of the transpassive couple, proposed to be $Co(OH)_2$/CoOOH by Mössbauer spectroscopy (25). If this is so, then the maximum at 500 nm would correspond to CoOOH, but at present the other candidate, $Co_2O_3$, cannot be excluded. The wavelength of the maximum decreased rapidly with decreasing potential, probably due to reduction to a nonstoichiometric oxycompound.

Summing up, the same compounds were observed in the anodic film on Co over the pH range 9.2–14, which is at variance with the results reported by other workers (23).

## 4. ANODIC IRIDIUM OXIDE FILMS (AIROFs)

A thick oxide layer can be grown on Ir in $H_2SO_4$ by potential cycling between −0.05 and 1.5 V *vs* RHE. Depending on pH, up to four anodic peaks can be seen in the voltammogram (Fig. 10). The main redox process is reversible, and has electrochromic character, the electrode colour changing from metallic at low potentials to blue–black at

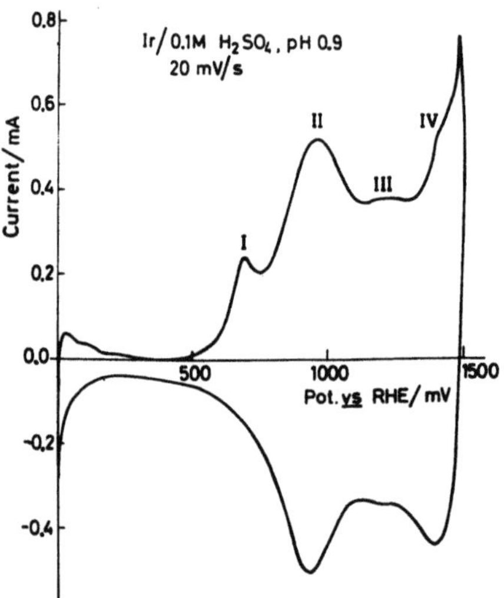


**Fig.10.** Cyclic voltammogram at 20 mV/s of an Ir electrode in 0.1M $H_2SO_4$. The electrochromic process is associated with the main voltammetric peak at 0.95 V *vs* RHE.

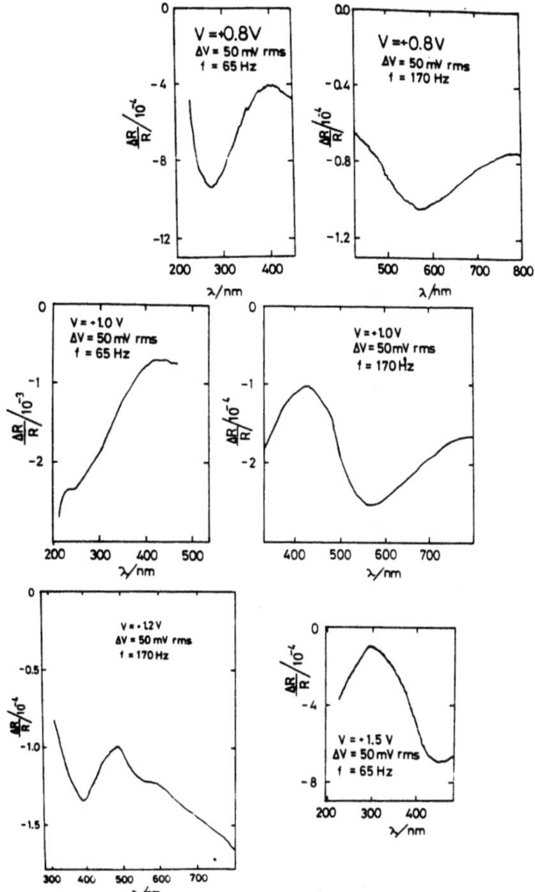


**Fig.11.** PMR spectra of an uncycled Ir electrode in 0.1M $H_2SO_4$ at several potentials.

high potentials. For this reason much effort has been devoted to the development of anodic Ir oxide films (AIROFs) for display applications. Much effort has been devoted to the identification of the electrochemical reaction involved in the electrochromic process, postulated to be the oxidation of a transparent, insulating Ir(III) oxyhydroxide to a blue-black, conducting Ir(IV) oxyhydroxide (26).

The density of AIROFs is 2 g.cm$^{-3}$ (27), only one sixth that of bulk $IrO_2$, due to a high proportion of water in the film, which probably has an amorphous structure. However, the electrochromic process is believed to involve a simple solid-state mechanism, in which raising the anodic potential decreases the Fermi level in the AIROF, from the top of the conduction band of $IrO_2$ (transparent, insulating state) to its middle (blue-black, conducting state). As is well known, bulk $IrO_2$ (and also $RuO_2$) have metallic conductivity and lustre, which is due to their having a half-empty conduction band, as true metals do. Strong support for this simple electrochromic mechanism has been obtained by UPS (see Section 3.2 in the chapter on Photoelectron Spectroscopy).

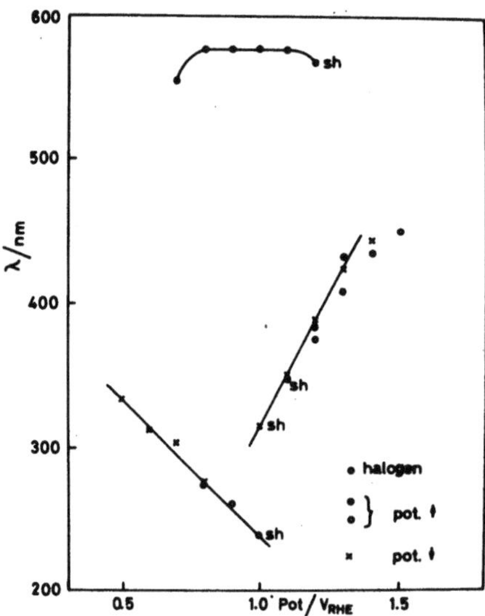


**Fig.12**. Ir electrode in 0.1M $H_2SO_4$. Dependence of the wavelengths of PMRS maxima on electrode potential. The Ir electrode had been previously cycled 5 sec at 2 V/sec, which resulted in a capacitance at the electrochromic peak of 1.3 mF; after recording all the PMR spectra the capacity had increased to 1.75 mF.

We have studied AIROFs by PMRS. The results are very interesting, but their interpretation is far from clear. At the low frequencies (≈ 100 Hz) used in PMRS, AIROFs behave as large capacitors in series with the ohmic electrolyte resistance, since the charge-transfer resistance is very low, and the pseudocapacitance of an adsorbed film (as a matter of fact, even of an adsorbed monolayer) is much higher than that of the double layer in parallel with it, which can therefore be neglected. Therefore, the PMRS signal must proceed from the electrooxidation and electroreduction of the compounds present in the AIROF.

Typical PMR spectra of an uncycled Ir electrode are shown in Fig. 11 and the dependence of the wavelengths of the PMRS maxima on electrode potential is depicted in Fig. 12. A peak at ≈ 580 nm is present from 0.8 to 1.1 V, the potential range corresponding to the electrochromic process, whose peak potential lies at 1.0 V. An absorption maximum at 610 nm was also observed in the transmission spectrum of the coloured form of an AIROF obtained by potential cycling a thin Ir film deposited on a metal electrode (28). We assign these two maxima to the transition at 620 nm found in the reflection spectrum of single-crystal $IrO_2$ (29). If this is so, it is remarkable that the transition energy is about the same for the Ir(IV) ion in bulk $IrO_2$ and in a hydrous, amorphous AIROF whose density is one-sixth that of the crystal.

The PMR spectrum of AIROFs in the range 250–450 nm shows a very curious behaviour (Fig. 12): from 0.5 to 1.0 V the wavelength of the maximum at ≈ 300 nm decreases linearly with increasing positive potential, with a slope of −0.18 nm/mV. At 1.0 V a new structure at ≈ 320 nm appears, whose wavelength increases linearly with potential (slope = +0.33 nm/mV), while the structure observed at lower potentials disappears.

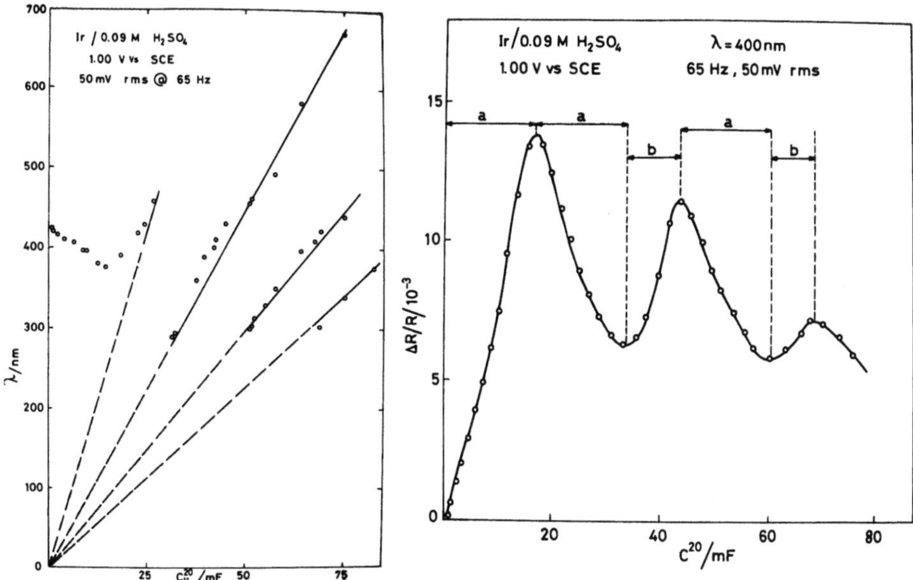


**Fig.13.** Ir electrode in 0.1M $H_2SO_4$. Dependence of the wavelengths of PMRS maxima on the pseudocapacitance of the AIROFs as determined from a voltammogram at 20 mV/s. Pseudocapacitance is assumed to be proportional to thickness of the AIROF.

**Fig.14.** Ir electrode in 0.1M $H_2SO_4$. Dependence of the PMR signal at 400 nm on the pseudocapacitance (assumed to be proportional to the thickness) of AIROFs.

The simplest explanation of a linear change with potential of the wavelength of a maximum would be that of interference, since it is often found that the thickness of passive films increases linearly with potential (10). This is not the case here, since: *1)* the oxide film on an uncycled AIROF should be thinner than the minimum thickness ($\sim\lambda$ /4n, where n = refraction index) necessary to observe an interference minimum. *2)* the same linear potential dependence has been found for AIROFs of many different thicknesses.

It has recently been discovered that there exist linear relationships between the electrode potential and the energy (i.e. frequency) of several types of transitions:

*1.* Surface plasmons, which are collectively propagating charge density waves confined to the surface and travelling parallel to it (30).

*2.* Transitions from the bulk metal to metal surface states located ≈ 0.1 nm away from the metal surface (31).

We think that neither of these two transition types can occur here, since the potential dependence we have observed is linear with the wavelength, not the frequency. Besides, surface plasmons are observed mostly in metals having free electron character (*viz.* Cu, Ag and Au), and surface states require the use of single crystal surfaces for their study. Therefore, the two potential-dependent transitions shown in Fig. 12 remain unassigned.

It could be thought that PMRS of passive layers is free from interference effects, since the magnitude measured is not the reflectance, R, but the change, ΔR, in reflectance brought about by a change in the electrode potential. However, we have found that PMR spectra of thick AIROFs show the interference expected for (non modulated) reflectance of film-covered metals (32). In Fig. 13 we have plotted the wavelengths of the maxima of PMR spectra as a function of the pseudocapacitance of the electrochromic peak, which is assumed to be proportional to the thickness of the AIROF. It can be seen that for films thicker than 25 mF the points can be grouped in four straight lines, all passing through the origin, and whose slopes decrease in the ratio 1:1/2:1/3:1/4. It can easily be calculated that this ratio corresponds to successive interference maxima, or to successive maxima and minima. From the ellipsometric measurements of Gottesfeld (33) we estimate that the first case holds here.

Interference effects can also be detected by measuring the reflectance at a single wavelength and increasing film thickness, the method having been used *e.g.* to determine the thickness of the oxide film formed by anodic oxidation of n-GaAs (34). It can be seen in Fig 14 that the intensity of the PMR signal at 400 nm of increasingly thicker AIROFs shows successive maxima and minima. Two differences appear with respect to interference in non-modulated reflectance. First, in the absence of film (zero thickness) the reflectance should obviously be maximum, while in this case the PMR signal is zero. This confirms that the PMR signal is due to the AIROF, and therefore initially increases linearly with increasing thickness, until interference effects begin to appear at 17 mF. Second, the distance from a maximum to the following minimum ("a" in Fig. 4) is not the same as that between a minimum and the following maximum ("b" in Fig. 14), but double, which remains unexplained.

## REFERENCES


(1) A. Bewick and A.M. Tuxford, Faraday Soc. Symp. No. 4, "Optical Studies of Adsorbed Layers at Interfaces", The Faraday Society, London, 1970, p.114.
(2) E.D. Palik, Ed., "Handbook of Optical Constants of Solids", Academic Press, 1985.
(3) N. Hara and K. Sugimoto, J. Electrochem. Soc. **126** 1328 (1979).
(4) C. Gutiérrez and M.A. Martínez, *ibid.* **133** 1873 (1986).
(5) K.L. Shaklee, F.M. Pollak and M. Cardona, Phys. Rev. Lett. **15** 883 (1965).
(6) D.E. Aspnes, Surf. Sci. **37** 418 (1973).
(7) L.M. Abrantes, R. Peat, L.M. Peter and A. Hamnett, Ber. Bunsenges. **91** 369 (1987).
(8) W. Höpfner and W.J. Plieth, Werkstoffe u. Korros. **36** 373 (1985).
(9) G. Larramona and C. Gutiérrez, J. Electrochem. Soc. **136** 2171 (1989).
(10) S. Haupt and H.-H. Strehblow, Langmuir **3** 873 (1987).
(11) K.J. Vetter, Z. Elektrochem. **62** 642 (1958).
(12) M.-i. Nagayama and M. Cohen, J. Electrochem. Soc. **109** 781 (1962).
(13) W. Paatsch, Surf. Sci. **37** 59 (1973).
(14) A.G. Akimov, M.G. Astaf'ev and I.L. Rozenfel'd, Electrokhimiya **11** 924 (1975).
(15) N. Hara and K. Sugimoto, a) J. Japan Inst. Metals **47** 31 (1983) b) *ibid.* **47** 39 (1983) c) Trans. Japan Inst. Metals **24** 236 (1983).
(16) B.P. Löchel and H.-H. Strehblow, J. Electrochem. Soc. **131** 713 (1984).
(17) R.E. Hummel, R.J. Smith and E.D. Verink, Corros. Sci. **27** 803 (1987).
(18) P. Delichère, S. Joiret, A. Hugot-Le Goff, K. Bange and B. Hetz, J. Electrochem. Soc. **135** 1856 (1988).
(19) T. Dickinson, A.F. Povey and P.M.A. Sherwood, J. Chem. Soc. Faraday Trans. 1, **73** 327 (1977).
(20) N. Sato and G. Okamoto, J. Electrochem. Soc. **110** 605 (1963).
(21) M.R. Barbosa, S.G. Real, J.R. Vilche and A.J. Arvia, *ibid.* **135** 1077 (1988).

(22) C. Zhang and S.-M. Park, *ibid.* **134** 2966 (1987).
(23) N. Hara and K. Sugimoto, J. Japan Inst. Metals **44** 915 (1980).
(24) P. Benson, G.W.D. Briggs and W.F.K. Wynne-Jones, Electrochim. Acta **9** 275, 281 (1964).
(25) G.W. Simmons, A. Vértes, M.L. Varsányi and H. Leidheiser, Jr., J. Electrochem. Soc. **126** 187 (1979).
(26) J.D.E. McIntyre, S. Basu, W.F. Peck (Jr.), W.L. Brown and W.M. Augustyniak, Phys. Rev. **B25** 7242 (1982).
(27) J.D.E. McIntyre, W.F. Peck (Jr.) and S. Nakahara, J. Electrochem. Soc. **127** 1264 (1980).
(28) K.S. Kang and J.L. Shay, J. Electrochem. Soc. **130** 766 (1983).
(29) A.K. Goel, G. Skorinko and F.H. Pollak, Phys. Rev.**B24** 7342 (1981).
(30) D.M. Kolb and R. Kötz, Surf. Sci. **64** 96 (1977).
(31) D.M. Kolb, W. Boeck, K.-M. Ho and S.H. Liu, Phys. Rev. Lett. **47** 1921 (1981).
(32) K.B. Blodgett and I. Langmuir, Phys. Rev. **51** 964 (1937).
(33) S. Gottesfeld and S. Srinivasan, J. Electroanal. Chem. **86** 89 (1978).
(34) R.P. Silberstein and F.H. Pollak, Surf. Sci. **101** 269 (1980).

# NONLINEAR OPTICAL TECHNIQUES FOR SURFACE STUDIES

Y. R. Shen
Department of Physics, University of California
Materials and Chemical Sciences Div., Lawrence Berkeley Lab.
Berkeley, California 94720 USA

ABSTRACT. Second-order nonlinear optical effects have recently been proven to be an extremely versatile analytical tool for surface studies. They are forbidden in media with centrosymmetry and hence, highly surface specific. Being coherent optical techniques, they have high temporal, spatial, and spectral resolution, and are useful for experiments requiring in-situ probing and remote sensing. They also have the advantage of being applicable to any interface accessible by light. In these lectures, we shall give a general review on the development of such techniques. We first present the underlying theory of a second-order optical process at an interface. We then discuss, with examples, the various possible applications of second harmonic and sum-frequency generation for surface studies.

## 1. INTRODUCTION

The progress of surface science and technology relies heavily on our ability to probe and characertize surfaces and interfaces. For this purpose, many techniques have been developed in the past,[1] but they are often limited in their applicabilities. New surface probes are always in great demand. Among the existing techniques,[1] many involve emission, adsorption, or scattering of massive particles; only a few rely on optical probing. The optical methods have, in general, the advantages of being nondestructive, capable of in-situ remote sensing with high spatial and temporal resolution, and applicable to any interface accessible by light. The difficulty of these methods often lies in the lack of surface specificity and sensitivity. Linear optical absorption and reflection, for example, can be employed for surface studies only if the dominant bulk contribution to the signal can be discriminated against.[2] The same is true with light scattering.

Recently, laser methods for material studies have advanced to a highly sophisticated level. One therefore wonders if they can also be applied to surface studies. Indeed, there have been a number of very interesting recent discoveries in this area. Laser-induced



*C. Gutiérrez and C. Melendres (eds.),*
*Spectroscopic and Diffraction Techniques in Interfacial Electrochemistry*, 281–311.

desorption,[3] photoacoustic spectroscopy,[4] and photothermal deflection spectroscopy[5] have been developed for spectroscopic studies of adsorbates and surface states. Resonance fluorescence[6] and multi-photon ionization[7] have been used to measure energy redistribution in molecules scattered or desorbed from a surface. Various surface nonlinear optical effects have also been seriously considered for surface studies. Stimulated Raman gain spectroscopy using continuous wave (CW) mode-locked lasers has been shown to be sensitive enough to yield the Raman spectrum of a molecular monolayer.[8] Coherent anti-Stokes Raman scattering that employs surface or guided waves is also very sensitive, and should be capable of detecting a monolayer.[9,10] Both techniques, however, suffer from the undesired contribution from the bulk; spectral selectivity is usually not sufficient to recover the surface signal from the bulk background.

Second-order nonlinear optical effects, such as second harmonic and sum frequency generation, are generally forbidden (under the electric-dipole approximation) in a medium with inversion symmetry. Since the inversion symmetry is necessarily broken at a surface, such effects appear to be highly surface-specific for interfaces between two centrosymmetric media.[11] As second-order effects, they also have a much stronger signal than third harmonic generation or four-wave mixing. In addition, they possess other advantages: the experimental arrangement is simple, the output can be easily separated from the input spectrally, and luminescence and light scattering background can be easily discriminated against by spatial filtering. As a result, submonolayers of molecules adsorbed on surfaces can be readily detected. Being optical, these probes are applicable to all interfaces accessible by light. They have inherently high spatial, temporal, and spectral resolutions, and are capable of nondestructive and remote sensing of surfaces or interfaces.

Optical second harmonic generation (SHG) appears to be most attractive because of its simplicity. The surface sensitivity of SHG was recognized in earlier experiments,[12] but the exploitation of SHG for surface probing did not begin until recently.[13] In a number of experiments,[14-20] it was demonstrated that SHG is extremely versatile as a surface tool. It can be used to study surfaces and molecular adsorbates at any interface between centrosymmetric media. The many input and output polarization combinations and the high spectral and time resolution of the technique provide some unique possibilities for surface studies. In some cases, even a CW semiconductor diode laser is intense enough to allow the detection of a submonolayer of molecular adsorbates.[21] The only shortcoming of SHG is its incapability for surface infrared spectroscopy because of the lack of sensitive photodectors in the IR range. This, however, can be overcome by replacing SHG with IR-visible sum-frequency generation (SFG), as has been successfully demonstrated recently.[22]

Historically, Bloembergen and Pershan[27] first set the foundation of surface nonlinear optical effects in their classical paper on "Light Waves at the Boundary of Nonlinear Media". They showed how harmonic generation from a medium can be affected by the boundary conditions on the surfaces. Later, the physical origins of surface nonlinearities were discussed[28,29] and experiments of SHG from media with inversion symmetry emphasizing the surface effects were carried out.[29-31] The submonolayer sensitivity of the surface SHG in detecting adsorbates on metals was actually observed.[32,33] Unfortunately, the work went unnoticed for many subsequent years. In 1974, surface enhanced Raman scattering on noble metals was discovered.[34] It soon caught the fancy of many researchers.[35] In the course of development, it was recognized that Raman scattering can be considered as a two-photon process, which is a nonlinear optical process; if one nonlinear optical process shows surface enhancement because of enhanced local fields on a metal surface, other nonlinear optical processes should experience similar surface enhancement. This was readily demonstrated by SHG from a roughened Ag surface.[36] Monitoring of electrochemical cycling on a Ag surface in an electrolytical solution indicated that SHG has a sensitivity of detecting far less than a monolayer of molecules.[14] A quantitative estimate then revealed that even without surface enhancement, SHG should be capable of detecting molecular monolayers at any surface or interface (certainly not limited to metal surfaces).[15] Thus, SHG as a possible surface probe was rediscovered. Today, we already witnessed successful applications of the technique in almost all disciplines of surface science.

In these lectures, we shall give a general survey on the development of SHG and SFG for surface studies. We shall first provide a theoretical background underlying the technique. This is important because of some persistent errors and confusion existing in the literature on the subject. We shall then discuss briefly the applications of the technique to many different surfaces, illustrating the power of the technique as an effective surface tool.

## 2. BASIC THEORY

The general theory of surface-second order nonlinear optical processes follows the early work of Bloembergen and Pershan.[27] A more rigorous and correct treatment of surface SHG has been developed more recently and can be found in Refs. 37-39. Here, we focus on the essential points using SHG in reflection as the representative case. The experimental geometry is shown in Fig. 1. We shall assume, for simplicity of discussion, only medium 2 is nonlinear.

Let $\vec{P}^{NLS}(2\omega)$ be the nonlinear polarization induced in medium 2 by the incoming laser beam. The radiation field generated by $\vec{P}^{NLS}$ can be obtained from the solution of the wave equation

$$\vec{\nabla} \times \vec{\nabla} \times \vec{E} + (\varepsilon/c^2)(\partial^2/\partial t^2)\vec{E} = (-4\pi/c^2)(\partial^2/\partial t^2)\vec{P}^{NLS} \qquad (1)$$

subject to

$$\vec{\nabla} \cdot \varepsilon\vec{E} = -4\pi\, \vec{\nabla} \cdot \vec{P}^{NLS} .$$

With $\vec{P}_s^{NLS}$ and $\vec{P}_b^{NLS}$ denoting surface and bulk nonlinear polarizations,

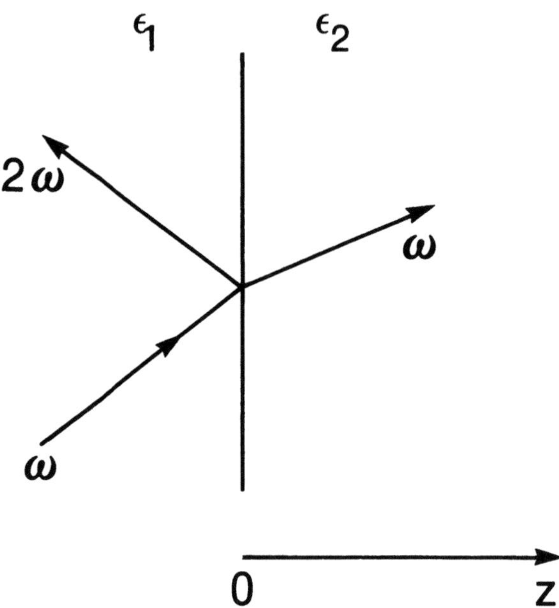


Figure 1. Geometry of second harmonic generation from an interface in the reflected direction.

respectively, we have

$$\vec{P}^{NLS} = \begin{cases} 0 & z < 0 \\ [\vec{P}_s^{NLS}\delta(z) + \vec{P}_b^{NLS}]\exp(i\vec{k}_s \cdot \vec{r} - i\omega_s t) & z \geq 0 \end{cases} \tag{2}$$

It can be shown that the radiation field $\vec{E}(2\omega)$ calculated using $\vec{P}^{NLS}$ in Eq. (2) is the same as that from an effective surface nonlinear polarization $\vec{P}^S_{eff}\delta(z)\exp(i\vec{k}_s \cdot \vec{r} - i\omega t)$,[40] where

$$P^S_{eff,j} = P^{NLS}_{s,j} + i(k_{2,z} + k_{s,z})^{-1}\, P^{NLS}_{b,j} \qquad \text{for } j = x,y$$

$$P^S_{eff,z} = P^{NLS}_{s,z} + i(1/\varepsilon_2)(k_{2,z} + k_{s,z})^{-1}\, P^{NLS}_{b,z}\ , \tag{3}$$

where $\vec{k}_1$ and $\vec{k}_2$ are the wavevectors of the freely propagating SH waves reflected into medium 1 and transmitted into medium 2, respectively, and $\vec{k}_s$ is taken to be in the x-z plane. The x-component of all the wavevectors is constrained by the continuity of the electric field at z = 0 such that $k_{1,x} = k_{2,x} = k_{s,x} \equiv k_x$. The bulk and surface nonlinear polarizations $\vec{P}_b^{NLS}$ and $\vec{P}_s^{NLS}$ are naturally related as we shall discuss explicitly later. The form of Eq. (3) is reasonable since one should be able to construct the solution using the superposition of thin slabs of polarization $\vec{P}_b^{NLS}$. Each slab will radiate just like the top surface slab with polarization $\vec{P}_s^{NLS}$ except they will have slightly different phase factors. The factor $(k_{2,z} + k_{s,z})^{-1}$ is the effective thickness of the bulk media over which the radiation is phase-matched.

The radiation from a surface layer with source polarization $P^S \times \exp(i\vec{k}_s \cdot \vec{r} - i\omega_s t)$ and with isotropic local dielectric $\varepsilon'$ can be derived using the boundary conditions:[41]

$$\Delta B_x = -4\pi i(\omega_s/c)P_y^S \quad ; \quad \Delta B_y = +4\pi i(\omega_s/c)P_x^S$$

$$\Delta E_x = -4\pi(\partial P_z^S/\partial x) \quad ; \quad \Delta E_y = -4\pi(\partial P_z^S/\partial y)$$

$$\Delta B_z = 0 \quad ; \quad \Delta D_z = -4\pi[(\partial P_x^S/\partial x) + (\partial P_y^S/\partial y)] \ . \tag{4}$$

Denoting the $\hat{p}$- and $\hat{s}$-polarized components of the electric field with subscripts, the electric field radiated back into medium 1 due to the surface polarization $\vec{P}^S$ is:[27,31,40]

$$E_p(\omega_s) = \frac{i4\pi k_1}{\varepsilon_2 k_{1,z} + \varepsilon_1 k_{2,z}} (k_{2,z}P_x^S + \varepsilon_2 k_x P_z^S)$$

$$E_s(\omega_s) = \frac{i4\pi k_1}{k_{1,z} + k_{2,z}} [(\varepsilon_1)^{-1} k_1 P_y^S] \ , \tag{5}$$

where $k_1$ is the magnitude of $\vec{k}_1$ and the wavevector components are as defined above. Inserting $\vec{P}_{eff}^{NLS}$ defined in Eq. (3) for $\vec{P}^S$ in Eq. (5), one can calculate the radiation reflected from an arbitrary nonlinear medium. We now consider the specific form of $\vec{P}_{eff}^{NLS}$ for centrosymmetric media.

Let us now focus on the nonlinear polarization $\vec{P}^{NLS}(2\omega)$ induced by $\vec{E}(\omega)$ in the medium with inversion symmetry. It is well known that the bulk nonlinear polarization $\vec{P}_b^{NLS}$ can be expanded into a series of multipole terms[42]

$$\vec{P}_b^{NLS} = \vec{P}^{(2)}(2\omega) - \nabla \cdot \overleftrightarrow{Q}^{(2)}(2\omega) + \frac{c}{i2\omega} \nabla \times \vec{M}^{(2)}(2\omega) + \dots \ , \tag{6}$$

where $\vec{P}$, $\overleftrightarrow{Q}$, and $\vec{M}$ denote electric-dipole polarization, electric-quadrupole polarization, and magnetization, respectively. All terms are quadratic functions of the field and/or their derivatives. Up to the first derivative of the field, we can write

$$\vec{P}^{(2)}(2\omega) = \overleftrightarrow{\chi}^D : \vec{E}(\omega)\vec{E}(\omega) + \overleftrightarrow{\chi}^P : \vec{E}(\omega)\nabla\vec{E}(\omega) \tag{7a}$$

$$\overleftrightarrow{Q}^{(2)}(2\omega) = \overleftrightarrow{\chi}^Q : \vec{E}(\omega)\vec{E}(\omega) \tag{7b}$$

$$\vec{M}^{(2)}(2\omega) = \overleftrightarrow{\chi}^M : \vec{E}(\omega)\vec{E}(\omega) \ . \tag{7c}$$

In a centrosymmetric medium, the first term in Eq. (7a), which is a pure electric-dipole term, vanishes identically. The nonlinear susceptibilities $\overleftrightarrow{\chi}$'s are constant in a uniform bulk. We then have

$$\vec{P}_b^{NLS} = \overleftrightarrow{\chi}_{bq}^{(2)}:\vec{E}(\omega)\nabla\vec{E}(\omega) \ , \tag{8}$$

which, for a cubic medium, can be written as[17,31]

$$P_{b,i}^{NLS} = (\delta - \beta - 2\gamma)[(\vec{E}(\omega) \cdot \nabla]E_i(\omega) + \beta E_i(\omega)[\nabla \cdot \vec{E}(\omega)] + \gamma\nabla_i \times [\vec{E}(\omega) \cdot \vec{E}(\omega)] + \zeta E_i(\omega)\nabla_i E_i(\omega) \ , \tag{9}$$

where $\delta$, $\beta$, $\gamma$, and $\zeta$ are constants. The same expression holds for an isotropic medium with $\zeta = 0$.

Extending $\vec{P}_b^{NLS}$ to an interface, we notice that because the inversion symmetry is necessarily broken at the interface, $\overleftrightarrow{\chi}^D$ is always nonvanishing. Furthermore, the rapid structural variation across an interface implies that the derivatives of $\overleftrightarrow{\chi}$'s along the surface normal are nonnegligible. The surface nonlinear polarization $\vec{P}_s^{NLS}$ in Eq. (2) is actually related to $\vec{P}_b^{NLS}$ by the integral[37-39]

$$P_{s,i}^{NLS}(2\omega) = \int_I P_{b,i}^{NLS}\, dz \qquad \text{for } i = x,y$$

$$= \int_I s(z,2\omega)P_{b,z}^{NLS}\, dz \quad \text{for } i = z \ , \tag{10}$$

where I denotes the interface layer over which the integration extends, and $s(z,2\omega) \equiv E_z(2\omega)/D_z(z,2\omega)$ describes the rapid variation of the normal field component $E_z$ across the interface. Note that the field components $E_x$, $E_y$ and the displacement current $D_z$ are continuous across a boundary.

From Eqs. (6) and (7), we have[37-39]

$$\begin{aligned}\vec{P}^{(2)}(2\omega,z) &= \vec{P}^D(z) - \vec{\nabla} \cdot \overleftrightarrow{Q}(z) + c\vec{\nabla} \times \vec{M}(z) + \ldots \\ &= \int \overleftrightarrow{\chi}^D(z,z',z''):\vec{E}(z')\vec{E}(z'')dz'dz'' \\ &\quad - \vec{\nabla} \cdot \int \overleftrightarrow{\chi}^Q(z,z',z''):\vec{E}(z')\vec{E}(z'')dz'dz'' \\ &\quad + c\vec{\nabla} \times \int \overleftrightarrow{\chi}^M(z,z',z''):\vec{E}(z')\vec{E}(z'')dz'dz'' + \ldots,\end{aligned} \tag{11}$$

(we have omitted the $\overleftrightarrow{\chi}^P$ term here for simplicity, as it can be combined into the $\overleftrightarrow{\chi}^Q$ term). Since the susceptibilities $\overleftrightarrow{\chi}^\mu(z,z',z'')$ ($\mu$ = D,M,Q) are response functions of the medium and are expected to be fairly local, they can be expanded into a power series

$$\chi^\mu(z,z',z'') = \overleftrightarrow{\chi}^\mu(z)\delta(z' - z)\delta(z'' - z) + \overleftrightarrow{\chi}_a^\mu[\vec{\nabla}\delta(z' - z)]\delta(z'' - z) + \overleftrightarrow{\chi}_b^\mu\delta(z' - z)[\vec{\nabla}\delta(z'' - z)] + \ldots \tag{12}$$

and hence

$$\int \overleftrightarrow{\chi}^{\mu}(z,z',z''):\vec{E}(z')\vec{E}(z'')dz'dz'' = \overleftrightarrow{\chi}^{\mu}(z):\vec{E}(z)\vec{E}(z) + \overleftrightarrow{\chi}^{\mu}_{a}(z):[\vec{\nabla}\vec{E}(z)]\vec{E}(z) + \overleftrightarrow{\chi}^{\mu}_{b}:\vec{E}(z)[\vec{\nabla}\vec{E}(z)] + \ldots \quad (13)$$

By defining the surface nonlinear susceptibility $\overleftrightarrow{\chi}_s^{(2)}$ by the relation

$$\vec{P}_s^{NLS} = \overleftrightarrow{\chi}_s^{(2)}:E(\omega,z = 0^+)E(\omega,z = 0^+) , \quad (14)$$

we find, from Eq. (10),

$$\chi^{(2)}_{s,ijk}E_j(\omega,z = 0^+)E_k(\omega,z = 0^+) = \int_I s(2\omega,z)\{\chi^D_{ijk}E_j(\omega,z) - (\partial/\partial z)[\chi^Q_{zijk}E_j(\omega,z)]E_k(\omega,z) + \chi^D_{a,izjk}[(\partial/\partial z)E_j(\omega,z)]E_k(w,z)\} \times dz . \quad (15)$$

If we let $S_{ijk} = s_i(2\omega,z)s_j(\omega,z)s_k(\omega,z)$ and $s_i(\Omega,z) = 1$ for $i = x,y$ and $s_z(\Omega,z) = E_z(\Omega)/D_z(\omega)$, then we obtain

$$\chi^{(2)}_{s,ijk} = \int_I \{\chi^D_{ijk}S_{ijk} - (\partial/\partial z)[\chi^Q_{zijk}s_j(\omega,z)]s_i(2\omega,z)s_k(\omega,z) + \chi^D_{a,izjk}[(\partial/\partial z)s_j(\omega,z)]s_i(2\omega,z)s_k(\omega,z)\}dz . \quad (16)$$

As an example, consider $\chi^{(2)}_{s,yzz}$. We hae

$$\chi^{(2)}_{s,yzy} = \int_I \left(\chi^D_{yzy}(z)s(z) - (\partial/\partial z)[\chi^Q_{zyzy}(z)s(z)] + \chi^D_{a,yzzy}(z) \times (\partial/\partial z)s(z)\right)dz . \quad (17)$$

Notice that $s(z) = 1/\varepsilon_i$ ($i = 1$ or 2) away from the interface so that $s(z)$ varies from $1/\varepsilon_1$ to $1/\varepsilon_2$ across the interface.

We can identify the physical origins of the various terms in $\chi^{(2)}_{s,yzy}$ in Eq. (17). The first term in the integral is the electric-dipole term which is nonvanishing because the interface layer, different from the isotropic bulk, has a structure with no inversion symmetry. The second term actually leads to electric-quadrupole radiation at $2\omega$ from the interface. This is a term that has so far been forgotten in the earlier papers on surface SHG. Its integration across the interface gives

$$\int_I (\partial/\partial z)[\chi^Q_{zyzy}(z)s(z)]dz = [\chi^Q_{zyzy}(0^-)]/\varepsilon_1 - [\chi^Q_{zyzy}(0^+)]/\varepsilon_2 , \quad (18)$$

which is often nonvanishing if there is a structural disparity between the two bounding media. This term is uniquely determined by the bulk parameters. It is insensitive to the surface treatment, and should in fact be called a bulk contribution. The third term in the integral in Eq. (17) comes from the field discontinuity, i.e., the rapid field variations across the interface. This is usually regarded in the literature as the electric-quadrupole (nonlocal) contribution to the surface nonlinearity. If the dielectric constants of the two media are matched, then the field becomes continuous across the interface and this particular term should vanish.

If we now write

$$\vec{P}^S_{eff}(2\omega) = \overleftrightarrow{\chi}^{(2)}_{S,eff}:\vec{E}(\omega)\vec{E}(\omega) \ , \tag{19}$$

the solution in Eq. (5) with $\vec{P}^S$ replaced by $\vec{P}^S_{eff}$ leads to a SH output (photons/pulse)[41]

$$S(2\omega) = \frac{32\pi^3\omega\sec^2\theta}{\hbar c^3\varepsilon_2(\omega)\varepsilon_2^{1/2}(2\omega)} |\vec{e}\,'(2\omega) \cdot \overleftrightarrow{\chi}^{(2)}_{S,eff}:\vec{e}\,'(\omega)\vec{e}\,'(\omega)|^2 I^2(\omega)AT \tag{20}$$

for an input pulse with intensity $I(\omega)$, cross-section A, and pulsewidth T. Here $\theta$ is the SH reflection angle. The vectors $\vec{e}\,'(\Omega)$ are related to the unit polarization vectors $\hat{e}$ in medium 1 by the Fresnel coefficients $\overleftrightarrow{L}$, in the case of a plane interface:

$$\vec{e}\,'(\Omega) = \overleftrightarrow{L} \cdot \hat{e}(\Omega) \ . \tag{21}$$

Here, L assumes a diagonal form with

$$\begin{aligned} L_{xx}(\Omega) &= 2\varepsilon_1 k_{2z}/(\varepsilon_2 k_{1z} + \varepsilon_1 k_{2z}) \\ L_{yy}(\Omega) &= 2k_{1z}/(k_{1z} + k_{2z}) \\ L_{zz}(\Omega) &= 2\varepsilon_2\varepsilon_1 k_{1z}/(\varepsilon_2 k_{1z} + \varepsilon_1 k_{2z}) \ . \end{aligned} \tag{22}$$

More generally, $\overleftrightarrow{L}$ acts as a local-field correction factor. On a rough metal surface with structure of the order of a wavelength, $\overleftrightarrow{L}$ can be very large, leading to a surface enhancement of SHG several orders of magnitude stronger than that from a smooth surface.[14,43] As a coherent process, the output from a plane interface is highly collimated in a well-defined direction determined by $k_{1x}(2\omega) = k_{2x}(2\omega) = k_{Sx}(2\omega) = 2k_x(\omega)$.

Equation (3) shows that in order for SHG to be surface-specific, we must have $|P_S^{NLS}| \gtrsim |P_b^{NLS}|/2k_2$. This would mean $|\chi_S^{(2)}| \gtrsim |\chi_b^{(2)}|/2k_2$ or $|\chi_S^{(2)}|/|\chi_b^{(2)}|d \gtrsim 1/2k_2d$ if the medium has no inversion symmetry with $P_b^{NLS} = \overleftrightarrow{\chi}_b^{(2)}:\vec{E}\vec{E}$, d being the thickness of the interface layer. Such an inequality is unlikely to hold since usually $k_2d \ll 1$. Therefore, SHG is generally not suitable for surface studies of a medium without inversion symmetry. If, however, the medium is centrosymmetric, then $\vec{P}_b^{(2)}$ from electric-quadrupole and magnetic-dipole contributions is ka times weaker than $\overleftrightarrow{\chi}_b^{(2)}:\vec{E}\vec{E}$ from the electric-dipole contribution, where a is the lattice constant or the size of the atoms or molecules in the medium. The condition $|P_S^{(2)}| \gtrsim (1/2k_2)|P_b^{(2)}|$ then becomes $|\chi_S^{(2)}|/|\chi_b^{(2)}|d \gtrsim a/d$ which, with $a \lesssim d$, can be satisfied in many cases. This happens, for example, with metal surfaces where free electrons give rise to large surface nonlinearities, with semiconductor surfaces where dangling bonds with lone-pair electrons yield large surface nonlinearities, and with surface layers of molecular adsorbates that possess large nonlinearities. As we shall discuss later, the surface layer can dominate over the bulk in SHG in some cases; in other cases, the surface nonlinearity can be determined separately by appropriate measurements.

As an order-of-magnitude estimate, we assume $|\chi_s^{(2)}| \sim |\chi_b^{(2)}|d$, which is normally in the $10^{-6}$-$10^{-8}$ esu range. With $d \sim 10^{-7}$ cm, we have, for a surface monolayer of atoms or molecules, $|\chi_s^{(2)}| \sim 10^{-13}$-$10^{-15}$ esu. One can also use the second-order perturbation calculation to give a rough estimate of $|\chi_s^{(2)}|$. Assuming the absence of local-field correction and assuming the laser excitation far of resonance, we have

$$|\chi_s^{(2)}| \sim N_s \langle (er)^3 \rangle / (\hbar\bar{\omega}_o)^2 \ . \tag{23}$$

With $N_s \sim 10^{14}$-$10^{15}/cm^2$ for a molecular monolayer, $r \sim 1$ Å, and $\bar{\omega}_o \sim 3 \times 10^{15}$/sec, we find $|\chi_s^{(2)}| \sim 10^{-14}$-$10^{-15}$ esu. Resonant enhancement, of course, can greatly enhance the value of $|\chi_s^{(2)}|$.

Knowing the value of $|\chi_s^{(2)}|$, we can estimate the signal strength of SHG from a surface layer using Eq. (20) and assuming $P_s^{(2)}$ dominating. If we take $|\chi_s^{(2)}| = 10^{-15}$ esu, $I(\omega) \sim 10$ MW/cm$^2$, $A \sim 0.2$ cm$^2$, $T \sim 10$ nsec, and $\theta \sim 45°$, the signal strength S is of the order of $10^4$ photons/pulse. Such a large signal should be readily detectable. This then predicts the submonolayer sensitivity of surface SHG. Measurement of the surface SHG allows us to deduce $\overleftrightarrow{\chi}_s^{(2)}$ for the surface layer.

From the surface science point of view, it is of course most important that the macroscopic $\overleftrightarrow{\chi}_s^{(2)}$ can be related to the surface properties. Unfortunately, the detailed microscopic understanding of $\overleftrightarrow{\chi}_s^{(2)}$ is still lacking. We can offer here only the following general picture. For a molecular monolayer adsorbed on a surface, the surface nonlinear susceptibility can be written as

$$\overleftrightarrow{\chi}_s^{(2)} = \overleftrightarrow{\chi}_{ss}^{(2)} + \overleftrightarrow{\chi}_m^{(2)} + \overleftrightarrow{\chi}_i^{(2)} \ . \tag{24}$$

Here, $\overleftrightarrow{\chi}_{ss}^{(2)}$ is the contribution from the bare substrate surface, $\overleftrightarrow{\chi}_m^{(2)}$ is the contribution from the molecular layer isolated from the substrate, and $\overleftrightarrow{\chi}_i^{(2)}$ arises from interaction between the molecule and the substrate that changes the properties of both the molecule and the surface of the substrate. Whether a particular term dominates in Eq. (24) or not depends on the material system. In some cases, asymmetric molecules with large second-order nonlinear polarizabilities can yield a dominating $\overleftrightarrow{\chi}_m^{(2)}$ if they are aligned on the surface. In other cases, $\overleftrightarrow{\chi}_m^{(2)}$ may be negligible and the only effect of the adsorbates on $\overleftrightarrow{\chi}_s^{(2)}$ is a modification of $\overleftrightarrow{\chi}_{ss}^{(2)}$ by $\overleftrightarrow{\chi}_i^{(2)}$. While it is possible to make a reasonable guess at the physical origins of the various terms in Eq. (24) for a given surface system, a more quantitative picture would require a rather thorough understanding of the microscopic properties of the surface system. It is possible, for example, to use the bond theory to calculate $\overleftrightarrow{\chi}_s^{(2)}$ for semiconductor and insulator surfaces, but one needs prior knowledge of the surface structure.

## 3. EXPERIMENTAL CONSIDERATIONS

One of the advantages of SHG as a surface tool is the simplicity of the experimental arrangement. As shown in Fig. 2, the basic elements in the setup are nothing but a laser and a photodetection system with proper spectral filtering. The SH signal from a smooth surface is coherent and highly directional, and therefore, collection of the signal is a trivial matter.

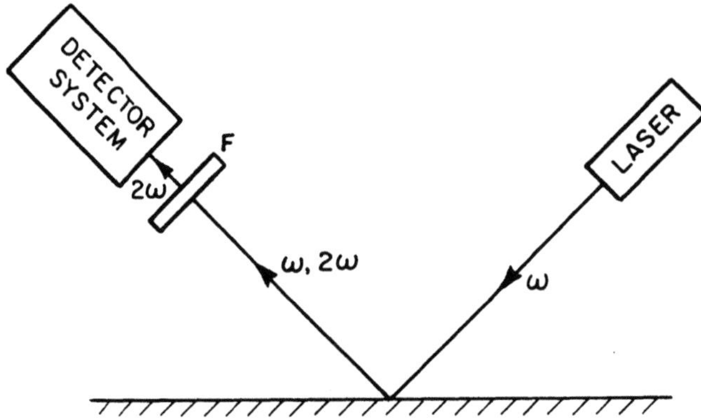


Figure 2. Experimental arrangement of second-harmonic generation from a surface. F denotes the filtering system that rejects the fundamental but passes the second harmonic.

Equation (20) shows that the SH signal should increase with increases in $I_p$, A, and T; however, if the pump pulse energy $W_p = I_pAT$ is fixed, then the signal should increase with decreases in A and T. If we assume that the breakdown is due to laser-induced melting, and the threshold energy for a short-pulse excitation is $I_p^{th}T \sim 1$ J/cm$^2$, the maximum SH signal is[24]

$$S_{max} \sim 10^{27} f |\chi_s^{(2)}|^2 A/T \text{ photons/sec},$$

where f is the pulse repetition rate. Let us consider a Q-switched pump laser with $W_p \sim 100$ mJ, $T \sim 10$ nsec, and $f \sim 10$/sec. Assuming that the minimum detectable signal is 10 photons/sec, we should have a minimum detectable $|\chi_s^{(2)}|$ of $\sim 3 \times 10^{-18}$ esu. Let $\chi_s^{(2)} = N_s\alpha^{(2)}$, where $N_s$ is the number of surface atoms or molecules per cm$^2$, and $\alpha^{(2)}$ is the nonlinear polarizability. Then, for typical values of $\alpha^{(2)}$ in the range between $10^{-31}$ esu (for weakly nonlinear molecules) and $10^{-29}$ esu (for more strongly nonlinear molecules, metals, and semiconductors), the minimum surface density of atoms or molecules detectable is between $3 \times 10^{13}$/cm$^2$ and $3 \times 10^{11}$/cm$^2$. Since $N_s$ for a monolayer is typically larger than $10^{14}/cm^2$, the submonolayer sensitivity of surface SHG is obvious. The detectability can be further improved by using a mode-locked laser. If we take $W_p \sim 10$ μJ, $T \sim 1$ psec, and $f \sim 10^3$/sec, we find a minimum detectable $|\chi_s^{(2)}|$ of $\sim 3 \times 10^{-19}$ esu.

An estimate for the CW case is also interesting. Because of thermal diffusion, the threshold power $P_{th}$ for laser-induced melting is roughly proportional to $A^{1/2}$ for sufficiently small A. If we assume $P_{th}/A^{1/2}$ to be $\sim 10^4$ W/cm (as for Si), the maximum SH signal is

$$S_{max} \sim 10^{36}|\chi_s^{(2)}|^2 \text{ photons/sec,}$$

which is independent of A. With $A \sim 10^{-8}$ $cm^2$ from tight focusing, $P_{th}$ is only about 1 W, a power readily obtainable from a CW laser. The minimum detectable $|\chi_s^{(2)}|$ is around $10^{-18}$ esu.

## 4. APPLICATIONS

We now discuss briefly the various applications of SHG and SFG to many different interfaces with demonstrative examples.

### 4.1. Applications to Electrochemistry on Electrodes in an Electrolytic Cell

It is possible to use surface SHG to monitor in-situ electrochemical processes at an electrode in an electrolytical solution.[14] When a metal electrode is roughened by the electrochemical processes, surface enfhancement of SHG due to the local field effect can greatly improve the sensitivity of the technique.[36] Then, even a CW diode laser could be intense enough to detect submonolayers of molecules.[21] An example is shown in Fig. 3, where surface SHG from a Ag electrode in a 0.1 m KCℓ solution is found to respond to the appearance and disappearance of AgCℓ on the electrode. The signal increases rapidly when the first monolayer of AgCℓ is formed at the roughened electrode, as judged from the amount of charge transfer through the electrode, and decreases precipitously when the last monolayer of AgCℓ is reduced back to Ag and Cℓ. In between, although hundreds of AgCℓ monolayers are formed and reduced, the signal varies only slightly, mostly in response to the change in surface roughness. This is the result of the surface-specific nature of SHG. If 0.05 M pyridine is added into the solution and a sufficiently negative bias is put on the Ag electrode, it is known that pyridine molecules would adsorb to the Ag surface at some active sites with no more than a monolayer. Yet the surface SHG signal in Fig. 3 shows that it can clearly respond to the adsorption of pyridine.[14] When a laser pulse of 0.2 mJ of energy at 1.06 μm and 6 nsec in pulsewidth was focused to a spot of 0.2 $cm^2$ on the pyridine-adsorbed electrode, a signal level of $8 \times 10^5$ photons/pulse was detected. Knowing that the enhancement of SHG on such a surface is $\sim 10^4$, and assuming that the surface coverage of pyridine is $4 \times$

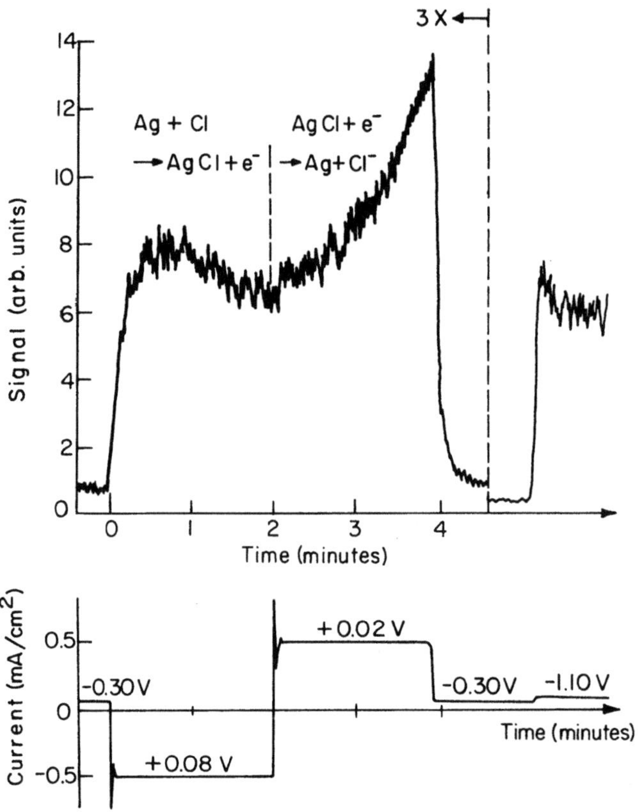


Figure 3. Current and diffuse second-harmonic signal as functions of time during and after an electrolytic cycle. The voltages listed on the lower curve are the potential of the Ag electrode with respect to a saturated calomel electrode. 0.05 M pyridine was added to the 0.1 M KCℓ solution following the completion of the electrolytic cycle.

$10^{14}/cm^2$, we can deduce from Eq. (20) a value of the nonlinear susceptibility per adsorbed pyridine molecule to be ~ $8 \times 10^{-30}$ esu.

Unless a tunable laser is used, the SHG technique is not spectrally selective. This is certainly a disadvantage of the technique. Another difficulty is the lack of microscopic understanding of the surface nonlinearity incurred by the adsorbates. For example, it is not understood why the SHG signal increases when pyridine molecules are adsorbed on the Ag electrode in the electrolytical solution[14] while it decreases when the same molecules are adsorbed on Rh(111) in an ultrahigh vacuum.[18] It may happen that the adsorption of pyridine on Ag induces a charge-transfer band in the green at which the SHG is resonantly enhanced.

The possible applications of SHG to electrochemistry are yet to be explored. As long as the adsorbed species on an electrode can be identified by some means (for example, SERS), surface SHG can be used to study the time dependence of adsorption or desorption of that species, and to help monitor molecular adsorption and desorption during an electrolytical cycle. Thus, for example, in-situ real-time studies of corrosion processes would be possible.

Recently, Richmond and coworkers have used SHG to study adsorption of anions on electrodes from aqueous solution and to monitor surface reconstruction of single-crystal electrodes resulting from electrolytical processes.[44]

## 4.2. Study of Adsorbates at a Liquid/Solid Interface

SHG can also be a viable method to detect molecular adosrbates at a smooth liquid/solid interface, where the solid is certainly not restricted to metals. As an example, Fig. 4 shows how SHG can be used to measure the adsorption isotherm of paranitrobenzoic acid (PNBA) on fused quartz from an ethanol solution.[16] From the slope of the isotherm at low concentraiton, we can estimate an adsorption free energy at infinite dilution of ~ 8 kcal/mole for PNBA on fused quartz in ethanol.[14]

The SH signal from adsorbed molecules on a substrate generally includes a contribution from the substrate, which may be comparable to that of the adsorbed layer. In this case, it must be measured separately and subtracted before the analysis described above can be applied. Because of the coherent nature of the SHG process, all of the signals have well-defined phases, permitting full compensation for the influence of the substrate. With this correction taken into account, it is then possible to study adsorbates at quite low fractional monolayer coverages by means of SHG.

## 4.3. Measurements of Orientation of Molecules at an Interface

The determination of the detailed arrangement of atoms and molecules at an interface is one of the central problems in surface science. The surface-specific SHG technique can provide some of the answers. The nonlinear susceptibility $\overleftrightarrow{\chi}_s^{(2)}$ can be related to the surface structure in two ways. First, the symmetries of $\overleftrightarrow{\chi}_s^{(2)}$ should correspond to macroscopic structural symmetries in the arrangement of the adsorbed molecules. A system lacking any preferred direction in the surface plane, for example, would give rise to a $\overleftrightarrow{\chi}_s^{(2)}$ with equal elements under $x \leftrightarrow y$, i.e., a relabeling of the coordinate axes in the plane. Second, knowledge of $\overleftrightarrow{\chi}_s^{(2)}$ and of the molecular polarizability $\overleftrightarrow{\alpha}^{(2)}$ can lead to information about the average orientation of the adsorbates.[16] This becomes clear if we write the relation between these two quantities in coordinate form:

$$(\chi_s^{(2)})_{ijk} = N_s \langle T_{ijk}^{\lambda\mu\nu} \rangle \alpha_{\lambda\mu\nu}^{(2)} .$$

In this expression $T_{ijk}^{\lambda\mu\nu}$ represents the coordinate transformation between the molecular ($\xi$, $\eta$, $\zeta$) system and the fixed (x, y, z)

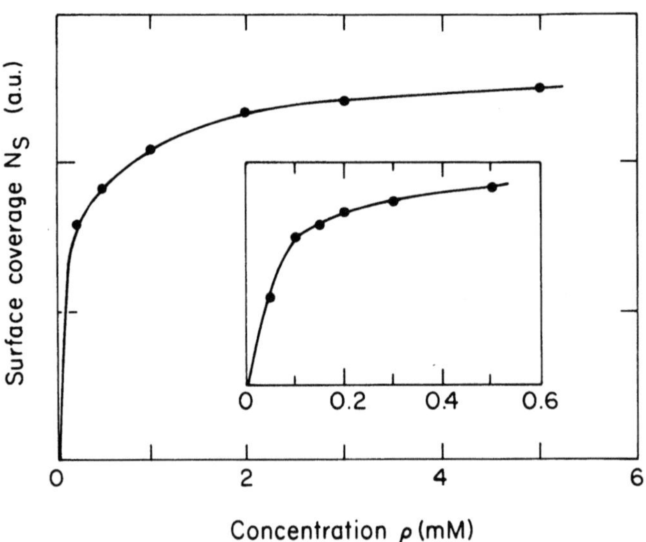


Figure 4. Isotherm for adsorption of PNBA on fused silica out of an ethanol solution. The results were deduced from the SH intensity under s-polarized excitation at 532 nm.

system. It describes the molecular orientation and is the only factor that must be retained in the angular average. Measurements of $\langle T_{ijk}^{\lambda\mu\nu} \rangle$ or, in some cases, of just the ratios of various terms should therefore give us the average orientation of the adsorbed molecules.

A study of adsorbed p-nitrobenzoic acid (PNBA) by surface SHG illustrates these points.[16] The structure of free PNBA and of the molecule adsorbed to fused silica is depicted in Fig. 5. From the observed symmetries of $\overleftrightarrow{\chi}_S^{(2)}$, we have found the molecular distribution in the surface plane to be isotropic. It has also been established that the nonlinearity of such molecules is dominated by $\alpha_{\zeta\zeta\zeta}^{(2)}$. We can therefore consider the molecules to be long rods defined by the $\zeta$-axis. Their orientation is specified by the angle $\theta$ between the molecular $\zeta$-axis and the surface normal. The ratio of any two independent components of $\overleftrightarrow{\chi}_S^{(2)}$ must then give a value for a weighted average of $\theta$. Experimentally, such a ratio can be conveniently obtained through a null measurement of the SH radiation, for which $|\hat{e}_{2\omega} \cdot \overleftrightarrow{\chi}_S^{(2)} : \hat{e}_\omega \hat{e}_\omega| = 0$ holds. The average orientation of PNBA adsorbed on fused silica at fractional monolayer coverages has been determined in this fashion both at solid/air and at solid/ethanol interfaces. Taking the orientational distribution to be sharply peaked, $\theta$ was found to be ~ 40° in the liquid and ~ 70° in air. This difference might be explained by the solvation energy of the PNBA in the liquid.

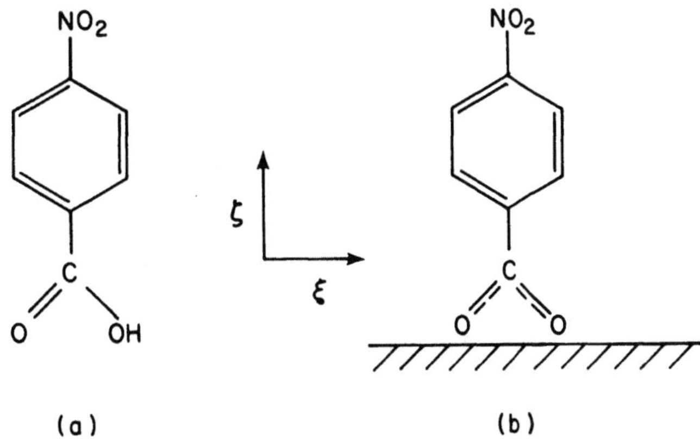


Figure 5. Structure of p-nitrobenzoic acid (PNBA) as a free molecule (a) and in its chemisorbed form (b).

## 4.4. Studies of Langmuir-Type Molecular Monolayers

Insoluble molecular monolayers at liquid/gas or liquid/liquid interfaces have been the subect of numerous studies in many fields of basic and applied research. We can use SHG to study the structure, optical nonlinearities and polymerization of Langmuir films of organic molecules. All the materials studied are spreadable on a water surface, which has the advantage of a low background SHG level and an easily controllable surface density of adsorbate molecules.

To measure the nonlinear polarizability $\overleftrightarrow{\alpha}^{(2)}$ of molecules, one usually relies on SHG measurements from powders[45] or dc electric field induced SHG (EFISH) from molecules in a liquid.[46] The problem with the first method is that one has to know the crystal structure and powder size distribution. With EFISH one actually measures an effective third-order nonlinear susceptibility, and to obtain $\overleftrightarrow{\alpha}^{(2)}$, a knowledge of solvent-solute interactions and local fields is essential. We now have an alternative method by measuring SHG from an orientationally ordered monolayer on a water surface.[47] This works well for amphiphilic molecules, i.e. molecules with one hydrophobic and one hydrophilic part, and has the advantage of an easily controllable surface density of molecules and a low background. The technique has been applied to a number of cyanobiphenyl molecules $C_nH_{2n+1}(C_6H_4)_2CN$ (nCB, n = 8,9,10,12), several derivatives of this cyanobiphenyl structure, a number of fatty acids $C_nH_{2n+1}COOH$ (nFA, n = 14, 17,22), and many others.

We can also use SHG to study polymerization of monolayers of monomers spread at a water/air interface. Monolayer polymerizations are of fundamental interest since their reactivity and kinetics may be studied under controllable and variable conditions of molecular separation and orientation.[48] Figure 6 shows an example of how SHG

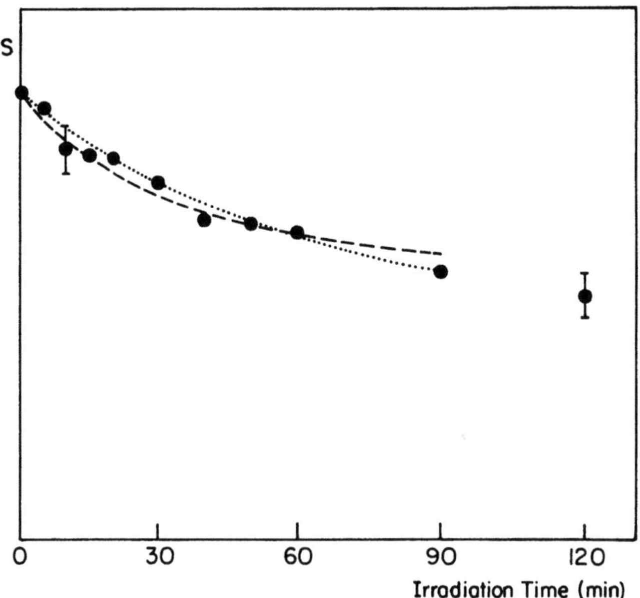


Figure 6. The relative SHG intensity (S) is plotted against irradiation time for UV polymerization of ODMA. The data (•) can be fitted satisfactorily by both first order (...) and second order (---) kinetics (see the text). All experimental data points have the same uncertainty, although error bars for most points have been omitted for clarity.

can follow the kinetics of polymerization of a monolayer of monomers.[49] The signal from a monolayer of octadecyl methacrylate (ODMA) decreases continuously with irradiation by a uv lamp. Unfortunately, due to the low values of $\alpha^{(2)}$ and the relatively small changes in SHG intensity during polymerization our measurements were not accurate enough to unequivocably distinguish between first and second order polymerization kinetics.

The observed decrease in SH intensity upon polymerization can be understood as follows: the second order optical nonlinearity mainly arises from chemical bonds in which the electron distributions are more readily distorted by optical excitations. In ODMA the $\pi$ electrons in the double bonds are likely to dominate the nonlinearity. Since the polymerization breaks a carbon-carbon double bond, $\alpha^{(2)}$ will decrease.

SHG can also be employed to probe the nature of the two-dimensional phase transition of a Langmuir monolayer.[50] An example is described in Figs. 7 and 8. Figure 7 shows the measured surface pressure, $\pi$, as a function of the surface area per molecule, A, for a monolayer of pentadecanoic acid (PDA) on a pH = 2 water substrate at various temperatures. The sharp kink in the middle of each $\pi$-A curve

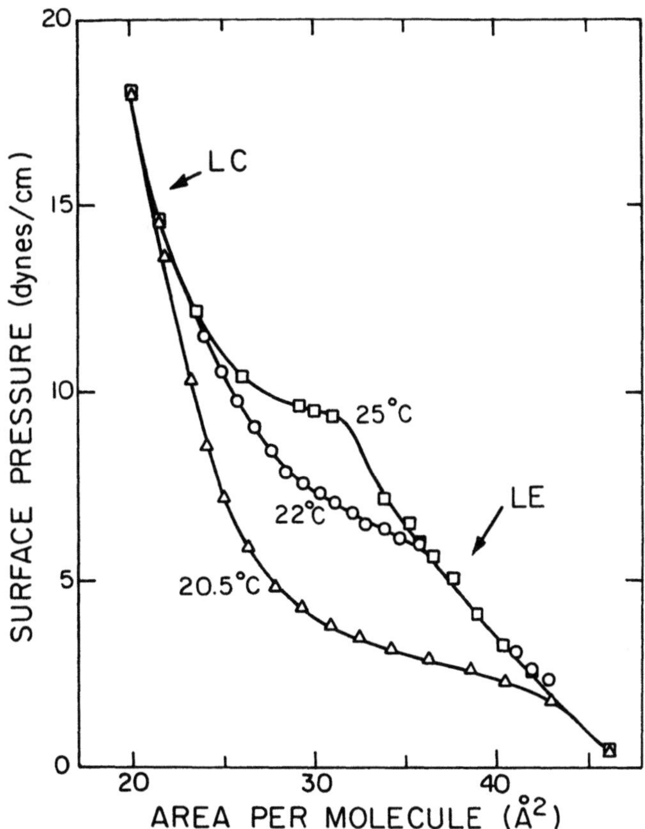


Figure 7. Surface pressure of PDA as a function of the area per molecule on a water surface of pH = 2 for various temperatures.

signals the onset of the transition between the so-called liquid-expanded (LE) and liquid-condensed (LC) phases. Though observed and intensively studied in many similar systems, the nature of the LE-LC transition is still controversial. Experimental data are almost exclusively limited to $\pi$-A measurements, and for their interpretation, various assumptions about the molecular orientation are made but, have never been checked experimentally. Figure 8 shows the results of our orientational measurements for PDA at 25°C as obtained from the SHG data. In the LE phase, $\theta$ rapidly increases with increasing density $N_s$ until the LE-LC transition is reached, whereafter it changes more slowly and linearly with $N_s$. Here, $\theta$ refers to the angle between the surface normal and the polar C-OH bond, which appears to make the dominant contribution to the SHG from this molecule. Intuitively, one expects this polar bond to align normal to the water (and hence the molecular chains would tilt away from the surface normal) as indeed is observed at lower densities

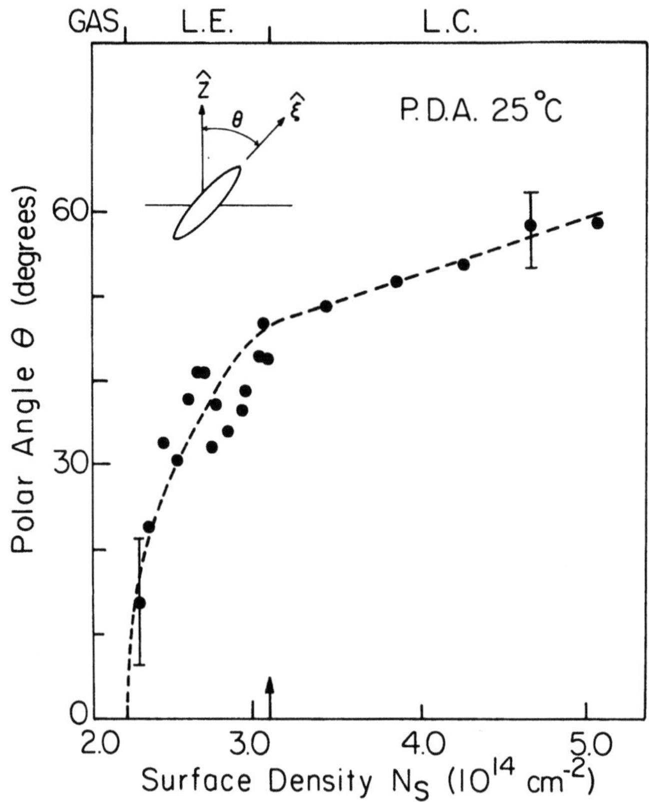


Figure 8. Tilt angle θ between the molecular axis and the surface normal as a function of the surface density of PDA on water at 25°C. The dashed line is an extrapolation through the data points.

where θ approaches 0°. When $N_s$ increases, the steric interaction between the hydrocarbon chains of neighboring molecules tends to align them towards, and hence forcing the C-OH orientation away from, the surface normal. At $N_s = 3 \times 10^{14}$ $cm^{-2}$ a phase transition to an oriented liquid occurs. By measuring θ just below the LE-LC transition (in the LE phase) we found θ = 45° ± 3° for all temperatures, though the transition point itself is very temperature dependent (Fig. 7).

## 4.5. Study of Adsorbates on Metals

The second-order nonlinearity of a metal surface is generally much stronger than that of an insulator. This is presumably because of the high density gradient and strong delocalization of electrons at a metal surface. Chemisorption of atoms or molecules on the surface may modify the surface electron distribution and change the surface nonlinearity. As examples, we consider here the adsorption of oxygen, CO, and alkali metals on transition metals studied by SHG.[18] Such adsorption processes are of practical interest because they are involved in the catalytic reactions of hydrocarbon formation and oxidation. To study adsorption on a well-defined surface, the SHG experiments were carried out in an ultrahigh vacuum (UHV) chamber equipped with low energy electron diffraction (LEED), Auger, and mass spectrometer probes.

The SH signal from Rh(111) as a function of exposure to $O_2$ is depicted in Fig. 9. It drops monotonically from the bare metal value to the final level of approximately 12% at about 1.8 L (Langmuir units) of $O_2$. This immediately suggests that oxygen must adsorb strongly to Rh with a large sticking coefficient. As shown in Fig. 9, the data can be fit theoretically by assuming that the adsorption obeys the simple Langmuir kinetics with only one adsorption site.

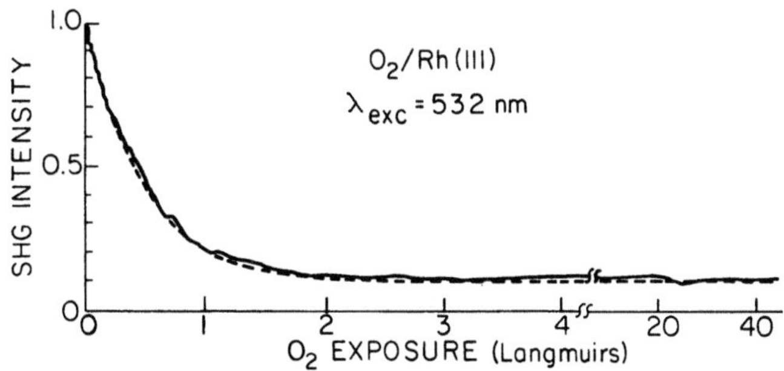


Figure 9. Second-harmonic signal from the Rh(111) surface during $O_2$ exposure [(——) = experimental result; (---) = theoretical fit].

The SH signal also decreases rapidly upon adsorption of CO on Rh(111), as seen in Fig. 10. The result shows a sudden change in slope at a surface coverage of one third of a monolayer ($\theta = 1/3$), and is an indication of the opening of a new adsorption site on Rh(111) for $\theta \geq 1/3$. Indeed, the data in Fig. 10 can be fit very well by a model that assumes a single adsorption site for $\theta \leq 1/3$ but two for $\theta \geq 1/3$. This is consistent with the results of electron energy loss spectroscopy and proves that SHG as a surface probe can even be site-specific.

4.6. Study of Semiconductor Surfaces

Atomic or molecular adsorption on Si and Ge can also be monitored by SHG. The sensitivity is quite high in the case of chemisorption. It is known that dangling bonds exist on clean Si and Ge surfaces. Because of asymmetry, these bonds are highly nonlinear. As a result, SHG from bare Si or Ge is dominated by the surface contribution. When adsorption quenches the dangling bonds, the surface nonlinearity should decrease accordingly.

We consider here oxidation of a Si(111) surface as an example.[51]

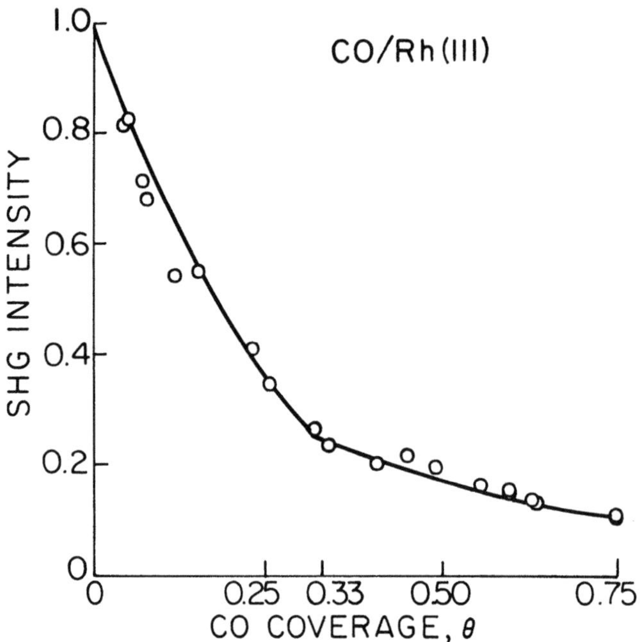


Figure 10. Second-harmonic generation from CO/Rh(111) as a function of CO fractional coverage. Circles are experimental points and solid curve is a theoretical fit.

Figure 11 shows how the SH signal varies as a clean Si(111) sample is exposed to oxygen. The decrease in the SH signal was found to be dominated by the adsorption of the first monolayer of oxygen. The data can be fit by the simple Langmuir model assuming a single adsorption site, as illustrated in Fig. 11. The residual SH signal at high oxygen exposure is due to bulk Si contribution. This background is quite appreciable when the pump laser beam is at 0.53 μm because of strong resonant enhancement, but reduces to a very small value at 1.06 μm.

A similar variation of SHG with adsorption of Pd on Si(111) was also observed.[52] This indicates that a chemical bond must have been established between the adsorbed Pd and the surface Si atoms. The monolayer of Pd on Si(111) appeared to be amorphous, as suggested by the observed azimuthal symmetry of SHG. After annealing, the SHG exhibited a threefold symmetry about the surface normal, which indicates that the surface Pd silicide layer crystalized following an epitaxial rearrangement.

Surface SHG can also be used to study bare surfaces.[20] A freshly cleaved Si(111) surface is known to recontruct to form a (2 × 1) structure as seen by LEED. Annealing changes the structure into (7 × 7). The (2 × 1) structure should yield a twofold symmetry in the

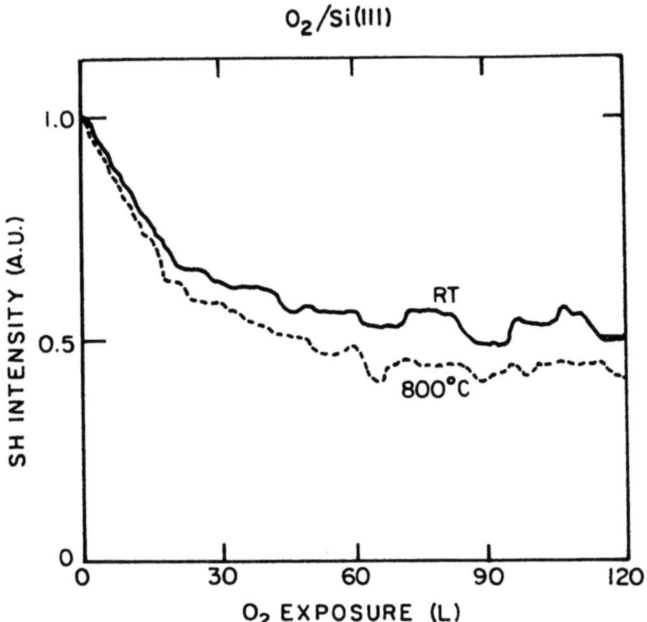


Figure 11. Second-harmonic signal from the Si(111) surface at (a) room temperature (RT) and (b) 800°C during $O_2$ exposure. The signal for the clean surface is normalized to 1.0. The dashed curves are the theoretical fits using the Langmuir model.

variation of SHG with rotation about the surface normal, while the (7 × 7) structure should lead to a result with full 3m symmetry. This is indeed what was observed by Heinz et al.,[20] as depicted in Fig. 12. The result also confirmed the π-bonded chain model description for the (2 × 1)-Si(111) surface. Continuous monitoring of the surface by SHG allowed a real-time study of the surface transformation during annealing.

Surface melting can also be monitored by SHG. Shank et al. have used the technique to study the dynamics of laser-induced melting of Si(111).[53] Before melting, the SH signal exhibited threefold symmetry in the surface plane; after melting, it became isotorpic. Using a 90-fsec pulsed laser system, they were able to monitor the melting in real time and concluded that the melting time was ~ 1 psec. Using the same technique, Akhmanov et al.[54] and Bloembergen et al.[55] have investigated laser-induced melting of a GaAs surface and found similar results.

### 4.7. Surface Spectroscopy of Electronic Transitions

SHG from an adsorbed molecular layer should be resonantly enhanced when either the fundamental or the SH frequency coincides with an

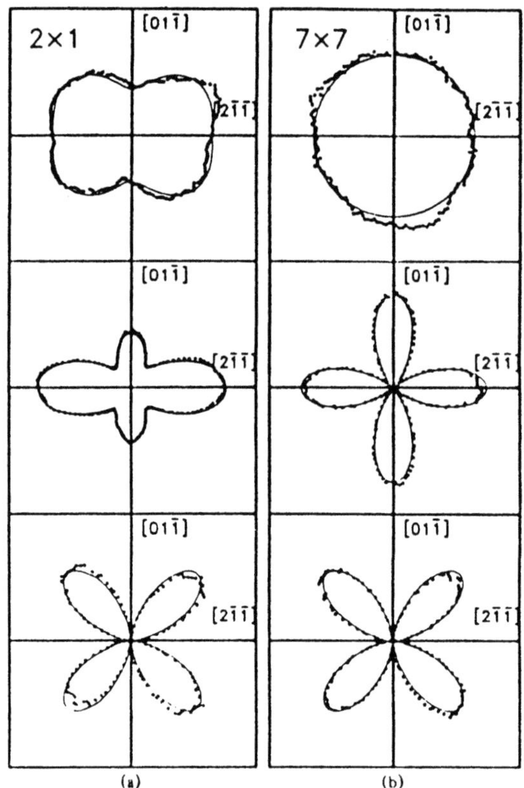


Figure 12. Second-harmonic intensity from (a) Si(111)(2 × 1) and (b) Si(111)(7 × 7) surfaces as a function of the polarization of the normally incident pump beam. The top panels display the total SH signal; the middle and lower panels show, respectively, the SH signal polarized along [2$\bar{1}\bar{1}$] and [01$\bar{1}$]. [Expermental = (---), theory = (——).]

absorption band of the molecules. Surface SHG with a tunable laser thus can yield a resonant spectrum of the molecular adsorbates. We use here dye molecules adsorbed at an air/fused quartz interface as an example.

Figure 13 shows the SH signal versus the SH frequency $2\omega$ for two samples on fused quartz, one with a half monolayer of Rhodamine 6G and the other with a half monolayer of Rhodamine 110.[15] The observed resonant structure in SHG reflects the $S_0 \rightarrow S_2$ electronic transitions in the molecules. Because of the slight difference in their molecular structures, the resonant peaks of the two dye molecules are shifted slightly from each other. The signal rise towards the low-frequency side indicates the appearance of another resonant peak as $\omega$ approaches the $S_0 - S_1$ transition.

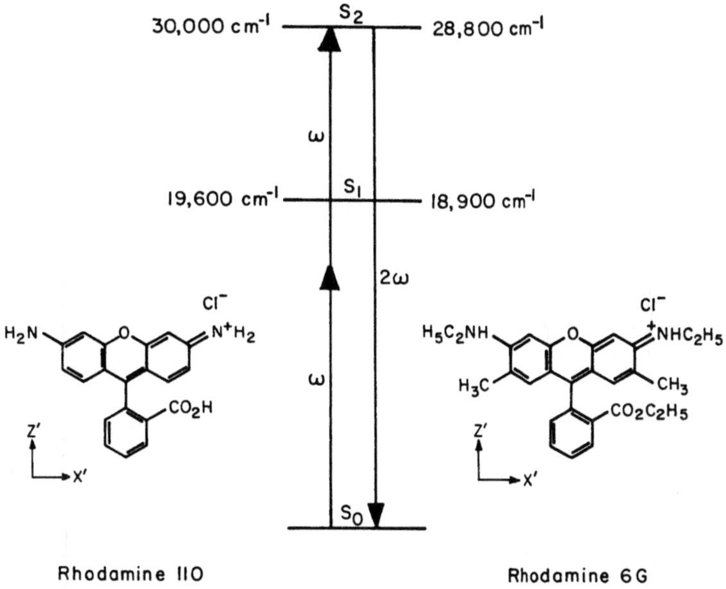

30,000 cm⁻¹ $S_2$ 28,800 cm⁻¹

Figure 13(a).

The SH signal from dye molecules was very strong. With a pump laser energy of 1 mJ in a pulsewidth of 5 nsec focused to a spot of $10^{-3}$ cm$^2$, the observed signal from a half monolayer of Rhodamine 6G ($N \sim 5 \times 10^{13}$/cm$^2$) was $\sim 10^4$ photons/pulse on resonance, corresponding to a nonlinear susceptibility $\chi^{(2)} \sim 7 \times 10^{-29}$ esu/molecule. Such a large nonlinear susceptibility was the combined effort of the resonance of $2\omega$ with the $S_0 \sim S_2$ transition, the near resonance of $\omega$ with the $S_0 \rightarrow S_1$ transition, and the large dipole moments of the transitions involved. In fact, in one type of coumarine dye molecules, one can find both $\omega$ and $2\omega$ simultaneously resonant with the $S_0 \rightarrow S_1$ and $S_0 \rightarrow S_2$ transitions, respectively. As a result, the SH signal from such molecules is $\sim 10^3$ times larger than that from Rhodamine 6G.[55]

The discussion here shows that surface SHG can be used to do spectroscopy on adsorbed molecules. The method is, of course, not limited to adsorbates on insulator surfaces. It can be used, for exmaple, to study the charge-transfer band created by adsorption of molecules on metals and the surface or interface states of semiconductors or metals. The latter has recently been demonstrated.[56]

## 4.8. Surface Vibrational Spectroscopy

So far, with tunable dye lasers, only the electronic transitions of adsorbates can be probed. In surface science, vibrational spectroscopy of adsorbates is of tantamount importance since the

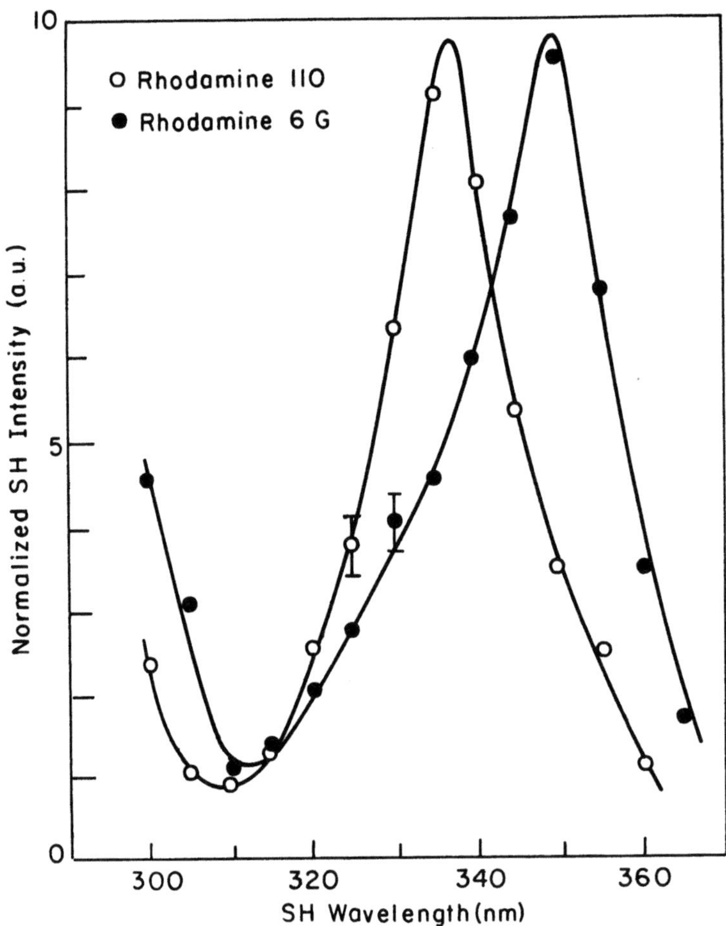


Figure 13(b). Resonant second-harmonic generation in rhodamine 110 and rhodamine 6G: (a) The resonant process in the two dyes with energy levels corresponding to the absorption line center for the molecules dissolved in ethanol; (b) the experimental SH spectrum in the region of the $S_0 \rightarrow S_2$ transition for submonolayers of the dye molecules adsorbed on fused silica.

spectra allow one to deduce the molecule-substrate interaction. One would think that this could be achieved by surface SHG using a tunable infrared laser. However, infrared detectors are generally much less sensitive than photomultipliers in the visible, so that infrared SHG is not likely to have the necessary sensitivity to detect submonolayer adsorbates. It is, however, possible to use yet another second-order process, sum-frequency generation, instead of SHG for vibrational spectroscopy. In this case, a tunable infrared laser is used to probe the vibrational transitions of the adsorbates, and the resonant spectrum can be up-converted to the visible by sum-frequency generation using a visible laser. Recent experiments show that this is indeed a viable method.[22-26]

The experimental setup is shown in Fig. 14. The beam from a mode-locked Nd:YAG laser was split into two. One was used to generate a second-harmonic output of 1 mJ/pulse at 532 nm from a KDP crystal, and the other was used to produce a tunable IR of 0.2 mJ/pulse in the range of 3.2-3.6 μm from an optical parametric amplifier. (The tunable range of the optical parametric amplifier is actually between 1.4 and 4.0 μm.) Both the green and the IR pulses had a width of ≈ 15 ps. They were focused to a common focal spot of ≈ 250 μm on the interface to be studied.

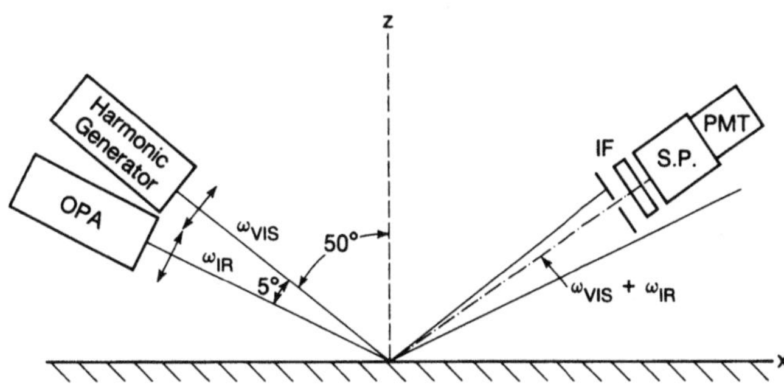


Figure 14. Experimental apparatus.

For both the air/solid and the air/liquid interfaces, we used a nearly collinear input beam geometry. The 5° angular separation between the input beams was sufficient for us to spatially filter the reflected input light from the SF signal. To calibrate the output, surface SFG from a quartz crystal was used.

As a demonstration of SFG as a surface vibrational spectroscopic tool, we have measured the C-H vibrational spectra of alcohol, methanol, and isopropanol evaporated on glass in the 3.3 to 3.5 μm wavelength range. Figure 15 displays the SFG spectra of the three adsorbed molecular species together with their Raman spectra from the bulk liquid.[57] The close agreement between the Raman and the SFG spectra allows us to assign unambiguously the observed peaks in the SFG spectra.

In Fig. 15(a) for methanol $CH_3OH$ on glass, two peaks at 2840 $cm^{-1}$ and at 2960 $cm^{-1}$ correspond to the $CH_3$ symmetric and asymmetric stretch vibrations, respectively. In Fig. 15(b) for ethylene glycol, $C_2H_4(OH)_2$, the symmetric and asymmetric $CH_2$ stretches appear at 2875 and 2935 $cm^{-1}$, respectively, Finally, for isopropyl alcohol, $(CH_3)_2CHOH$, we assign the peak at 2885 $cm^{-1}$ as the $CH_3$ symmetric stretch, the two peaks at 2950 and 2980 $cm^{-1}$ as the degeneracy-lifted $CH_3$ asymmetric stretches, and the little bump at 2920 $cm^{-1}$ as the CH stretch. All the spectra were taken with the visible light

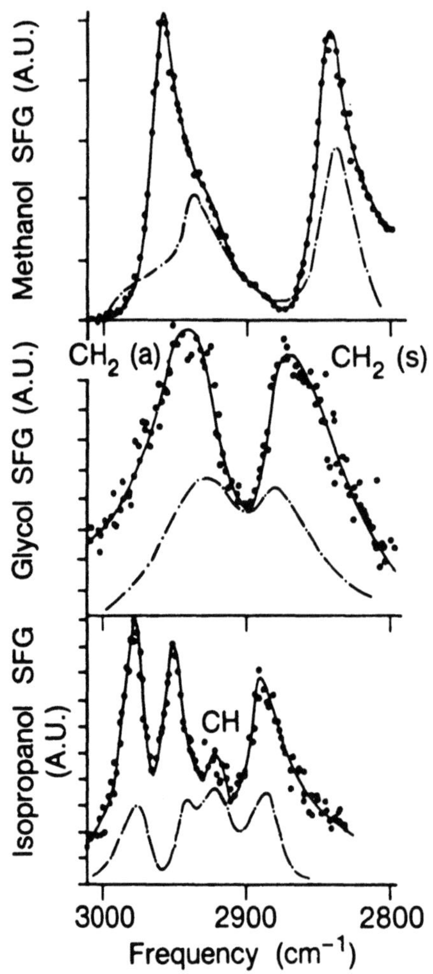


Figure 15. Sum-frequency generation spectra as a function of infrared input frequency. The bold lines are the bulk liquid Raman spectra.

s-polarized and the infrared light unpolarized. It is seen that the spectra for the three alcohols are distinctly different, and various CH stretches are clearly distinguishable.

SFG has also been used to obtain the vibrational spectra of a monolayer of pentadecanoic acid (PDA) on a water surface in the CH stretch region.[24] Figure 16 depicts the SF spectra of the $CH_3$ end group of PDA at full monolayer for three different polarization combinations. They are obviously different, indicating that one should be able to deduce information about the orientation of the $CH_3$ group and hence that of the molecule. It was found, from Fig. 16,

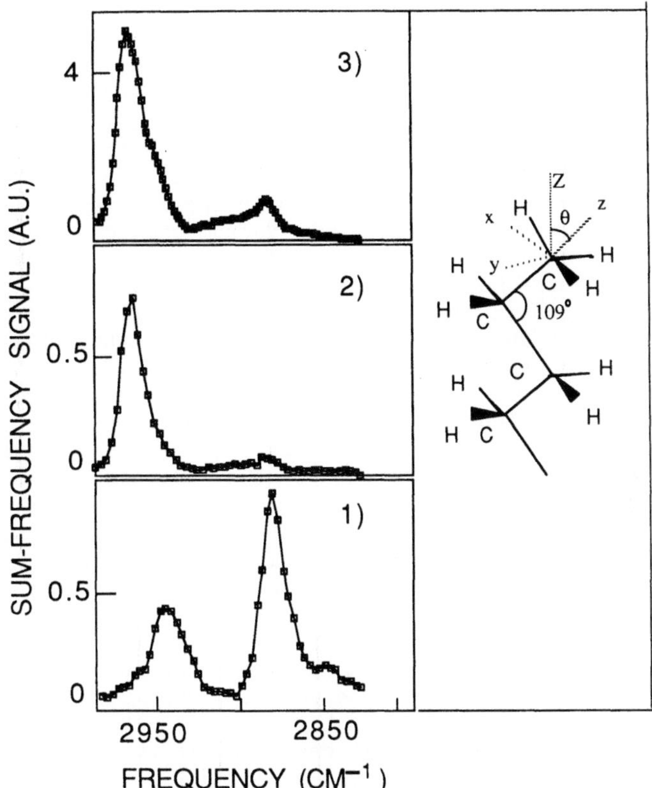


Figure 16. SFG spectra of PDA at full monolayer coverage with various (vis; IR) polarization combinations. 1) s visible, p IR; 2) p visible, s IR; 3) p visible, p IR. The lines simply connect the experimental data points (squares). Inset: coordinate axes chosen for the methyl group.

that the average angle θ between the C-$CH_3$ axis and the surface normal (see inset in Fig. 16) is about 35°, corresponding to a vertical alignment of the long alkane chain of PDA on the water surface.

Applications of surface vibrational spectroscopy using SFG to molecules adsorbed on metals and semiconductors are also possible, as has been recently demonstrated. Presently, the main limitation of the SFG technique is in the availability of an infrared laser tunable over a wide range. The advance of infrared free electron lasers would undoubtedly remove such a limitation.

## 5. CONCLUSION

Because of limitation of scope, we have omitted many other possible applications of SHG and SFG demonstrated or discussed in the recent literature. Hopefully, the many examples described in the previous section already provide enough convincing evidence to the power of the techniques as surface tools. The strength of the technqiues is in their great flexibility. While few existing surface probes can be used for in-situ measurements outside a vacuum chamber, SHG (or SFG) is applicable to essentially all interfaces that are accessible to light. That the technique can be very sensitive, nondestructive, and capable of remote sensing makes it highly promising. Yet among the many advantages that SHG (SFG) can offer, the most attractive one is its time-resolution capability.

SHG (SFG) is a process arising from the electronic response of the medium to the applied field, and should therefore have a response time of the order of $10^{-14}$-$10^{-15}$ sec away from resonance. Thus, one could in principle use the technique to study changes on a surface with a femto-second time resolution. This is most exciting because it would allow us to investigate many interesting problems of surface dynamics and reactions in real time, a field that has been largely neglected because of the limitation of the existing techniques.

The spatial resolution of SHG (SFG) is also worth commenting on. Since SHG (SFG) can have a monolayer sensitivity with a submicrometer resolution, it has the potential of being used as a true surface optical microscope, i.e. a microscope that will allow scrutiny of surface monolayers. In this respect, SHG (SFG) can also be used to read a monolayer optical recording if there is ever such an interest for future optical data processing.[21]

As a surface analytical tool, SHG may suffer from insufficient spectral selectivity, but SFG as a laser spectroscopic technique, certainly has an inherent spectral resolution better than any other conventional surface tool. Together with its subpicosecond time-resolution capability, SFG is likely to open up many new exciting areas in surface science. Transient surface vibrational spectroscopy, in-situ real-time mapping of surface species, and probing intermediate species in catalytical reactions are three obvious examples.

Exploitation of SHG and SFG for surface studies has just begun, but the potential of these new surface tools is already well recognized. These techniques provide a unique opportunity for us to look at surface science in a new light.

## ACKNOWLEDGMENT

This work was supported by the Director, Office of Energy Research, Office of Basic Energy Sciences, Materials Sciences Division of the U.S. Department of Energy under Contract No. DE-AC03-76SF00098.

## REFERENCES


1. G. Somorjai, Chemistry in Two Dimensions: Surfaces (Cornell Univ. Press, Ithaca, NY, 1981).
2. D. E. Aspnes and A. A. Studna, Phys. Rev. Lett. 54, 1956 (1985).
3. T. J. Chuang and H. Seki, Phys. Rev. Lett. 49, 382 (1982).
4. F. Trager, H. Coufal, and T. J. Chuang, Phys. Rev. Lett. 49, 1720 (1982).
5. A. C. Boccara, D. Fournier, and J. Badoz, Appl. Phys. Lett. 36, 130 (1980); M. A. Olmstead and N. M. Amer, Phys. Rev. Lett. 52, 1148 (1984).
6. H. Zacharias, M. M. T. Loy, and P. A. Roland, Phys. Rev. Lett. 49, 1790 (1982).
7. J. Haeger, Y. R. Shen, and H. Walther, Phys. Rev. A 31, 1962 (1984).
8. J. P. Heritage and D. L. Allara, Chem. Phys. Lett. 74, 507 (1980).
9. C. K. Chen, A. R. B. de Castro, Y. R. Shen, and F. DeMartini, Phys. Rev. Lett. 43, 946 (1979).
10. W. M. Hetherington, N. E. Van Wyck, E. W. Koenig, G. I. Stegeman, and R. M. Fortenberry, Optics Lett. 9, 88 (1984); W. M. Hetherington, Z. Z. Ho, E. W. Koenig, G. I. Stegeman, and R. M. Fortenberry, Chem. Phys. Lett. 128, 150 (1986); W. M. K. P. Wijikoon, Z. Z. Ho, and W. M. Hetherington, J. Chem. Phys. 86, 4384 (1987).
11. Y. R. Shen, The Principles of Nonlinear Optics (J. Wiley, New York, 1984), Ch. 25, p. 479.
12. J. M. Chen, J. R. Bower, C. S. Wang, and C. H. Lee, Optics Commun. 9, 132 (1973).
13. T. F. Heinz, H. W. K. Tom, and Y. R. Shen, Laser Focus 19, 101 (1983).
14. C. K. Chen, T. F. Heinz, D. Ricard, and Y. R. Shen, Phys. Rev. Lett. 46, 1010 (1981); Phys. Rev. B 27, 1965 (1983).
15. T. F. Heinz, C. K. Chen, D. Ricard, and Y. R. Shen, Phys. Rev. Lett. 48, 478 (1982).
16. T. F. Heinz, H. W. K. Tom, and Y. R. Shen, Phys. Rev. A 28, 1883 (1983).
17. H. W. K. Tom, T. F. Heinz, and Y. R. Shen, Phys. Rev. Lett. 51, 1983 (1983).
18. H. W. K. Tom, C. M. Mate, X. D. Zhu, J. E. Crowell, T. F. Heinz, G. A. Somorjai, and Y. R. Shen, Phys. Rev. Lett. 52, 348 (1984).
19. Th. Rasing, Y. R. Shen, M. W. Kim, P. Valint, and J. Bock, Phys. Rev. A 31, 537 (1985).
20. T. F. Heinz, M. M. T. Loy, and W. A. Thompson, Phys. Rev. Lett. 54, 63 (1985).
21. G. T. Boyd, Y. R. Shen, and T. W. Hansch, Optics Lett. 11, 96 (1986).
22. X. D. Zhu, H. Suhr, and Y. R. Shen, Phys. Rev. B 35, 3047 (1987).
23. J. H. Hunt, P. Guyot-Sionnest, and Y. R. Shen, Chem. Phys. Lett. 133, 189 (1987).

24. P. Guyot-Sionnest, J. H. Hunt, and Y. R. Shen, Phys. Rev. Lett. 59, 1597 (1987).
25. P. Guyot-Sionnest, R. Superfine, J. H. Hunt, and Y. R. Shen, Chem. Phys. Lett. 144, 1 (1988).
26. R. Superfine, P. Guyot-Sionnest, J. H. Hunt, C. T. Kao, and Y. R. Shen, Surf. Sci. (Letters), (in press).
27. N. Bloembergen and P. S. Pershan, Phys. Rev. 128, 606 (1962).
28. N. Bloembergen and Y. R. Shen, Proceedings of the Physics of Quantum Electronics Conference, Puerto Rico, 1965, p. 119.
29. N. Bloembergen, R. K. Chang, and C. H. Lee, Phys. Rev. Lett. 16, 986 (1966).
30. C. H. Lee, R. K. Chang, and N. Bloembergen, Phys. Rev. Lett. 18, 167 (1967).
31. N. Bloembergen, R. K. Chang, S. S. Jha, and C. H. Lee, Phys. Rev. 174, 813 (1968).
32. F. Brown and M. Matsuoka, Phys. Rev. 185, 985 (1969).
33. J. M. Chen, J. R. Bower, C. S. Wang, and C. H. Lee, Optics. Commun. 9, 132 (1973).
34. M. Fleischmann, P. J. Hendra, and A. J. McQuillan, Chem. Phys. Lett. 26, 123 (1974).
35. See, for example, Surface Enhance Raman Scattering, eds. R. K. Chang and T. E. Furtak (Plenum, New York, 1982).
36. C. K. Chen, A. R. B. de Castro, and Y. R. Shen, Phys. Rev. Lett. 46, 145 (1981).
37. P. Guyot-Sionnest, W. Chen, and Y. R. Shen, Phys. Rev. B 33, 8254 (1986).
38. P. Guyot-Sionnest and Y. R. Shen, Phys. Rev. B 35, 4420 (1987).
39. P. Guyot-Sionnest and Y. R. Shen, Phys. Rev. A (to be published).
40. C. C. Wang, Phys. Rev. 178, 1475 (1969).
41. A detailed derivation can be found in the Ph.D. Thesis of T. F. Heinz (Univ. of Calif., Berkeley, 1982).
42. See, for example, P. S. Pershan, Phys. Rev. 130, 910 (1963).
43. G. T. Boyd, Th. Rasing, J. R. R. Leite, and Y. R. Shen, Phys. Rev. B. 30, 519 (1984).
44. G. L. Richmond, H. M. Rojhantalab, J. M. Robinson, and V. L. Shannon, J. Opt. Soc. Amer. B 4, 228 (1987).
45. See, for example, A. F. Garito, K. D. Singer, and C. C. Teng, in Nonlinear Optical Properties of Organic and Polymeric Materials, ed. D. J. Williams (Amer. Chem. Soc., Washington, 1983), pp. 1-26.
46. See, for example, G. R. Meredith, in Nonlinear Optical Properties of Organic and Polymeric Materials, ed. D. J. Williams (Amer. Chem. Soc., Washington, 1983), pp. 27-56.
47. Th. Rasing, G. Berkoic, Y. R. Shen, S. S. Grubb, and M. W. Kim, Chem. Phys. Lett. 130, 1 (1986).
48. S. A. Letts, T. Fort, and J. B. Lando, J. Colloid. Int. Sci. 56, 64 (1976).
49. G. Berkovic, Th. Rasing, and Y. R. Shen, J. Chem. Phys. 85, 7374 (1986).
50. Th. Rasing, Y. R. Shen, M. W. Kim, and S. Grubb, Phys. Rev. Lett. 55, 2903 (1985).

51. H. W. K. Tom, X. D. Zhu, Y. R. Shen, and G. A. Somorjai, Surf. Sci. 167, 167 (1986).
52. T. F. Heinz, M. M. T. Loy, and W. A. Thompson, J. Vac. Sci. Technol. B3, 1467 (1985).
53. S. A. Akhmanov, N. I. Koroteev, G. A. Paitian, I. L. Shumay, M. F. Galjautdinov, I. B. Khaibullin, and E. I. Shtyrkov, Optics Commun. 47, 202 (1983); J. Opt. Soc. Amer. B 2, 282 (1985).
54. A. M. Malvezzi, J. M. Liu, and N. Bloembergen, Appl. Phys. Lett. 45, 1019 (1984).
55. G. A. Marowsky, A. Gierulski, and B. Dick, Optics Commun. 52, 339 (1985).
56. E. Palange, T. H. Heinz, F. J. Himpsel, and E. Burstein, Bull. Am. Phys. Soc. 33, 298 (1988).
57. Selected Raman Spectral Data, ed. L. B. Beach (Thermal Research Center Hydrocarbon Project Publications, College Station, TX, 1983), Ser. No. 807,653,55.

# X-RAY DIFFRACTION AT THE ELECTRODE–SOLUTION INTERFACE

J. ROBINSON
*Department of Physics*
*University of Warwick*
*Coventry, CV4 7AL*
*Great Britain*

ABSTRACT. The interaction of X-rays with condensed matter and the scattering of X-rays by an ordered lattice to give a diffraction pattern are described. A brief description of X-ray generation, X-ray detection and diffraction experiments is followed by a discussion of the problems associated with cell design for *in situ* diffraction studies of electrochemical systems. A number of examples are given, with particular emphasis on the methods used to improve surface sensitivity and selectivity. Finally the related technique of X-ray standing wave generation is briefly described, and possible future developments are suggested.

## 1. INTRODUCTION

As will be apparent from the other chapters in this book the last 20 years have seen the development of a very wide range of spectroscopic techniques for studying the electrode/electrolyte solution interfacial region, and many of these techniques have now moved on from the development phase to almost routine application. The earliest developments were based on the interaction of UV and visible light with an electrochemical system, and these yielded information about electronic structure of the electrode surface, surface phases, and the solution part of the interface. Subsequently routes to probing molecular structure at the interface, using infra-red and Raman spectroscopies, were developed. The missing 'weapons' in this electrochemist's 'armoury' were techniques that could reveal information about the structural order (both long and short range) at the interface, at the atomic level, i.e. techniques that could yield the type of information now routinely provided by X-ray diffraction (either powder or single crystal) in conventional solid state studies.

Surface scientists, studying systems under ultra high vacuum conditions (UHV), are faced with similar problems to those of electrochemists interested in interfacial structure. In recent years they have developed a number of techniques, based on electron probes, which have proved invaluable, and great progress has been made. Many electrochemists have sought ways of using these UHV based techniques to study electrochemical systems, an approach described elsewhere in this book. The major problem associated with this is that it requires removal of the solution side of the interface, and transfer of the sample to UHV conditions, before measurements can be made. Whilst elaborate procedures have



*C. Gutiérrez and C. Melendres (eds.),*
*Spectroscopic and Diffraction Techniques in Interfacial Electrochemistry*, 313–341.

been developed for this transfer, for many systems it will still inevitably lead to changes in structure, and even composition. The desirability of a structural technique capable of probing interfacial structure, *in situ*, is clear.

The traditional method for determining the structure of solid crystalline matter is of course X-ray diffraction. Recently the increased availability of synchrotron radiation has resulted in the development of extended X-ray absorption fine structure (EXAFS) as a very useful probe of local structure in both ordered, and disordered, solids. Electrochemists have in the past be put off using X-rays to study structure at the electrode solution interface, firstly because at the energies conventionally used in diffraction studies X-ray photons will not penetrate far in solution, and secondly because X-rays are not surface selective. Despite these problems the possibility of using both EXAFS and X-ray diffraction has now been recognised by a number of electrochemists, and work in this area is proceeding in several laboratories around the world.

The application of X-ray diffraction to the study of structure in electrochemical systems is the subject of this chapter, whilst EXAFS is discussed in the chapter by Professor Hoffman. The theory of the interaction of X-rays with condensed matter is discussed, and in particular the scattering of X-rays to give a diffraction pattern. Then follow experimental considerations such as X-ray generation, X-ray detectors and cell design. Finally a number of examples of the application of X-ray diffraction to the study of electrochemical systems, of both fundamental and practical interest, will be discussed, with particular attention to describing ways of improving surface sensitivity and selectivity.

## 2. THE INTERACTION OF X-RAYS WITH CONDENSED MATTER

### 2.1. INTRODUCTION

On encountering a thin slab of condensed matter an X-ray photon may suffer one of three possible fates. It may be transmitted through the sample unchanged, it may be scattered, either elastically or inelastically, or it may be absorbed, thereby exciting a core electron associated with one of the atoms in the material. In this section some of the key conclusions of X-ray scattering theory will be presented, and a brief description of how an analysis of the scattered radiation can lead to the determination of the structure of the scattering phase will be given. Before this problem is tackled, however, we will briefly look at the absorption process. This process is discussed fully in the chapter on EXAFS but for our purposes here a much simpler approach will suffice.

### 2.2. THE ABSORPTION OF X-RAYS

If we consider an X-ray beam of intensity $I_0$, incident at a thin slab of material of thickness $x$, then the intensity of the transmitted beam, $I$, is readily obtained by the application of the Beer-Lambert law,

$$I = I_0 \exp(-\mu x)_\lambda = I_0 exp((-\mu/\rho)_\lambda x\rho) \tag{1}$$

where $\mu$ is the linear absorption coefficient, $\rho$ is the density and $(\mu/\rho)_\lambda$ is the mass absorption coefficient, which is independent of the state of the matter (gas, solid or liquid). The values of mass absorption coefficients as a function of X-ray energy are readily available in reference books.

Whilst we have no direct interest here in the absorption of X-rays this absorption has major implications for the practicability of *in situ* X-ray diffraction experiments. If we consider copper $K_\alpha$ X-rays (wavelength, $\lambda = 1.54$ Å), which are those most commonly used in conventional X-ray diffraction, and we calculate the extent of absorption of these photons by a layer of water, then we find that a 100 $\mu$m layer will absorb nearly 20 % of the incident photons, whilst for a 1 mm layer this figure rises to nearly 90 %. Clearly it will be impractical to conduct a diffraction experiment in which the vast majority of the incident photons are absorbed before they have a chance to be scattered at the electrode surface, or where the scattered photons are themselves largely absorbed. Therefore this absorption by water (or other solvents and materials such as cell windows) must be taken into account when designing experiments. Increasing the X-ray energy decreases the absorption; so, for molybdenum $K_\alpha$ X-rays ($\lambda = 0.71$ Å) path lengths through solution of several centimetres become feasible. However, as will be seen later, the use of such high energy photons raises other problems, particularly with detectors.

The problem of fluorescence, which will appear if the incident photon energy is just above the absorption edge of one of the metals in the system, should always be borne in mind when designing experiments. The tunability of synchrotron radiation can be used to great advantage here.

## 2.3. SCATTERING OF X-RAYS

As we saw above when X-rays interact with matter they can be scattered. The origin of this scattering is the electrons surrounding the nucleus which are caused to oscillate by the oscillating electric field of the electromagnetic radiation (X-rays), these oscillating electrons acting as sources of electromagnetic radiation. For unpolarized X-rays the scattered intensity at any point, $I_e$ is given by the famous Thomson equation,

$$I_e = \frac{I_0 e^4}{(4\pi\epsilon_0 r m c^2)^2}\left(\frac{1+\cos^2\phi}{2}\right) \tag{2}$$

where $I_0$ is the incident intensity, $e$ is the proton charge, $r$ is the distance from the scattering electron, $c$ is the velocity of light, $m$ is the rest mass of an electron, $\epsilon_0$ is the permittivity of free space and $\phi$ is the scattering angle. From this equation, and assuming that each electron scatters X-rays independently of all the other electrons, we can calculate the fraction of the incident photons that are scattered per unit length of sample, $\sigma$, as,

$$\sigma = \frac{8\pi}{3}\left(\frac{e^2}{4\pi\epsilon_0 c^2 m}\right)^2 n = 6.6\text{x}10^{-29}\,.\,n\ \ [\text{m}^{-1}] \tag{3}$$

where $n$ is the electron density in the sample.

It is interesting to consider the implications of this equation for our proposed studies. If we have a typical material with an electron density of 3 x $10^{29}$ m$^{-3}$, then $\sigma = 20$ m$^{-1}$, and so for a 1 mm thick slice of this material 2 % of the incident X-ray photons will be scattered. If we now consider scattering by a monolayer of this material the situation is much less favourable, as this factor decreases to about $10^{-6}$ %. This immediately suggests

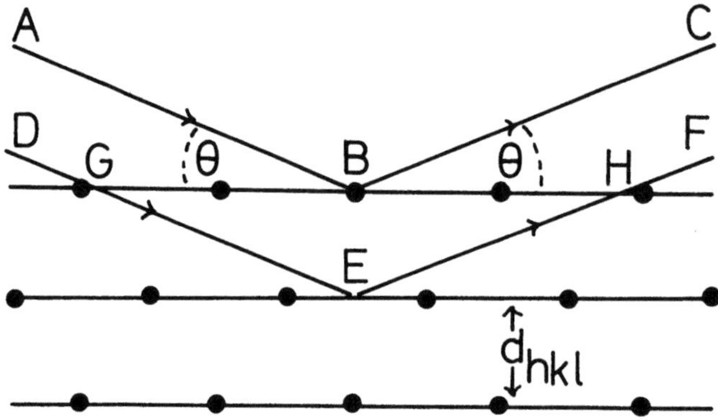


Figure 1: Scattering of X-rays by lattice planes according to the Bragg formulation.

to us that in using X-ray scattering to study surfaces we are likely to face some serious problems, and as we will see later this is indeed the case.

The scattering that was referred to in Eqn. 2 is coherent and therefore the scattered photons may interfere. When the electrons are associated with an ordered array of atoms this interference leads to the formation of diffraction patterns that can be analysed to provide structural information. The simplest description of the scattering of X-rays by an ordered array of atoms ordered in a lattice formed by repetition of a unit cell in the three space directions is that due to Bragg. Fig. 1 shows an end-on view of a set of parallel equidistant planes defined by the Miller indices $(hkl)$, on which an X-ray beam of wavelength, $\lambda$, comparable with the lattice spacing, is incident at an angle $\theta$. The path length difference between the paths $ABC$ and $DEF$ is $GEH$, and from Fig. 1 it can be seen that,

$$GE = EH = d_{hkl} \sin \theta$$

If this distance $GEH$ is an integral number of wavelengths then the waves $ABC$ and $DEF$ reinforce, leading to a diffraction maximum, the condition for this being

$$n\lambda = 2d_{hkl} \sin \theta \tag{4}$$

where $d_{hkl}$ is the lattice spacing. This is the famous Bragg equation for diffraction.

One of the virtues of this Bragg formulation is its simplicity, but whilst it is useful for describing simple systems its limitations are soon exposed when more complex ones are considered, e.g. those containing several atom types. For such systems we must use a more complex formulation based on the concept of the reciprocal lattice. Whilst we do not need to go into great detail about the reciprocal lattice here it is useful to remind ourselves of how it is related to the real lattice. To each set of parallel, equidistant planes in the real lattice we can assign a reciprocal lattice vector, whose direction is normal to the lattice planes whilst its magnitude is given by the inverse of the interlayer spacing. The reciprocal lattice vector associated with the planes $(hkl)$ is represented as $\mathbf{r}^*{}_{hkl}$. For each real lattice the set of all possible reciprocal lattice vectors constitutes its reciprocal lattice.

To see how we can use the concept of the reciprocal lattice to predict the form of the diffraction pattern arising from a sample we must introduce vector notation and a construction known as the Ewald sphere. As shown schematically in Fig. 2 this sphere is centred at the sample and has a radius of $1/\lambda$, where $\lambda$ is the wavelength of the X-rays. We can then define unit vectors in the incident and scattered directions as $\mathbf{S_0}$ and $\mathbf{S}$ respectively, such that $OS_0 = \mathbf{S_0}/\lambda$. It can then be shown that if a point of the reciprocal lattice lies on this sphere there will be a diffracted beam in the direction of $\mathbf{S}$ such that $\mathbf{S}/\lambda = OR_{hkl}$. The vector $(\mathbf{S} - \mathbf{S_0})/\lambda$ is called the scattering vector, $\mathbf{s}$, and it plays a fundamental rôle in scattering theory. All scattering studies on an object involve determining the scattering intensity for all values of $\mathbf{s}$. The discussion above shows us that when considering X-ray scattering by an atomic array diffraction maxima will occur for scattering vectors equal to reciprocal lattice vectors. Put briefly diffraction occurs when a sphere of reflection of radius $1/\lambda$ intersects a reciprocal lattice point. Whilst the Ewald sphere construction and the reciprocal lattice can be used to predict the directions in which scattering maxima will occur, they provide no information on the intensities of the beams. These are determined by what are known as the structure factors which can be calculated from a knowledge of the atomic structure. Detailed discussions of these calculations are given in any textbook on X-ray crystallography, and of course these also give full derivations of X-ray scattering theory for those who are interested.

In our discussion we have indicated how the Ewald sphere construction can be used to determine in which directions diffracted beams will go, these directions depending only on the shape (i.e. crystal system) and size of the unit cell. However, it is the positions and nature of the atoms in the unit cell (i.e. the crystalline structure) which determine the intensities of these beams, which can be calculated by Fourier synthesis. Thus, given any structure we can calculate both the diffraction pattern and the beam intensities. However, in practice of course it is generally the inverse of this operation that we wish to perform, i.e. given a diffraction pattern we wish to determine the structure. While it is very easy to determine the unit cell from the diffraction pattern, calculating the crystal structure requires the accurate measurement of beam intensities, and can be very difficult.

## 2.4. STRUCTURE DETERMINATION

As we will see later much of the work that has been done so far in applying X-ray diffraction to electrochemical systems has been concerned with powder diffraction. As most readers will be aware powder diffractograms consist of measurements of the intensity of the scattered radiation from a random powdered sample (or one that can be regarded as such), as a function of the scattering angle. On such a diffractogram there will usually be a number of narrow peaks. These diffraction patterns are not well suited to *ab initio* structure determinations, but only to identifying the phase, or phases, present. This is usually done

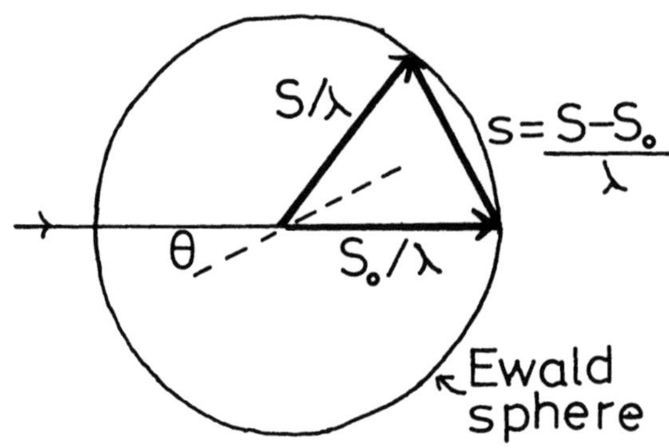


Figure 2: Ewald sphere construction for determining diffraction conditions

by comparison with powder diffraction patterns of known materials, either recorded by the experimenter, or on file with the Joint Committee on Powder Diffraction Standards (JCPDS, 1601 Park Lane, Swathmore, Pennsylvania). This database, which is maintained on cards, and has also recently been made available on optical disc for access by computer, contains references to approaching 50,000 substances and is now very widely used. In relying on this database, however, it should always be remembered that it is quite possible that thin surface phases of a material may differ significantly in structure from bulk material of nominally the same phase, and also that it is possible to form hitherto unknown phases electrochemically.

From the time of the discovery of X-ray diffraction the most notable achievements of this technique have been the structural determinations of crystalline materials using single crystal diffraction. One thinks in particular of the unravelling of the structure of DNA, and more recently of certain viruses. Whilst it is unlikely that electrochemists will be confronted with this type of problem, there is one area where single crystal studies are made, and that is in the investigation of adsorbates on single crystal substrates. Here the electrode is a three dimensional single crystal, and the adsorbate phase may be a two dimensional one. In solving the structures of such systems one is interested both in the in-plane structure of the adsorbate and also in its registry with the substrate. This is essentially the same problem encountered by surface scientists using low energy electron diffraction (LEED) to study systems under UHV, and the approaches used are well covered in the literature of this subject. The techniques conventionally used by crystallographers are not particularly suited to this type of problem.

# 3. EXPERIMENTAL CONSIDERATIONS

## 3.1. X-RAY GENERATION

The discovery of X-rays is due to Röntgen, who in 1895 observed that, when a voltage was applied between two metal plates in an evacuated tube, rays (subsequently called X-rays) emerged that penetrated solid matter and 'fogged' photographic film. But it was not until the early part of this century that, with the development of quantum mechanics, the true nature of these rays as electromagnetic radiation became established.

The operation of these early X-ray tubes depended first on the ionisation of the residual gas in the tube, followed by the release of electrons from the cathode, as the result of its bombardment by the positive ions. The electrons were then accelerated to the anode, with which they collided, releasing X-rays. The use of this type of X-ray tube was beset with many problems, and they were anyway not particularly efficient. In the modern tube, which is a development of this early version, the source of electrons is not gas ionisation but a heated tungsten filament. The electrons released by this filament are accelerated by a voltage of about 40 kV, to collide with a water cooled target anode releasing the X-ray photons. The electron current in these devices would typically be about 20 mA, but by rotating the anode it is possible to reduce the effects of target heating, and thus to raise the electron current to perhaps as much as 2 A, and thereby to greatly increase the X-ray flux. The vast majority of routine X-ray diffraction work is undertaken with a conventional fixed anode source. Such sources are available from many suppliers, though in fact there are few anode manufacturers. With the benefit of modern electronics these sources can be made highly stable over long periods, and are very reliable. Rotating anode sources are significantly more expensive both to purchase and to operate, they are therefore usually only used in applications that require their increased intensity. As we saw earlier the scattering from surfaces is very weak and absorption by solution may be significant, therefore, for electrochemical applications a case can readily be made for the desirability of such a high intensity source.

The spectrum of the X-ray radiation arising from an X-ray tube is shown schematically in Fig. 3. It can immediately be seen that this consists of two distinct components: a broad featureless part which has a high energy cut-off and a long low energy tail, and superimposed on this some narrow lines. It is found that altering the tube voltage does not affect the energies of the lines, but that the high energy cut-off varies with the voltage. We saw earlier in our discussion of X-ray scattering (section 2.3) that an accelerating (oscillating) charge acts as a source of radiation. This is the cause of the X-rays giving rise to the broad envelope in the X-ray emission spectrum: the electrons are rapidly brought to a halt as they collide with the target and whilst most of their kinetic energy is converted to heat, the remainder is released as X-ray photons. The high energy cut-off corresponds to the conversion of all the electron's kinetic energy into an X-ray photon. In view of the origin of these X-rays they are called Bremsstrahlung (or braking) radiation. For our purposes the lines in the spectrum are in fact more interesting, as it is these that are generally used in diffraction experiments.

As we have seen the energies of the characteristic lines do not vary with tube voltage; however, they are found to vary with the target metal. This suggests that they are in some

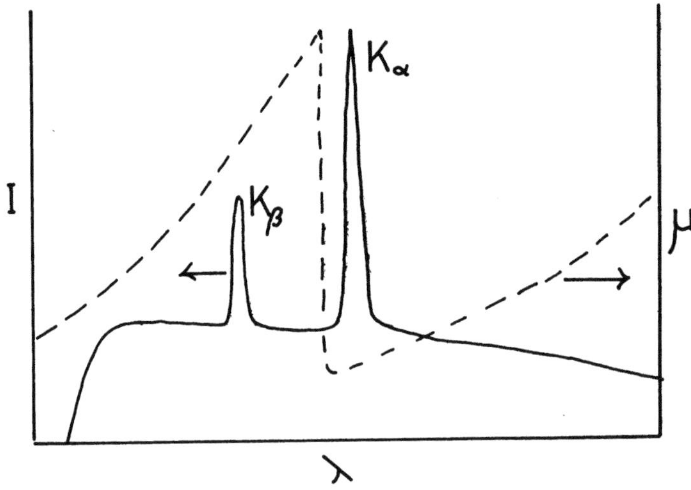


Figure 3: A schematic plot of the X-ray emission spectrum of an X-ray tube and of the absorption spectrum of a (Z-1) filter.

way related to the electronic structure of the target, and this is indeed found to be the case. If the incident electron has sufficient energy it can eject one of the core electrons in a target atom, ionising the atom and leaving a core hole. This situation is of course not stable and the core hole is filled by an electron from a higher shell with the emission of an X-ray photon, whose energy is given by the difference in energies of the two electronic states involved. If the electron is ejected from a 1s state then the X-ray is called a $K$ photon whilst subscripts $\alpha$ and $\beta$ indicate from which shell ($L$ or $M$) the electron filling the core hole came. For X-ray diffraction work $K$ photons are almost exclusively used as, for the most convenient target materials, these have wavelengths comparable with atomic spacings. $L$ or $M$ photons are usually of much longer wavelength. For most types of diffraction experiment it is necessary to work with a monochromatic X-ray beam or else interpretation of diffraction patterns becomes impossible. The most effective way of achieving this is by using a crystal monochromator, typically made of Si or pyrolitic graphite, which is capable of selecting a very narrow band pass of X-rays, in much the same way as a diffraction grating is used at optical frequencies. These crystal monochromators are not however very efficient and therefore the purity of the beam is achieved at the expense of intensity. For most laboratory based work the refinement of a monochromator is rarely required, instead we take advantage of the fact that the presence of the characteristic lines in the spectrum has in part done the job for us. In general there will be two significant characteristic lines, $K_\alpha$ and $K_\beta$, and these will be very much more intense than the Bremsstrahlung. If we can remove one of these lines we will then effectively have a monochromatic beam. Fortunately, an absorption edge of the element one atomic number less than the target material usually lies at an energy between the $K_\alpha$ and $K_\beta$ lines as shown in Fig. 3. By using an appropriate thickness of a foil of this (Z-1) material as a filter, it is possible to reduce greatly the intensity of the $K_\beta$ line whilst having only a slight effect on the intensity of the $K_\alpha$ one. So it is possible to reduce the intensity of the $K_\beta$ line to 1/500 of its initial value whilst only reducing the intensity of the $K_\alpha$ one by 1/2. In fact most laboratory based diffraction work is conducted with a fixed anode source using (Z-1) filtering.

Fixed anode sources are available commercially from a range of suppliers and with a variety of target materials. By far the most commonly used is copper, whose $K_\alpha$ line has a wavelength of 1.54 Å. The other materials are used for special purposes. For electrochemical applications one might want to consider molybdenum, whose $K_\alpha$ radiation has a wavelength of 0.71 Å, in view of its greater penetrating power for getting through solutions.

In addition to being monochromatic the X-ray beam must also be well collimated. At optical frequencies we use focusing optics, i.e. lenses, for this purpose but with X-rays we can only achieve focusing through the use of bent metal coated mirrors, or bent monochromator crystals. The technology of X-ray mirrors is rather complex and their use is usually restricted to synchrotrons. For laboratory work collimation is usually achieved with slits, i.e. we simply select a part of the beam that is travelling in the correct direction and prevent the rest from reaching our sample. Some care has to be taken in designing these slits to avoid scattering effects from their edges. Typical designs are given in literature and collimators are also available commercially.

The other significant, and increasingly important, source of X-rays is synchrotron radiation from synchrotrons or storage rings. In these devices bunches of electrons (or positrons) circulate in a vacuum at speeds near that of light. The electrons are kept in this path by electromagnets which bend the beam, and each time the beam is bent the electrons experience an acceleration, and hence emit radiation. This radiation is broad band, very intense, highly collimated and polarised. Storage rings, or sections of storage rings, are optimised for different regions of the electromagnetic spectrum, and X-ray sources delivering photon fluxes at the sample several orders of magnitude greater than those attainable with conventional sources are now routinely available. In addition to this increased intensity synchrotron radiation has the advantage that it is 'white', and therefore can be tuned, using a crystal monochromator, over a wide range of wavelengths. We are therefore no longer restricted to characteristic lines. The disadvantage of course is that most of us do not have a synchrotron radiation facility in our back yard. We have to travel to one, and once there we only get limited access. Despite these difficulties the use of synchrotron radiation is becoming more widespread, and as we will see later it makes possible a range of experiments which would otherwise have been very difficult to perform.

## 3.2. CELL DESIGN

The problem of cell design has been alluded to earlier. It is perhaps helpful in this context if we list some of the cell requirements, first from the electrochemical standpoint and then from the X-ray one.

*Electrochemical requirements.*

**(a)** uniform current density at the electrode surface

(b) uniform potential distribution across the electrode

(c) low $iR_u$ drop and $i^2R$ heat dissipation

(d) ease of assembly and disassembly

(e) deoxygenating facilities

*X-ray requirements.*

(a) an X-ray transparent window

(b) low X-ray absorption and scattering by the electrolyte solution

(c) convenience of alignment

(d) clear access to the sample for the X-rays

When trying to design a cell one immediately realises that reconciling these various requirements is extremely difficult, particularly when working with relatively low energy radiation, and any design is therefore inevitably a compromise. For the body of the cell glass and machinable plastics seem to be the most appropriate materials, whilst plastics are favoured for windows. Mylar and Kapton have been extensively used as windows and have been shown to be very effective, having good mechanical stability, resistance to radiation damage and good chemical resistance. Thicknesses as low as 10 $\mu$m have been shown to be adequate. Where some flexibility in the window is desirable polyethylene has been used. With the cell and window materials chosen the problem becomes one of designing a suitable cell configuration for the type of electrode being investigated, and the system being studied. The main design criterion must be one of reducing the solution path length of the X-rays without compromising the electrochemical behaviour of the cell significantly. Fig. 4 shows an example of a cell design that has been used for a range of systems with disc shaped electrodes sealed into movable syringe barrels [1]. This cell was mainly used with synchrotron radiation and was mounted on a precision goniometer. This goniometer, and various other positions, e.g. electrode height, were adjustable via computer controlled stepper motors so that the alignment could be optimised with the X-ray beam on.

Another type of design that has found favour [2] is one in which the window (here usually polyethylene) can be collapsed on to the electrode, trapping a thin layer of electrolyte, thereby minimising absorption by the solution. This design however requires that the system is at equilibrium with negligible current flow, or $iR_u$ drop and other problems will be severe. This design is therefore particularly suited to studies of adsorption and underpotential deposition.

## 3.3. X-RAY DIFFRACTION AND DETECTORS

As we saw earlier a complete scattering experiment on any object involves determining the scattered intensity as a function of the scattering vector for all values of **s**. However, it is not usual to carry out such a complete experiment. A number of different types of experiment have been developed, the choice depending on the type of sample available and the nature of the information that is sought. Most of the experimental results to be described later in this chapter are from powder diffraction studies, which as we have seen are useful for

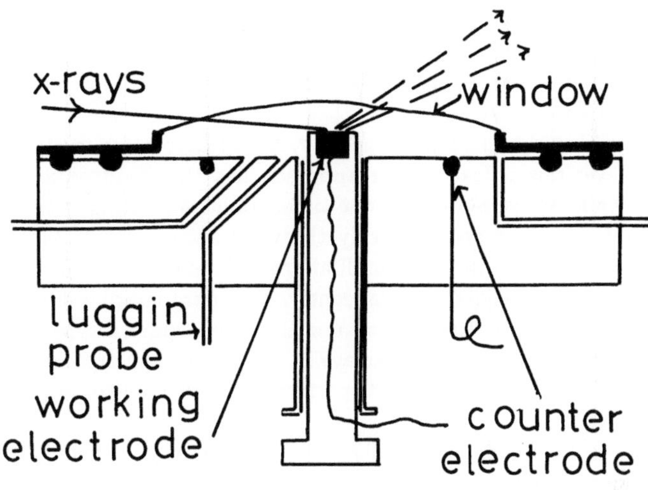


Figure 4: An electrochemical cell for *in situ* electrochemical X-ray diffraction studies.

phase identification. This discussion of the diffraction experiment will therefore be largely restricted to this type of study.

The conventional powder sample consists of many small crystals, or grains, randomly oriented and either packed in a thin walled capillary tube or onto a flat plate. This sample is then illuminated with collimated monochromatic radiation. One way of thinking of this configuration is by assuming, as usual, a fixed Ewald sphere centred at the sample. Since each reciprocal lattice vector $\mathbf{r^*}_{hkl}$ can take on all possible orientations, because of the random nature of the sample, the reciprocal space for this powdered sample is a set of concentric spheres of radius $r^*_{hkl}$. Each of these intersects the Ewald sphere in a circle, leading to a cone of diffracted radiation with the direct beam passing along its axis, the angle between the cone and the direct beam being $2\theta$. Probably the simplest way of recording the diffraction pattern is to wrap a photographic film in a circle around the sample to intersect the diffraction cones. For electrochemical work, however, this is not a particularly satisfactory approach for a number of reasons, e.g. the poor contrast, and therefore other detection methods are used.

For high resolution, high contrast, powder diffraction measurements a diffractometer is generally used. With a capillary sample the detector is slowly rotated about the axis of the sample and the count rate recorded as a function of the scattering angle. With flat plate samples it is necessary to use a $\theta/2\theta$ drive where the sample is rotated about the same axis as the detector, but at half the rate, in order to maintain the scattering conditions. For this type of work a simple Geiger-Muller tube could be used as the detector, but it is more usual to use a proportional detector. This can operate at higher count rates but, perhaps more importantly, since the output pulse height is proportional to the incident photon energy pulse height discrimination can be used to discriminate against fluorescent photons, thus reducing background signals.

We will see later that it is possible to use a $\theta/2\theta$ diffractometer to study electrochemical systems. However this diffractometer approach has the disadvantage that only one scattering angle is being measured at any one time, whilst scattering is occurring in many directions. In other words we are effectively throwing away a lot of the available data, which for weakly scattering samples we can ill afford. We therefore need some way of looking at a range of scattering angles simultaneously, as a photographic film does, but with the advantages of an electronic detection system, i.e. we require a type of parallel detection such as is achieved with photodiode arrays in UV-visible spectroscopy. There are two possible approaches to this: to use energy dispersive techniques, or to use position-sensitive detection.

The energy dispersive technique is really limited to use with synchrotron radiation as it relies on an incident beam that is 'white' over a wide energy range. The experimental configuration is shown in Fig. 5. The 'white' collimated beam is incident on the sample and there is a solid state detector mounted at a fixed scattering angle. The solid state detector is effectively a high resolution proportional counter in that it produces an output pulse whose height is proportional to the photon energy. These photons can then be sorted by height, and counted, using a multi-channel analyser (MCA) to produce a spectrum of counts as a function of pulse height, and thus photon energy. Since the photon energy is inversely proportional to its wavelength we can see from the Bragg equation, Eqn. 4, that we are effectively recording a powder diffraction pattern of our sample, with the whole of the pattern being acquired all of the time. Diffraction maxima appear at those energies, E, fulfilling the condition

$$E = \frac{hc}{2d_{hkl}\sin\theta} \tag{5}$$

In addition to the advantage of speed this parallel detection approach has the advantage that we can take 'snapshots' (or time slices) of the diffraction pattern, which can be particularly valuable if the sample is undergoing a transformation, i.e. we can make kinetic, albeit rather slow, studies.

Position-sensitive detectors are not restricted to use with synchrotron radiation. They rely on the use of monochromatic radiation and the detector is effectively the electronic equivalent of photographic film. Various types of position-sensitive detector have been used in X-ray work, e.g. vidicons with image intensifiers (largely for medical studies), photodiode arrays, and charge coupled devices, but for diffraction work the most successful types have undoubtedly been those based on wire detectors with delay line readout. Perhaps the simplest form of this type of detector, both to construct and to understand, is that due to Borkowski and Kopp [3] and this is shown schematically in Fig. 6. The detector is to a large extent a conventional proportional counter. The X-rays enter the active volume through a window of metallised plastic, or beryllium, in the side of the detector. The active volume contains an ionisable gas in which the X-ray photon may ionise a gas molecule and liberate an electron. This electron then accelerates towards the central anode wire

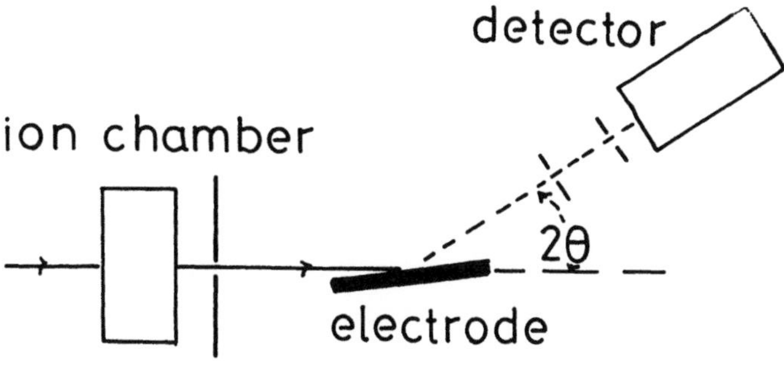


Figure 5: The experimental configuration for energy dispersive X-ray diffraction.

causing an avalanche of charge, and a charge pulse forms on the wire. The only difference from a conventional proportional detector is that this central wire is highly resistive, e.g. a resistivity of 8 k $\Omega$ $mm^{-1}$, being made of carbon coated quartz or possibly nichrome. Since the wire is resistive it behaves as part of a distributed RC delay line, and the charge pulse therefore takes different times to reach the two ends of the wire. By determining this time difference we can locate the position of the ionisation, and hence the point at which the X-ray entered the detector. This type of timing experiment is very common in high energy physics and can be conducted with high precision. Suitable electronic modules are available commercially, built usually to either the NIM or CAMAC standards. The basic requirements for this application are shaping amplifiers, cross over detectors, a time to amplitude converter and an MCA. This set of electronic modules converts the time difference to a pulse height and then the pulses can be sorted, counted and displayed with the MCA, as with the energy dispersive method. Again the diffraction pattern can be displayed in real time as it is being acquired and we can make 'snapshot' measurements.

Detectors based on resistive anodes have a limited maximum count rate of about $10^4$ counts $s^{-1}$ and they are also rather fragile. These problems can be overcome by using another type of detector where the readout is via the cathode plane, which is either a segmented cathode where the segments are linked with discrete L and C components [4], or the cathode plane is itself a distributed LC delay line [5]. The operation of these detectors is essentially the same as that of the resistive anode variety, except that it is the induced charge pulse on the cathode that is timed. The problem with all these detectors is that they are linear, which is not the ideal form for a detector for powder diffraction work. A detector that formed a segment of a circle would be much better as the output of

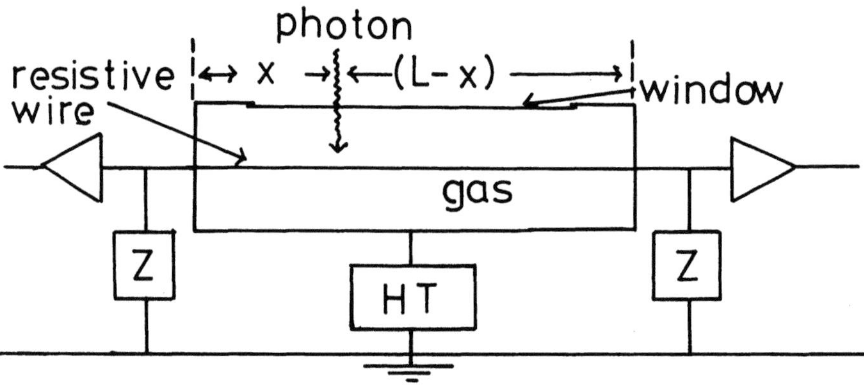


Figure 6: A schematic diagram of a resistive anode position-sensitive X-ray detector.

the detection system would be linear in scattering angle, and more importantly, it would not suffer from parallax problems that lead to smearing of diffraction peaks as all photons would enter the detector along a radius. Detectors of this type have now been developed. In one design the anode wire is held in the form of an arc by using magnets and passing a current through the anode wire [6], whilst in the other type the anode wire is replaced by a curved knife edge [7]. Detectors of both these designs are available commercially. They can be used with both laboratory and synchrotron sources and produce high quality diffraction patterns with a resolution approaching that of most diffractometers.

We have seen that all wire detectors require an ionisable gas to be present, and also a quencher gas. These gases are available commercially, ready mixed, and for copper $K_\alpha$ work 90:10 $Ar/CH_4$ and $Ar/C_2H_6$ mixtures are used with the hydrocarbon as the quencher. These mixtures are not suitable for harder radiation, such as molybdenum $K_\alpha$, as the ionisation cross sections are too small. Hard radiation requires the use of xenon or krypton mixtures, often at high pressures. Most position-sensitive detectors operate in a flow mode to reduce contamination effects, but of course xenon and krypton are too expensive to use in this way. For position-sensitive detector work with hard X-rays one must either accept a low counting efficiency, which partly negates the reason for using this radiation, or work with a sealed detector, which requires great care in the choice of construction materials and great cleanliness in assembly, if it is to operate reliably over a long period. Thus whilst the use of hard radiation does overcome the problem of solution penetration it introduces new problems. Perhaps a compromise is to use radiation of wavelength around 1 Å, which although quite penetrating can be detected efficiently without recourse to expensive detector gases.

Finally before considering some applications a brief comment about safety is in order. X-rays can undoubtedly be dangerous, particularly with high intensity sources such as synchrotrons. At synchrotron radiation facilities safe working practices have been developed but in the laboratory it will be up to the user to determine these, in line with local safety requirements. In view of the nature of the electrochemical cell as a sample the safest approach is to conduct the experiments in a shielded, interlocked room or cabinet.

## 4. CASE STUDIES

A number of examples of *in situ* X-ray diffraction studies will now be discussed. This overview will in no way attempt to be exhaustive but rather the examples have been chosen to give some indication of the range of the studies that are possible and of the applicability of the various experimental approaches. Possible future developments will also be identified.

### 4.1. POWDER DIFFRACTION

4.1.1. *Diffractometer Studies.* As was pointed out earlier it is sometimes possible to obtain reliable *in situ* powder diffraction data for electrochemical systems by using a conventional laboratory diffractometer. Indeed, on occasion these data can even be obtained dynamically. For this approach to work there must be a large amount of the phase of interest present and therefore it is topics such as phase transitions in compressed paste electrodes, of the type used in batteries, that have been studied in this way. We will now look at two examples of X-ray studies that have used this simple laboratory diffractometer approach. The first of these is a study of phase transitions in a battery system, the other is perhaps a rather surprising application, being concerned with the use of X-ray diffraction to elucidate the mechanism of an organic electrode reaction.

The battery study we will consider was made by Chianelli *et al* [8] and was an investigation of the structural changes that accompany the discharge of a $TiS_2$ electrode forming part of a Li/$TiS_2$ battery system. The electrochemical cell, which is shown schematically in Fig. 7, was of a simple two-electrode type. The lithium foil anode and the $TiS_2$ paste cathode, which was loaded on a Ni grid, were kept apart by a porous glass fibre separator, and this combination was mounted in a cell constructed from Teflon and aluminium. As shown in Fig. 7 the $TiS_2$ cathode was placed next to a beryllium or polyethylene X-ray transparent window, and the cell was then filled with an electrolyte of $LiClO_4$ in dioxolane. It was then mounted on a Philips diffractometer and diffraction patterns were recorded, using Cu $K_\alpha$ radiation from a fixed anode source, as the cell was discharged.

Alignment of electrochemical cells in X-ray beams can be rather difficult because of their size and therefore it is rather important that, if possible, there is an internal method of calibrating the diffraction angles. In this particular case the diffraction pattern prior to discharge served this purpose, as the crystallography of $TiS_2$ is well defined. As was pointed out earlier this diffractometer approach requires a significant amount of material to be present if interpretable diffraction patterns are to be obtained. The $TiS_2$ electrodes were prepared as 700 $\mu$m thick films, but the authors estimate that only 50 $\mu$m of this was being sampled by the X-ray beam, and thus contributed to the X-ray diffraction pattern.

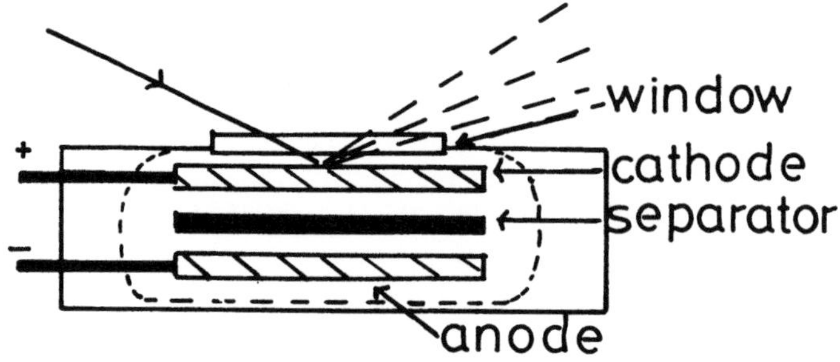


Figure 7: Schematic diagram of the two-electrode X-ray diffraction cell.

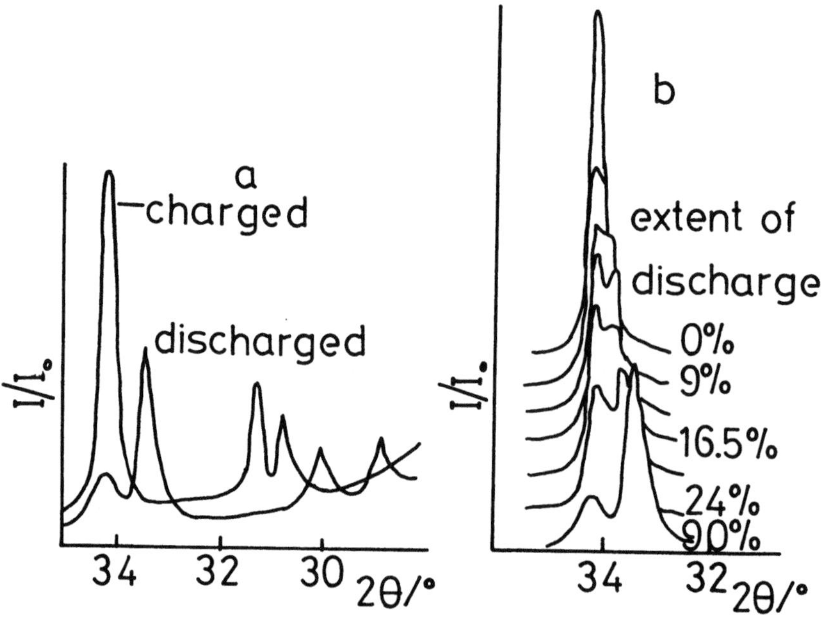


Figure 8: X-ray diffraction patterns obtained during discharge of $TiS_2$ electrode to form $LiTiS_2$. (a) wide angle scan; (b) narrow angle scan. After reference 8.

It might therefore have been better, though technically more difficult, to have prepared thinner electrodes so that the changes observed could be attributed to a larger fraction of the electrode This is particularly relevant as the region sampled was the back of the electrode where changes may have been occurring less rapidly due, for example, to potential variations through the sample. The results reported, however, assumed that the electrode was in a quasi-equilibrium state, i.e. that the volume sampled was representative of the bulk of the material.

The discharge of the cell corresponds to the intercalation of Li into the $TiS_2$ lattice by the following electrochemical process.

$$xLi + TiS_2 + xe^- \rightleftharpoons Li_xTiS_2 \quad 0 < x < 1$$

This leads to a change in the lattice parameters, and hence by monitoring the diffraction peaks of the $TiS_2$ it should be possible to follow the state of charge of the cell and to identify the phase changes. Fig. 8a shows how the (101), (002) and (100) Bragg peaks of the $TiS_2$ change as the cell is discharged. It can be seen that all the peaks shift to lower angles on discharge, the change being greatest for the (002) peak. However, the (101) peak was used for further dynamical studies because of its greater amplitude, and hence greater ease of detection. Fig. 8b shows scans over a narrow angular range in the region of the (101) peak at various depths of discharge. It can be seen that as soon as discharge commences the peak broadens, due to the nucleation of $Li_xTiS_2$, and at about 15% discharge two separate peaks are resolved. The lower-angle peak then shifts to smaller angles as the discharge proceeds. From a detailed analysis of peak widths and areas the authors proposed that intercalation of Li initially leads to buckling of the $TiS_2$ layer structure which, however, slowly anneals out. They also proposed that there was a certain amount of ordering in the intercalated Li layers. A detailed discussion of these points is however beyond the scope of this overview.

A topic that has been fairly extensively investigated by *in situ* X-ray diffraction is the formation of hydrides, particularly of palladium, during cathodic hydrogen evolution. Our second topic for consideration here is related to the structure of palladium hydrides, but with the structural information being used as an indicator of electrocatalytic activity.

The oxidation of small organic molecules is of major interest to electrochemists because of the possible use of these reactions in fuel cells. One possible such fuel that has received considerable attention is formaldehyde. In alkaline media formaldehyde is rapidly oxidised at Group IB metal electrodes (Cu, Ag and Au) but unfortunately the oxidation is incomplete, yielding formate and hydrogen. The following overall reaction scheme involving the hydroxymethanolate anion has been proposed to explain this incomplete oxidation:

$$HCHO + OH^- \rightleftharpoons [HOCH_2O]^- \tag{6}$$

$$[HOCH_2O]^- \rightleftharpoons [HOCH_2O]_{ads} + e^- \tag{7}$$

$$[HOCH_2O]_{ads} + OH^- \rightleftharpoons HCOO^- + H_{ads} + H_2O \tag{8}$$

$$2H_{ads} \rightleftharpoons H_2 \tag{9}$$

$$H_{ads} + OH^- \rightleftharpoons H_2O + e^- \tag{10}$$

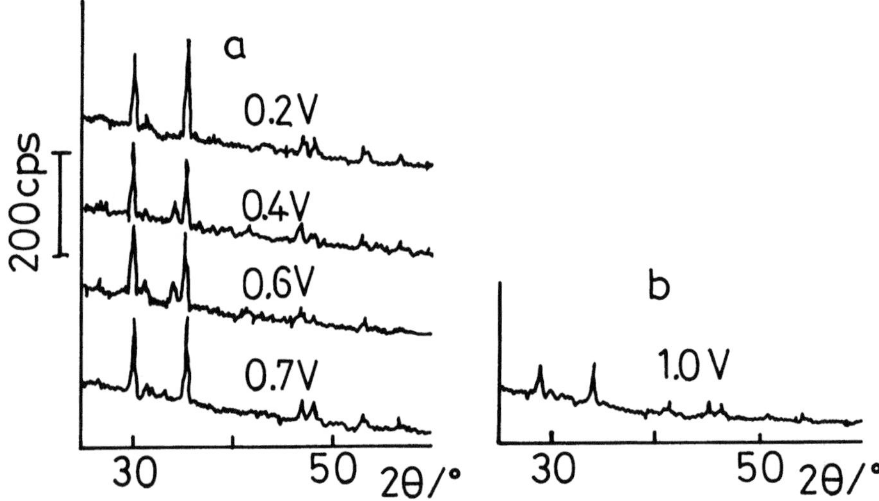


Figure 9: X-ray diffraction patterns obtained after three hours of anodic HCHO oxidation at the potentials shown on (a) a Pd electrode; (b) a Au/Pd electrode. After reference 9.

Since reaction 10 is not favoured on these Group IB metals the overall reaction is the one electron oxidation given in reaction 11, and therefore hydrogen evolution is even possible at potentials positive of the reversible hydrogen potential.

$$HCHO + 2OH^- \rightleftharpoons HCOO^- + 0.5H_2 + H_2O + e^- \tag{11}$$

To achieve a two electron reaction an electrode material that is more active to the $H_{ads}$ oxidation (reaction 10) must be used. The Group VIII metals are suitable for this but, except for Pt, are less active to the first electron transfer. One line of research that has been followed to overcome this problem has been to investigate the possibility of using alloys such as Au/Pd. Machida and Enyo [9] have investigated such systems using X-ray diffraction.

The basis of the study is that as hydrogen is evolved at a palladium surface it enters into the lattice to form the $\alpha PdH_x$ phase, until at a certain composition there is a transition to the $\beta$ phase, which is isostructural but has a larger lattice constant. This change in lattice constant leads to a change in the diffraction pattern which can be used to monitor the state of the palladium, and thus the activity of hydrogen in the region of the electrode. Fig. 9 shows diffraction patterns that were obtained for this system after three hours at the potentials shown. It can be seen that on Pd, where appreciable oxidation of formaldehyde occurs at 0.4V, peaks due to the $\beta$ phase become visible at this potential and persist up to 0.7V. This shows that hydrogen is being incorporated into the lattice, despite the oxidising environment, and thus confirms that $H_{ads}$ atoms are present as intermediates in the oxidation. On the Au/Pd alloy electrode the hydride peaks persist up to potentials as high as 1.0V (Fig. 9), showing, as expected, that this alloy is less efficient than Pd at oxidising the $H_{ads}$ atoms. The loss of the diffraction peaks due to the $\beta$ phase at more anodic potentials shows that reaction 10 has become more favoured, and therefore there is no significant build up of $H_{ads}$ on the electrode.

This study confirms the suggested reaction mechanism outlined earlier, and also demonstrates the possible benefits that may be accrued through the use of an alloy electrode. Such a confirmation of an electro-organic reaction mechanism is certainly a novel application for a solid state structural technique!

4.1.2. *Position-sensitive detection.* As we have already seen conventional diffractometers, whilst providing high resolution, suffer from the wastage of photons in that at any time they only sample a small fraction of the total scattered photons. Position-sensitive detectors, however, have the ability to sample a range of scattering angles simultaneously, and may therefore be better suited to the study of dilute systems.

A particular area of interest in electrochemical studies is monolayer adsorption, such as metal underpotential deposition. Systems of this type are effectively very dilute, and therefore position-sensitive detection will clearly be of benefit if X-ray diffraction is to be used in their study. Even so the detection of diffraction from a single monolayer will be difficult, but it may be possible to improve the situation by arranging for the X-ray beam to intersect a large number of surfaces. A study by Fleischmann and co-workers [10,11] of the adsorption of iodine on graphite, whilst not electrochemical, illustrates this point.

Iodine was adsorbed on graphite from $KI_3$ solutions and the structure of the adsorbed phase was studied *in situ* by X-ray diffraction, using position-sensitive detection. Two important features of this study relate to the form of the graphite used, 'Papyex'. This graphite has a high surface area of 20 $m^2/g$ and the covalently-bound basal surfaces are oriented largely parallel to one another, and to the surface of the material. By directing the X-ray beam perpendicularly to these surfaces it was possible to probe correlations parallel to the surface. The background-subtracted diffraction pattern obtained in this configuration with, as for all the examples in this sub-section, nickel filtered Cu $K_\alpha$ radiation from a fixed anode source is given in Fig. 10. The diffraction peak at $2\theta = 20^o$, which is the (10) diffraction of the 2-dimensional array of hexagonally close packed $I_2$ molecules (the ad-layer), exhibits the shape expected for scattering from a 2-dimensional layer. By contrast no diffraction features attributable to the ad-layer were obtained in the other configuration with the scattering vector perpendicular to the surface, i.e. with the incident beam parallel to the surface. The coherence length of the $I_2$ ad-layer was obtained from the half width of the (10) diffraction peak, and was found to be comparable to the graphite crystallite size and to be independent of the degree of coverage, as interestingly was the peak position. As expected the peak area was found to be proportional to coverage. These results clearly show that each graphite crystallite was either entirely covered by a monolayer of iodine or was empty. From classical wet chemical techniques (iodometric titration) this system was shown to follow a Langmuir isotherm, which suggests that in the Langmuir sense the sites are the crystallites, which are either fully covered or fully uncovered.

Whilst this use of high surface area materials was highly successful in improving surface

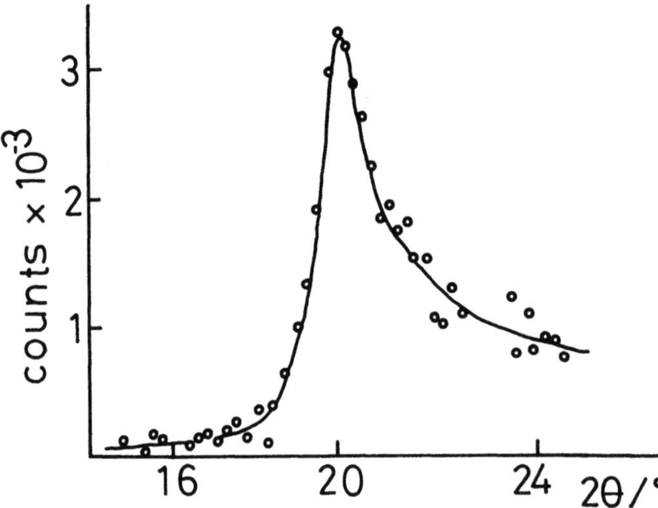


Figure 10: Background subtracted diffractogram of Papyex in 0.02M $I_2$ in 0.1M KI in the in-plane geometry.

sensitivity in the above example, it is not clear that this approach will be particularly appropriate in electrochemical cases, due to the probable unsuitability of these high surface area materials as electrodes, e.g. non uniform current and potential distribution. In studies of electrochemical systems Fleischmann and co-workers [11-13] have used a different approach to the improvement of surface sensitivity, namely the use of potential modulation. In this technique, which is applicable only to chemically reversible systems, the potential of the electrode is modulated repeatedly between two potentials corresponding to two different states of the electrode, e.g. a clean electrode surface and one covered by a thin oxide film. The diffraction patterns at the two potentials are then accumulated separately and after a suitable period of data acquisition, they are differenced. In this way constant contributions, such as scattering by the solution and cell window, are subtracted out and the difference diffractogram contains only contributions related to structural changes caused by the potential modulation. We will consider here two examples of this type of investigation.

The first of these studies relates to a system of significant technological importance, the nickel hydroxide/oxyhydroxide system as used in nickel-cadmium rechargeable batteries [11]. The electrodes consisted of thin films of nickel hydroxide that had been cathodically deposited on a polished nickel electrode, and aqueous KOH was used as the electrolyte. The films were kept thin ($\sim$ 1500Å) to ensure rapid charge and discharge during the potential modulation. The potential of these electrodes was then modulated between values corresponding to the charged and discharged states and the respective diffraction patterns acquired. Fig. 11 shows one such difference diffractogram for an aged electrode (i.e. one that had been exposed to the electrolyte for several hours) that resulted from six hours potential modulation. In a difference diffractogram positive-going peaks correspond to fea-

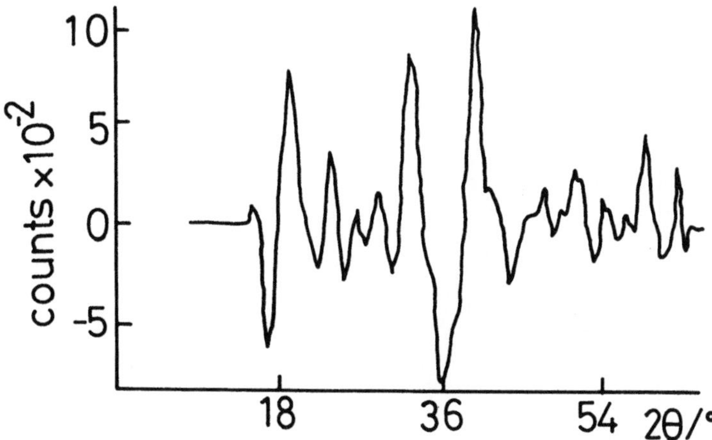


Figure 11: Difference diffractogram for the $Ni(OH)_2/NiOOH$ system in 1M KOH. Six hours acquisition time at a modulation frequency of 5 x $10^{-3}$ Hz. Data are presented as the diffractogram at +100mV minus that at +400mV.

tures present at the first potential and not at the other. Thus in the present case the positive-going peaks relate to structure present only at a potential of 100mV and were attributed to $\alpha$ $Ni(OH)_2$, whilst the negative-going ones related to the structure at 400mV and were attributed to $\gamma$ NiOOH. Further studies of this system showed that for the unaged electrode the transition was between the $\beta$ phases of the hydroxide and oxyhydroxide, whilst for incompletely aged electrodes a mixture of both types of behaviour was observed

The first example of the use of potential modulation was relatively straightforward as it related to changes in quite thick layers, and as a result could probably even have been conducted with a conventional diffractometer system, though in this case a significantly thicker layer would probably have been required. The second example we are going to consider, namely oxide formation and hydrogen adsorption on platinum, is experimentally much more demanding.

Fleischmann and Mao [12] have recorded difference diffractograms obtained by modulation of the potential of a platinum electrode in various potential regions. Whilst potential modulation was essential for this study, this of itself was not enough to obtain adequate discrimination between surface and background signals. The surface contribution was therefore enhanced by using a platinum-black electrode whose area was about 10 times the geometrical area. As an example of their results Fig. 12 shows how the difference diffractogram for potential modulation between the double layer and oxide formation regions varies as chloride ions are added to the sulphuric acid electrolyte. If we first consider the diffractogram obtained in the absence of added $Cl^-$ ions the main features are two negative-going peaks at positions corresponding to the (111) and (200) diffractions of Pt. Thus the sole result of forming the oxide seems to be a loss of Pt-structure. No features directly attributable to the

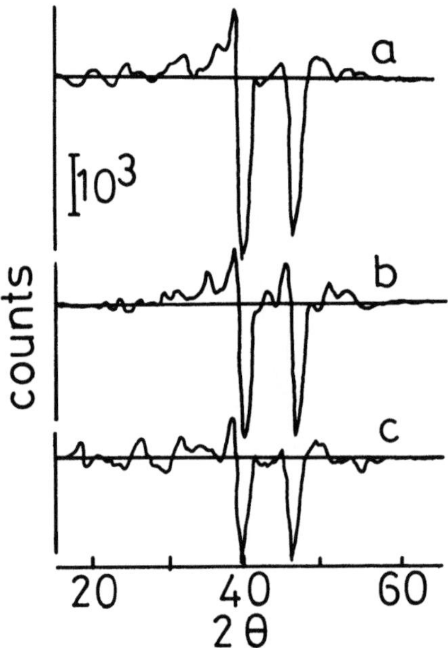


Figure 12: Difference X-ray diffractograms for platinized Pt in 1M $H_2SO_4$ modulated between +0.2V and +1.0V obtained after $10^4$ s at a modulation frequency of 4 x $10^{-3}$ Hz. The solution contained (a) 0 (b) 0.01M and (c) 0.1M KCl. After reference 12.

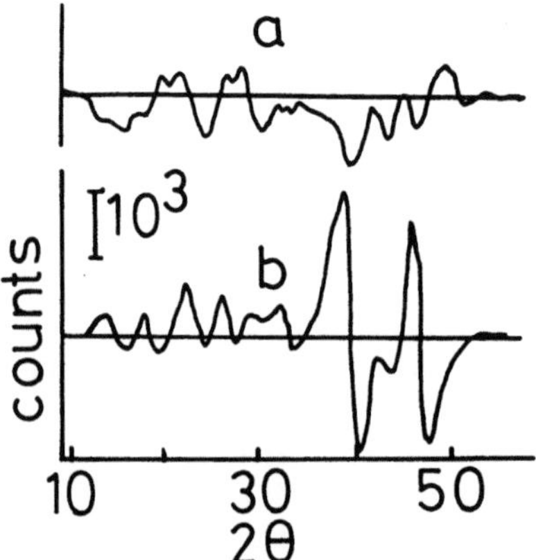


Figure 13: Difference X-ray diffractogram for platinized Pt in 1M $H_2SO_4$ in the hydrogen adsorption region. Modulation between (a) +0.2V and -0.05V and (b) +0.2V and -0.2V. After ref^rence 12.

oxide are observed (there are no positive going peaks in the difference diffractogram) and therefore it was concluded that the oxide was of low crystallinity. The addition of chloride had the effect of reducing the amplitude of the negative-going peaks. This suggests that less crystalline Pt is being lost, and therefore that less oxide is being formed, which is in agreement with the voltammetric data.

A rather more surprising result is obtained when the modulation is between the double layer and hydrogen adsorption regions. Whilst modulation to the strongly bound hydrogen region results in a featureless difference diffractogram, large effects are seen when the modulation potential is extended into the weakly bound region, as shown in Fig. 13. Again there are negative-going peaks, corresponding to the loss of Pt-structure on forming the weakly bound hydrogen, but these are accompanied by positive-going peaks which the authors attribute to scattering by an incommensurate surface layer. In other words, the adsorption of the weakly bound hydrogen is causing a reconstruction of the surface. As the authors themselves point out these results are rather surprising, as one would probably expect strongly bound hydrogen to have had the greater effect, as it is thought to enter the Pt surface, whereas the weakly adsorbed hydrogen is generally regarded as a simple chemisorbed layer. Further work is probably required before this matter can be fully resolved. What these results do clearly show is that in favourable cases structural changes induced by monolayer adsorption can be studied using the potential-modulation technique.

## 4.2. GRAZING INCIDENCE DIFFRACTION

In recent years interfacial electrochemical studies have increasingly been conducted on single crystal substrates, as opposed to polycrystals, because the surface state of such a material is much better defined and interpretation of data is more straightforward and precise. This change has brought about significant advances in our understanding of electrochemical surface processes. The use of this type of electrode in X-ray diffraction studies requires a rather different approach from that used in powder diffraction studies, being more like the single crystal studies common in protein crystallography, or LEED studies in surface science.

When conducting surface structural studies on single crystal substrates using X-ray diffraction, it is usual to use a glancing angle of incidence configuration to achieve surface sensitivity. Simple geometrical considerations show that by using glancing angles the effective path length through a surface will increase, thereby increasing the surface sensitivity. However, by decreasing the angle of incidence to below the value of the critical angle (it must be remembered that for X-rays the refractive index of metals is less than unity) a significant further enhancement of the surface selectivity also occurs. Since the X-ray beam undergoes total external reflection only a thin layer of the substrate is penetrated, and therefore the contribution from the substrate to the total scattered signal is greatly reduced. Under these conditions it can be shown that the ratio of the signal from the surface to that from the substrate is enhanced by a factor of $2/\theta_c$ (where $\theta_c$ is the critical angle in radians). For metals at X-ray frequencies the value of the critical angle is typically around $0.5^o$ (measured with respect to the surface plane), i.e. about 9 milli-radians, for transparent incident phases, e.g. air or water, and thus it can be seen that this enhancement can be highly significant. Fig. 14 shows schematically the sample geometry for this glancing inci-

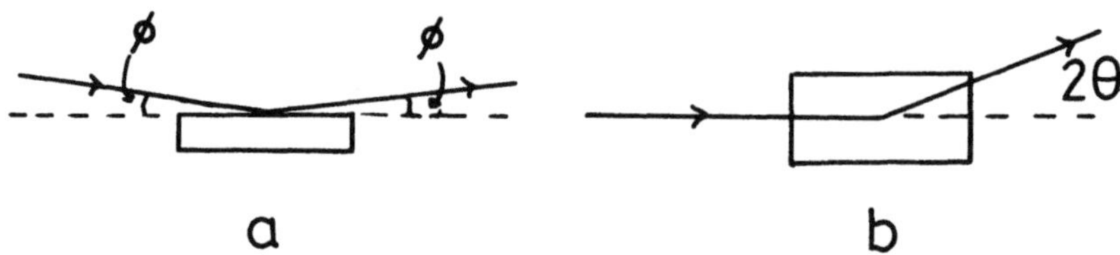


Figure 14: Schematic diagram showing sample geometry in the TRBD technique. (a) plane view and (b) top view.

dence approach, which is known as total reflection Bragg diffraction (TRBD). For a fuller description of the technique the reader is referred to a paper by Marra *et al* [14].

The TRBD technique has been increasingly used in the last few years in studies of semiconductor surfaces, particularly in problems related to growth using molecular beam epitaxy. Applications in electrochemistry are particularly problematic because of the presence of a layer of electrolyte solution over the electrode. At the very small angles of incidence that are required by this technique the path length through the solution will inevitably be large, leading to absorption problems. The only way to overcome this is to collapse a thin window onto the electrode, trapping a very thin electrolyte layer. Thus whilst potential control is maintained dynamic measurements are effectively impossible. To further reduce the absorption problem hard X-ray radiation is generally used, and the high flux and high brightness of synchrotron radiation has proved helpful. Despite these experimental difficulties some exciting studies are beginning to appear in the literature.

An example of this type is the study by Samant *et al* [2] of Pb monolayers on Ag(111) electrodes. These electrodes were prepared by the evaporation of Ag onto cleaved mica, and the lead monolayers were deposited at an underpotential from a lead acetate solution. Once the monolayer had been deposited, and prior to any X-ray studies, the polypropylene cell window was collapsed onto the electrode. For control of the scattering geometry the cell was mounted on a four-circle diffractometer with the sample placed vertically. Synchrotron radiation with a wavelength of 1.53 Å was used at an incidence angle of $0.54^o$, which was chosen to maximise the signal.

(10), (11) and (20) reflections from the Pb overlayer were detected, and Fig. 15 shows both a rocking scan for the (10) reflection and also a radial scan. The identification of these peaks was confirmed from their disappearance when the electrode potential was increased to a value where the lead oxidised. The rocking scan was referenced to the Ag($\bar{2}11$) direction, and hence it can be seen from Fig. 15 that the Pb lattice is rotated by $4.4^o$ from the Ag lattice. It was found that the radial scan could be repeated at $60^o$ intervals, indicating

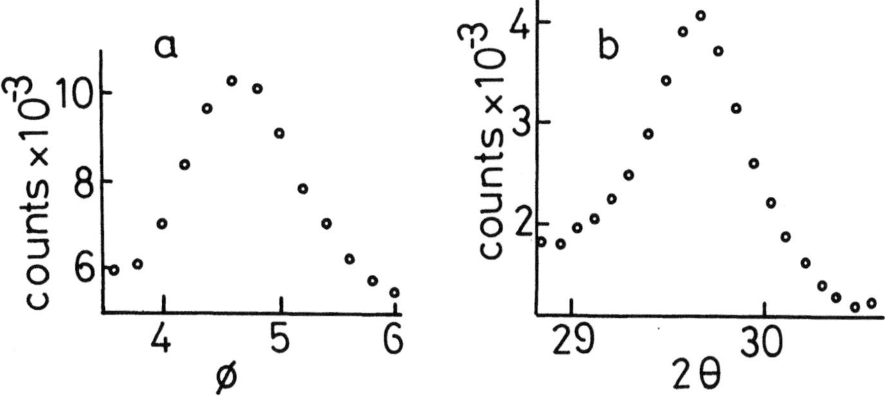


Figure 15: (a) Rocking curve scan of the Pb (10) reflection at $2\theta$=29.7° and (b) radial scan of the (10) refection at $\phi$=4.4°. After reference 2.

that the ad-layer had six-fold symmetry. From these results it was concluded that the ad-layer formed a hexagonally close packed layer, and that the ad-layer and substrate lattice were rotated with respect to one another. Further, more detailed, studies of relative peak intensities showed that the ad-layer was almost completely ordered. This study clearly demonstrates the power of this technique, and points to very significant developments in the next few years.

## 4.3. STANDING WAVE STUDIES

For the study of adsorption on single crystals the grazing incidence method can be used to study in-plane structure, but it does not lend itself so readily to the determination of the spacing between the ad-layer and the substrate. Another technique, which is not strictly a scattering method though it relies on scattering, has been developed to obtain this information. This technique is the standing wave method which works in the following way.

When an X-ray beam is incident on a crystal lattice under Bragg diffraction conditions, the coherently related incident and diffracted beams can interfere with one another to set up a standing wave-field. This wave-field, which has the same periodicity as the lattice planes that are scattering, also extends into the incident medium. As the incident beam is scanned through the Bragg peak the position of the standing wave-field moves with respect to the lattice. On the low angle side of the Bragg peak the anti-nodal planes of the wave-field lie halfway between the $(hkl)$ diffracting planes whilst on the high angle side the anti-nodal planes and the lattice planes are coincident. Thus, if there is an ad-layer on the surface the standing wave can be made to sample it. If an anti-nodal plane coincides with the atom plane then the ad-layer may fluoresce, and the fluorescence yield will be modulated

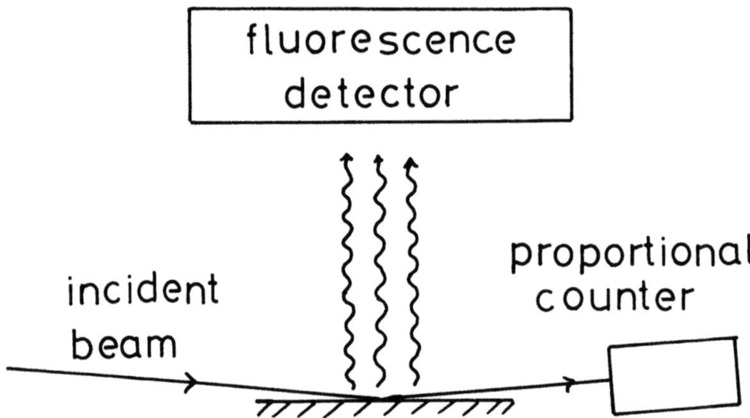


Figure 16: Schematic representation of the experimental configuration for standing wave experiments.

as the Bragg peak is scanned through. From the phase and amplitude of this modulation the position, and width of the distribution, of the adatoms can be found. The experimental configuration is shown schematically in Fig. 16. In order to penetrate electrolyte solutions relatively hard radiation ($\sim$ 15 KeV) is typically used, and hence for conventional electrode materials, e.g. Cu or Au, strong diffractions are to be found at relatively low incidence angles (measured with respect to the electrode surface). An unfortunate consequence of this is that perfect, or nearly perfect crystals are required. Whilst this is not too problematic with materials like Si, it is difficult to make, and more importantly preserve, single crystals of soft metals of the required quality. As a result few studies of this type have been made, but Materlik and co-workers [15] have shown that with great care a Cu(111) crystal of the required quality can be prepared, which they used to study Tl underpotential deposition. Fig. 17 shows how the reflectivity and flúorescence signal, both in the absence and presence of oxygen, varied for this system as a function of the angle of incidence. An analysis of these data showed that in the absence of dissolved oxygen the Tl layer lay 2.58 Å above the Cu surface which can be compared with the value of 2.67 Å obtained from studies of an emersed electrode. On exposure to oxygen this value contracted to 2.27 Å for both *in situ* or *ex situ* studies. The authors suggest that these results are consistent with the Tl occupying a two-fold site *ex situ*, but a three-fold one in the *in situ* case. The effect of oxygen is thought to be *via* an interaction with the copper.

Despite the success obtained for Tl on Cu(111) it is unlikely that there will be many further studies of this type, in view of the difficulties with crystal preparation. In order to lift the requirement for high quality crystals one needs to work with Bragg reflections where the rocking curve width is larger, and hence the mosaic spread of the substrate has little effect. There are two ways in which this can be achieved, the first is to work at near normal

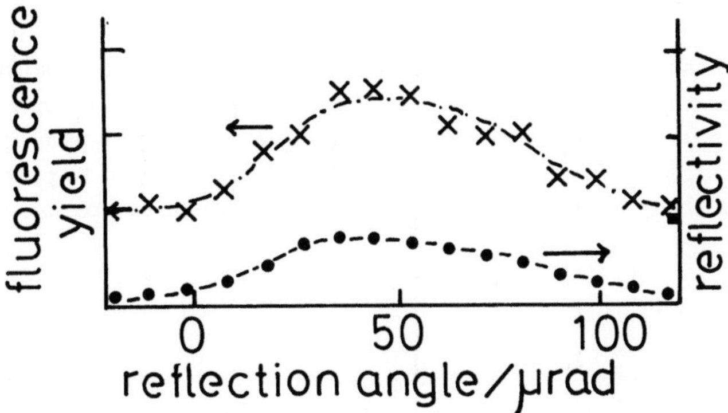


Figure 17: The Tl fluorescence yield and the reflectivity as a function of the incidence angle in the region of the Bragg peak for the Tl on Cu(111) system. After reference 15.

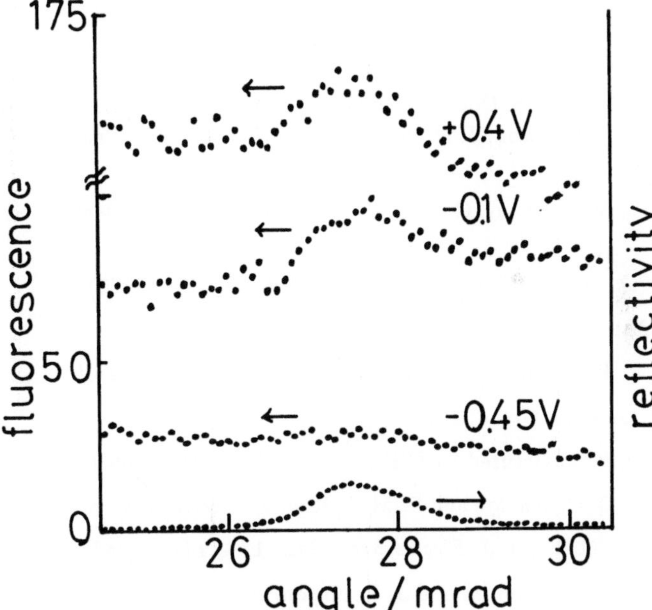


Figure 18: Normalised iodide fluorescence intensity for the Pt/C layer structure electrode immersed in 50 $\mu$M NaI in 0.1M $Na_2SO_4$ as a function of incidence angle for various potentials. Also shown is the reflectivity curve of the layer structure. After reference 16.

incidence, and the second is to use materials of large layer spacing as the electrode. Abruña and co-workers [16] have adopted the latter approach, using artificial Pt/C layer structures with d spacings in the range 20 to 100 Å as electrodes to look at iodide adsorption on Pt. The success of this approach is shown in Fig. 18, which shows the I fluorescence signal as a function of angle for various electrode potentials. These data were used to construct an adsorption isotherm, but do not as yet appear to have been analysed to provide structural information. From the quality of the data, however, this should be possible.

### 5. CLOSURE

It is really early days now for X-ray scattering studies of electrochemical interfaces, but the examples outlined above should give some indication of the type of study that is possible. In the next few years one can expect the number of applications to increase and the range of systems studied to broaden. In particular, we can expect powder diffraction studies to be extended into the area of time-resolved studies, and glancing incidence and standing wave studies to be taken up by a wider range of laboratories.

## References

[1] N. Barlow, C. Brennan, S. Doyle, G. N. Greaves, M. Miller, A. H. Nahle, R. J. Oldham, K. J. Roberts, J. Robinson, J. N. Sherwood and F.C. Walsh, *Rev. Sci. Inst.*, **60**, 2386 (1989).

[2] M. G. Samant, M. F. Toney, G. L. Borges, L. Blum and O. R. Melroy. *Surf. Sci.*, **193**, L29 (1988)

[3] C. J. Borkowski and M. K. Kopp, *Rev. Sci. Inst.*, **46**, 951 (1975)

[4] A. Gabriel, *Rev. Sci. Inst.*, **48**, 1303 (1977)

[5] P. Lecomte, V. Perez-Mendez and G. Stokes, *Nucl. Instrum. Meth.*, **153**, 543 (1978)

[6] E. R. Wolfel, *J. Appl. Cryst.*, **16**, 341 (1983)

[7] INEL CPS120, Manufactured by INEL, 126 rue L. Blériot, 78530 BUC, France.

[8] R. R. Chianelli, J. C. Scanlon and B. M. L. Rao, *J. Electrochem. Soc.*, **125**. 1472 (1987)

[9] K. Machida and M. Enyo, *J. Electrochem. Soc.*, **134**, 1472 (1987)

[10] M. Fleischmann, P. J. Hendra and J. Robinson, *Nature*, **288**, 152 (1980)

[11] M. Fleischmann, A. Oliver and J. Robinson, *Electrochim. Acta*, **31**, 899 (1986)

[12] M. Fleischmann and B. W. Mao, *J. Electroanal. Chem.*, **229**, 125 (1987)

[13] M. Fleischmann and B. W. Mao, *J. Electroanal. Chem.*, **247**, 297 (1988)

[14] W. C. Marra, P. Eisenberger and A. Y. Cho, *J. Appl. Phys.*, **50**, 6927 (1979)

[15] G. Materlik, M. Schmah, J. Zegenhagen and W. Uelhoff, *Ber. Bunsen-Ges. Phys. Chem.*, **91**, 292 (1987)

[16] H. D. Abruña, J. H. White, M. J. Albarelli, G. M. Bommarito, M. J. Bedzyk and M. McMillan, *J. Phys. Chem.*, **92**, 7045 (1988)

# X-RAY REFLECTIVITY AND SURFACE ROUGHNESS

B.M. Ocko

*Physics Department*
*Brookhaven National Laboratory*
*Upton, New York 11973*

**ABSTRACT.** Since the advent of high brightness synchrotron radiation sources there has been a phenomenal growth in the use of x-rays as a probe of surface structure. The technique of x-ray reflectivity is particularly relevant to electrochemists since it is capable of probing the structure normal to an electrode surface in situ. In this paper the theoretical framework for x-ray reflectivity is reviewed and the results from previous non-electrochemistry measurements are summarized. These measurements are from the liquid/air interface ($CCl_4$), the metal crystal vacuum interface (Au(100)), and from the liquid/solid interface(liquid crystal/silicon).

## 1. INTRODUCTION

The surface structure of electrodes in electrolyte solutions is a fundamental problem in electrochemistry, catalysis and surface science. The arrangement of the atoms at the surface has a pronounced impact on the electronic structure and reactivity of the surface. Direct structural studies of electrode interfaces in-situ are not possible with conventional structural surface science techniques such as Low Energy Electron Diffraction (LEED). Recently several groups have realized the importance of x-rays as a direct structural probe in-situ and these efforts have utilized EXAFS[1, 2], Standing wave[3] and Glancing Incident Angle Diffraction[4] techniques at a number of synchrotron sources. X-ray reflectivity techniques probe the structure along the surface normal direction, including the effects of roughness. Since reflectivity techniques can also be applied in situ, they should play an important role in understanding electrochemical systems. These techniques will be applicable to the problems of surface reconstruction, underpotential deposition, and surface oxidation.

It is important to define the coordinate system before proceeding. We will choose the surface normal direction to be $\hat{z}$ and the two directions in the plane of the surface as $\hat{x}$ and $\hat{y}$. For specular reflectivity the scattering is aligned along the $\hat{z}$ direction. The momentum transfer $Q_z=(4\pi/\lambda)\sin\theta$ where $\lambda$ is the x-ray wavelength and $\theta$ is the incident angle with respect to the surface plane. Since specular scattering is confined to the $\hat{z}$ direction the



*C. Gutiérrez and C. Melendres (eds.),*
*Spectroscopic and Diffraction Techniques in Interfacial Electrochemistry,* 343–359.

reflectivity is often referred to as a rod of scattering[5]. By measuring the specular reflectivity starting at very small $\theta$ and possibly extending out to Bragg peaks information on the structure along the surface normal direction is obtained. Fitting the data to simple surface models allows us to deduce the surface roughness, mass, and expansion(contraction) of the surface. For small incident angles the data can be modeled as a perturbation from a perfect interface and the corresponding deviation from the Fresnel Law of Optics.

Specular reflectivity techniques have been applied to a large variety of problems at solid and liquid interfaces. X-ray studies have been performed at the liquids surfaces of mercury[6, 7, 8], water[9], carbon tetrachloride and methanol[10], liquid crystals[11, 12, 13], microemulsions[14], on Langmuir films on water[15, 16, 17] and on thin water layers on solids. Experiments on solid surfaces include Au(100)[18], Cu(110)[19], silicon oxide interfaces[20] and on silanes coated silicon[21, 22] These same classes of experiments could also be explored at the air/electrolyte surface and on the metal/electrolyte surface.

This paper is divided into four sections. In this section the scattering geometry is outlined. In the next section we will consider a continuum model of the electron density. This approach is most applicable at small angles and provides a simple description of the reflectivity in terms of a deviation from the Fresnel Law. These deviations can be simply related to surface roughness effects and for simple liquids this roughness can be calculated from the surface tension of the liquid. In the third section, we will consider the scattering from discrete atoms. This is particularly useful in describing the structure of single crystal surfaces. In the final section, preliminary measurements at the liquid crystal/solid interface are presented.

Before turning to the continuum and discrete models, we will briefly discuss x-ray sources. X-rays are produced via a conventional fixed target tube source, via a rotating anode generator or by a synchrotron. The first two methods produce x-rays which exit from the source in all directions. A small collimated beam is chosen from slits which are located a distance from the source which is usually less than 1 meter. These slits are followed by a monochromator crystal, typically germanium or silicon, which is tuned to accept the intrinsic radiation from the source. The source is typically either copper or molybdenum and the characteristic lines are at 1.54 Å and 0.71 Å respectively. At a synchrotron source the divergence of the x-ray beam is controlled by the energy of the ring and is just $1/\gamma$. At NSLS this corresponds to $2.5\times10^{-4}$ radians. To provide a monochromatic beam, the x-rays are Bragg reflected from single crystals. The intensity from current synchrotron sources at about 1 Angstrom is about one thousand times more intense then from a rotating anode, which is typically ten times brighter then a conventional tube source. In order to measure the specular reflectivity it is convenient to use a two circle x-ray goniometer which allows the incident and outgoing angles to be controlled independently. For liquid interfaces, the sample can not be tilted and the incident angle is controlled by tilting the incident beam downward and adjusting the sample height.

## 2. CONTINUUM MODELS (Liquid Surfaces)

In the continuum model, the index of refraction for materials is given by the free electron model[23] where the dielectric constant is given by

$$\epsilon = 1 - \frac{\rho_e \lambda^2}{\pi} \left( \frac{e^2}{mc^2} \right) \tag{1}$$

where $\rho_e$ is the electron density, e is the electron charge, and m is the mass of an electron. The effect of absorption has been ignored which would give rise to an imaginary index of refraction. We will now consider a vacuum/solid interface in which the dielectric constant of the vacuum is unity and the dielectric constant of the solid is given by Eq. 1. For angles below the critical angle $\theta_c$ the x-rays are totally externally reflected according to Snell's Law. Solutions to Maxwell's Equations gives the intensity of the reflected field divided by the incident field. This is often referred to as the Fresnel Law of Optics. In the small angle approximation it is given by

$$R = \frac{|E_R|^2}{|E_0|^2} = \left| \frac{\theta_c}{\theta^2 + \sqrt{\theta^2 - \theta_c^2}} \right|^4 \tag{2}$$

where the critical angle for total external reflection

$$\theta_c = \sqrt{\rho_e^2 \lambda^2 / \pi m c^2} \tag{3}$$

shown by the solid line in figure 1. At angles below $\theta_c$ there is total external reflection. The critical wavevector $Q_c = (4\pi/\lambda)\theta_c$ is independent of wavelength. Numerically, $Q_c = 3.75 \times 10^{-14} \rho_e^{1/2}$ Å$^{-1}$, where $\rho_e$ is in electrons/cm$^3$. It is convenient to rewrite Eq. 2 as

$$R = \left[ \frac{\theta_c}{2\theta} \right]^4 |T(\theta)|^4 \tag{4}$$

where the enhancement factor

$$T(\theta) = \frac{2\sin\theta}{\sin\theta + \sqrt{\cos^2\theta_c - \cos^2\theta}}. \tag{5}$$

expresses the amount by which the electric field is enhanced at the interface. For $\theta$ several times $\theta_c$ the enhancement factor is very close to unity and the reflectivity can be well approximated by

$$R = \left[\frac{\theta_c}{2\theta}\right]^4 \tag{6}$$

which is shown by the dashed line in figure 1. At five times the critical angle the reflected intensity is slightly less than $10^{-5}$.

For interfaces which are not ideally sharp ( i.e. a diffuse interface) the reflectivity falls off faster than the $\theta^{-4}$ dependence given by the Fresnel Law. This can be understood physically by considering the interface as discrete slabs of dielectric material which mimic the actual electron density profile. This is a generalized extension of the technique to measure film thickness with ellipsometry.

For smoothly varying interfaces, the actual reflectivity can be calculated as a perturbation from the ideal step interface. The theoretical expression for the ratio $R(\theta)/R_F(\theta)$ in terms of the average electron density gradient along the surface normal is given by

$$\frac{R(\theta)}{R_F(\theta)} = \left| \frac{1}{<\rho_e>} \int_{-\infty}^{\infty} \left\langle \frac{\partial \rho_e}{\partial z} \right\rangle e^{iQ_z z}\, dz \right|^2 \tag{7}$$

in the Born approximation limit, i.e. $\theta >> \theta_c$. This expression has been used extensively to model the reflectivity of liquid crystal/air interfaces where the density is modulated by the "smectic layers" induced by the surface. This same expression has been used to describe the interface of rough surfaces. The derivative of the density with respect to z depends on the length scale over which the roughness is integrated over. This length scale depends on the resolution of the measuring instrument and can vary between several hundred Angstroms to length scales approaching a micron.

Several simple forms for $\rho_e(z)$ and the resulting reflectivities, $R(\theta)/R_F(\theta)$, given by Eq. 7 will be considered.

(i) If $\rho_e$ is a step function, then the derivative is a delta function. Taking the fourier transform of a delta function gives unity and we are left with the desired result that $R(\theta) = R_F(\theta)$.

(ii) If $\rho_e$ is an error function, then the derivative is a Gaussian. In addition, the reflectivity for such an interface deviates from the Fresnel law as a Gaussian, i.e. $R(\theta)/R_F(\theta) = \exp(-\sigma^2 Q_z^2)$.

For many systems a single Gaussian may not be adequate to describe the derivative of the density profile. In such cases the system can be modeled by a density derivative which is the sum of several Gaussians which may be centered at different positions. As we shall see, for simple liquids, the expected reflectivity deviates from the Fresnel form very much like a Gaussian.

Up to now we have considered diffuse interfaces in which there is no in-plane structure. However, this is usually not the case since real systems are

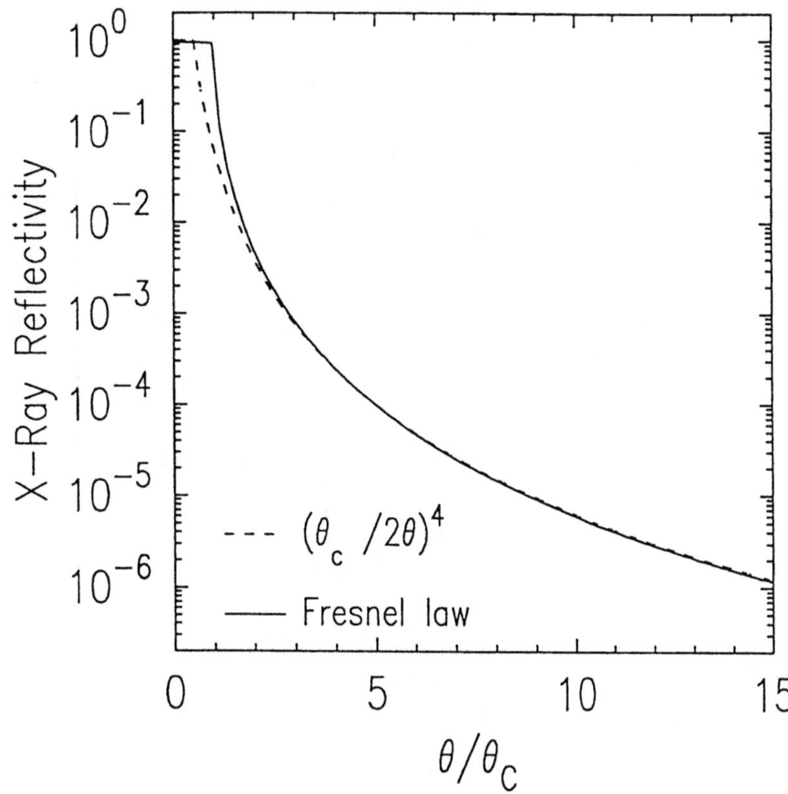


Figure 1. The solid line is the Fresnel law of classical optics and the dashed line is the approximation for $\theta >> \theta_c$: $R_F(\theta) \approx (\theta_c/2\theta)^4$.

"rough". By rough we mean that the height-height correlation is non-zero. In defining the height-height correlation function, the length scale over which we are considering defines the limits of integration. For most systems, the roughness increases as the area of the sampling is increased, hence a "universal" measure of the roughness is inappropriate.

For simple liquids, the height-height correlation function results from thermally excited capillary waves. The two-dimensional energy density for surface waves is made up of a surface tension term, $(\gamma/2)[\nabla_{xy} h(\mathbf{r}_{xy})]^2$, and a gravitational term (important only at long wavelengths),

$$U = \frac{\gamma}{2}\int d^2\mathbf{r}_{xy}\, |\nabla_{xy} h(\mathbf{r}_{xy})|^2 + \frac{\rho g}{2}\int d^2\mathbf{r}_{xy}\, h(\mathbf{r}_{xy})^2 . \tag{8}$$

Taking into account the density of states and applying the equipartition theorem one obtains the following expression for the root-mean-squared(r.m.s.) roughness of the free surface due to capillary waves,

$$\langle h(0)^2 \rangle = \frac{k_B T}{4\pi^2\gamma}\int d^2\mathbf{Q}_{xy} \frac{1}{|\mathbf{Q}_{xy}|^2 + k_g^2} \tag{9}$$

where $k_g^2 \equiv \rho g/\gamma$. The upper limit of this integral, $k_{\mathrm{max}}$, is on the order of $\pi/r_M$, where $r_M$ is the molecular radius. Without the gravity term, $\rho g/\gamma$, the integral would have an infrared divergence $\mathbf{Q}_{xy} \rightarrow 0$. Using these limits, integration of Eq. 9 results in the following equation for the intrinsic roughness of a liquid surface,

$$\langle h(0)^2 \rangle = \frac{k_B T}{4\pi\gamma} \log\left(\frac{k_{\mathrm{max}}^2 + k_g^2}{k_g^2}\right). \tag{10}$$

For water $\langle h(0)^2 \rangle^{1/2} \approx 3.98$ Å at room temperature[24]. The gravitational length scale, $1/k_g$, is inaccessible with realistic resolution volumes, and the measured, or effective, roughness is larger than the intrinsic roughness.

In this approximation, the reflectivity $R(\theta)$, is given by

$$R(Q_z) = e^{-\sigma_T^2(Q_z)Q_z^2} \tag{11}$$

where $\sigma_T^2 = \sigma_C^2 + \sigma_P^2$. The first term is from thermal capillary waves and the second term is from the finite size of the liquid molecules. Braslau et al[10] have shown that the capillary term is given by

$$\sigma_C^2(Q_z)Q_z^2 = \frac{k_B T}{2\pi\gamma} Q_z^2 \log\left(\frac{2\pi}{Q_z r_M(\Delta\theta_d)}\right) \tag{12}$$

where $r_M$ is the molecular radius and $\Delta\theta_d = h_d/L$ is the full width of the angular resolution in the plane of incidence.

X-ray Reflectivity studies have been carried out on the free surface of water, carbon tetrachloride, and methanol[10]. As predicted by the capillary wave model the roughness of the interface is higher for the liquids with the smaller surface tensions, i.e. it is easier to excite thermal roughness on surfaces with a smaller stiffness. All of these liquids are in reasonable agreement with the predictions of Eq. 10 and Eq. 11. The data has been fit to a single parameter, $\sigma_P$, whereas $\sigma_C$ has been calculated exactly.

For carbon tetrachloride the reflectivity normalized to $R_F$ is shown in Figure 2. The filled square data is taken with higher resolution then the filled triangle data, therefore, the effective $\sigma_C$ from Eq 10. is larger and the reflectivity is smaller. The results of the capillary model with $\sigma_P$=0 is shown as the solid lines. In order to fit the data, $\sigma_P$ has been adjusted. For the higher resolution data $\sigma_P$=2.5 Å and for the lower resolution data $\sigma_P$=3.3 Å. The model with these parameters is shown as dashed curves in figure 2. The data has also been equally fit to a simple Gaussian form

$$R(Q_z) = e^{-\sigma_M^2 Q_z^2} \tag{13}$$

This fit gives a value $\sigma_M$=5.13 Å for the higher resolution data and $\sigma_M$=5.93 Å for the lower resolution data.

## 3. DISCRETE MODELS (Au(100) Surface)

We will now turn to the subject of specular reflectivity from crystalline surfaces, and in particular from the (100) surface of gold. In this example, we will consider the "Fresnel" reflectivity which arises from the termination of the crystal as well as the nature of the "Bragg" reflectivity which results from the periodicity of the atomic arrangement. As we shall see, these effects are essentially the same and can not be easily separated from each other.

The Au(100) is believed to exhibit a buckled, slightly rotated triangular structure, incommensurate (in-plane) with the underlying square substrate. This picture has emerged from a variety of experiments and calculations. Initial LEED observations concluded that the overlayer reconstructs with a "(5×1)"[25, 26]. Shortly thereafter, Fedak and Gjostein resolved a splitting in the LEED pattern and proposed a "(20×5)" structure with a a hexagonal overlayer on a square substrate. This model corresponds to a ~20% contraction along the direction corresponding to the "(5×1)" reconstruction and a ~5% contraction along the orthogonal surface direction. More recent LEED measurements[27] have resolved additional spots. This same data has been reinterpreted by rotated domains which are nearly 14×5 which is more precisely described by a large unit cell given by c(26×68) reconstruction[28]. Unfortunately, multiple scattering effects have made a definitive determination of the structure impossible so far. The additional ~25% mass required by such an triangular overlayer has been detected by ion scattering[27].

In the kinematical approximation, the scattering can be calculated by summing the cross-section for electrons over the resolution volume. In the

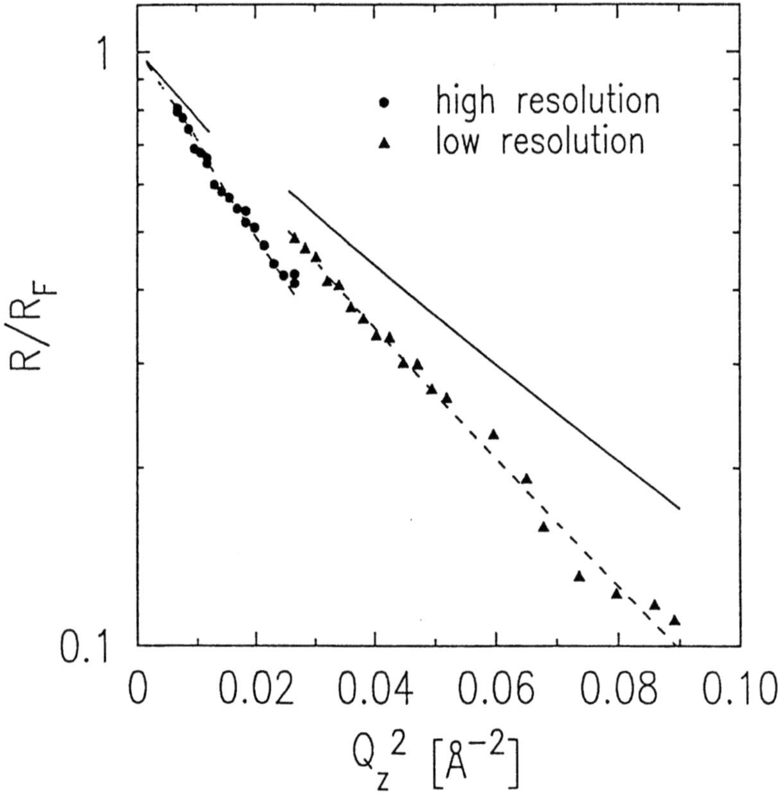


Figure 2. The measured reflectivity of carbon tetrachloride normalized to the Fresnel reflectivity is shown. The solid lines are the predictions of the capillary-wave model for the two resolutions. The dashed lines are the best fit to a surface profile width which is added in quadrature to the capillary roughness.

case of a perfectly terminated crystalline material, $R(Q_z)$ can be expressed as either a sum over the positions of each atomic layer

$$R = \frac{16\pi^2 r_0^2 T^4(\theta) \, |F(Q_z)|^2 e^{-2W(Q_z)}}{a^4 Q_z^2} \left| \sum_{n=0}^{\infty} e^{iQ_z dn} \right|^2 \tag{14}$$

or equivalently as a sum over Bragg peaks[29, 5, 30, 31, 32]

$$R = \frac{64\pi^2 r_0^2 T^4(\theta) \, |F(Q_z)|^2 e^{-2W(Q_z)}}{a^6 Q_z^2} \left| \sum_{(L\ even)} \frac{(-1)^{L/2}}{Q_z - Lc^*} \right|^2. \tag{15}$$

In these expressions $d = a/2$ is the layer spacing. The Debye-Waller factor is given by $W(Q_z)$ and the form factor is given by $F(Q_z)$. The $(Q_z - Lc^*)^{-2}$ tails evident in Eq. (15) are familiar in the context of Darwin theory of Bragg reflectivity. Because of the factor $(-1)^{L/2}$, the amplitudes of neighboring peaks add constructively.

Figure 3 displays the specular reflectivity of Au(100) from $Q_z/c^* = 0.15$ to $Q_z/c^* = 3.5$ at T=310K. The inset shows a typical rocking curve obtained at the anti-Bragg position ($Q_z/c^* = 1.0$) at this temperature. Before turning to a more detailed analysis, it is worth noting that the measured reflectivity in Figure 3 extends over a range of nearly seven decades and over scattering angles from $2\theta = 1^\circ$ to $75^\circ$, smoothly joining the (002) reflection to the origin. This is all the more impressive in view of the finite reflectivity obtained at the "forbidden" (001) reflection, where successive planes scatter exactly out of phase. Together with the absence of any measurable diffuse scattering in the transverse direction this suggests that on the length scales sampled by the resolution volume, the surface may be both atomically flat and perfectly aligned with the crystallographic planes. To estimate the in-plane coherence area we note that the width of the $\theta$-scan(inset to Figure 3) is only $0.07^\circ$ FWHM. This corresponds to the $2\theta$ resolution and gives a width in reciprocal space of $\Delta Q_x = Q_z \Delta\omega = 0.001$ Å$^{-1}$. Because it is very unlikely that a Au surface can be cut to these specifications, we conclude that upon annealing it must rearrange itself to expose smooth (100) facets several thousand Angstrom across. It seems certain that the several thousand Angstrom correlation length is a lower limit because of the contributions to $\Delta\omega$ of "extrinsic" factors, such as the macroscopic sample mosaic ($0.02^\circ$) and the finite $2\theta$-resolution.

The most striking feature of the data in Figure 3 is the pronounced asymmetry evident in the reflectivity about the (002) reflection. The solid line in Figure 3 was obtained from Eq. 14 using the known atomic and crystal structure of bulk Au. It is clear that the model's prediction is substantially too low for $Q_z/c^* < 2$ and too high for $Q_z/c^* > 2$. Nor can the asymmetry be qualitatively understood as resulting from surface damage or contamination. The reflectivity at the anti-Bragg position from a rough surface is generally reduced from that of a smooth surface[29, 30, 31], which is clearly not

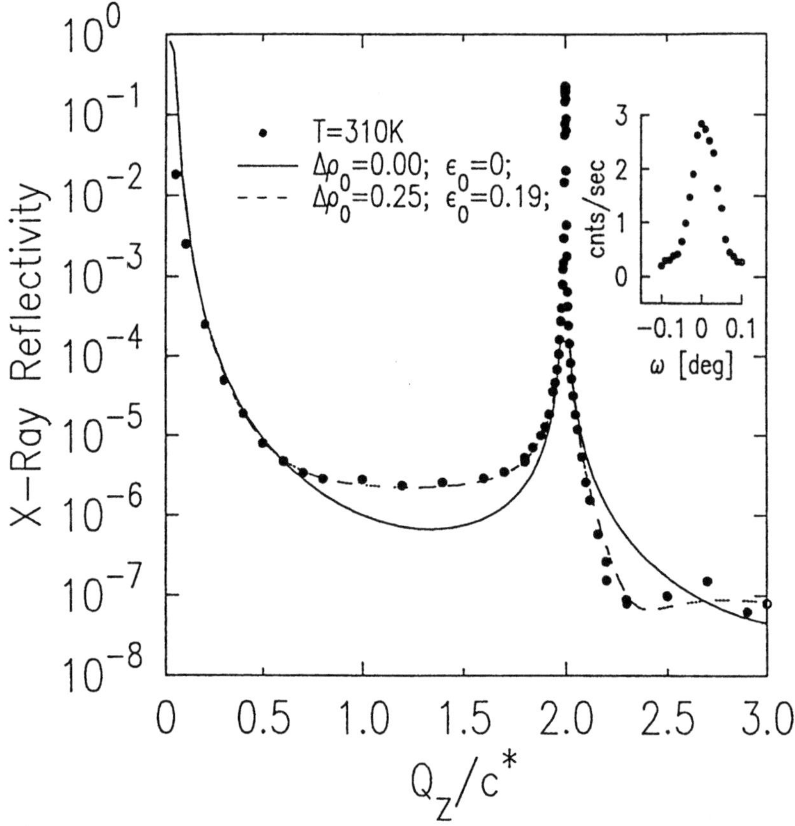


Figure. 3. Specular reflectivity of the clean Au(100) surface normalized to the incident flux at T=310K. The solid line corresponds to a real space model with an ideally terminated bulk as given by Eq. 14. The dashed line correspond to $\epsilon_0$=19%, an increased mass of $\Delta\rho_0$=25%, and a buckling amplitude of $\varsigma_0$=20% as described by Eqn(16). The inset corresponds to the rocking curve at $Q_z/c^*$=1.0

supported by the data.

The room temperature data can be described with simple real space models of the surface reconstruction where only the top layer has been modified. In the following we consider three different parameters:

- $\epsilon_0$ - spacing between the top and next layers
- $\Delta\rho_0$ - top layer excess mass density
- $F_0(Q_z)$ - Buckling induced roughness

We modify Eqn. (14) by rewriting the sum as

$$|f_0(Q_z)(1+\Delta\rho_0)e^{iQ_z d\epsilon_0} + \sum_{n=1}^{\infty} e^{iQ_z dn}|^2 \tag{16}$$

Here $f_0(Q_z)$ represent the Fourier transform of the charge distribution for the top layer. For a buckled top layer in the specular geometry

$$f_n(Q_z) = J_0(Q_z\varsigma_n), \tag{17}$$

where $J_0$ is the zeroth order Bessel function and $\varsigma_n$ is the buckling amplitude[33, 34]. This form distributes the charge along the surface normal direction by introducing a sinusoidal mass distribution in z into the nth layer, while preserving the center of mass of the layer. For small $\varsigma_n$, ie. $Q_z << \varsigma_n^{-1}$, the Bessel function and a Gaussian ($f_n(Q_z)=e^{-Q_z^2\varsigma_n^2/2}$) are both quadratic and differ only by a scale factor. Best fits to both forms are indistinguishable over the $Q_z$-range of the existing data. Since the Bessel function form is motivated by a simple real space model, we have chosen to use this form in the analysis. From the present data it is not possible to determine the sine wave period. We have also included a bulk Debye-Waller factor $W(Q_z) = -Q_z{}^2 <u^2>/2$ with $\sqrt{<u^2>}$ varying between 0.1 Å and 0.2 Å. Such values are appropriate for bulk Au in the temperature range studied.

The most important result from the least squares fitting is that of the three parameters $\epsilon_0$, $\Delta\rho_0$, and $\varsigma_0$ ,only the variation of $\epsilon_0$ can produce the observed asymmetry. The dashed line in Figure 3 shows the best fit obtained by varying $\Delta\rho_0$, $\epsilon_0$, and $\varsigma_0$ (and including all the data). This model, which gives an excellent fit, has an expansion of $\epsilon_0$=19%, an increased mass of $\Delta\rho_0$=25%, and a buckling amplitude of $\varsigma_0$=20%. It is noteworthy that the increased mass is consistent with the 5X20 model of Zehner[27]. The measured expansion, combined with the increased mass of the top layer, gives a mass density which is close to the bulk value.

It is interesting that no surface roughness has been required to fit the data, supporting the suggestion that the clean, annealed surface is atomically smooth over regions of at least several thousands of Angstroms in extent. In contrast, one would expect roughness to be important in a description of damaged or contaminated surfaces. Examples of these are shown in Figure 4. The reflectivity is very much lower near the anti-phase position than for the clean room temperature data. Near the Bragg peak the three data sets agree more

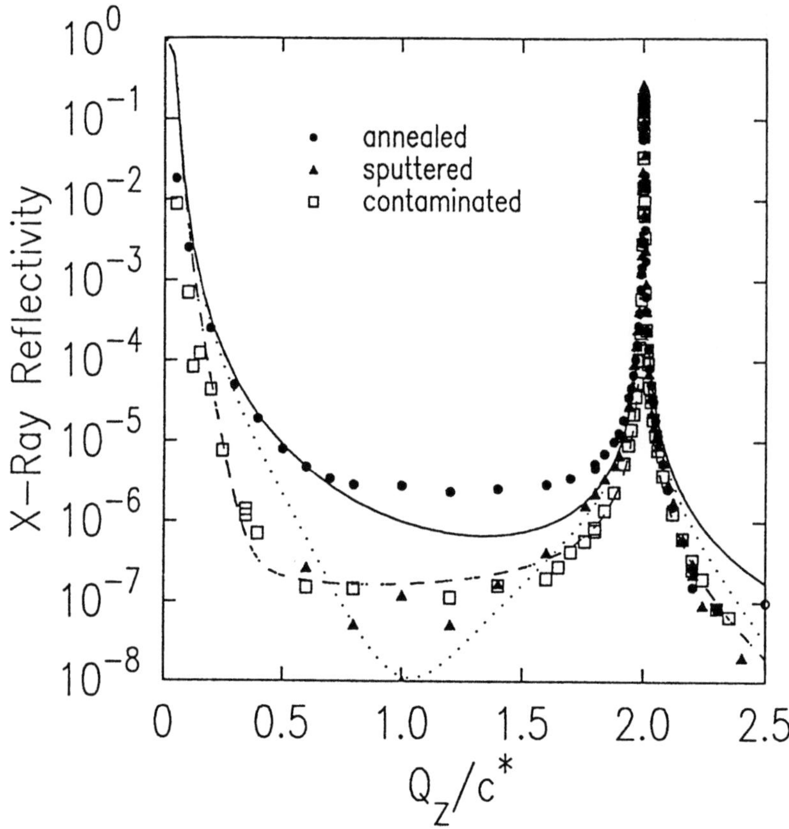


Figure 4. Specular reflectivity of the clean Au(100) surface at room temperature (filled circles), of a surface exposed to atmosphere for > 24 hours and then heated to 690K (open squares), and of a clean surface sputtered for 50 minutes at T=300K (filled triangles). The solid line corresponds to the ideally terminated bulk. The dotted line is the result of a roughness model given by Eq. 18 as described in the text. The dashed line is the result of a fit to Eq. 19.

closely, although not perfectly.

Several models have been considered to fit the annealed and sputtered data shown in Figure 4. In the first model, we have allowed for a variable occupation in the density of successive layers according to an error-function distribution. This is the discrete equivalent to the Gaussian model discussed in the previous section. To accomplish this, the sum in Eq. 14 is replaced by

$$\left| \sum_{n=-\infty}^{\infty} 1 - erf\,(\sigma(dn - z_0))e^{iQ_z dn} \right|^2 \tag{18}$$

where $\sigma$ is the width of the distribution and $z_0$ is the position. For this model it is assumed that all of the atomic layers are uniformly spaced. The result of this model with $\sigma$=3 Å and $z_0$=1.5 Å is shown as the dotted curve in Figure 4. This model does not adequately describe the contaminated or sputtered data.

Rather than construct an elaborate real space model, we have chosen to modify the sum of Eq. 15 to include two charge-density waves,

$$\left| \frac{e^{-\sigma_0^2 Q_z^2/2}}{Q_z} - \frac{e^{2\pi i \phi} e^{-\sigma_2^2 (Q_z - 2c^*)^2/2}}{Q_z - 2c^*} \right| . \tag{19}$$

One density wave corresponds to $Q_z$=0 and the other corresponds to $Q_z=2c^*$ which should be sufficient in the region relevant to our experiments. Each, however, is modified by its own Gaussian of widths, $\sigma_0$ and $\sigma_2$, respectively. In simple models[10], $\sigma_0$ can be identified with the r.m.s. roughness of the interface between the metal and the vacuum, while $\sigma_2$ is the r.m.s. roughness of the interface between the ideally terminated crystal and the overlayer above it. We allow for a variable phase, $\phi$,which for perfect termination is zero (Eq. (18)). The fit of the reflectivity data from a surface exposed to air for > 24 hours (open squares) to this form (dashed line) is shown in Figure 4. The best fit parameters are $\phi$=0, $\sigma_0$=4.5 Å, $\sigma_2$=0.5 Å, and $<u^2>^{1/2}$=0.25 Å. Thus, we arrive at the physically appealing picture of a damaged, rough surface layer (r.m.s. roughness 4.5 Å), in which Au atoms are not positioned in lattice planes and are separated from the undamaged bulk by a relatively smooth interface (0.5 Å r.m.s. roughness).

In summary, by the use of absolute reflectivity techniques we have shown that atomically smooth Au(100) facets may be prepared over length scales of several thousand Angstroms. The asymmetric angular dependence of clean annealed surfaces is consistent with an expanded, possibly buckled, hexagonal surface structure. A more sophisticated analysis of the out-of-plane structure will accompany future in-plane studies.

## 4. LIQUID/SOLID INTERFACES (8OCB)

Due to the penetrating nature of x-rays, the characteristics of a liquid/solid interfaces can also be investigated using x-ray reflectivity techniques. In this case, the scattering/reflectivity must be considered upon entering the fluid layer and upon impinging on the liquid/solid interface. In order to account for absorption losses in the liquid layer, it is important to utilize a geometry where these effects can be understood simply. Therefore, we have chosen a cylindrical "transmission" geometry for these reflectivity measurements. In this geometry, the liquid is placed in a cylinder which is of the order of the x-ray absorption length. For molybdenum radiation($\lambda$=0.71 Å) and for hydrocarbons the absorption length is ~1 cm. Within the liquid immersed cylindrical sample cell there is a concentric silicon disk. The x-rays enter the liquid and are then reflected from the liquid/silicon interface because of the change in the index of refraction in going across the interface.

Since the sample is rather small by reflectivity standards, we were not able to limit the incident beam size to avoid the extremes of the x-ray beam from spilling over the sample. Instead, it was assumed that the incident flux scaled as the incident angle and the reflectivity was appropriately scaled. The intensity scale factor was obtained by least-squares analysis.

In figure 5(circles), the measured reflectivity is shown for the liquid crystal material 4-cyano-4'octyloxybiphenol(8OCB) at the interface of an alkylsiloxane monolayer on a silicon substrate at T= 66.6 °C. We have chosen to study 8OCB since surface effects have been very well characterized at the air/liquid interface by Pershan and coworkers[11]. At this temperature, 0.3 °C above the bulk smectic phase, there is smectic layering induced by the interactions with the alkylsiloxane surface. The dashed line in figure 5 is the expected reflectivity from Eq. 4 using the critical angle for the silicon substrate($q=4\pi/\lambda \sin[\theta]$). This is what would be expected at small q-vectors from the air/silicon interface. This model does not agree with the data since the dielectric constant of the liquid is not unity.

The actual change in the dielectric constant(free electron density) at the liquid crystal/silicon interface is considerably smaller then for the air/silicon interface. Since the critical angle scales as the square root of the density change, it can be shown that the critical angle at the liquid/solid interface is given by

$$q_c^2[l/s]=q_c^2[s]-q_c^2[l] \tag{20}$$

where $q_c[l/s]$ is the critical wavevector of the liquid/solid interface, $q_c[s]$ is the critical wavevector of the solid, and $q_c[l]$ is the critical wavevector of the liquid. For the silicon/8OCB interface, $q_c$[silicon]=0.0316 Å$^{-1}$, $q_c$[8OCB]=0.0217 Å$^{-1}$ and $q_c$[8OCB/silicon]=0.0230 Å$^{-1}$. The Fresnel law, given by Eq. 4 with $\theta_c=q_c$[8OCB/silicon]$(\lambda/4\pi)$ is shown as the dotted line in figure 5. At small q-vectors the data agrees very well with the predictions of the model. At q-vectors of order 0.2 Å$^{-1}$ there are deviations between the Fresnel law and the data. This can be understood in terms of the smectic layering induced by the broken symmetry at the interface. As the nematic to smectic A transition temperature is approached this peak increases in intensity and narrows corresponding to an increasing number of surface smectic layers. This result clearly demonstrates that the x-ray reflectivity technique is applicable at a liquid/solid interface.

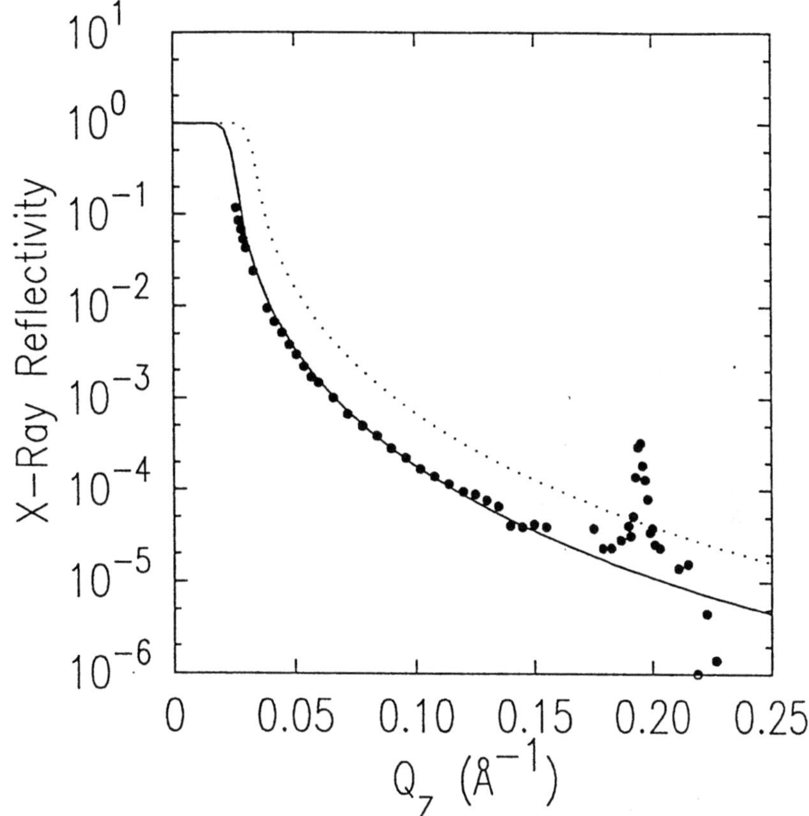


Figure 5. Specular reflectivity of the interface between the liquid crystal 8OCB and an alkylsiloxane coated silicon surface(circles). The dashed line is the Fresnel law with the critical angle of silicon($q_c$=0.317 Å$^{-1}$). The solid line is the Fresnel law with the critical angle for the 8OCB/silicon interface($q_c$=0.230 Å$^{-1}$). The deviation between the solid line and the data at ~0.2 Å$^{-1}$ is a consequence of smectic layering at the interface.

## 5. SUMMARY

The technique of X-ray reflectivity provides detailed information about the structure along the surface normal direction. In this paper we have considered air/liquid, vacuum/solid, and liquid/solid interfaces. The ability to probe liquid/solid interfaces should be particularly valuable in understanding the electrode/electrolyte interface in the future.

The measurements on $CCl_4$ were done in collaboration with A. Braslau, P.S. Pershan, G. Swislow, and Jens Als-Nielsen. These measurements were in part supported by grants from the U.S. National Science Foundation through grants DMR-85-13523 and DMR-86-14003, from the Joint Services Electronics Program through grant N00014-75-C-0648, the Danish National Science Foundation, and the RIS National Laboratory. The measurements on Gold were done in collaboration with Doon Gibbs, S.G.J. Mochrie, and D.M. Zehner. The experiments on the liquid crystal/solid interfaces were done in collaboration with Larry Sorensen. Work performed at Brookhaven National Laboratory is supported by the Division of Materials Research, U.S. Department of Energy, under Contract No. DE-AC02-76CH00016.

## References

1. Richard Hoffman, NATO Advanced Study Institute on Spectroscopic and Diffraction Techniques in Interfacial Electrochemistry.
2. L. Blum, H.D. Abruna, J. White, J.G. Gordon, G.L. Borges, M.G. Samant, and O.R. Melroy, *J. Chem. Phys.*, vol. 85, p. 6732, 1986.
3. G. Materlik, J. Zagenhagen, and W. Uelhogg, *Phys. Rev. B*, vol. 32, p. 5502, 1985.
4. Mahesh G. Samant, Michael F. Toney, Gary L. Borges, Lesser Blum, and Owen Melroy, *J. Phys. Chem*, vol. 92, p. 220, 1988.
5. I. K. Robinson, *Phys. Rev. B*, vol. 33, p. 3830, 1986.
6. L. Bosio and M. Oumezine, *J. Chem. Phys.*, vol. 80, p. 959, 1984.
7. B.C. Lu and S.A Rice, *J. Chem. Phys.* , vol. 68, p. 5558, 1978.
8. D. Sluis and S.A. Rice, *J. Chem. Phys.* , vol. 79, p. 5658, 1983.
9. A. Braslau, M. Deutsch, P.S. Pershan, A.H. Weiss, J. Als-Nielsen , and J. Bohr, *Phys. Rev. Lett.*, vol. 54 , p. 114 , 1985.
10. A. Braslau, P.S. Pershan , G. Swislow, B.M. Ocko, and J. Als-Nielsen , *Phys. Rev. A*, vol. 38, p. 2457, 1988.
11. P.S. Pershan, A. Braslau, A.H. Weiss, and J. Als-Nielsen , *Phys. Rev. A.*, vol. 35, p. 4800, 1987.
12. B.M. Ocko, A. Braslau, P.S. Pershan , J. Als-Nielsen , and M. Deutsch, *Phys. Rev. Lett.* , vol. 57, p. 94, 1986.
13. E.F. Gramsbergen, W.H. de Jeu, and J. Als-Nielsen, *J. Phys.(paris)*, vol. 47, p. 711, 1986.

14. D.K. Schwartz, A. Braslau, B.M. Ocko, P.S. Perhsan, J. Als-Nielsen, and J.S. Huang, *Phys. Rev. A*, 1988. accepted for publication
15. C.A. Helm, H. Mohwald, K. Kjaer, and J. ALs-Nielsen, *Europhys. Lett.*, vol. 4, p. 697, 1987.
16. S.G. Wolf, L. Leiserowitz, M. Lahev, M. Deutsch, K. Kjaer, and J. Als-Nielsen, *Nature*, vol. 328, p. 63, 1987.
17. R.M. Richardson and S.J. Roser, *Liquid Crystals*, vol. 2, p. 797, 1987.
18. Doon Gibbs, B. M. Ocko, D. M. Zehner, and S. G. J. Mochrie, *Phys. Rev. B*, vol. 38, p. 7303, 1988.
19. B. M. Ocko and S. G. J. Mochrie, *Phys. Rev. B*, vol. 38, p. 7378, 1988.
20. R.A. Cowley and T.W. Ryan, *J. Phys. D; Appl. Phys*, vol. 20, p. 61, 1987.
21. M. Pomerantz, A. Segmuller, L. Netzer, and J. Sagiv, *Thin Solid Films*, vol. 32, p. 153, 1985.
22. I.M. Tidswell, B.M. Ocko, P.S. Pershan, S.R. Wasserman, G.W. Whitesides, and J.D. Axe. submitted to Phys. Rev. B
23. J.D. Jackson, in *Classical Electrodynamics*, p. 281, John Wiley & Sons(New York), 1975.
24. R. Loudon, "Riples on Liquids Interface," in *Surface Excitations*, ed. V.M. Agranovich and R. Louden, p. 589, Elsevier Science Publishers B.V., 1984.
25. D.G. Fedak and N.A. Gjostein, *Phys. Rev. Lett.*, vol. 16, p. 171, 1966.
26. A.M. Mattera, R.M. Goodman, and G.A. Somorjai, *Surf. Sci.*, vol. 7, p. 26, 1967.
27. D. M. Zehner, B. R. Appleton, T. S. Noggle, J. W. Miller, J. H. Barret, L. H. Jenkins, and O. E. Schow III, *J. Vac. Sci. Technol.*, vol. 12, p. 454, 1975.
28. M.A Van Hove, R.J. Koestner, P.C. Stair, J.B. Biberian, L.L. Kesmodel, I. Bartos, and G.A. Somorjai, *Surf. Sci.*, vol. 103, p. 189, 1981.
29. S.R. Andrews and R.A. Cowley, *J. Phys C*, vol. 18, p. 6427, 1985.
30. D.A. Bruce, *J. Phys. Chem C*, vol. 14, p. 5195, 1981.
31. A.M. Afanas'ev, P.A. Aleksandrov, S.S. Fanchenko, V.A. Chaplanov, and S.S. Yakimov, *Acta. Cryst. A*, vol. 42, p. 116, 1986.
32. I. S. Gradshteyn and I. M. Ryzhik, *Table of Integrals, Series, and Products*, p. 36, Academic Press, New York, 1965. By including an infinitesimal imaginary component to $Q_z$ this sum can be performed exactly: $|\sum_{n=0}^{n=\infty} e^{iQ_z dn}|^2 = |1/\sin(Q_z d/2)|^2$. Then using Eqn. (1.422.3) of this reference i.e. $1/\sin\pi z = \frac{1}{\pi}\sum_{n=-\infty}^{n=\infty}(-1)^n/(z-n)$, we obtain our Eq. (15).
33. J.D. Axe, *Phys. Rev. B*, vol. 21, p. 4181, 1980.
34. One obtains a zeroth order Bessel function if $Q_x = Q_y = 0$.

# SURFACE STRUCTURAL INVESTIGATIONS BY ELECTRON DIFFRACTION TECHNIQUES

D. M. KOLB, G. LEHMPFUHL AND M. S. ZEI
*Fritz-Haber-Institut der Max-Planck-Gesellschaft*
*Faradayweg 4-6*
*D-1000 Berlin 33*
*West Germany*

**ABSTRACT. Detailed information on surface structures can be obtained by the use of electron diffraction techniques, such as LEED or RHEED. Their application to electrochemical problems, however, requires that the electrode is transferred from the electrochemical environment into a UHV chamber. Since the controlled emersion of electrodes is the basis of any successful ex-situ study, this process will be discussed first, followed by a brief introduction to the basic principles of electron diffraction at surfaces. Several selected examples of electrode surface studies are then presented, which demonstrate the power of electron diffraction techniques in assessing the structures of bare and adsorbate-covered surfaces.**

## 1. Introduction

In recent years single-crystal electrodes have been increasingly employed in electrochemical studies [1-4] for a number of reasons. In order to derive a microscopic picture of the electrode-electrolyte interface, the use of single-crystal electrodes with their atomically flat and well-defined surface structure is indispensable. Furthermore, many electrochemical reactions are found to depend strongly on the surface structure, especially when the adsorption of reaction intermediates or products is involved. The use of single-crystal material allows a detailed study of the structure specificity of reactions and hence provides the key to a mechanistic interpretation.

Electrochemistry with single-crystal material, however, requires surface structure determination not only before the electrochemical experiment, i.e., for surface preparation, but also afterwards in order to detect structural changes such as restructuring, facetting or roughening which might have occurred in the course of electrochemical manipulation. When studying adsorbate-covered surfaces, structural information about short- or long-range order is essential for a complete characterization of adlayers or thin film deposits, but is almost impossible to obtain by conventional electrochemical techniques. The obvious way to proceed in structure studies is to employ electron diffraction methods, which, however, require ultrahigh vacuum (UHV) conditions and hence cannot be used in-situ. In order that these techniques can be applied, the electrode has to be transferred from the electrochemical cell into a vacuum chamber.



*C. Gutiérrez and C. Melendres (eds.),*
*Spectroscopic and Diffraction Techniques in Interfacial Electrochemistry*, 361–382.

## 2. Electrode Emersion: The Crucial Link between Electrochemistry and UHV Techniques

A successful emersion process should meet the following two requirements: (i) the condition of the electric double layer of the emersed electrode (with or without adsorbates) and the surface structure must be the same as in-situ; (ii) there should be no noticeable amount of bulk electrolyte on the emersed electrode. This is called "dry emersion", where the break between electrode and electrolyte is very close to the outer Helmholtz plane. (For a thorough discussion of the emersion process, see ref. 5.)

In order to ascertain the fulfillment of the first requirement, we need criteria by which the state of the emersed double layer can be tested against the in-situ state. The electric double layer is usually characterized by the electrode potential U, the surface charge $q_m$ of the metal, and the chemical composition, given by the surface concentrations $\Gamma_{\pm}$. The potential U of the emersed electrode can be determined by its work function $\Phi$, since U and $\Phi$ represent one and the same quantity: the electrochemical potential of the electrons in the metal. For example, a change of the electrode potential U by 1 V corresponds to a $\Delta\Phi$ of exactly 1 eV. Secondly, the surface charge $q_m$ of an electrode in the emersed state can be determined either by surface resistance measurements [6] or, less conveniently and less accurately, by electroreflectance [7]. Both effects have been shown to depend linearly on $q_m$ [8,9]. And finally, the chemical composition of the emersed double layer is readily determined by ESCA measurements. In Fig. 1 is shown the work function $\Phi$ of an emersed gold electrode as a function of emersion potential [10]. In this case $\Phi$ was obtained under UHV conditions by ultraviolet photoelectron spectroscopy

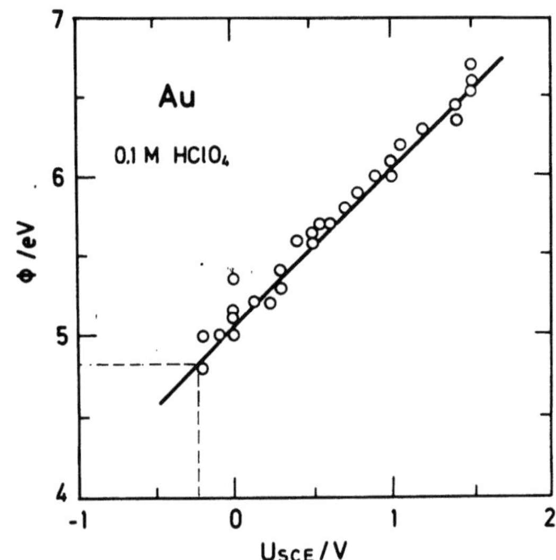


Figure 1. Work function $\Phi$ of an emersed gold electrode in UHV as a function of emersion potential. 0.1 M $HClO_4$. Dashed lines indicate $\Phi$-value for emersion at 0 V vs. NHE. After ref. [10].

(UPS) from the cut-off energy of the secondary electrons. The strictly linear correlation of unity slope between $\Phi$ and U demonstrates that indeed the potential drop across the double layer is retained during and after emersion (see also refs. [11,12]), and even after transfer of the electrode into a UHV chamber. A second point is noteworthy: the measurement of the (absolute) work function by UPS allows a determination of the absolute electrode potential $U^{o}_{NHE}$, which corresponds to the work function of an electrode emersed at 0 V vs NHE, corrected for the dipole contribution of the electrolyte-air interface. The latter quantity is not included in $\Phi$ if the electrode is emersed dry, i.e., without a layer of bulk electrolyte.

Fig. 2 shows the result of an ESCA study on emersed gold electrodes, where the surface concentrations of cations and anions in the case of halide adsorption were determined ex-situ [13]. Chloride, bromide and iodide ions are known to adsorb "specifically" on gold [14], i.e., these ions form a direct chemical bond with the electrode surface under partial loss of their solvation shell. This is sometimes called contact adsorption. The surface concentrations of the halide ions on polycrystalline gold as a function of emersion potential are given in Fig. 2b. These curves represent directly the respective adsorption isotherms for a 0.01 M CsX solution, with $X^-$ = $Cl^-$, $Br^-$ or $I^-$. An interesting and well established consequence of specific adsorption is demonstrated in Fig. 2a, where the concentration of the counter ions (in our case $Cs^+$) in the double layer is shown as a function of emersion potential and compared with the situation for non-specific adsorption (dashed curve in Fig. 2a). In the latter case, the $Cs^+$ concentration in the double layer is strictly governed by electrostatics, which causes the surface concentration of the cation to increase continuously with negative charging of the electrode and to vanish for positive charging. This situation is distinctly different for specific adsorption, where the chemical interaction between halide ion and Au substrate partly overrules the electrostatic repulsion. Hence the anions can adsorb already at potentials negative of the pzc (potential of zero charge) which for polycrystalline gold was found to be around -0.05 V vs. SCE. The potentials at which halide adsorption from 0.01 M solution starts are about -0.2, -0.6 and -1.2 V for $Cl^-$, $Br^-$ and $I^-$, respectively, the values reflecting the increasing bond strength when going from chloride to iodide. As a direct consequence of chemical bonding, more negative charge is brought into the double layer by the specifically adsorbed anions than is required by electrostatics. Therefore, specific adsorption is always accompanied by the accumulation of counter ions in the double layer for charge compensation [15]. This effect is clearly seen in Fig. 2, where the $Cs^+$ concentration increases - relative to the case of non-specific adsorption (dashed curve in Fig. 2a) - as soon as the halide adsorption starts (e.g., at -0.6 V for $Br^-$ adsorption). In the absence of any specific anion adsorption at very negative potentials, the $Cs^+$ concentration becomes independent of the anion and all the $\Gamma_{Cs^+}$ curves merge with the curve for non-specific adsorption. From a visual inspection of the extra amount of $Cs^+$ brought into the double layer by the halide, it appears as if the number of counter ions per adsorbed halide decreases after a certain coverage has been reached. This is especially clearly seen for $I^-$ adsorption, but also for the $Br^-$ case. This behaviour of $\Gamma_{Cs^+(X^-)} - \Gamma_{Cs^+(SO_4^{=})}$ can be understood by the decreasing ionicity of the adsorbed halide with increasing coverage, a fact which is also reflected in the coverage dependence of the respective electrosorption valencies [16].

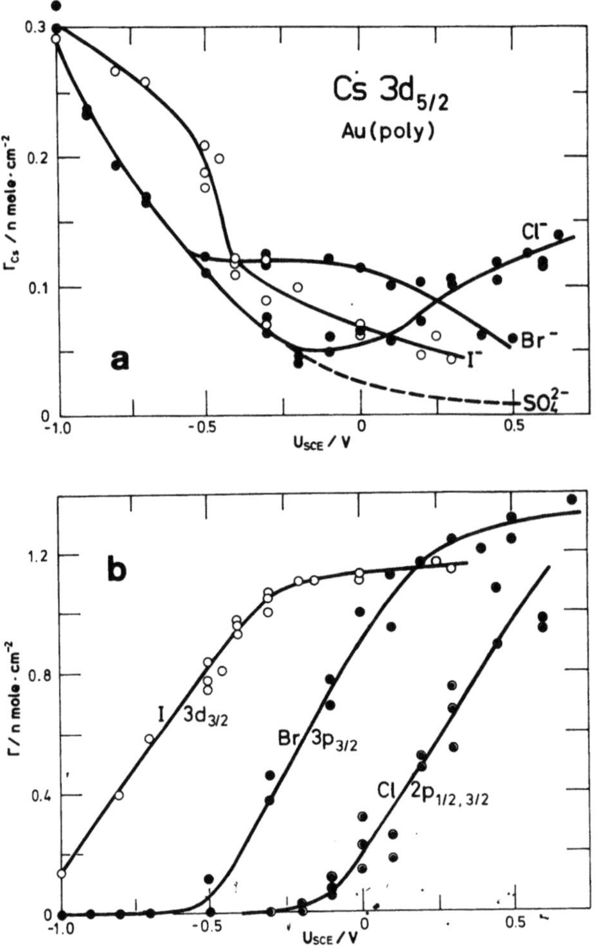


Figure 2. Surface concentrations of (a) $Cs^+$ and (b) $Cl^-$, $Br^-$ and $I^-$ on gold, derived from ESCA measurements, as a function of emersion potential. 0.01 M cesium halide solutions. For comparison, the result for $Cs^+$ in the case of $Cs_2SO_4$ is also shown in (a). After ref. [13].

It having been shown that an electrode can be emersed with its double layer apparently intact, the question immediately arises regarding the stability of the emersed double layer. This question can be answered most conveniently by work function measurements, which allow monitoring of the electrode potential outside the electrochemical cell. A requirement for a stable electric double layer in the emersed state is the absence of faradaic reactions. The latter, if present during the emersion process, will continue to consume electric charge after loss of the potentiostatic control, and hence such reactions will cause the emersed double layer to collapse. This is demonstrated in Fig. 3 for

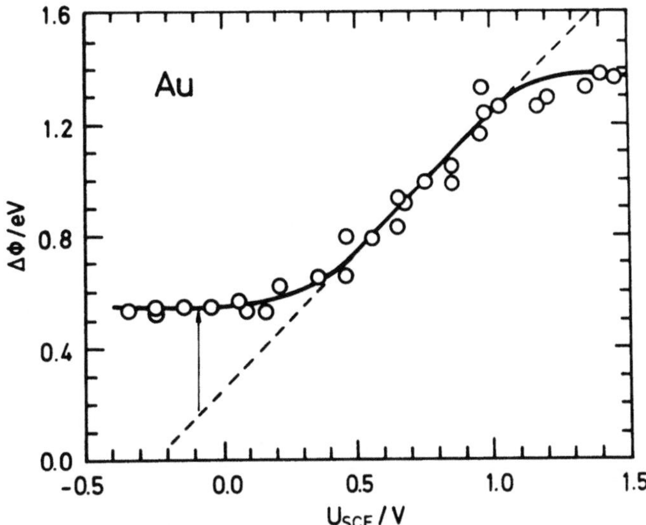


Figure 3. Work function change $\Delta\Phi$ of an emersed gold electrode with electrode potential. 0.1 M $NaClO_4$, air-saturated. The arrow illustrates the change in potential of the emersed electrode due to a collapse of the double layer by oxygen reduction. After ref. [5].

the emersion of Au from an air-saturated solution, where the reduction of oxygen for $U \leqq +0.4$ V is clearly seen to cause a deviation of the $\Delta\Phi(U)$ plot from a straight line [5,11,12]. In this case, the work function $\Delta\Phi$ of the emersed gold electrode was determined by a Kelvin probe (vibrating capacitor) which was mounted above the electrolyte level in an argon atmosphere [11]. The deviations of $\Delta\Phi(U)$ from the linear correlation of unity slope signal a drift of the electrode potential in the emersed state to a value at which the reaction ceases (rest potential). This drift at a given emersion potential is schematically indicated by the arrow in Fig. 3. A similar deviation is found at very positive potentials where obviously the oxidation of some impurities in the solution renders the double layer unstable for $U > 1.1$ V vs. SCE. It could be shown that in an oxygen-free solution the stability range of the emersed double layer is extended considerably to more negative potentials. Likewise the careful exclusion of oxidizable species in the electrolyte allows an extension of the double layer stability to very positive potentials (e.g., up to 1.5 V [10]). Simple adsorption reactions (including the surface oxide formation for $U \geqq 1.0$ V) do not cause deviations from the 1:1 correlation of $\Delta\Phi(U)$, because for a given potential the current drops to zero immediately after the desired coverage has been reached. It may be noteworthy that the double layer collapse occurs typically on a seconds or minutes time scale in the case of oxygen reduction [11]. This is surprisingly slow, and we may speculate that it is the absence of bulk water at the emersed surface, which would be needed for solvation of the reaction products, that causes the reduction process to occur at such a low rate. It has been demonstrated [12,17] that work function measurements are ideally suited to monitor electrochemical reactions outside a conventional cell, i.e., corrosion of a metal in a humid atmosphere.

The second important aspect of the emersion process is the so-called "dry emersion" [5]. It has been demonstrated that gold electrodes can be emersed dry from aqueous solutions, i.e., without noticeable amounts of bulk electrolyte, as judged by the unaided eye. In order to assess quantitatively the thickness of the residual water layer, various experiments were performed. Ellipsometric measurements were carried out for emersed gold in air which yielded an average thickness of the order of 1 nm [18]. A similar value was obtained by ESCA measurements for highly concentrated solutions (~ 5 M), where the thickness of the emersed electrolyte layer could be calculated from the amount of counter ions at the surface under potential conditions, where their concentration should be zero in the double layer [13]. Finally, surface plasmon excitation in the ATR configuration was employed to determine the thickness of the water layer on emersed Ag electrodes [19]. Again a value of about 1 nm was obtained under low-humidity conditions, the thickness value depending somewhat on the emersion potential. All these experiments support the view that for dry emersion the sheer plane is close to the outer Helmholtz layer, where a marked change in the water structure is expected to occur.

According to our experience, cleanliness and flatness (on an atomic scale) are the two most important prerequisites for the dry emersion of gold electrodes [5]. With respect to cleanliness, Auger electron spectroscopy (AES) should be able to give a decisive answer. Indeed, AES of gold single-crystal surfaces such as Au(111) and Au(100), obtained after electrochemical treatment, showed no carbon contamination, whereas the presence of the double layer was clearly seen in the respective Auger peaks (e.g., Cl peak in the case of perchlorate solution). Furthermore, the presence of the double layer caused only a slightly increased background in the corresponding LEED pattern, which was routinely obtained immediately after emersion and evacuation, with no further "cleaning" treatment. This very fact demonstrates the progress made in ex-situ studies because of a better understanding of the emersion process [5]. Several years back, it was quite frequently reported in the literature that a "mild" sputtering had to be applied to the emersed electrode in order to obtain a LEED pattern [20]. Needless to say, such a treatment removed all of the electrochemical information.

The importance of the surface topography for dry or wet emersion and for electrochemically induced transitions from the "hydrophobic" to the "hydrophilic" state of the electrode surface may be illustrated by the following observations. Gold electrodes which initially behaved hydrophobic (i.e., they were emersed dry), could be made hydrophilic by applying several oxidation-reduction cycles. Such a treatment is known to roughen the gold surface microscopically. Annealing the electrodes changed them back to hydrophobic [12]. Quite recently, it was shown by in-situ monitoring of electrode reactions with the quartz microbalance that a few oxidation-reduction cycles increased the coupling between Au and the electrolyte as seen by a marked decrease in the resonance frequency for the bare Au surface. This was interpreted as a roughness-induced transition of the Au surface from hydrophobic to hydrophilic [21]. A change in the surface topography (e.g., by roughening) obviously can cause a change of the interfacial tension such that the roughened surface becomes hydrophilic and therefore strongly couples to the liquid phase.

Although a clean smooth gold surface acts hydrophobic, other surfaces and other materials do not. We noticed, for example, that adsorbate-covered surfaces, particularly in the case of Ag, often behave hydrophilic. Wet emersion, which can cause substantial amounts of electrolyte to be withdrawn from the electrochemical cell, makes ex-situ studies always very difficult but not necessarily impossible. The electrolyte residues at the emersed surface usually form tiny crystallites after evacuation (see e.g., Fig. 11) which distort the LEED patterns (e.g., increased background intensity) and give rise to so-called transmission spots in RHEED. More seriously, a quantitative evaluation of surface concentrations by, e.g., ESCA, is no longer possible, because the bulk electrolyte now contributes substantially to the signal. In the case of persistent wet emersion, the following two procedures helped in obtaining reliable information from the electrochemical interface. Firstly, the electrolyte concentration was kept relatively low, typically 0.01 M, in order to reduce the amount of electrolyte residues. For the study of strongly bound adsorbates, such as metal atoms or halides, even much lower electrolyte concentrations are possible. However, since for dilute solutions only the compact double layer is retained in the emersion process, whereas the diffuse part is lost [11], double-layer studies will require electrode emersion from relatively concentrated electrolytes, i.e., $c \geqq 0.01$ M. Secondly, we found that droplets adhering to the electrode surface after emersion are conveniently removed by a glass capillary which was attached to the electrolyte vessel (see Fig. 4). Other groups reported a successful removal of droplets after emersion by blowing them off the surface with a stream of argon [22].

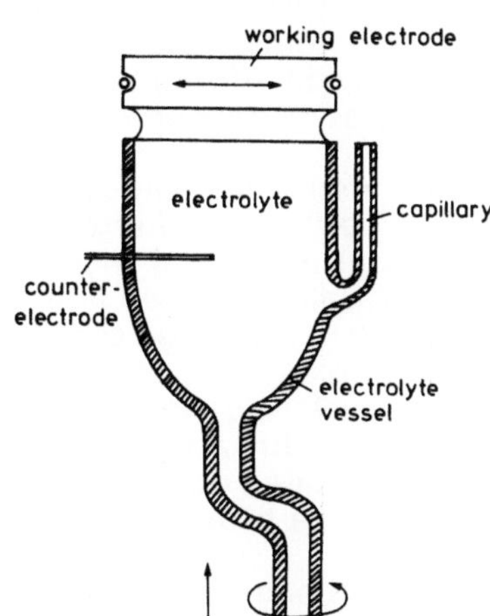


Figure 4. Sketch of the electrochemical cell for ex-situ studies of electrode surfaces. The capillary is used to remove droplets of electrolyte in case of wet emersion. The arrows indicate possible movements of electrolyte vessel and working electrode.

## 3. Basic Considerations of Electron Diffraction at Surfaces

Electron diffraction results from the wave nature of the electrons. According to the de Broglie equation, the wavelength λ of an electron with kinetic energy E is given approximately by [23]

$$\lambda\,[\text{Å}] = (150/E\,[\text{eV}])^{1/2} \tag{1}$$

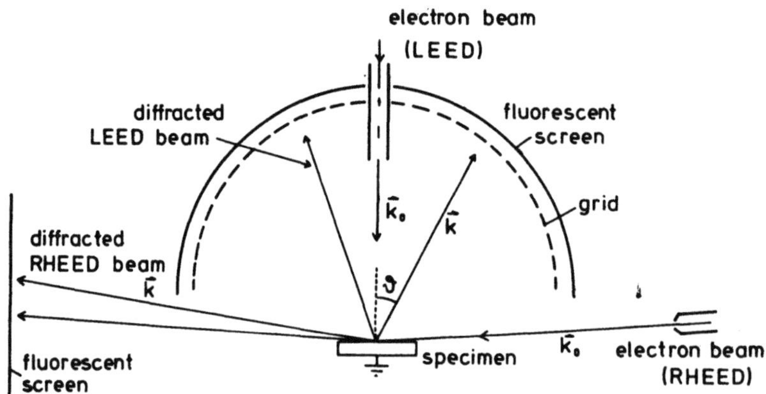


Figure 5. Schematic diagram of a set-up for surface studies by LEED and RHEED. $\vec{k}_0, \vec{k}$: wave vectors of incoming and diffracted beams.

In LEED experiments, the electrons impinging on the surface have energies typically in the range between 10 and 500 eV, which corresponds to wavelengths between 4 and 0.5 Å, i.e., of the order of interatomic distances. Thus, electrons can be diffracted from a crystal surface to provide information about periodic surface structures in very much the same way as diffraction of light at a grating does.

The experimental set-up for a LEED experiment is schematically shown in Fig. 5 [23]. A monoenergetic beam of electrons ($\Delta E \approx 1$ eV) in the energy range from 10 to 500 eV hits the surface of a single crystal at normal incidence. Roughly 1 - 5 % of the incoming electrons will be elastically scattered and give rise to an interference pattern on the fluorescence screen, the Bragg condition ( $a \sin \vartheta = n\lambda$, with interatomic distance a and scattering angle $\vartheta$) determining the directions of the interference maxima. Hence the surface geometry can be derived from the positions of the diffraction spots. Sharp diffraction spots are obtained only if the regions of well ordered atoms (so-called domains) are of the size 100 x 100 $Å^2$ or larger. Smaller domains result in a broadening of the spots.

Diffraction experiments, such as LEED, inherently image the surface lattice in reciprocal or *k*-space, and therefore the diffraction pattern has to be inverted in order to obtain the surface structure in real space. There are certain well established procedures for doing this [23-25]. In Fig. 6 surface structures and the corresponding LEED notations are shown for a few cases. The periodic structure of periodicity a (e.g., interatomic distance) of a surface or an overlayer is associated with the reciprocal lattice vector, which is proportional to 1/a; hence the larger the periodicity a, the closer the LEED spots.

In the case of an adsorbate-covered surface, ordered adsorption may give rise to so-called superstructure spots from which information about the lateral arrangement of the adatoms or admolecules in the overlayer can be derived. However, the LEED pattern proper can only give an answer about the "unit cell" (i.e., the lateral distance of the adatoms in units of the atomic distance of the underlying substrate, see Fig. 6) but not about the exact position. For example, the three possible arrangements of adsorbed particles on a (100) sur-

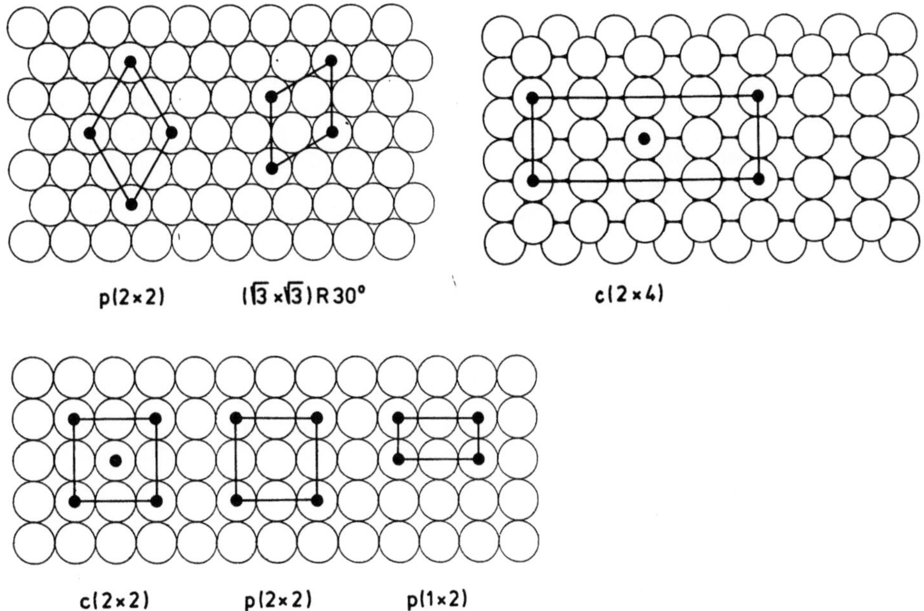


Figure 6. Sketches of commonly observed superstructures for adlayers on the three low-index faces of an fcc metal. Top:(111) and (110); bottom : (100). The corresponding unit cells are also shown.

face of a fcc crystal, as demonstrated in Fig. 7 (on-top, bridge position, fourfold hollow), all yield the same geometrical LEED pattern, namely the so-called c(2x2). The exact position of surface atoms can only be determined by analyzing LEED intensities as a function of the electron energy [23-25]. This, however, is far beyond the scope of this chapter.

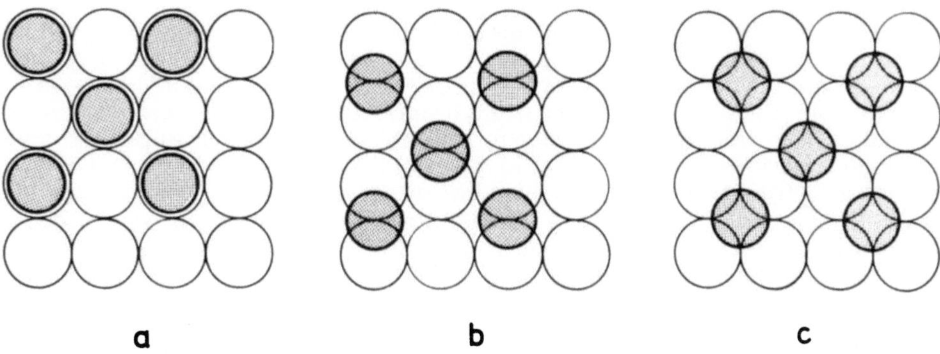


Figure 7. Three different positions for adatoms in a c(2x2)-superstructure (on-top, bridge and fourfold hollow), which yield identical diffraction patterns.

Besides LEED (Low Energy Electron Diffraction), a second powerful diffraction technique is RHEED (Reflection High Energy Electron Diffraction), where high-energy electrons of typically 40 keV are diffracted off the surface at grazing angles of incidence (see Fig. 5). In contrast to the two-dimensional image of the surface obtained from a LEED pattern, a RHEED pattern, in essence, yields information along one crystallographic surface direction, namely perpendicular to the direction of the incident beam. In order to obtain the complete information about the surface structure from RHEED, a number of patterns must be taken at various azimuthal angles. Despite the somewhat lower surface sensitivity caused by the greater penetration depth of the high-energy electrons, RHEED has a number of distinct advantages over LEED: (i) RHEED can yield information about the surface layer as well as the bulk (e.g., Kikuchi lines [26]). In addition, the lattice constants of a surface layer can be determined more precisely from RHEED than from LEED patterns. (ii) Surface roughness, epitaxially grown deposits or amorphous overlayers give rise to transmission spots or halo patterns and hence may be studied to some extent by RHEED, which would be impossible with LEED. (iii) A structural disorder (e.g., by thermal movement) reduces the intensity of Bragg reflections, but affects RHEED to a lesser extent than LEED. For example, a root-mean-square deviation of about 0.2 Å causes the LEED pattern of an overlayer to disappear, while in RHEED the overlayer reflections are still visible [27]. This may be an important aspect in the study of emersed electrodes, where partial solvation or thermal movement of weakly bound species could induce a disorder normal to the surface. (iv) Because of the weak interaction of the high-energy electrons with matter, multiple scattering can be neglected in RHEED and the kinematic theory can be applied to a first approximation for the calculation of diffraction intensities (similar to the case of X-ray diffraction). This is contrasted by LEED, where multiple scattering events require the use of dynamic theories, which are much more difficult to handle. For more details the interested reader is referred to the literature [23].

## 4. Case Studies

### 4.1 SURFACE RECONSTRUCTION OF GOLD SINGLE-CRYSTAL ELECTRODES

It is well established that certain metal and semiconductor surfaces can reconstruct, i.e., the surface atoms acquire a different position than expected from a mere termination of the bulk, in order to minimize their surface energy. For example, the (100) surface of Pt, Ir and Au rearranges itself into the so-called (5x20) structure [28,29] which represents a hexagonal close-packed overlayer on top of the (100) bulk (Fig. 8). The (110) surface of these metals reconstructs into the (1x2) structure which consists of narrow (111) microfacets in the [110] direction [30]; and for gold even the densely packed (111) surface has been found to reconstruct into the so-called (1x23), which indicates that the surface atoms are slightly compressed in [110] direction [31,32] so that every 23rd surface atom is in registry with the underlying bulk.

In the above-mentioned cases, the clean reconstructed surface is thermodynamically more stable than the unreconstructed (1x1) surface, although the former is obtained in UHV only at elevated temperatures because of an activation barrier for surface reconstruction, which in the case of a (100) face was

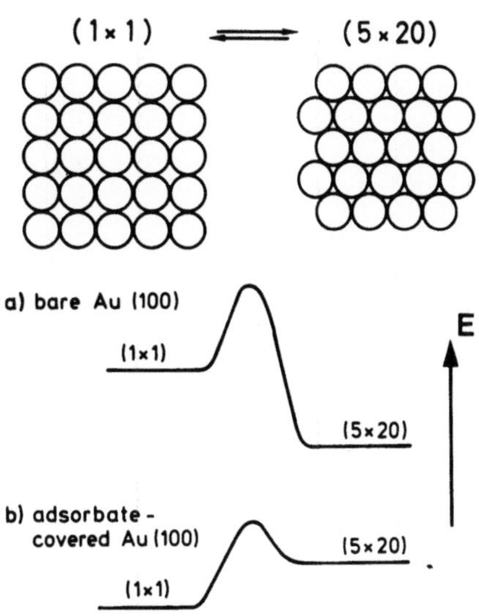


Figure 8. Surface structures for reconstructed (5x20) and unreconstructed (1x1) Au(100), with energy diagrams for the bare and the adsorbate-covered surface.

found to be of the order of 20 kcal/mol [33]. Adsorption of atoms and molecules, however, usually lifts the reconstruction, because very often adsorption is more favorable on the respective (1x1) surface, which is structurally more open than the reconstructed surface.

Because of the general observation made for surfaces in UHV that adsorbates tend to remove the reconstruction, it was not clear at all whether a reconstructed surface would be stable when in contact with an electrolyte. Although surface reconstruction had been postulated earlier to explain special features in cyclic voltammograms [34], it was not proven until recently that reconstructed surfaces exist in aqueous electrolytes. Yeager et al. [35] have shown by combining electrochemical studies with LEED investigations that a reconstructed Pt(110)-(1x2) surface, prepared in UHV, did withstand potential cycling in the double layer charging and hydrogen adsorption region, whereas Pt(100)-(5x20) apparently did not survive the contact with the electrolyte. More recently, we have demonstrated that all three reconstructed Au(hkl) surfaces are stable in aqueous electrolytes, and their properties can be studied over a surprisingly wide potential range [36-38].

Reconstructed Au(100)-(5x20) was prepared in UHV by sputtering and annealing, and tested by LEED and RHEED. The sample was then transferred to the electrochemical cell and subjected to different potential cycles. Finally, the electrode was emersed again and transferred back into the UHV chamber for a surface structural test by LEED and RHEED. The electron diffraction patterns obtained from the Au(100) surface before and after electrochemical manipulation provided conclusive evidence that the reconstructed Au(100) surface, the so-called (5x20) structure, is stable in an aqueous electrolyte as long as no

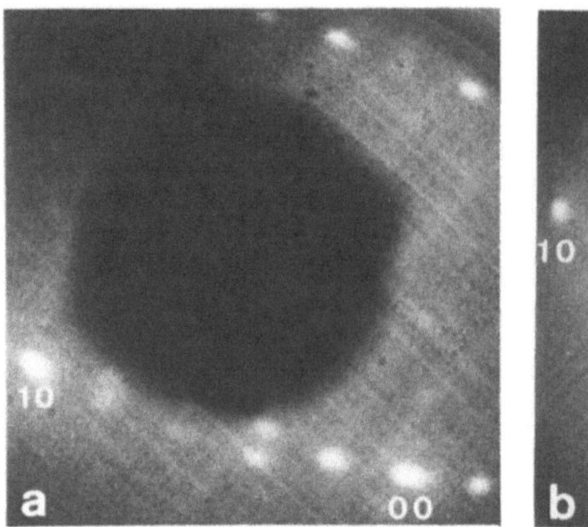


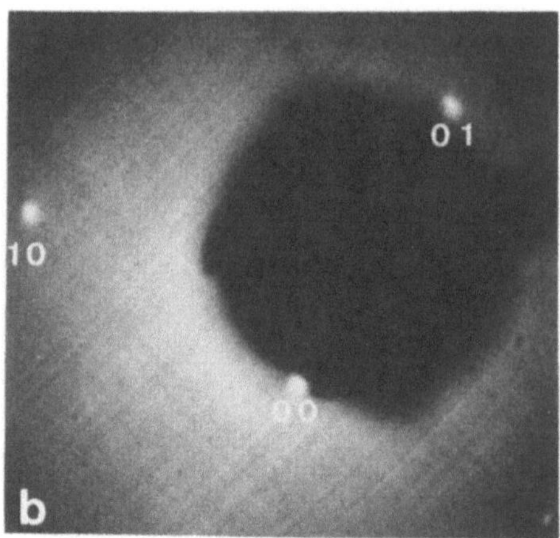


Figure 9. LEED patterns of an Au(100) electrode, which had been prepared in UHV to show the (5x20) reconstruction, after emersion from 0.01 M $HClO_4$. (a) After cycling the potential between -0.3 and +0.3 V vs. SCE. The surface is still (5x20). (b) After cycling between -0.2 and +0.7 V vs. SCE. The surface is now (1x1).

specific adsorption occurs [36]. This is illustrated in Fig. 9. In 0.01 M $HClO_4$ the stability range for Au(100)-(5x20) was limited by the onset of perchlorate adsorption around +0.55 V vs. SCE at the anodic side, while no such limit was found for the cathodic side of the potential cycle [37]. A single potential excursion beyond +0.55 V vs. SCE lifts the reconstruction due to specific adsorption of $ClO_4^-$, and the surface acquires the so-called (1x1) structure. After desorption of the perchlorate on the cathodic scan, the unreconstructed surface remains in its metastable state. Since the hexagonal close-packed (5x20) surface has about 20 % more surface atoms than the less densely packed unreconstructed Au(100)-(1x1), the (5x20) → (1x1) transition inevitably leads to a certain roughening of the surface.

It has been demonstrated by in-situ electrochemical and optical measurements that the surface properties of Au(100)-(5x20) and Au(100)-(1x1) differ markedly [37]. For example, it was observed that the pzc of the reconstructed Au(100) is 220 mV more positive than that of Au(100)-(1x1) and therefore close to - although not identical to - the pzc of Au(111). The electroreflectance spectra of reconstructed and unreconstructed Au(100) were also found to differ vastly from each other in their spectral shape. Optical transitions into empty electronic surface states, which are only seen for the unreconstructed Au(100) and not for the (5x20) structure, cause characteristic features for Au(100)-(1x1), which are completely absent in the spectrum for Au(100)-(5x20), the latter resembling closely that of Au(111). This is not unexpected, since both Au(111) and Au(100)-(5x20) have about the same surface structure, namely hcp.

We have recently shown that also flame-treated gold single-crystal surfaces are reconstructed [37]. This finding now makes possible the preparation of reconstructed gold electrodes simply by means of a Bunsen burner rather than by expensive UHV equipment which may not be available in every electrochemical laboratory. We should, however, mention that it is important to have the electrode connected to the potentiostat before contact with the electrolyte in order to retain the surface reconstruction, which requires a relatively negative electrode potential at the moment of immersion. Similarly, care has to be taken when emersing a reconstructed surface. Emersion at very negative potentials often causes the double layer to collapse due to reduction of spurious amounts of oxygen. This reduction process generates reaction products at the surface which obviously lift the (5x20) reconstruction. Therefore suitable emersion potentials for reconstructed Au(100) are typically slightly positive of 0 V vs. SCE, where the double-layer collapse is negligible and specific adsorption of anions not yet possible. In order to obtain high-quality reconstructed Au(100) surfaces by flame annealing, we found it advantageous if the crystal was not quenched too rapidly. As was demonstrated recently for Pt(111) by a reflection electron microscopy study [39], rapid quenching can introduce a considerable number of surface defects. For this reason, the gold crystal after flame annealing was allowed to cool down in air for 2 - 3 sec until the glowing had just disappeared, before it was quickly dipped into a beaker of high-purity water. With such a procedure the crystal is obviously still hot enough to prevent surface contamination by the laboratory air before it is immersed in water.

We recall that in UHV the bare Au(100)-(1x1) surface will transform into the (5x20) structure only after heating the crystal above 100 °C in order to overcome the activation barrier for reconstruction [28]. At the metal-electrolyte interface, however, we found that the Au(100) surface can reconstruct - at least partly - even at room temperature, if a sufficiently negative electrode potential is applied. Obviously,the activation barrier for surface reconstruction is influenced by the electrode potential and can be lowered significantly by charging the surface negatively. Such a potential-assisted surface reconstruction of Au(100), which has been studied recently in great detail by electrochemical and optical methods [40], is demonstrated by the sequence of LEED patterns in Fig. 10. A freshly prepared Au(100)-(5x20) electrode was transferred from the UHV chamber into the electrochemical cell containing 5 mM $H_2SO_4$. After potential cycling between -0.35 and +0.25 V, the surface was still reconstructed (Fig. 10a). The electrode was then transferred again into the cell, where potential excursions up to +0.45 V were applied, i.e., into the region of sulphate adsorption. The subsequent LEED examination (Fig. 10b) clearly revealed that the reconstruction was lifted and the surface had acquired the (1x1) structure. A potential of -0.35 V vs. SCE was then applied to the very same electrode for 3 min, which restored the (5x20) reconstruction as demonstrated unequivocally by the LEED pattern (Fig. 10c) [38]. This potential-assisted reconstruction has been explained tentatively by the surface excess charge which favors densely packed structures because of a stronger attraction between the surface atoms at increased electron densities [41].

We should like to stress the fact that only electron diffraction techniques such as LEED and RHEED could give decisive answers to questions about surface reconstruction in electrochemistry. Meanwhile we have obtained a substantial body of evidence for the existence of reconstructed gold surfaces in an electrochemical cell derived from ex-situ electron diffraction [36,38] as well as in-situ double-layer capacity [37], electroreflectance [37,40], and SHG [4] measurements, which should not be ignored completely when attempts are made to derive conclusions on surface structures merely from voltammetry [42].

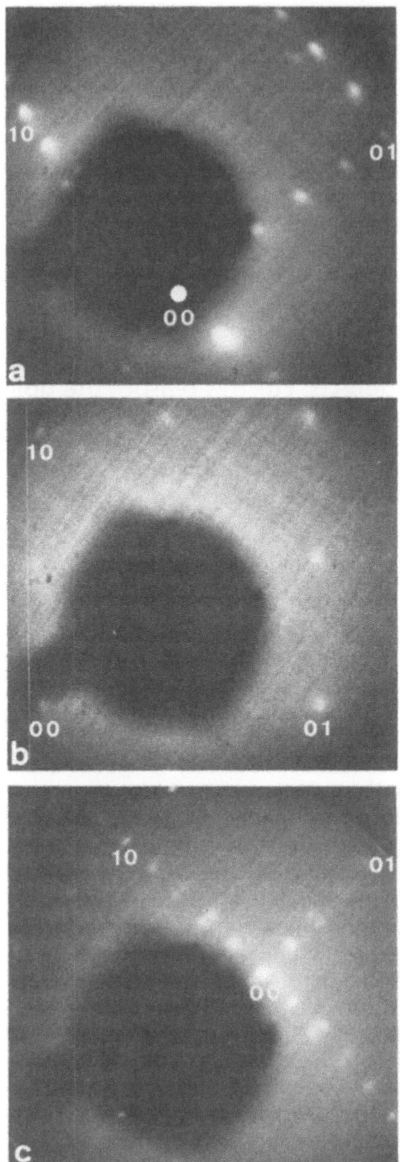


Figure 10. LEED patterns of an Au(100) electrode, which had been prepared in UHV to show the (5x20) reconstruction, after emersion from 5 mM $H_2SO_4$. (a) After cycling between -0.35 and +0.25 V vs. SCE. The surface is still (5x20). (b) After cycling between -0.35 and +0.45 V vs. SCE. The surface is now (1x1), with a p(2x2) superstructure due to an ordered $H_2SO_4$ overlayer (see text). (c) After applying a potential of -0.35 V vs. SCE for 3 minutes. The (5x20) pattern has reappeared. After ref. [38].

## 4.2 $H_2SO_4$ OVERLAYERS ON GOLD SINGLE-CRYSTAL ELECTRODES

As we mentioned in chapter 2, "dry" emersion (i.e., the absence of any bulk electrolyte on the electrode surface after emersion) is one of the prerequisites for successful ex-situ experiments. Such experiments were mostly performed with electrolytes containing either hydrofluoric or perchloric acid. Occasionally, small droplets do adhere to the electrode surface during emersion, which causes bulk electrolyte to be introduced into the vacuum chamber. Subsequent evacuation in most cases either causes the electrolyte to be pumped away, if it is volatile like HF, or leads to the formation of small crystallites of salt. The latter do not interfere seriously with the surface studies, because only a tiny fraction of the surface will be covered by these salt crystals (see Fig. 11).

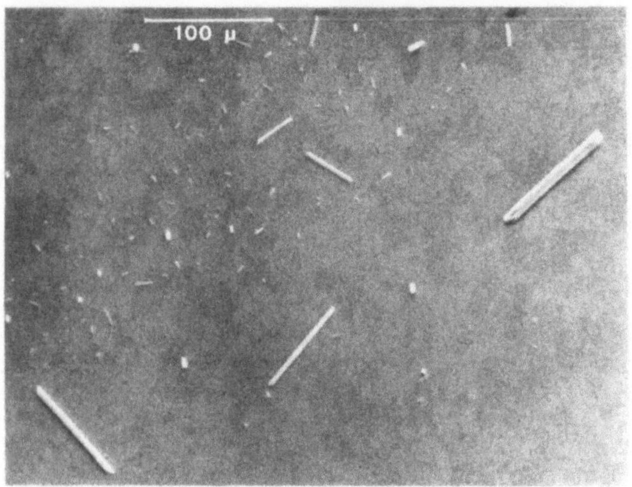


Figure 11. Scanning electron microscope image of Au(111) with 0.5 monolayers of Cu, after wet emersion from 0.01 M $HClO_4$. Small crystallites of hydronium perchlorate, formed after evacuation, are clearly seen. Their orientation indicates epitaxial growth.

A distinctly different behavior, however, was observed for emersion of gold single-crystal electrodes from sulphuric acid solutions, where wet emersion led to a superstructure pattern in LEED and RHEED due to ordered adsorption of the bulk electrolyte residue [43,44].

Reconstructed and unreconstructed Au(111), (100) and (110) surfaces were brought into contact with 5 mM $H_2SO_4$ and the interfacial response examined by cyclic voltammetry. The samples were then emersed from solution, always with a drop of solution adhering to the surface ("wet" emersion). Upon evacuation of the transfer chamber the droplet was no longer visible, indicating that most of the solution had evaporated. The specimens were then transferred into the main UHV chamber for structural studies by electron diffraction. In all four cases examined - Au(111)-(1x23), Au(100)-(5x20), Au(100)-(1x1), and Au(110)-(1x1) - superstructure patterns emerged which differed for the

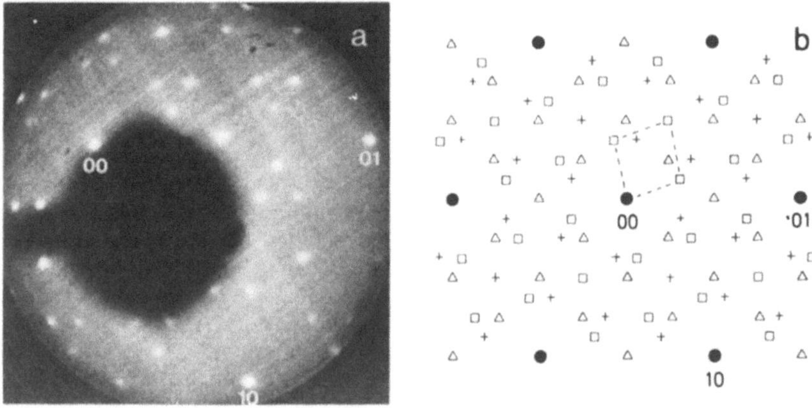


Figure 12. (a) LEED pattern (27 eV) of Au(111) after emersion at 0 V vs. SCE from 0.005 M $H_2SO_4$ showing a superstructure. (b) Schematic diagram of the LEED pattern. The spot pattern is a superposition of the diffraction patterns from three different domains ( +, Δ, □ ). After ref. [43].

various surfaces but were independent of the emersion potential. Even the mere contact with the sulphuric acid solution, without any potential applied, resulted in the same diffraction pattern, as tested with Au(111) [43]. The observation of identical diffraction patterns regardless of emersion potential clearly indicates that such a structure cannot be connected with the electrochemical double layer but, rather, is a consequence of wet emersion from sulphuric acid solution followed by evacuation. In other words, $H_2SO_4$ forms an ordered overlayer on gold after evacuation. This conclusion is corroborated by corresponding experiments with $Na_2SO_4$ solutions, where such superstructures were not observed. Besides, the electrodes were emersed dry from this solution [43].

In Fig. 12 is shown the LEED pattern of a Au(111) electrode after emersion at 0 V vs. SCE. In addition to the features associated with the reconstructed Au(111), a superstructure of the type "oblique $\sqrt{7}$x$\sqrt{7}$" has appeared due to the ordered $H_2SO_4$ overlayer. A detailed structure analysis of this pattern has been given elsewhere [43]. Real space images as derived from the corresponding LEED and RHEED patterns of the $H_2SO_4$ overlayers on the various gold surfaces are given in Fig. 13. For Au(100) and Au(111) it was observed that the sulphuric acid overlayer did not remove the reconstruction. In this respect the study of reconstructed and unreconstructed Au(100) is particularly interesting because of the large structure differences between (5x20) and (1x1) and of the close similarity of Au(100)-(5x20) and Au(111)-(1x23). And indeed, when the reconstructed Au(100)-(5x20) surface was emersed at -0.3 V, a "centered $\sqrt{7}$x$\sqrt{7}$" structure with respect to the hexagonal overlayer [(2x$\sqrt{3}$)-rectangular] was observed for the $H_2SO_4$ monolayer which is very similar - although not identical - to the result for Au(111)-(1x23) [44]. On the other hand, emersion of unreconstructed Au(100) led to a p(2x2) superstructure, i.e., the $H_2SO_4$ molecules arrange themselves according to the fourfold symmetry of the substrate. The various superstructures of $H_2SO_4$ overlayers on Au(hkl) are summarized in Table 1. Two different types of structural arrangements may be distinguished.

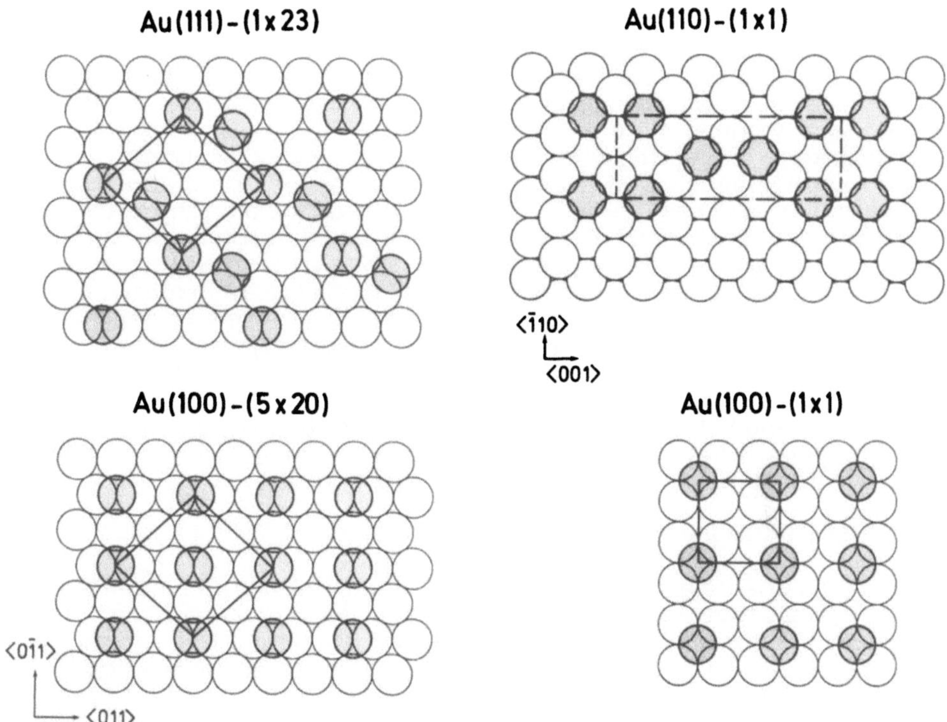


Figure 13. Models for the $H_2SO_4$ adlayer on reconstructed and unreconstructed Au surfaces, as derived from LEED and RHEED measurements. With the exception of Au(100)-(1x1), all structures contain two or more molecules per unit cell. After refs. [43,44].

**Tab. 1:** The structure of $H_2SO_4$ adlayers on reconstructed and unreconstructed Au surfaces. After refs. [43, 44]

| substrate | adlayer | no. of ad-molecules per unit cell |
|---|---|---|
| Au(111)-(1x23) | $(\sqrt{7}\text{x}\ \sqrt{7})$-oblique | 2 |
| Au(100)-(5x20) | centered $(\sqrt{7}\text{x}\ \sqrt{7})$ [or(2x $\sqrt{3}$)-rectangular] | 2 |
| Au(100)-(1x1) | p(2x2) | 1 |
| Au(110)-(1x1) | c(2x4) | 4 |

For Au(111)-(1x23) and Au(100)-(5x20) (densely packed surfaces in both cases), the interaction between adsorbate and substrate is very weak, allowing the $H_2SO_4$ overlayer to acquire a lateral structure ($\sqrt{7}$x$\sqrt{7}$) similar to that in solid $H_2SO_4$ [43,45]. The more open faces, however, to which adatoms and admolecules are usually bound more strongly, force the $H_2SO_4$ molecules into structures which are in full registry with the substrate (e.g., p(2x2) or c(2x4)).

The peculiar behavior of sulphuric acid residues forming ordered layers after evacuation has been rationalized by the fact that the monohydrate has a rather low melting point of 9 °C and therefore is liquid at room temperature, at which all our experiments were performed. This is in contrast to perchloric acid solutions which after evacuation form small crystallites of hydronium perchlorate with a melting point of about 50 °C. From the above it becomes clear that the superstructure patterns of ordered $H_2SO_4$ overlayers - although extremely interesting in their own right - are an artefact from the standpoint of in-situ electrochemistry, because these overlayers are formed only after emersion and evacuation. It is one of the rare cases where the ex-situ derived structure information is not fully congruent with the in-situ situation and hence may lead to misinterpretations of the double layer structure.

## 4.3 ORDERED ADSORPTION OF METAL DEPOSITS

The formation of a metal deposit in submonolayer or monolayer amounts on a foreign metal substrate (e.g., Cu on Au) is often found to occur within a potential range which is positive of the so-called Nernst potential for the respective bulk metal deposition. Such behavior is called underpotential deposition [46]. It is observed when the metal adatoms are bound more strongly to the foreign metal substrate than to a substrate of its own kind (e.g., Cu on Cu). Underpotential deposition has been studied extensively by cyclic current-potential curves which in most cases reflect directly the adsorption and desorption behavior of the metal deposit. Since the current is proportional to the change in coverage, $d\Theta/dt$, and the electrode potential is a direct measure of the free enthalpy of adsorption, current integration yields the adsorption isotherm $\Theta(U)$.

Cyclic voltammograms have revealed that the adsorption isotherms for metal monolayer formation in the underpotential region in general depend markedly on the surface crystallographic orientation [47,48]. In addition it was found that the formation of a complete monolayer often occurs in several distinctly different adsorption steps (as indicated by a multi-peak structure in the I-U curves). From such measurements it was inferred that metal monolayer formation in many cases proceeds via ordered structures [47,48] similar to the ones observed by LEED for metal evaporation in UHV [49].

In order to obtain conclusive information about the structure of metal adsorbates formed by underpotential deposition, electron diffraction methods such as LEED or RHEED were employed in a number of cases [31,50-54]. Cu on Au(111) was among the first systems for which ordered adsorption was proven to exist by ex-situ RHEED and LEED studies [31]. In Fig. 14a is shown the LEED pattern of Au(111) covered with approximately half a monolayer of Cu, which was emersed from a sulphate-containing electrolyte. A superstructure is observed which clearly indicates ordered adsorption of Cu on Au(111) of the type ($\sqrt{3}$x$\sqrt{3}$)R 30° (honeycomb structure) for medium coverages. For the complete Cu monolayer, a (1x1) structure was inferred from LEED and RHEED, i.e., the

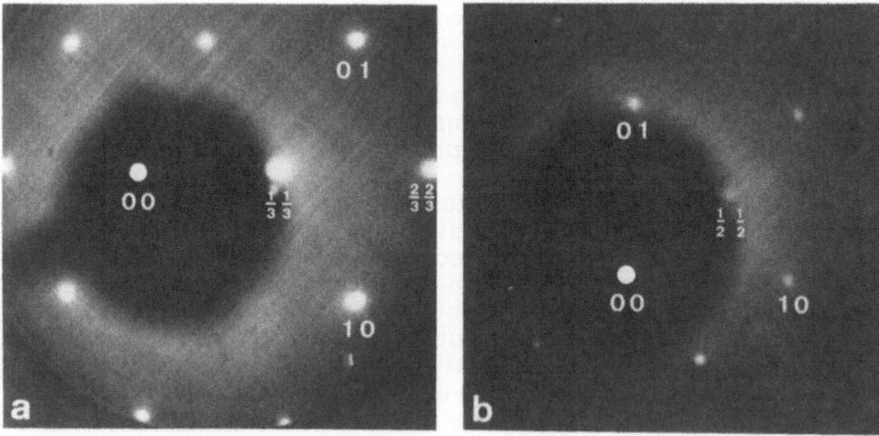


Figure 14. (a) LEED pattern (43 eV) of Au(111) with half a monolayer of Cu after emersion from 0.1 M $HClO_4$ + 1 mM $CuSO_4$. The Cu overlayer shows a (√3x√3)R 30° superstructure. After ref. [31].
(b) LEED pattern (75 eV) of Ag(100) with half a monolayer of Tl after emersion from 0.01 M $Na_2SO_4$ + 1 mM $Tl_2SO_4$. The Tl overlayer shows a c(2x2) superstructure. After ref. [38].

adlayer is in registry with the substrate [31]. We mention in passing that our findings on the structure of Cu upd on Au(111) derived from ex-situ studies have been confirmed recently in every detail by in-situ surface EXAFS measurements [55], a fact which emphasizes again the validity of the ex-situ approach.

Despite the strong interaction between metal adsorbate and substrate, a pronounced anion influence on the adsorption process has frequently been observed in underpotential deposition. In the case of Cu on Au(111) we discovered that the anions can play a structure-decisive role [54]. While in sulphate-containing electrolytes a (√3x√3)R30° structure was found for medium Cu coverages (see above), a rather complicated, non-commensurate (2.2x2.2) superstructure was observed for perchlorate solutions. From a quantitative analysis of the Auger intensities it was inferred that the (2.2x2.2) structure must contain two Cu atoms per unit cell [54]. Hence a maximum Cu coverage of about 0.4 is accommodated in such a structure, whereas the limiting $\Theta$ value for the √3 structure is 2/3 of a monolayer. In both cases, for sulphate and perchlorate solutions, Cu deposition in the underpotential region was accompanied by massive coadsorption of anions, as could be seen in the corresponding Auger spectra. The anion concentration at the interface increased almost linearly with the amount of adsorbed Cu and reached a ratio of about 1:3 at maximum Cu coverage. The two different surface structures, (√3x√3)R30° and (2.2x2.2), of electrochemically formed Cu on Au(111) at medium coverages may be compared with the growth mechanism for Cu evaporated onto Au(111) under UHV conditions. In the latter case, a pseudo-morphic growth (i.e., islands of (1x1) structure) was observed from the very beginning [31].

In Fig. 14b, the LEED pattern for Tl on Ag(100) is shown. The electrode had been emersed from a solution of 0.01 M $Na_2SO_4$ + 1 mM $Tl_2SO_4$ with about half a monolayer of Tl deposited at underpotentials. A c(2x2) superstructure due to the Tl deposit is clearly seen, which proves the existence of ordered adsorption under these conditions. This superstructure is also seen in the corresponding RHEED pattern by 1/2 superstructure streaks in [001] azimuth [38]. It is hoped that such detailed information on adlayer structures will help to elucidate the initial steps of metal deposition and their role in determining the further growth behavior.

In conclusion it seems only fair to say that electron diffraction techniques have contributed significantly towards our understanding of electrode surface structures and their influence on electrochemical processes. The widespread use of single-crystal electrodes, which has advanced tremendously our knowledge about the electrochemical double layer and about adsorption reactions, would not have been possible without the intensive use of LEED and RHEED in surface preparation and characterization. Even with new and powerful structure-sensitive in-situ techniques such as X-ray diffraction methods or scanning tunneling microscopy being available in the near future, the ex-situ approach still seems indispensable when it comes to a detailed characterization of electrode surfaces. This is particularly true for the use of electron diffraction techniques in studying long-range order.

**References**


[1] See, e.g., W. N. Hansen, D. M. Kolb and D. W. Lynch (eds.), *Electronic and Molecular Structure of Electrode-Electrolyte Interfaces*, J. Electroanal. Chem. **150** (1983).

[2] See, e.g., H. Gerischer and D. M. Kolb (eds.), *Structure and Dynamics of Solid-Electrolyte Interfaces*, Ber. Bunsenges. Phys. Chem. **91** (1987) 262-496.

[3] A. Hamelin, in: *Modern Aspects of Electrochemistry*, Vol. **16**, p. 1, eds. B. E. Conway, R. E. White and J. O'M. Bockris (Plenum, New York, 1985).

[4] D. M. Kolb, Ber. Bunsenges. Phys. Chem. **92** (1988) 1175.

[5] D. M. Kolb, Z. Phys. Chem. N.F. **154** (1987) 179.

[6] W. N. Hansen, C. L. Wang and T. W. Humpherys, J. Electroanal. Chem. **93** (1978) 87.

[7] D. M. Kolb and W. N. Hansen, Surface Sci. **79** (1979) 205.

[8] W. J. Anderson and W. N. Hansen, J. Electroanal. Chem. **43** (1973) 329.

[9] See, e.g., A. Bewick and J. Robinson, J. Electroanal. Chem. **60** (1975) 163.

[10] R. Kötz, H. Neff and K. Müller, J. Electroanal. Chem. **215** (1986) 331.

[11] W. N. Hansen and D. M. Kolb, J. Electroanal. Chem. **100** (1979) 493.

[12] D. L. Rath and D. M. Kolb, Surface Sci. **109** (1981) 641.

[13] D. M. Kolb, D. L. Rath, R. Wille and W. N. Hansen, Ber. Bunsenges. Phys. Chem. **87** (1983) 1108.

[14] See, e.g., W. Paik, M. A. Genshaw and J. O'M. Bockris, J. Phys. Chem. **74** (1970) 4266.

[15] D. M. Mohilner, in: *Electroanalytical Chemistry*, Vol. **1**, p. 241, ed. A. J. Bard (Dekker, New York, 1966).

[16] C. N. van Huong, C. Hinnen and A. Rousseau, J. Electroanal. Chem. **151** (1983) 149.

[17] M. Stratmann and H. Streckel, Ber. Bunsenges. Phys. Chem. **92** (1988) 1244.

[18] W. N. Hansen and D. M. Kolb, unpublished.
[19] J. G. Gordon, J. Phys. (Paris) **44** (1983) C10-171.
[20] P. N. Ross, Jr., J. Electrochem. Soc. **126** (1979) 67.
[21] W. Stöckel and R. Schumacher, Ber. Bunsenges. Phys. Chem. **91** (1987) 345.
[22] D. Aberdam, this conference.
[23] G. Ertl and J. Küppers, *Low Energy Electrons and Surface Chemistry* (VCH, Weinheim, 1985).
[24] G. A. Somorjai, *Chemistry in Two Dimensions: Surfaces* (Cornell Univ. Press, Ithaca, 1981).
[25] M. A. Van Hove, W. H. Weinberg and C. M. Chan, *Low Energy Electron Diffraction* (Springer, Berlin, 1986).
[26] H. Raether, in: *Structural Research,* Encyclopedia of Physics, Vol. **32**, p. 443, ed. S. Flügge (Springer, Berlin, 1957).
[27] M. S. Zei, Y. Nakai, D. Weick and G. Lehmpfuhl, Surface Sci. **152/153** (1985) 254.
[28] M. A. Van Hove, R. J. Koestner, P. C. Stair, J. P. Biberian, L. L. Kesmodel, I. Bartos and G. A. Somorjai, Surface Sci. **103** (1981) 189, 218.
[29] N. Ikarashi, K. Kobayashi, H. Koike, H. Hasegawa and K. Yagi, Ultramicroscopy **26** (1988) 195.
[30] G. Binnig, H. Rohrer, Ch. Gerber and E. Weibel, Surface Sci. **131** (1983) L379.
[31] Y. Nakai, M. S. Zei, D. M. Kolb and G. Lehmpfuhl, Ber. Bunsenges. Phys. Chem. **88** (1984) 340.
[32] K. Takayanagi, Ultramicroscopy **8** (1982) 145.
[33] G. Ertl, Surface Sci. **152/153** (1985) 328.
[34] A. Hamelin, J. Electroanal. Chem. **142** (1982) 299.
[35] E. Yeager, A. Homa, B. D. Cahan and D. Scherson, J. Vac. Sci. Technol. **20** (1982) 628.
[36] D. M. Kolb, G. Lehmpfuhl and M. S. Zei, J. Electroanal. Chem. **179** (1984) 289.
[37] D. M. Kolb and J. Schneider, Electrochim. Acta **31** (1986) 929.
[38] M. S. Zei, G. Lehmpfuhl and D. M. Kolb, to be published.
[39] Y. Uchida, G. Lehmpfuhl, T. Twomey and D. M. Kolb, to be published.
[40] J. Schneider and D. M. Kolb, Surface Sci. **193** (1988) 579.
[41] V. Heine and L. D. Marks, J. Electron Spectr. Rel. Phenom. **38** (1986) 229.
[42] A. Hamelin, J. Electroanal. Chem. **255** (1988) 281.
[43] M. S. Zei, D. Scherson, G. Lehmpfuhl and D. M. Kolb, J. Electroanal. Chem. **229** (1987) 99.
[44] M. S. Zei and D. Weick, Surface Sci. **208** (1989) 425.
[45] R. Pascard, C. R. Acad. Sci. **240** (1955) 2162.
[46] D. M. Kolb, in: *Advances in Electrochemistry and Electrochemical Engineering*, Vol. **11**, p. 125, eds. H. Gerischer and C. W. Tobias (Wiley, New York, 1978).
[47] J. W. Schultze and D. Dickertmann, Surface Sci. **54** (1976) 489.
[48] K. Jüttner and W. J. Lorenz, Z. Phys. Chem. N.F. **122** (1980) 163.
[49] K. Takayanagi, D. M. Kolb, K. Kambe and G. Lehmpfuhl, Surface Sci. **100** (1980) 407.
[50] D. Aberdam, R. Durand, R. Faure and F. El-Omar, Surface Sci. **162** (1985) 782.

[51] J. L. Stickney, S. D. Rosasco, D. Song, M. P. Soriaga and A. T. Hubbard, Surface Sci. **130** (1983) 326.
[52] B. C. Schardt, J. L. Stickney, D. A. Stern, A. Wieckowski, D. C. Zapien and A. T. Hubbard, Surface Sci. **175** (1986) 520.
[53] L. Laguren-Davidson, F. Lu, G. N. Salaita and A. T. Hubbard, Langmuir **4** (1988) 224.
[54] M. S. Zei, G. Qiao, G. Lehmpfuhl and D. M. Kolb, Ber. Bunsenges. Phys. Chem. **91** (1987) 349.
[55] O. R. Melroy, M. G. Samant, G. L. Borges, J. G. Gordon, L. Blum, J. H. White, M. J. Albarelli, M. McMillan, and H. D. Abruña, Langmuir **4** (1988) 728.

# AUGER ELECTRON SPECTROSCOPY AND THE ELECTROCHEMICAL INTERFACE

**Daniel J. ABERDAM**
Laboratoire de Spectrométrie Physique,
associé au CNRS (UA 8)
Université Joseph Fourier, Grenoble I
BP 87
F-38402 Saint Martin d'Hères, France

**ABSTRACT**. A brief introduction to Auger Electron Spectroscopy (AES) is given from a practical point of view. An arrangement for associating this spectroscopy with an electrochemical experiment is then schematically presented. Some examples from the author's work are described in order to illustrate what AES may teach us about the electrochemical interface. These examples concern the Under Potential Deposition (UPD) of lead and silver on the platinum (100) single crystal surface. Finally, the problem of electrode emersion is discussed, as well as the way AES can help us to handle it.

## 1. INTRODUCTION

The knowledge of the electrode surface composition during and after an electrochemical reaction, and of the electrode structure at the atomic scale, is most necessary to improve our understanding of a number of electrochemical processes. Most of the optical techniques which have the outstanding feature of being applicable 'in situ', i.e. directly on the solid-liquid interface, are not element-sensitive, so that the combination of various methods is necessary. Together with X-Ray Photoelectron Spectroscopy (XPS), also called Electron Spectroscopy for Chemical Analysis (ESCA), Auger Electron Spectroscopy is the easiest (and probably the cheapest) among the 'ex-situ' element-sensitive techniques. Obviously, element sensitivity is meaningful when the elements detected come only from the system under study, and not from exposure to a dirty liquid or gas phase. This requires 'ex-situ' analysis to be performed on samples which have been carefully emersed from the electrochemical cell and transferred to the Auger analyser without contact with the laboratory atmosphere. This rule has a major exception : gold electrodes seem sufficiently inert with respect to adsorption to accept transfer through air without major contamination (1). Otherwise, it is necessary to couple together an Ultra High Vacuum (UHV) chamber with an Electrochemical chamber, where electrochemistry is performed under an inert gas atmosphere. The



*C. Gutiérrez and C. Melendres (eds.),*
*Spectroscopic and Diffraction Techniques in Interfacial Electrochemistry*, 383–407.

transfer of the electrode from one chamber to the other is accomplished in vacuo with a transfer device, such as a magnetically actuated rod.

The chapter is organised as follows :**(i)** a short introduction to the principles and practice of AES is followed by a schematic description of one possible way of coupling a cell to an UHV chamber.**(ii)** two examples, chosen from the author's work, illustrate the type of information which can be gained on electrochemical systems with this technique, and **(iii)** results taken from the same examples demonstrate how careful one has to be when emersing an electrode, and how AES may help in this context.

## 2. AUGER ELECTRON SPECTROSCOPY.

### 2.1.Physical bases of Auger Electron Spectroscopy.

Discovered by Pierre Auger in 1925 (2), the Auger process is the non radiative de-excitation of an atom which has been ionized in an inner level, say level $E_1$. The way the atom has been first ionized is unimportant. The primary ionization may equally be produced by an energetic electron (0.5 to 10 keV), or by an X-Ray photon, or by a massive ion. Contrary to XPS, where one detects the electron ejected during this ionization process, the ejected electron is in general disregarded by Auger Electron spectroscopists. What they look at in the energy distribution of backscattered electrons are those secondary electrons resulting from a subsequent non-radiative de-excitation of the atom. Two de-excitation paths are offered the atom to reduce its excess energy. The first is a radiative process, fluorescence, where an electron from a higher energy level fills the deep hole, and the excess energy is released through the emission of a photon. The second one is the Auger process, which is a two-electron electromagnetic relativistic interaction. A crude way model of the process is to consider that an electron in level $E_2 > E_1$ fills the deep hole and that another electron in level $E_3 > E_2$ is ejected with a kinetic energy equal to the difference between the initial and final energies of the atom. The theory of the Auger process was first developed by Moller (3). The interaction operator is the sum of a coulombic repulsive term and of a relativistic magnetic term, which has been studied by Massey and Burhop (4) and by Rose (5). The interaction between the two electrons is weak, and so the lifetime of the excited state is long enough to guarantee that the initial states of the electrons are well defined stationary states. Chattarji (6) has studied the various approximations useful to calculate transition probabilities according to the atomic number of the excited atom. He found that for a small or moderate first ionization energy, the Auger process is favoured with respect to the radiative process. Bambynek *et al.* (7), have reviewed the calculations of radiative and non radiative transition probabilities

**2.1.1. Notation of Auger Lines.** Auger Lines are usually denoted according to the historical X-Ray notation, which is equivalent to the spectroscopic notation in jj coupling. The atomic energy levels (or shells) are labelled K,L,M,N, etc., which correspond to the principal quantum numbers n=1,2,3,4, etc. Subshells are labelled according to the jj coupling scheme, the $^2p_{1/2}$ level being designated as $L_2$. Similarly, the $^3d_{3/2}$ and $^3d_{5/2}$ levels are designated as $M_4$ and $M_5$, respectively. When calculating transition probabilities, it may happen that this coupling scheme is not adequate, in which case the notation is completed accordingly. However, if the aim of the calculation is to determine the total radiationless transition probability for a given final state configuration, the choice of the coupling scheme is unimportant as long as the initial vacancy is not in the final state configuration (7).

**2.1.2. Energy of Auger Electrons.** The energy of Auger electrons is their most important associated quantity, because it is element-dependent and constitutes a fingerprint of the emitting atom. It is given by

$$E(\text{Auger}) = E_1 - E_{23X}$$

where $E_1$ is the first ionisation energy of the atom ionized in level $E_1$ of the relevant shell, and $E_{23X}$ is the energy of the final state of the atom doubly ionized in levels $E_2$ end $E_3$ and in configuration X. $E_1$ is known, and $E_{23X}$ may be either calculated from first principles, or deduced semiempirically from binding energies and optical data. Gallon has discussed the various schemes of calculation of $E_{23X}$ (8). A very crude, but useful approximation to E(Auger) is obtained from the so called Z+1 formula, in which $E_{23X}$ is replaced by the sum of the first ionization energy of atom Z in shell $E_2$ and of the first ionization energy of atom Z+1 in shell $E_3$. This latter term approximates the second ionization energy of atom Z in shell $E_3$. Therefore,

$$E(\text{Auger}) \approx E_1(Z) - E_2(Z) - E_3(Z+1)$$

We shall not discuss further this question since a very abundant literature deals with Auger electron energy (8), and we are more interested in the Auger process in solids than in isolated atoms.

**2.1.3. Auger Electrons from Solids.** When the emitting atom is embedded in a solid, the energy, width, and line shape of the Auger transition are modified by several processes.

*Energy.* The energy of the Auger transition is subject to displacements due to band structure, chemical bonds, and relaxation of the electron gas in the field of the doubly ionized atom.

*Width*. The width of an Auger line is generally higher in a solid because the lifetime of the final state decreases due to delocalization in the energy bands.

*Shape*. The Auger line shape also suffers from solid state effects, viz. **(i)** the onset of transitions forbidden in isolated atoms, **(ii)** the hybridization of wave functions, **(iii)** electron-electron correlation energy.

These changes depend on the chemical nature of the solid and the type of bonding of the emitting atom in the solid. Of course, they are more prominent in Auger lines which involve valence electrons, i.e. CCV or CVV lines, C and V standing for Core and Valence, respectively. Fig. 1 illustrates the shift in energy and the change in lineshape of silicon Auger peaks with increasing electronegativity of the bond (9).

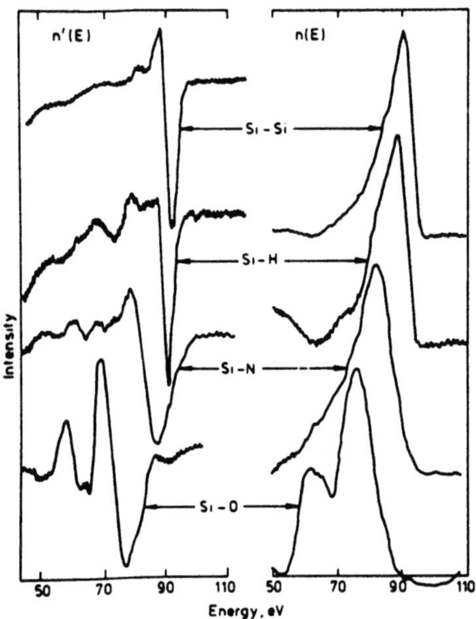


**Figure 1**. Auger electron peaks for silicon, showing the changes in shape and the progressive shift towards lower energies with increasing bond electronegativity. Taken from Madden (9). Left : peaks recorded in the d(n(E))/dE (or derivative) mode, which automatically performs background subtraction, thus making spectra acquisition much easier. Right : peaks recorded in the n(E) (or distribution) mode. It clearly appears here that quantitative analysis will be much easier in the distribution mode (see text).

This perturbation of Auger line characteristics by solid state chemical effects may be turned to the advantage of the experimentalist, because it involves mainly the outer electron shells, and thus brings information about the chemical state of the emitting atom. This subject has been reviewed by Madden (9).

**2.1.4. Surface sensitivity of AES**. AES is a surface sensitive technique because the strong interaction of electrons with solids limits their inelastic mean free path, $\lambda$, in the material. Consequently, only those electrons emitted close enough to the surface will escape without suffering any inelastic scattering event. Seah and

Dench (10), among others, have compiled from the literature known mean free paths of electrons in various materials. Fig. 2 (log-log scale) shows that the behaviour of $\lambda$ may be described as the sum of two empirical power laws of electron energy :

$$\lambda = A_m\, a/E^2 + B_m\, (a^3E)^{1/2}$$

where E is the electron kinetic energy, expressed in eV, and $A_m$, $B_m$ are constants fitted to the data which depend only on material type, viz. metal, oxide, or organic material. In this expression, a is a length, in nm, which characterizes the size of the atoms constituting the absorbing medium. It is simply derived from the molar mass M (g/mole), density $\rho$ (g/cm$^3$), and Avogadro's number through the following equation :

$$a\,(\text{nm}) = 10^7\ (M/(\rho N_a))^{1/3}$$

In order to express $\lambda$ in nm, for metals the following values must be assigned to $A_m$, $B_m$ (10) :

$$A_m = 538; \qquad B_m = 0.41;$$

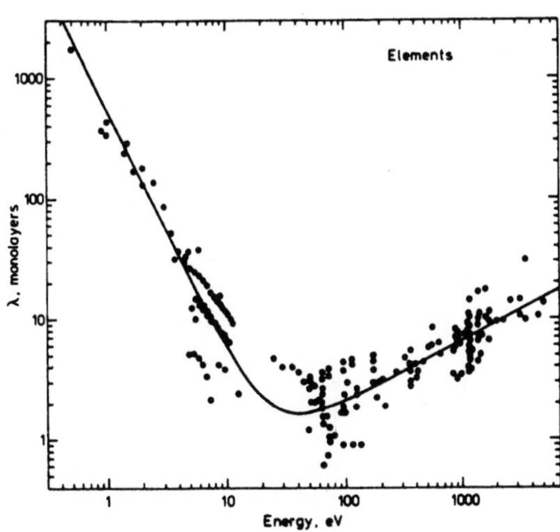


**Figure 2**. Electron Inelastic Mean Free Path, $\lambda$, expressed in units of material monolayer thickness as a function of electron energy, according to Seah and Dench (10). Note the double log scale.

Fig. 2 also shows that for any material $\lambda$ is minimum in the energy range 20 eV < E < 100 eV, and that this minimum value amounts to about 1 to 5 monolayers, depending on the material ; in this range the surface sensitivity of AES is maximum. It is possible, to some extent, to change this surface sensitivity by selecting, e.g., Auger lines for a given element with Auger energies well within and well outside the energy range for minimum $\lambda$ (11). Another way of changing surface sensitivity is by changing the angle of incidence of the exciting electron beam, which changes the depth distribution of excited atoms. This can be used as a method to gain information on the depth distribution of a given element in the first few layers of a material in a non destructive way. This will be illustrated below in the case of surface alloying of silver on platinum

## 2.2.Practising Auger Electron Spectroscopy.

**2.2.1. Primary ionization.** As already mentioned, the means by which primary ionization is obtained is unimportant in principle. However, electron beams are most commonly used in AES because they are easy to produce at low cost, and can achieve a good lateral resolution thanks to the possibility of focusing the beam of primary electrons over an area a few micrometers wide. Since backscattered electrons create a high-level background of electrons in the energy distribution n(E), it is necessary to use some tricks to extract the Auger signal from this background.

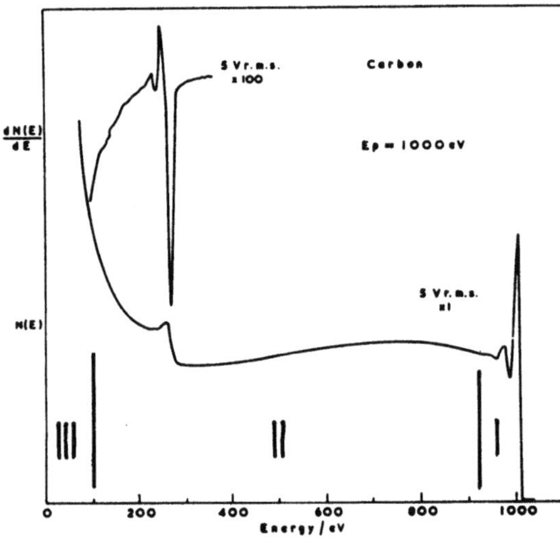


**Figure 3**. Distribution of energy of secondary electrons ejected from a graphite surface by incident electrons of energy 1000 eV. Taken from Briggs and Seah (20). Lower curve : Spectrum recorded in the n(E) mode. Three regions may be distinguished :
**(I)** the region of the quasi-elastically scattered electrons, which also includes electrons having lost a definite amount of energy exciting plasma oscillations of the electron gas in the solid.
**(II)** the region where Auger electrons appear at a fixed energy, independent of the primary beam energy, superimposed on a relatively flat but high background.
**(III)** at low energies, the peak of true secondary electrons, i.e. those electrons which have suffered some cascade of inelastic scattering, so that no sharp structure can be found in this peak, unless some low energy Auger line is superimposed.
Upper curve : differential distribution dn(E)/dE over the energy range containing the carbon KLL Auger peaks. By convention, in

the differential distribution the peak 'position' is taken to be that of the high energy minimum.

This is not so critical, for example, for Auger signals excited by X-Rays in an XPS experiment, in which case, however, the lateral resolution is rather poor. Fig. 3 shows the spectrum of electrons backscattered from a graphite surface (lower curve).

**2.2.2. Extracting Auger lines from the spectrum.** Various instruments have been developed to extract the Auger signal from the spectrum of backscattered electrons. Some of them provide good lateral resolution, together with the possibility of sputtering for depth profiling and signal processing for imaging, thus enhancing Auger microscopy. We will briefly describe the working principles of the two basic Auger analysers, namely the Retarding Field Analyser (RFA), and the Cylindrical Mirror Analyser (CMA) ; a detailed study may be found in ref. 12.

***Retarding Field Analyser.*** This analyser makes use of the filtering grid system of the Low Energy Electron Diffraction (LEED) apparatus, and thus requires only a modulating system and a phase sensitive detector to display Auger lines. Fig. 4 shows the principle of the detection. Let us denote by $E_p$ the kinetic energy of the primary (i.e., incident) electrons, and by $V_r$ the retarding potential applied to the intermediate grids 2 and 3. Electrons scattered by the sample will be slowed down by this retarding potential, and only those electrons scattered with a kinetic energy larger than $E = -eV_r$ will overcome the potential barrier and reach the collector, which is simply the LEED screen.

The current collected is thus proportional to the integral of the energy spectrum of the scattered electrons between $-eV_r$ and $E_p$.

$$I(RFA) = C \int_{-eV_r}^{E_p} n(E)dE$$

Modulation of the electron energy at the frequency $\omega$ allows one to obtain the energy distribution n(E) by lock-in detection of the fundamental frequency $\omega$, and its first derivative dn(E)/dE by detection at the second harmonic $2\omega$. A nice feature of the RFA is its large solid angle of acceptance, usually higher than $\pi$ steradians, which guarantees a good angular integration. This may be of importance when working with single crystal surfaces, because diffraction effects may act strongly on Auger electrons, giving rise to a strong anisotropy in the Auger emission pattern. It is, however, worth noting that such diffraction effects have been used to determine the adsorption site of sulfur on a nickel single crystal surface, thus making AES a crystallographic tool (13).

The main drawback of the RFA comes from the fact that it detects a very large number of electrons, viz. the integral of the energy spectrum in the window $-eV_r$ to $E_p$, whereas one is interested only in the electron current in a narrow window including the Auger line.

Consequently, the signal to noise ratio in this narrow window is rather poor since the noise level is proportional to the square root of the number of collected electrons. This is the reason why dispersive analysers, whose prototype is the CMA, have been developed.

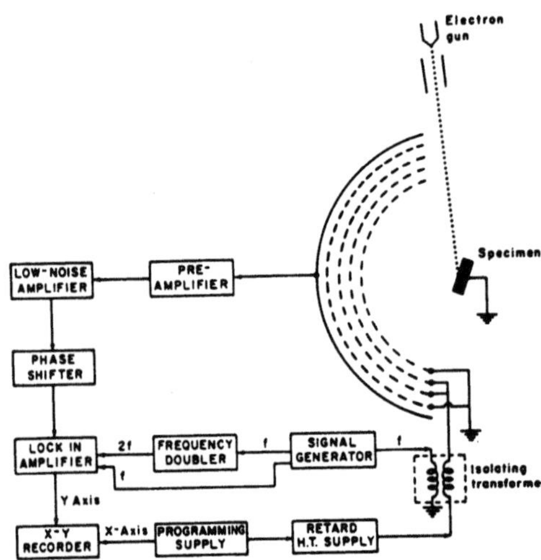


**Figure 4**. Retarding Field Auger Analyser using LEED display system. Taken from Briggs and Seah (20). The primary electrons are accelerated between the emitting filament set at a potential $V_p$ negative to ground and the grounded anode. The first grid is also grounded to ensure a field free propagation of electrons. The fourth grid is also grounded to ensure electrostatic shielding between grids 2,3 and collector. Electrons scattered by the sample are retarded by a potential $V_r$ negative to ground applied to both grids 2 and 3, in order to minimize equipotential distortions by grid nets. A sine wave modulation is superimposed to the retarding potential, which is scanned in the range of interest. The modulated part of the output current is detected by a phase sensitive amplifier, locked either to the modulation frequency $\omega$ or to its second harmonic $2\omega$. In the former mode, the output signal is proportional to n(E). In the latter mode, it is proportional to dn(E)/dE, which improves the display of Auger lines.

***Cylindrical Mirror Analyser.*** This analyser selects electrons according to their velocity, so that only those electrons in a narrow range of velocity (energy) reach the collector. The selection of electrons according to energy is similar to that of light by a prism monochromator. The selection of energies is achieved electrostatically by means of a cylindrically symmetric electric field applied between two coaxial cylinders, the innner one being grounded, and the outer one maintained at the analysing potential V, negative to ground. The geometry of the apparatus is schematically illustrated in Fig. 5 (14). The current to the collector is :

$$I(CMA) = C\ n(E)\ \Delta E = (C\ R)\ E\ n(E).$$

where the width of the energy window, $\Delta E$, is proportional to the energy of the analysed electrons, and the constant $R = \Delta E/E$ is the resolution of the apparatus, usually in the range $(2\text{-}5).10^{-3}$. Since the Auger current is proportional to E, the readability of Auger lines at high energies is improved. If a sine wave is superimposed to the analysing potential V on the outer cylinder, then a phase sensitive amplifier locked to the modulation frequency allows us to record d[E n(E)]/dE, related to the derivative of the energy spectrum by

$$d[n(E)]/dE = (1/E)\ \{d[E\ n(E)]/dE - n(E)\}$$

However, since the second term within square brackets is small and its variation with respect to the first one is also small, many spectra recorded with the CMA are published in the form of d[E n(E)]/dE

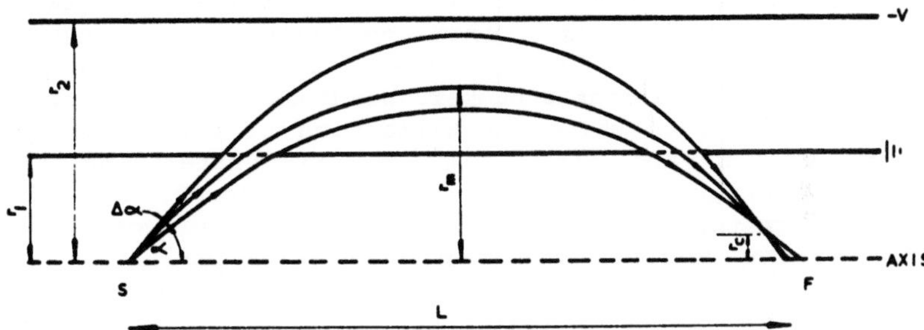


**Figure 5**. Cylindrical Mirror Analyser, after Bishop, Coad, and Rivière (14). The apparatus consists of an electron gun located on the common axis of two cylinders, which act as an electrostatic velocity selector device. The inner cylinder with radius $r_1$ is grounded and contains two ring-shaped apertures, through which electrons are admitted in the velocity selector, and escape to the collector located on the axis of the apparatus, behind the electron gun. This collector is generally associated with an electron multiplier. The outer cylinder with radius $r_2$ is set at a potential V negative to ground, thus creating a cylindrically symmetric electric field. Let L be the distance along the axis between the electron source (the sample) and the collector, and $\alpha$ the mean angle of acceptance of the entrance aperture. It may be shown (12) that focusing on the collector is second order with respect to $\alpha$ provided that $\alpha = 42.18°$ and the following relation is satisfied :

$$(E/V)\ \ln(r_2/r_1) = 1.31$$

In this relation E is the kinetic electron energy in eV, and V is the voltage applied to the outer cylinder. Under these conditions, $L = 6.13\ r_1$. Due to the second order focusing condition, the divergence of the incoming electrons around $\alpha$ may be relatively large, say ± 3 deg, so that the transmission efficiency of the apparatus is good.

Figure 3 shows the derivative of the carbon KLL Auger peaks (upper curve). The signal to noise ratio is much better than for the RFA. The n(E) spectrum with the advantages of the phase sensitive amplifier can also be obtained by modulating the voltage of the Wehnelt electrode of the electron gun, which results in a modulation of the incoming beam intensity.

One drawback of the CMA with respect to the RFA is that the distance from the sample to the front of the apparatus is very critical, so that lengthy adjustments may be necessary.

**2.2.3. Identifying elements.** Auger spectroscopy is sensitive to all elements, except Hydrogen and Helium, which do not have enough energy levels for an Auger process to occur. For a given element transition probabilities may differ largely from one Auger line to another, and for a given type of Auger line the transition probability is a non monotonic function of element Z number. Handbooks of Auger line energy and spectra exist, from which identification of the elements is in general straightforward (15). Fig. 6 illustrates the relative sensitivity of various Auger lines for the elements. This data are available in computer data bases, which, with the help of a specially designed algorithm, may help in element identification.

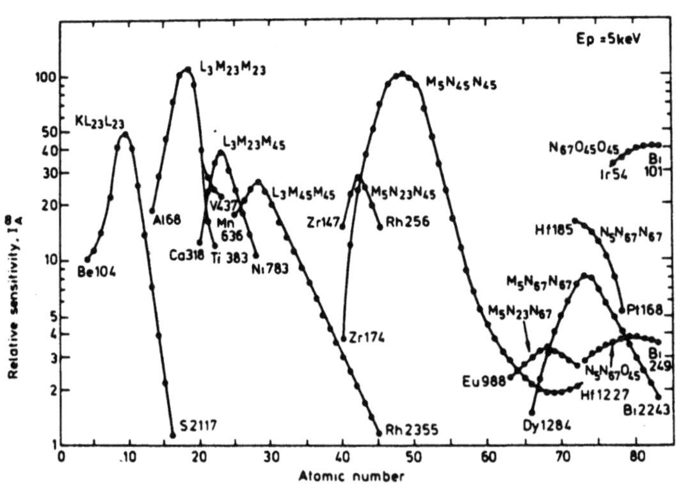


**Figure 6.** Relative sensitivity factors for derivative spectra in AES, taken from Davis *et al.* (15). The numbers at the end of the curves indicate the energies of the peaks (in eV) and the elements at these points. Taken from Briggs and Seah (20).

**2.3. Towards a quantitative analysis.**

Quantitative Auger analysis is a difficult problem, and requires much care in recording and interpreting data. As already mentioned, Auger spectra may be recorded in two modes : the distribution or n(E) mode, and the distribution derivative or dn(E)/dE mode, both modes allowing quantitative analysis. However, conditions to perform quantitative analysis are more restrictive in the derivative mode, which is much easier to practise because it removes the high background of electrons. On the contrary, using the distribution mode requires to solve properly the difficult problem of background subtraction. After giving general rules applying in both cases, we shall specify the restrictive conditions in each mode.

***General rules for quantitative analysis.***

**(i)** A quantitative analysis requires the knowledge of the Auger spectrum of reference materials in order to perform relative measurements. Absolute measurements are difficult, both experimentally, because the surface topography and the instrumental specifications should be known accurately, and theoretically, because Auger transition probabilities are difficult to calculate, particularly in the solid state.

**(ii)** The depth distribution of atoms actually excited in the sample, supposed homogeneous, weighs the contribution to the Auger signal of atoms belonging to a layer located at a given depth z. Let us denote this function as the excitation function, $R(z,\alpha)$, where $\alpha$ stands for the angle of incidence of the primary beam. $R(z,\alpha)$ differs from the attenuation law of the primary beam because some energetic electrons are backscattered deep in the solid and contribute to the excitation of atoms close to the surface. Monte-Carlo calculations (16,17) show that, in conditions where the penetration depth of the exciting beam is reasonably larger than the Auger escape depth, the excitation function is roughly a constant over the range of Auger electron escape depth. This 'constant' is usually written as $R = 1 + r_M(E_{AX},E_p,\alpha)$ where $r_M$ is the backscattering factor. The backscattering factor depends on the material M, on the primary electron beam energy, $E_p$, and angle, $\alpha$, and on the ionization energy for Auger excitation, $E_{AX}$. Ishimura and Shimizu (18) have computed backscattering factors by Monte-Carlo techniques. Values of R=1+r range from 1 to 2.5.

**(iii)** The mean free path $\lambda$ of electrons has to be known for the various solid phases they will traverse in their escape path, in order to calculate the contribution to the Auger signal of emitting atoms located at a given depth. If $\lambda$s are not available from a direct measurement, they may be estimated from plots similar to that of Fig. 2, with the help of empirical formulae for their energy dependence. If $\theta$ is the escape angle of the electrons, the attenuation law for Auger intensity is assumed to be of the form :

$$I(z)/I(0) = \exp(-z/\lambda\cos\theta).$$

**(iv)** It is necessary to know the depth distribution of the element analysed, which can be a very difficult problem when the element of interest has not been simply deposited on top of the solid surface. One example of such a situation is given below, and one way to gain information on that distribution is described.

***Quantitative analysis in the derivative mode.*** This procedure has three restrictions.
**(i)** Firstly, one must be sure that Auger line shapes are identical in the reference and in the analysis spectra, and furthermore that these line shapes do not change in the course of the analysis. This is a critical requirement, as illustrated in Fig.1.
**(ii)** Secondly, the analysers used for recording reference spectra and analysis spectra must have the same resolution, and
**(iii)** Thirdly, the same modulation amplitude must be used in both cases. Conditions (ii) and (iii) can be met by recording reference and analysis spectra with the same analyser at the same settings.

***Quantitative analysis in the distribution mode.*** This procedure has the great advantage of getting rid of the problem of line shape. However, due to the high level of the electron background, it is sometimes difficult to obtain a signal well separated from this background. A related problem is background subtraction, which is rather complex, because the results are very sensitive to boundary conditions. Seah (19) has discussed this problem carefully. Usually, the background under an Auger peak is not smooth : it contains a step-like contribution which appears clearly in Auger peaks excited by X-Rays.
It is not the purpose of this introduction to AES to dwell into details about principles, instrumentation or quantification techniques The interested reader is referred to the comprehensive and very interesting book by Briggs and Seah, where these aspects are developed (20).

## 2.4. Coupling AES to Electrochemistry.

The necessity of minimizing electrode contamination during transfer from an electrochemical cell to a surface analysis UHV chamber has motivated a few groups to build especially designed transfer systems. The number of such systems was relatively small until 1987, but seems presently to explode. Pioneering work was done by Hubbard's group (21,22). Early work was also done by Yeager's group (23,24), and by Ross's (25). More recently transfer systems were described by Bellier *et al.* in Paris (26), Aberdam *et al.* in Grenoble (27), Solomun *et al.* in Berlin (28), and others. The transfer may be performed in a controlled atmosphere (26) or in vacuum (21,25,27,28). The electrochemical cell may either be of the thin layer type (25) or of the classical three electrode type, using the dipping technique (27,28).

**2.4.1. System design**. Fig. 7 illustrates schematically the transfer system built in Grenoble (27). It is a combination of a UHV chamber equipped with LEED, CMA, RFA, gas analyser, etc., with an electrochemical (EC) chamber in which an electrochemical cell may be introduced and operated in a pure argon atmosphere.

Both parts are coupled through a gate valve, and may be pumped down independently, an ion pump being attached to the UHV chamber and a turbo-molecular pump to the EC chamber.

A sample manipulator with a horizontal axis and X,Y,Z,θ degrees of freedom allows sample holder positioning in the UHV chamber, and a vertical axis sample manipulator is used to adjust the dipping of the electrode in the electrolytic solution.

A magnetically actuated transfer rod transports the crystal holder from one manipulator to the other. Several cold traps are installed in order to maintain a better cleanliness of the residual atmosphere during the various stages of the experiment.

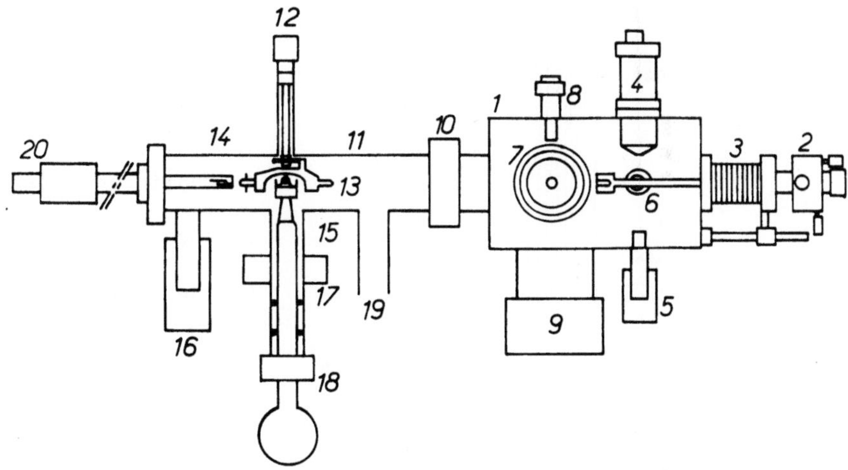


**Figure 7**. Schematic diagram of the tranfer system.
1: UHV chamber. 2: sample manipulator. 3: quick bellows translation system. 4: CMA. 5: gas analyser. 6: ion gun. 7: LEED. 8: auxiliary electron gun. 9: ion pump. 10: gate valve. 11: EC chamber. 12: vertical sample manipulator. 13: crystal holder. 14: sample. 15: electrochemical cell. 16: gas analyser. 17: gate valve. 18: cell holder. 19: gate valve to turbo-molecular pump (not shown). 20: transfer rod.

**2.4.2. Electrochemical cell**. In order to guarantee a controlled argon atmosphere in the EC chamber a pyrex glass cell has been especially designed to ensure air-tightness, and to allow manipulation of the solutions, and elimination of the residual electrolyte from the electrode surface after emersion. Fig. 8 is a schematic view of this cell. It is mounted on top of a cylindrical pyrex glass tube which slips in 'O' rings assembled in the EC chamber. This 'O' rings ensure both the mechanical alignment of the cell assembly and air-tightness. Solutions are pushed from the bottom tanks to the cell by means of pressurised argon gas. Residual electrolyte film or droplet are eliminated from the electrode surface after emersion by blowing argon gas, which disperses the liquid at once.

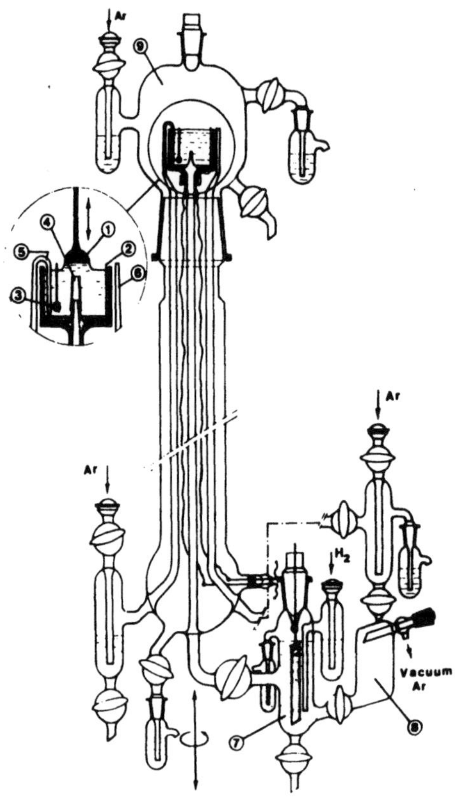


**Figure 8.**
Schematic view of the electrochemical cell.

1:working electrode.

2: counter electrode.

3: secondary electrode.

4: Luggin capillary.

5: argon bubbling tube.

6: capillary to blow argon onto the electrode.

7: Hydrogen reference electrode.

8: tank for vacuum-degassing of solutions.

9: protecting cover.

## 3. EXAMPLES OF AES STUDY OF ELECTRODES.

### 3.1. UPD of lead on a platinum (100) electrode.

The platinum (100) surface, when modified by UPD of lead atoms, exhibits improved properties for the electro-oxidation of formic acid (29). The question then arises about the mechanism of this improvement: is it of a purely chemical nature, or is it due to some particular structural feature of the adsorption layer ? A step towards the answer is to study the structure and composition of the deposited layer of lead. This has been done by LEED and AES.

**3.1.1. Voltammetry.** Electrosorption of lead on Pt(100) from a 0.1 M perchloric acid solution millimolar in $Pb^{2+}$ ion is characterized in voltammetry by a very sharp, nearly reversible UPD peak located at about 0.9 V positive with respect to bulk Pb deposition, at about +0.75 V(RHE) (30). Fig. 9 shows the corresponding voltammogram. The amount of charge under the peak is 220 μC/cm$^2$, from which the double layer contribution must be subtracted. With the assumption that the lead ion completely discharges on adsorption, this corresponds to a coverage of 0.45 lead atoms per platinum surface atom.

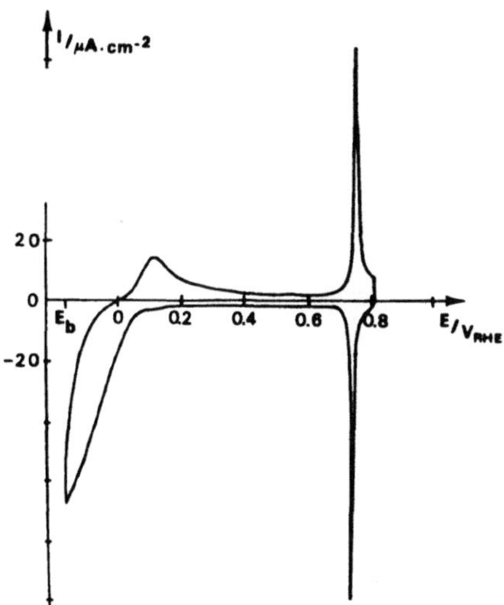


**Figure 9.** Voltammogram showing the electrosorption of lead on Pt(100) in perchloric acid aqueous solution (29). The electrode was prepared in the UHV chamber, and vacuum tranferred in the EC chamber prior to immersion in the 0.1M $HClO_4$, $10^{-3}$M $Pb^{2+}$ solution. Scan rate : 50 mV/s.

**3.1.2. LEED.** It is not possible in this paper to develop the details of this LEED study. Let us, however, summarize the results as follows.
**(i)** The Pt(100) surface prepared in UHV exhibits a reconstruction, characterized by a complex LEED pattern, subject to subtle variations, and known as the (5x20) pattern. Fig. 10 shows this pattern, which is the result of the superposition of a hexagonal compact layer of platinum, slightly distorted, on the regular Pt(100) square lattice.

The (5x20) LEED pattern is due to the Moiré, or coincidence, pattern between these two lattices. The notation (5x20) is a short notation for the (2x2) matrix **[M]** which determines the two basis vectors **A**, **B**, of the new surface net from the basis vectors **a**, **b**, of the non reconstructed surface net.
**(ii)** When immersed in an aqueous solution, it seems that the Pt(100) surface loses its characteristic reconstruction and converts to the regular termination denoted by (1x1) (Fig. 10).

**(iii)** Now, if the electrode is emersed with a layer of lead electroadsorbed on its surface, then the LEED pattern shows that the deposit has long range order, and that the new surface has a c(2x2) mesh, or, in matrix notation, a (1,-1,1,1) mesh (Fig.10).

**a.**
"(5x20)" pattern of the clean, reconstructed Pt(100)
Primary energy:55.8 eV
Angle of incidence:10 deg.

**b.**
(1x1) pattern of the Pt(100) after voltammetry in $HClO_4$.
Primary energy:177 eV
Normal incidence.

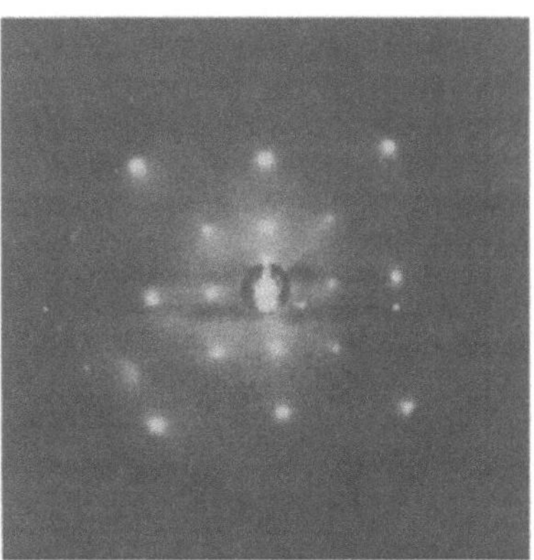

**c.**
c(2x2) pattern of lead electroadsorbed on Pt(100).Emersion at 0.65V(RHE). Primary energy: 173 eV. Normal incidence.

**Figure 10**. LEED patterns for UPD of lead on Pt(100).

This new surface mesh is twice as large as the (1x1) mesh. Since the metallic radius of the lead atom is much larger than that of platinum, it is tempting to assume that the new mesh contains one lead atom, which would be consistent with the electric charge measurement.

**3.1.3. AES.** The only way to check the above assumption is to determine directly the lead coverage by a quantitative Auger analysis. This has been performed by recording the Auger spectra of clean and lead-covered platinum surfaces with the CMA. Fig. 11 shows the Auger spectrum of the lead-covered platinum surface; the only elements detected were lead, platinum, and the elements of the anion, i.e. chlorine and oxygen. The result of the analysis was a coverage of 0.4 ± 0.04 lead atoms per platinum, in reasonable agreement with the electric charge measurements.

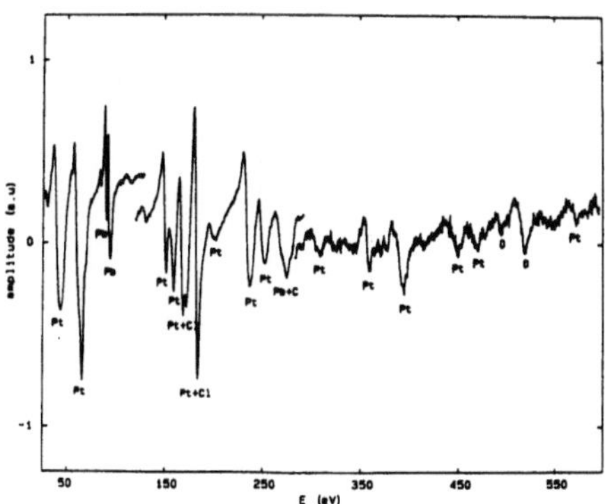


**Figure 11.** E d[n(E)]/dE Auger spectrum for the lead-covered Pt(100) surface. Primary beam energy : 3 keV. Target current : 10 μA. Electron multiplier voltage : 1.65 kV. Modulation parameters : frequency: 1.2 kHz. amplitude : 0.1 V(rms) below 100 eV, 0.5 V(rms) above.100 eV. Lock-in amplifier gain : 4000 below 290 eV, $10^4$ above.290 eV.

Once the atomic structure and composition of the lead adlayer are known, one can now face the electrocatalysis mechanisn in a better position to elucidate it.

### 3.2. UPD of silver on a platinum (100) electrode.

Improvement of the electrocatalytic properties of a metal surface might be related not only to the activity or the selectivity of the surface for the reaction of interest, but also to the stability of these properties with time. The UPD of silver on Pt(100) favours the electro-oxidation of small organic molecules (30). It has been reported, in addition, from silver adsorption on polycrystalline platinum that silver could alloy to platinum (31), a property which looks interesting with regard to stability of the silver deposit. A LEED and AES study of this system has been undertaken to check whether this assumption is correct on a single crystal surface (33).

**3.2.1. Voltammetry.** The voltammetric behaviour of silver UPD on Pt(100) is rather complex. Fig. 12 summarizes the situation. A first

UPD peak occurs close to 1.1 V (RHE), well within the potential range of oxygen adsorption. The electric charge of this peak corresponds, assuming that silver completely discharges on adsorption, to about one silver atom per platinum surface atom. This UPD peak is quite sensitive to a pre-existing oxygen adsorption on the surface, in which case the electrosorption of silver seems to occur without a well defined UPD peak. At a more cathodic potential, very close to the Nernst potential for bulk deposition of silver (about 0.62 V(RHE)), a second UPD peak occurs, with a smaller electric charge than the first peak. The corresponding desorption peak exhibits fine structure. Fig. 12 also shows that this voltammogram changes gradually when the number of cycles increases, which indicates some gradual change of the surface structure and/or composition.

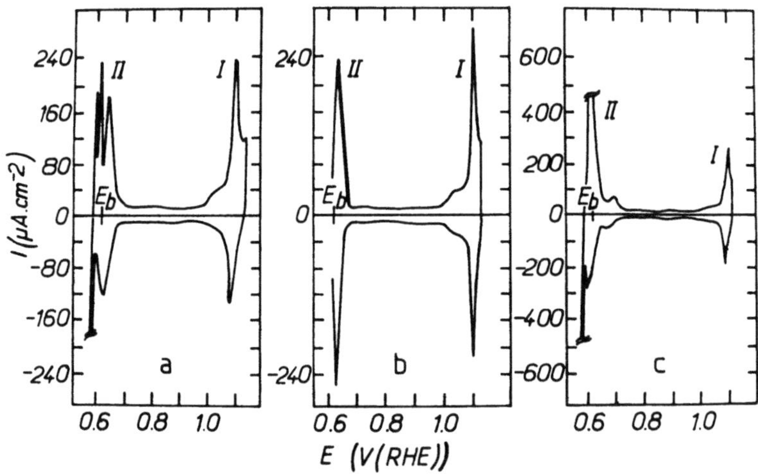


**Figure 12.** Voltammograms for UPD of silver on Pt(100) from an 1M $H_2SO_4$, $10^{-3}$M $Ag^+$ solution after **a**: 5 cycles; **b**: 12 cycles; **c**: ≈ 50 cycles. Note the disappearance of the fine structure of peak II, and the growth of a satellite on the high potential side of this peak. Scan rate: 50 mV/s.

**3.2.2. LEED.** LEED examination of the Pt(100) surface, either covered with silver or from which silver has been at least partially stripped, always shows a (1x1) type of pattern. The conclusion is that whatever occurs on the surface, the long range periodicity is saved, at least in a statistic sense. Substitutional alloying at random is not incompatible with this observation.

**3.2.3. AES and alloying.** The problem with alloying is not so much to perform an Auger quantitative analysis, since some of the most necessary information is missing, namely the depth distribution of silver in platinum, but rather to gain information about this depth distribution. Only then can a quantitative analysis be carried out.

***Measuring depth distribution by non destructive AES.*** As mentioned in section 2.3, the excitation function, $R(z,\alpha)$, is roughly a constant for a given matrix material and a given ratio between primary beam energy and ionization energy, provided the penetration depth of the primary electrons is large enough with respect to the escape depth of Auger electrons. If this is not so, the backscattering factor is no longer a meaningful quantity : the excitation function is no longer a constant, and it will weigh the contribution of the various layers of the material in a different way. Information will be gained about the depth distribution of silver if the following conditions can be met :
**(i)** The penetration depth of primary electrons must be close to the escape depth of Auger electrons.
**(ii)** Have enough knowledge about the excitation function to be able to build a model of the contribution of the different layers 1, 2,… with silver concentrations C1, C2,…respectively.
***The first condition*** is satisfied if the primary beam energy is moderate, say between 0.8 and 2 keV, and the polar angle of incidence is larger than about 70 deg. This configuration requires an auxiliary electron gun shooting on the sample at grazing incidence when the sample faces the Auger collector of the RFA analyser.
***The second condition*** was solved by Segaud (32). He found that at large angles of incidence, very close to the surface, the excitation function takes the simple form :

$$R(z,\alpha)/R(0,\alpha) = \exp(-z/\lambda_p\cos\alpha)$$

where $\lambda_p$ is an effective mean free path, which according to fitting procedures is close to the primary beam MFP, $\lambda$.
With this knowledge we are in a position to compute the contribution to the Auger signal of silver atoms in concentration C1, C2, in layers 1, 2, of the platinum matrix (it is not realistic to introduce more than two concentrations as unknowns). In practice we have to make relative measurements, which requires a reference object. This reference is chosen as a complete UPD monolayer of silver _on_ Pt(100), i.e. without any cycling. This reference should be a good one because it is very similar in structure and composition to our unknown object. Let $y(\alpha)$ and $m(\alpha)$ be the angle dependences of the calculated and measured, respectively, relative Auger amplitudes. $y(\alpha)$ takes the following form :

$$y(\alpha, C1, C2) = \frac{(I(Ag)/I(Pt))_{sample}}{(I(Ag)/I(Pt))_{reference}}$$

Fig. 13 shows $m(\alpha)$, measured on an electrode cycled about 50 times in the range of peaks I+II, at a sweep rate of 50 mV/s, and the best fit

of y($\alpha$), using a Least Mean Squares fitting program working with a gradient algorithm. Due to a problem at the stage of electrode emersion, which we shall discuss below because it is illustrative, the output we get are not numbers for C1, C2, but only ranges. The results are the following :
*Electrode cycled in the range of peak I (Fig. 12) :* ***no alloying.***
*Electrode cycled in the ranges of peak II or I+II :* ***some alloying***
***Penetration of silver in layers number 1 and 2 was detected*** *The larger the number of cycles, the larger the alloying.*

To summarize, the procedure briefly described here is efficient for determining the element concentration in the very first layers of a binary system. This may be of great value in the field of electrochemistry and in surface science. We have learnt that silver penetrates in the platinum lattice under some definite conditions. We have also learnt that this penetration occurs likely through site substitution, so that the mean lattice is preserved.

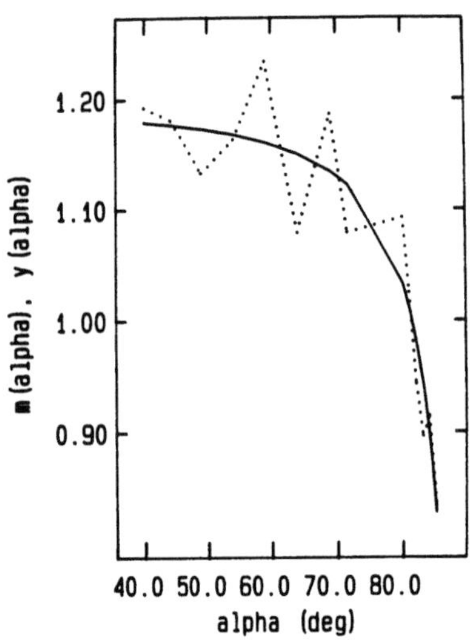


**Figure 13.**
Dependence of the Auger intensity ratio I(Ag)/I(Pt) of the sample relative to that of a reference object, on the angle of incidence $\alpha$ of the primary beam.
*Dotted line* : Dependence of m($\alpha$), the measured Auger intensity ratio.
*Full line* : Best-fit y($\alpha$), the calculated Auger intensity ratio, from which concentrations of silver in the first layers may be found.

## 4. AES AND THE PROBLEM OF ELECTRODE EMERSION.

The problem of electrode emersion with a view to 'ex-situ' measurements was addressed for the first time by Hansen (34). Kolb, in the present volume, explains how an emersion can be successful in the sense that the double layer remains intact and the electrode keeps its charge and potential. He points out the importance of dry emersion, which, unfortunately, is not expected to be the general case. In the case of wet emersion it is necessary to eliminate the excess of electrolyte from the surface. Various techniques are possible, but two questions arise at once :

**(i)** Has the excess of electrolyte actually gone away?

**(ii)** Is the interface stable during the time elapsed between emersion and electrolyte elimination ? The following examples show how AES may help answering these questions.

## 4.1. Elimination of the excess of electrolyte after a 'wet' emersion.

The following example is taken from work relative to the UPD of lead on Pt(100) in a perchloric solution (30). The excess of electrolyte left on the electrode after emersion from a perchloric solution was dispersed by blowing pressurised argon on the surface. The AES analysis performed was by no means quantitative, since it was based on calibrations of Auger signals for oxygen and chlorine published in the literature (35, 36). However, it showed that chlorine and oxygen were present in proportions close to $ClO_4$. It also showed that the amount of '$ClO_4$' on the surface was slightly smaller than 5% if the electrode was emersed at 0.65 V(RHE) from a 0.1M perchloric solution, and slightly larger than 5% if the solution is 1M instead of 0.1M. The amount of anions electrostatically adsorbed in the double layer may be estimated, knowing that the electrode potential was roughly 0.5 V positive with respect to the pzc, and that the double layer capacity is $\approx 20\ \mu F/cm^2$. The negative countercharge in the outer Helmholtz plane is then equal to $10\ \mu C/cm^2$, i.e. $6.2 \times 10^{-13}$ '$ClO_4$'$/cm^2$ or 5% $ClO_4$ molecule per platinum atom on Pt(100). This result shows that on the surface the amount of elements which constitute the anions roughly corresponds to that electrostatically adsorbed in the double layer, and that it does not depend much on solution concentration. The conclusion is that the thickness of the liquid layer after blowing argon on the electrode cannot be greater than a few times the double layer thickness.

## 4.2. Stability of the interface after emersion.

As pointed out by Kolb, the stability of the double layer after emersion is limited to the potential range where no electrochemical reaction can take place. This range may be narrowed **(i)** by the presence of dissolved molecular oxygen in the solution, which produces corrosion after emersion, or **(ii)** by slow oxygen adsorption after emersion, coupled to metal deposition.

**4.2.1. Corrosion by dissolved oxygen.** This example is again taken from the UPD of lead on Pt(100) (30). Sometimes, following UPD of lead and emersion of the electrode at 0.65 V(RHE), it happened that LEED showed an (1x1) type of pattern. However, AES showed a lead-free surface, which was in contradiction with expectations. On the other hand, when the electrolytic solution was degassed by vacuum pumping instead of by bubbling argon, a plausible amount of lead was found on the surface. It was concluded that after emersion the following reactions occurred in the first case :

$$Pb_{ads} \rightarrow Pb^{2+} + 2\ e^-$$
$$1/2\ O_2 + 2\ H^+ + 2\ e^- \rightarrow H_2O$$
$$\overline{Pb_{ads} + 1/2\ O_2 + 2H^+ \rightarrow Pb^{2+} + H_2O}$$

up to the point where no lead was left on the surface. To prevent this occurence, the concentration of molecular oxygen dissolved in the solution should be $< 10^{-6}$M.

**4.2.2. Metal deposition produced by slow oxygen adsorption.**

This example is taken from the UPD of silver on Pt(100) (33). In order to make more accurate the AES analysis of the penetration of silver in the lattice of Pt(100), it was decided to strip anodically the first layer of silver adsorbed *on* the surface by emersing the electrode at 1.125 V(RHE). Unfortunately, at this potential electroadsorption of oxygen-containing species is well under way, and it is known to be a slow, irreversible process. AES analysis of the electrode emersed at this potential always shows an excess of silver (less than, close to, or even larger than one monolayer), and a rather large oxygen peak. It is likely that reactions of the following type occur on the surface after emersion :

$$H_2O \rightarrow OH_{ads} + H^+ + e^-$$
$$Ag^+ + e^- \rightarrow Ag_{ads}$$
$$\overline{Ag^+ + H_2O \rightarrow Ag_{ads} + OH_{ads} + e^-}$$

Thus, instead of having no silver *on top* of the surface, the actual concentration $C_0$ of silver on top was a new unknown. This is the reason why we obtained only ranges for $C_1$ and $C_2$. This problem can probably be overcome by emersing the electrode at a potential where no oxygen-containing species can adsorb.

## 5. CONCLUSION.

In this very short introduction to Auger spectroscopy, a few starting keys have been given, with some emphasis on problems related with quantitative analysis. Quantification involves reference spectra, and a choice between the distribution or the derivative mode, the former requiring sophisticated background subtraction, and the latter requiring great care about line shapes, which may change because of changes in the chemical environment. In addition, with single crystals care must be taken of possible diffraction effects on the escape path of Auger electrons. Two examples taken from the author's work show how AES may contribute to a better knowledge of some electrochemical processes, namely the ordering of a UPD layer of lead on Pt(100), and the alloying of silver in the very top layers of Pt(100) under voltammetric cycling. However it must be stressed that the stability of the double layer after electrode emersion must be checked very carefully. In this respect, a careful approach to the problem, such as that initiated by Hansen, is necessary.

## REFERENCES

**1**. D. M. KOLB. 'UHV techniques in the study of electrodes surfaces.' *Z. Physik. Chem. N. F.* **154**, 179 (1987).

**2**. P. AUGER. *C. R. Acad. Sc. Paris* **180**, 65 (1925).

**3**. C. MOLLER. *Z. Phys.,* **70**, 786.(1931)

**4**. H. S. W. MASSEY and E. H. S. BURHOP. *Proc. Roy. Soc. London* **A 153**, 661 (1936).

**5**. M. E. ROSE. *Multipole Fields*. Wiley Interscience, (1955).

**6**. D. CHATTARJI. *The theory of Auger transitions.* Academic Press (1976).

**7**. W. BAMBYNEK, B. CRASEMAN, R. W. FINK, H. U. FREUND, H. MARK, C. D. SWIFT, R. E. PRICE and P. VENUGOPALA RAO. 'X-Ray Fluorescence Yields, Auger, and Coster-Kronig Transition probabilities.' *Rev. Mod. Phys.,* **44**, 716 (1972).

**8**. T. E. GALLON. 'Current problems in Auger Electron Spectroscopy', in *Electron and Ion Spectroscopies of Solids.* Ed. Fiermans. Plenum (1978).

**9**. H. H. MADDEN. *J. Vac. Sci. Technol.* **18**, 677 (1981).

**10**. M. P. SEAH and W. A. DENCH. *Surf. Interf. Anal.,* **1**, 2 (1979).

**11**. T. S. KING and R. G. DONNELLY. *Surf. Sci.* **151**, 374 (1985).

**12**. D. ROY and J. D. CARETTE,. 'Design of Electron Spectrometers for Surface Analysis', in *Topics in Current Physics : Electron Spectroscopy for Surface Analysis.* Ed. H. IBACH. Springer Verlag, **4**, 12 (1977).

**13**. D. ABERDAM, R. BAUDOING, E. BLANC and C. GAUBERT.
*Surf. Sci.* **57**, 306 (1976).
*Surf. Sci.* **71**, 279 (1978).

**14**. H. E. BISHOP, J. P. COAD and J. C. RIVIERE.
*J. Electron Spec.* **1**, 389 (1972-1973).

**15**. L. E. DAVIS, N. C. MacDONALD, P. W. PALMBERG and G. E. RIACH, R. E. WEBER. *Handbook of Auger Electron Spectroscopy.* 2nd ed., Physical Electronics Division, Perkin-Elmer Corp., Minnesota, (1976).

**16**. M. CAILLER and J. P. GANACHAUD. *Surf. Sci.* **154**, 524 (1985).

**17**. J. Le HERICY. *Le Vide, Les Couches Minces.*
**Special Issue**, 307 (1979).

**18**. S. ISCHIMURA and P. R. SHIMIZU. *Surf. Sci.* **112**, 386 (1981).

**19**. M. P. SEAH. *Surf. Interf. Anal.* **1**, 86 (1979).

**20**. *Practical Surface Analysis.* Eds. D. BRIGGS and M. P. SEAH. Wiley, (1983).

**21**. R. M. ISHIKAWA and A. T. HUBBARD.
*J. Electroanal. Chem.* **69**, 317 (1976).

**22**. A. T. HUBBARD, R. M. ISHIKAWA and J. KATEKARU.
*J. Electroanal. Chem.* **86**, 271 (1978).

**23**. W. E. O'GRADY, M. Y. C. WOO, P. L. HAGANS and E. YEAGER.
*J. Vac. Sci. Technol.* **14**, 365 (1977).

**24**. A. S. HOME, E. YEAGER and B. D. CAHAN.
*J. Electroanal. Chem.* **150**, 181 (1983).

**25**. P. N. ROSS. *Surf. Sci.* **102**, 463 (1981).

**26**. J. P. BELLIER, J. LECOEUR and A. ROUSSEAU.
*J. Electroanal. Chem.* **200**, 55 (1985).

**27**. D. ABERDAM, R. DURAND, R. FAURE and F. El-OMAR.
*Surf. Sci.* **171**, 303 (1986).

**28**. T. SOLOMUN, W. RICHTERING and H. GERISCHER.
*Ber. Busenges. Phys. Chem.* **91**, 412 (1987).

**29**. F. El-OMAR. Ph.D. Thesis. Université Grenoble, France (1986).

**30**. D. ABERDAM, S. TRAORE, R. DURAND and R. FAURE.
*Surf. Sci.* **180**, 319 (1987).

**31**. R. C. SALVAREZZA, D. V. VASQUEZ MOLL, M. C. GIORDANO and A. J. ARVIA. *J. Electroanal. Chem.* **213**, 301 (1986).

**32**. J. P. SEGAUD, E. BLANC, C. LAUROZ and R. BAUDOING. *Surf. Sci.* **206**, 297 (1988).

**33**. D. ABERDAM, S. TRAORE, C. SALEM, R. DURAND and R. FAURE. In preparation.

**34**. W. N. HANSEN, C. L. WANG and T. W. HUMPHERYS. *J. Electroanal. Chem.* **93**, 87 (1978).

**35**. G. A. GARWOOD, A. T. HUBBARD. *Surf. Sci.* **112**, 281 (1981).

**36**. J. L. GLAND. *Surf. Sci.* **93**, 487 (1980).

# PHOTOELECTRON SPECTROSCOPY (XPS AND UPS) OF ELECTRODE SURFACES

R. KÖTZ
*Asea Brown Boveri*
*Corporate Research*
*CH-5405 Baden*
*Switzerland*

ABSTRACT. A survey is given of the various applications of x-ray and ultra-violet photoelectron spectroscopy (XPS and UPS) to the investigation of electrochemical phenomena. After a short introduction to the physics and experimental techniques of XPS and UPS a number of specific examples demonstrate the type of information available with photoelectron spectroscopy when used for ex-situ studies of the electrode/electolyte interface. XPS was first applied to the study of anodic oxide layers on noble metal electrodes and of underpotential deposition of Cu and Ag on Pt. These XPS investigations are discussed in detail together with recent UPS data. In addition, XPS studies of corrosion systems and nonmetallic electrodes are briefly mentioned. Finally, recent XPS and work function studies of the emersed double layer are presented.

## 1. Introduction

X-Ray Photoelectron Spectroscopy (usually designated as XPS), also known as Electron Spectroscopy for Chemical Analysis (ESCA), has become a standard surface analytical tool during the last two decades. A major step towards the success of XPS were the developments of Siegbahn /1/ who was awarded the Nobel prize in 1981.

The development of XPS has to be correlated to the rapid progress of Photoelectron Spectroscopy (PS) in general. Experimental and theoretical premises for this rapid development were provided in the late 50s when reliable ultrahigh vacuum (UHV) control was achieved, and by the success of the three-step model of photoemission /2/ around 1970. Excellent overviews of the development of Photoemission Spectroscopy have been given by Spicer /3/ and by Feuerbacher et al. /4/.

In view of the tremendous success of PS it is not surprising that already in 1971 XPS was first applied to the study of electrode surfaces by Kim and Winograd /5/, who investigated the chemical and electrochemical oxidation of platinum and, later, the underpotential deposition of Cu and Ag on Pt /6, 7/. However, compared to other fields, XPS and especially Ultra-violet Photoemission Spectroscopy (UPS), were only scarcely applied in electrochemistry thereafter.

A major drawback of PS applied to electrochemical surfaces is, of course, the fact that XPS or UPS are ex-situ techniques, since they require UHV. Whenever there is the possibility to gain the same information with an in-situ technique this is to be preferred. Recently, however, it has been demonstrated clearly that under certain circumstances the electrochemical interface can be removed intact from the electrolyte and transferred to the



*C. Gutiérrez and C. Melendres (eds.),*
*Spectroscopic and Diffraction Techniques in Interfacial Electrochemistry,* 409–438.

ultra high vacuum of an XPS or UPS system (see the chapter by D.M. Kolb). In addition, more or less sophisticated transfer systems have been developed recently /8 - 13/ in order to effect the electrode transfer under an inert gas and avoid any contact with air.

The scope of the present chapter is to bring XPS (and also UPS) nearer to those who deal with electrochemical processes. The information provided by XPS - although being ex-situ - is of utmost importance for the control and understanding of interfacial electrochemical processes. The approach will be rather pragmatic, and fundamental theoretical and experimental aspects of photoemission spectroscopy will be treated only in so far as they are necessary to understand and interpret measured data. Several articles in the literature review the fundamentals and new developments in PS in an extensive and competent manner /3,4,14,15/. In this chapter several examples from different areas of electrochemistry will demonstrate the use and usefulness of XPS and also UPS in electrochemical studies.

## 2. X-Ray and UV Photoelectron Spectroscopy

In the following a brief survey of the fundamental aspects of the photoemission process will be given. For further reading the books by Feuerbacher et al /15/ and by Cardona and Ley /14/ are recommended. Then the information provided by XPS, and special aspects of the data evaluation, will be discussed. Finally, some experimental aspects especially the sample transfer will be briefly reviewed.

The basic arrangement of an XPS or UPS experiment is sketched in Fig. 1. Photons from a light source hit the sample and cause emission of photo electrons. Electrons travelling within a certain emission cone are analysed with respect to their kinetic energy, and the number of electrons with a defined kinetic energy per time interval is measured.

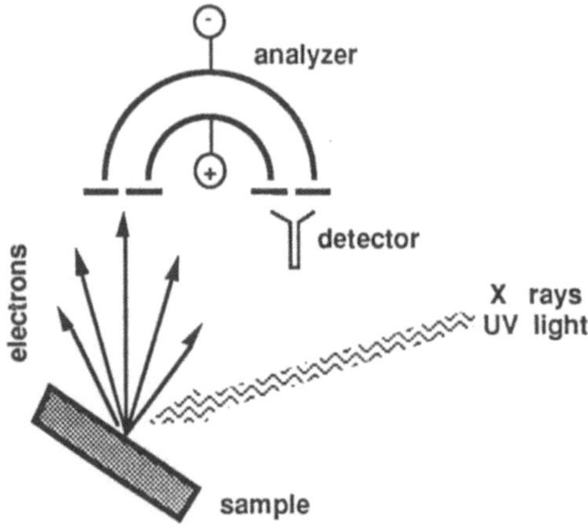


Figure 1. Sketch of the basic setup of a photoelectron spectrometer for XPS or UPS.

### 2.1. FUNDAMENTAL ASPECTS

Within a single particle approximation the total energy of a photon absorbed in the sample is used to excite an electron from an initial state energy $E_i$ to some empty final state with energy $E_f$. The probability for this optical excitation is proportional to the combined density of states. Conservation of energy requires that

$$E_f = E_i + h\nu \quad (1)$$

Conservation of momentum is provided by a lattice vector or, in some cases, by a surface vector. The optical excitation is the first step in the above mentioned three-step model. The second step is the transport of the electron to the surface, where the third step, escape of the photoelectron, occurs, provided the energy of the electron is higher than the vacuum level $E_{vac}$. During transport to the surface the electron may lose energy as a consequence of scattering events. Only those electrons which are excited within a distance from the surface which is smaller than the mean free path will reach the surface without a scattering event, and will retain their energetic information; the other so-called secondary electrons contribute to a structureless background only. This is what makes XPS surface sensitive, the information depth being in the order of 20 Å for Mg or Al x-ray sources.
At the surface the electrons have to overcome the surface potential barrier which is the work function $\phi$ of the sample. Therefore their kinetic energy in vacuum will be reduced by the work function when compared to the final state energy $E_f$ within the sample.

$$E_{kin} = E_f - \phi \quad (2)$$

The kinetic energy of the emitted electron is detected at the analyzer, which is in most cases a cylindrical mirror analyzer (CMA). The projection of the density of states (DOS) below the Fermi energy of the solid sample to the kinetic energy measured at the analyzer of a photoelectron spectrometer is sketched in Fig. 2, from which it becomes evident that only filled states are reflected in the measured spectrum. The increase of background intensity

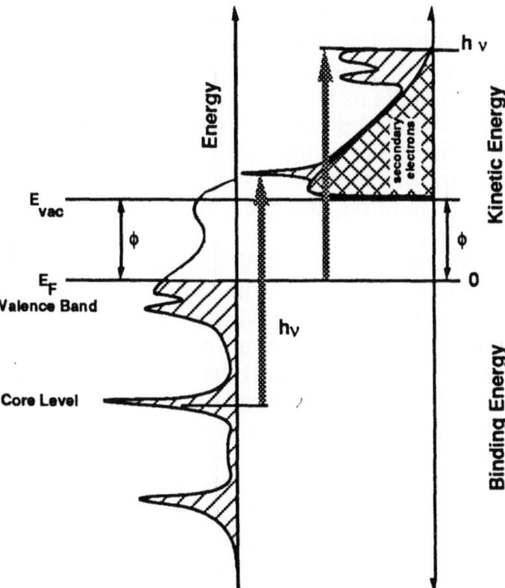


Figure 2. Sketch of the projection of the density of electronic states (DOS) in a solid onto the kinetic energy distribution measured at the analyzer of a photoelectron spectrometer.

with decreasing kinetic energy originates from secondary electrons which have lost their information due to scattering events.

The energy range between $E_{kin}= 0$ and $E_{kin}= \phi$ exhibits zero emission intensity because the surface barrier prevents electrons with energies $< \phi$ from leaving the sample, and this effect can be used to measure the work function of the sample. The measured spectrum reflects the kinetic energy distribution of emitted electrons, the highest kinetic energy corresponding to the excitation energy hν. The binding energy $E_b$ (energy of the initial state with respect to the Fermi level) of an electron is, obviously

$$E_b = h\nu - E_{kin} \quad (3)$$

The $\phi$ term is missing, because in practice the analyzer is calibrated such that $E_{kin} = h\nu$ corresponds to $E_b = 0$. The excitation energy hν is constant during the course of an experiment. Different energies can be achieved by changing the light source. As indicated in Fig. 2 the electronic states in a solid are divided into two regions, the core levels and the valence band. While XPS with a photon energy of typically > 1000 eV covers both regions, UPS is also called valence band spectroscopy, because the energy of the generally used He I and II gas discharge lamps is only about 20 and 40 eV, respectively. With modern synchrotron radiation sources the borderline between XPS and UPS fades away.

## 2.2. INFORMATION PROVIDED BY XPS AND UPS

2.2.1. *Chemical Analysis.* As discussed in the above paragraph, XPS provides information about the energetic position of localized core levels, which is the reason why XPS can be used to determine the elemental composition of solid samples. Each element has a characteristic distribution of core levels, which can be used as a fingerprint for the presence of this element. These spectra are tabulated /16/. In practice it is sufficient to monitor the strongest level, whose position is characteristic for the element, within ±5 eV. These levels are also tabulated /16/. Occasionally levels of different elements overlap, so that one has to switch to the second strongest emission peak. Monitoring of the characteristic XPS peak of an element tells us whether this element is a component of the sample or not. If one has no idea about the composition of the sample, an overview spectrum has to be obtained, covering the total kinetic energy range from 200 eV to the incident photon energy (1253.6 eV for Mg and 1486.6 eV for Al). These spectra are either time consuming or of relatively low resolution, and just serve as a starting point for further analysis.

In addition to qualitative elemental analysis, XPS also provides information on the relative amounts of the different elements present. Although the emission intensity of a core level as detected by XPS is independent of the environment of the respective element, in order to determine the amount of a certain element on an absolute scale too many parameters concerning the light source, the sample and the analyzer have to be known. If, however, the total number of components or the amount of one component is known, the amount of additional elements can be determined relatively to this known parameter. The amount of an element within the sample is directly proportional to the signal strength of the respective core level divided by the relative ionization cross section (also tabulated /16,17/). For a rough estimate it is sometimes sufficient to measure the peak height of the respective core level. For a more accurate determination the emission peak has to be integrated, and possible satelites have to be taken into account, but even so the analysis is accurate only within 10%.

The energetic position of core levels and, therefore, the emission energy is strongly dependent on the environment of the respective element. This is what makes XPS a technique for chemical analysis. The so called "chemical shift" of core level binding energies as a function of the environment is a consequence of several physical processes during the excitation and transport of the photoelectron whose treatment lies beyond the scope of this chapter. Chemical shifts depend on the charge on the atom, on the charge on the surrounding atoms, and on intra- and inter-atomic relaxation processes. In a rough approximation one can say that the higher the oxidation state of a certain element, the higher the respective core level binding energy. For Pt the binding energy of the 4f level for the metal is 71.0 eV while it is found at 73.8 eV for PtO and at 74.7 for $PtO_2$. The chemical shift is much smaller for lead, where the binding energy of the Pb4f level is 136.5 eV in the metal, and those of PbO, $Pb_3O_4$, and $PbO_2$ are all in the range 137.0 - 138.5 eV.

For the identification of the chemical environment of the elements the corresponding chemical shifts have been tabulated /16/, and are commonly measured with respect to standard reference samples. In the case of insulating samples charging effects, which increase the binding energy due to positive charging of the sample, have to be considered. In electrochemistry a new effect, the so-called electrochemical shift, has to be taken into account. As will be discussed in more detail in 3.6, it has been observed by D´Agostino and Hansen /18/ and also by Kötz et al. /19/ that the binding energy of adsorbed counter ions depends on the potential at which the electrode was emersed before transfer to the UHV system.

Oxygen is of particular importance for the investigation of electrode surfaces treated electrochemically in aqueous electrolytes. A particularly complicated situation is encountered when interpreting binding energies of the oxygen O1s level because the O1s signal may originate from oxide layers, from adsorbed oxygen-containing counter ions or from water and hydroxyls. In addition, an O1s signal may also arise from organic contaminants. To complicate matters even further, the binding energy of many oxygenated species, like $H_2O$, depends on the substrate on which they are adsorbed. Fuggle /20/ has investigated this aspect in detail and gives a collection of O1s binding energies for several oxygen species on different metal surfaces. Binding energies of chemisorbed oxygen, for instance, cover the range from 528.8 eV to 533.6 eV, and for water the O1s level may occur between 532.5 eV and 535.7 eV. Thus O1s levels of electrode surfaces should be interpreted with special care and, if possible, a second complementary surface technique like thermal desorption or EELS should be used.

2.2.2. *Depth Profiling*. Depth profiling in XPS can be achieved in two different ways, (i) destructively and (ii) non-destructively. The former technique is the one applied most commonly and uses an ion etching technique (argon ion bombardment) for surface abrasion. A consecutive sequence of XPS measurements and etching processes yields the desired information about the chemical composition of the sample down to some μm below the surface. Typical sputtering rates are 10 ... 100 Å/min, depending on energy and current of the ions. Disadvantages of sputtering are the destruction of the surface and the danger of preferential sputtering, which occurs when one of several elements has a different sputtering rate, and consequently the composition of the surface is altered by the etching process. This is particularly problematic when examining oxide surfaces: in many cases the oxygen is sputtered preferentially so that e.g. $RuO_2$ turns into metallic Ru even after mild sputtering. A list of some oxides and their tendency toward preferential sputtering of oxygen has been given by Kim et al. /21/.

Non-destructive depth profiling within the information depth of the XPS technique is achieved via angle dependent measurements. If the the sample normal is tilted with respect to the direction of detection by an angle $\varepsilon$, the effective information depth will be reduced by $\cos \varepsilon$. Thus, in the case of an oxide layer on top of a metal substrate the relative contribution of the oxide as compared to the substrate will be increased by tilting the sample (see fig. 1). This technique tells immediately whether a certain species is on top of the sample or evenly distributed within the bulk. Pijolat and Hollinger /22/ have used angle-dependent XPS measurements for quantitative depth profiling assuming a stratified layer model and smooth sample surfaces. Angle dependent measurements of XPS peaks originating from a layer on top of a substrate can be used to determine the layer thickness according to /23/

$$I_o / I_s = I_o^{\infty} / I_s^{\infty} (1 - e^{-d/(\lambda \sin \theta)}) / e^{-d/(\lambda \sin \theta)} \tag{4}$$

where d is the thickness of the overlayer, $\lambda$ the mean free path of electrons in the overlayer, $\theta$ the angle between surface plane and direction of electron detection, $I_o/I_s$ is the experimentally determined intensity ratio of overlayer and substrate peaks, and $I_o^{\infty}$ and $I_s^{\infty}$ are the respective values for infinite thickness. In the case of a metal oxide on top of the same metal the ratio $I_o^{\infty}/I_s^{\infty}$ is mainly determined by the atomic densities of the overlayer and substrate /21,23/.

A second possibility of non destructive depth profiling within the sampling depth is the use of different excitation sources with varying photon energies. This technique is based on the fact that the mean free path of the electrons in the solid, and thus the information depth, is a function of the electron energy. While for MgKα excitation with a photon energy of 1253.6 eV the sampling depth is roughly 20 Å, a TiKα (4510 eV) source would provide information from a surface layer of 60 Å thickness.

2.2.3. *Sensitivity*. According to the three types of information provided by XPS, sensitivity has to be discussed in terms of (i) elemental sensitivity, (ii) energetic resolution, and (iii) surface sensitivity.

(i) The answer to how much of a certain element can be detected by XPS depends very much on its specific scattering cross section. On a relative scale, these cross sections are tabulated with respect to either the cross section of carbon, that of sodium /17/ or of fluorine /16/. Scattering cross sections vary by a factor of 30, from that of 0.5 for B to that of 16 for Sb, both compared to carbon. Detection limits can be easily determined by adsorption measurements, either from the gas phase or by metal underpotential deposition (UPD) from the liquid electrolyte phase (including sample transfer). As a thumb rule one can assume that 10% of a monolayer is easily detected. Modern instruments are capable of detecting $10^{12}$ atoms (0.1% of a monolayer) of gold (scattering cross section 10 vs. C) on a substrate /24/.

(ii) With respect to the determination of chemical shifts´the limiting factor is the energetic resolution of XPS, which is determined by the natural line width of the X-ray excitation source of about 1.2 eV, and not by instrumental resolution. Thus it is difficult to resolve small shifted peaks from an unshifted main peak. In order to reduce the line width of the excitation source X-ray monochromators may be used. They yield a line width of the order of about 0.3 eV /17/, but source intensity is reduced. The low resolution can in some cases be compensated by a deconvolution procedure. Assuming Gaussian or Lorentzian line shapes the measured asymmetric emission intensities may be deconvoluted into several subspectra which would otherwise be hidden beneath the strong main peak. These deconvolution procedures are available with the software of modern XPS machines, but should be used with care, since the theory of XPS line shapes is rather complicated /25/, and asymmetries can be explained in many different ways.

(iii) The surface sensitivity of PS is a function of the excitation energy, the emission angle, and also of the element under investigation. These parameters strongly influence the mean free path of an electron traveling in the solid. The mean escape depth exhibits a minimum of about 7 Å at 100 eV excitation energy. For UPS with an excitation energy of

40.8 eV (He II) the escape depth is roughly 10 Å, and for XPS with MgKα excitation about 20 Å. These values can be reduced by tilting the sample.

2.2.4. *Valence Band Spectroscopy.* Valence band spectroscopy is traditionally performed using He I or He II gas discharge lamps with an excitation energy of 21.22 eV and 40.82 eV and a line width of 3 meV and 17 meV respectively, and is primarily used for the verification of band structure calculations. Such excitation energies allow to obtain the projection of the delocalized electronic states close to the Fermi level (see Fig. 2) with a resolution determined by the spectrometer. Although valence band spectra can also be obtained by XPS, as a consequence of the rather low resolution of this technique (without monochromator) most of the information is lost. With the use of a monochromator, however, XPS may be used to scan the valence band with an energetic resolution of 0.3 eV.

Unlike XPS, UPS has not been used for the characterization of electrochemically treated electrode surfaces until very recently /19,26,27/. As a consequence of its extreme sensitivity towards surface contamination, it was regarded as almost impossible to record a reliable UP spectrum of a surface treated ex-situ. In contrast to XPS, where the sample can readily be analyzed through a thin contamination layer - provided substrate elements are not identical to the contamination elements - any contamination shows up in the valence band region on top of the desired spectrum; especially the O2p line at 6 eV binding energy has a very high UPS strength, and therefore rather small amounts of oxygen contamination are sufficient to dominate the UPS spectrum. Kötz and Neff /19,27/ have clearly demonstrated that UP spectra of polycrystalline noble metal electrodes like Au, Pt and Ir can be obtained after electrochemical treatment when using a controlled transfer system. Although to obtain the complete band structure information available from UPS measurements well characterized single crystal samples are required, this technique applied to polycrystalline electrodes allows the determination of the electronic density of states at the Fermi level (see 3.1.), the position of the Fermi level with respect to the valence band, and the electronic distribution curves which may be compared with those calculated from optical in-situ measurements (see 3.4.)

## 2.3. THE TRANSFER PROBLEM

The application of XPS to the investigation of electrode surfaces is often criticized because of the ex-situ character of this technique. The interface studied under UHV conditions after emersion of the electrode from the electrolyte and transfer to the XPS system may differ from the interface in-situ. This situation has led to significant efforts in the development of controlled electrode transfer systems /8-13/ and in the study of the emersion process /18,28-30/ (see Chapter by D.M. Kolb).

The danger of a significant contamination is of course extremely great when carrying the electrode through air. In some cases the contamination layer may not affect the information of interest, but such a layer may well reduce the intensity of the signal under investigation, and thus reduce the quality of the analysis. The situation may be improved when covering the surface with a protecting layer like a drop of $H_2O$ /31/ or an oxygen scavenging layer /32/.

The next step towards transfer control is the use of glove boxes, running the electrochemical experiment in an inert, deoxygenated atmosphere. Either the glove box is attached to the spectrometer, or especially designed containers are used for the electrode transfer to the UHV system.

The best approach for electrode transfer is the use of an electrochemical cell directly attached to the UHV spectrometer, enabling transfer of the electrode after UHV cleaning procedures to the electrochemical cell and back to the spectrometer after electrochemical treatment. Such cells have been designed by several groups /8-13/; only two examples, from O´Grady /8/ and from the authors laboratory, are given in Fig. 3 and 4, respectively.

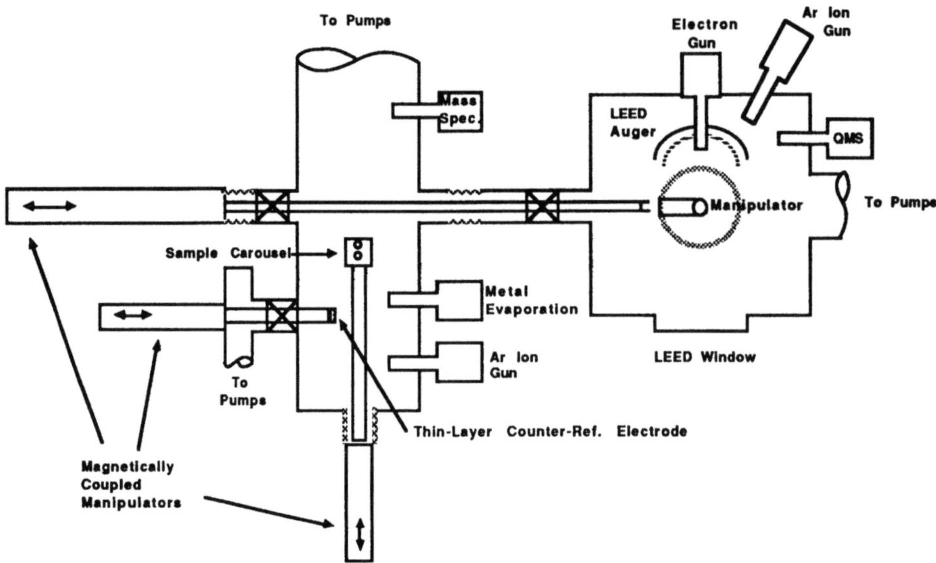


Figure 3. Sketch of the LEED Auger electrochemical UHV system, including an especially designed electrochemical thin layer cell. After O´Grady et al./33/.

The key feature of the design in Fig. 3 is a combined Pd-hydrogen reference and counter electrode, which is brought close to the sample with the help of magnetically coupled manipulators to form an electrochemical thin layer cell. During electrochemistry the electrochemical chamber is flooded with ultra pure argon. The advantage of this arrangement is the extremely small volume of electrolyte, which makes evacuation after electrochemistry easy and reduces possible contaminations to a minimum. This kind of cell is especially appropriate for double layer and surface studies on single crystal electrodes, but cannot be used for corrosion experiments or for reactions at high current densities.

A full size electrochemical cell with the standard three electrode arrangement directly attached to the fast insertion lock of the UHV spectrometer has been designed by Neff et al. /12/. This cell is constantly overpressurized with ultra pure argon, and electrochemistry is performed by bringing the sample at the end of the manipulator rod in contact with the meniscus of the electrolyte in the vertically arranged glas tube. Contact and interruption is achieved by rising and lowering, respectively, the electrolyte level in the vertical glas tube. This cell has been used for double layer studies, as well as for corrosion and $O_2$ evolution experiments.

The level of contamination in the course of a transfer experiment can be readily checked by monitoring the carbon or oxygen peaks in AES or XPS after in-situ cleaning and after exposure of the sample to the electrochemical chamber. Another approach was chosen by Neff et al. /12/, who recorded the UPS spectrum of a clean Au electrode before and after exposure to the argon atmosphere. Although some reduction in intensity of the

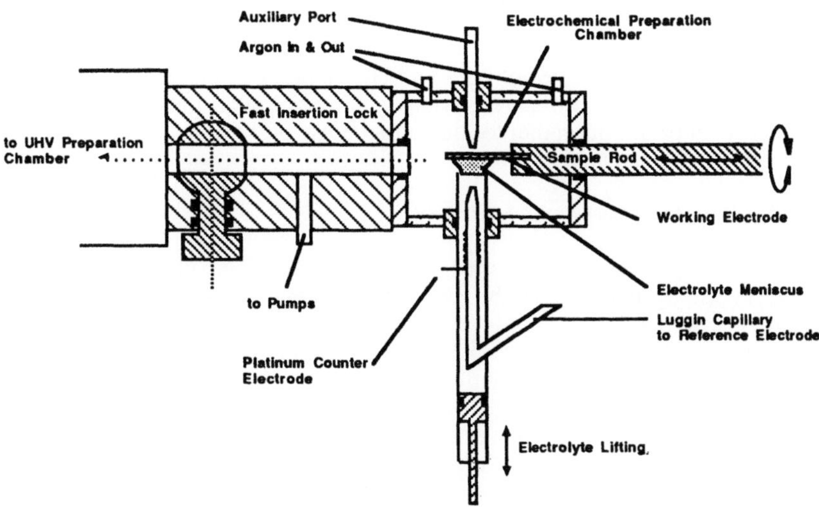


Figure 4. Sketch of an electrochemical preparation chamber for the Kratos ES 300 photoelectron spectrometer with fast insertion lock. After Neff et al. /12/.

UPS spectrum after 30 sec in the electrochemical preparation chamber was observed, no distortion of the spectral features could be detected, which in view of the extreme surface sensitivity of UPS proves the quality of the transfer system /12/. The quality standards of the transfer system have of course to be met by the electrolyte. Only ultra pure water, either purchased or produced by reverse osmosis or combined ultrafiltration and UV treatment, in combination with ultra pure chemicals should be used. Contact of organic impurities concentrated at the surface of the electrolyte with the electrode can be avoided by placing a drop of ultra clean water on the electrode surface before contacting the electrolyte.

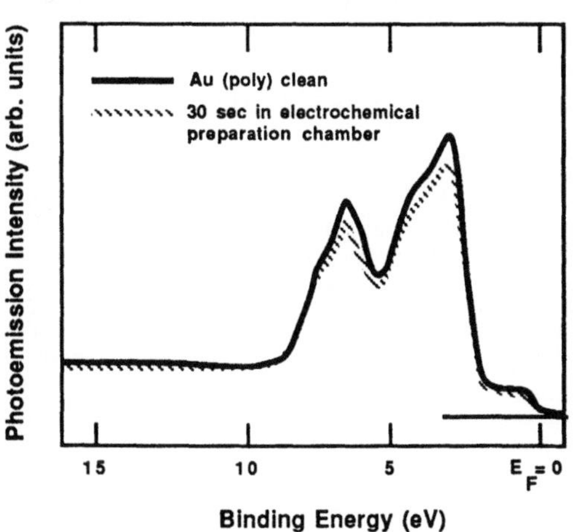


Figure 5. UPS spectrum of a clean polycrystalline Au surface before and after exposure to the electrochemical preparation chamber of Fig. 4. After Neff et al. /12/.

## 3. Application of XPS and UPS to Electrochemical Processes

In this paragraph several examples of XPS and UPS studies of electrode surfaces will be given. The intention is not to discuss several electrochemical processes and fields of electrochemistry in detail, but rather to illustrate the use of XPS and UPS and the kind of information that can be obtained.

### 3.1. OXIDE FORMATION ON NOBLE METALS

3.1.1. *Platinum* As already mentioned, oxide formation on platinum was the first electrochemical process to be studied by XPS. Kim, Winograd and Davis /5/ attributed the chemical shifts observed for the Pt 4f level during electrochemical oxide formation to PtO and $PtO_2$, with binding energies of 73.3 eV and 74.1 eV, respectively ($E_b$ of Pt metal = 70.7 eV). In addition they observed chemisorbed oxygen at 71.6 eV. This detailed analysis was achieved using a deconvolution process. The electrode was rinsed after emersion with deoxygenated water and transfered to the electron spectrometer through air. Although this analysis looks rather conclusive, later XPS studies of the oxidation of Pt by other workers came to different conclusions about the composition of the anodic oxide.

Peuckert et al /31,34/ conducted several XPS investigations of the anodic oxide formation on noble metals, valence band analysis, thermodesorption /35/ and EELS /31/. For Pt they put forward that the oxide formed on Pt in $H_2SO_4$ during polarization at 3.0 Vshe for 18 h is $Pt(OH)_4$, this conclusion being based on the analysis of Pt 4f and O1s levels with respect to binding energies and relative intensities. The formation of $Pt(OH)_4$ was confirmed by recent XPS studies performed by Rach and Heitbaum /36/ for thin layers of the anodic oxide.

Although the anodic oxide on Pt is one of the most important catalytic materials its electronic properties are still a matter of discussion. Based on the optical data by McIntyre and Kolb, it is generally assumed that the oxide has n-type semiconducting properties with a band gap of 1.3 eV /37/. This is in disagreement with UP spectra, which, as has been demonstrated recently, can be obtained with reasonable cleanliness for electrochemically treated electrodes /19/ (see 3.1.2.). UPS data of Pt electrodes anodized at a current density of 0.1 $mA/cm^2$ (1.45 Vsce) show /38/ that the top of the valence band coincides with the Fermi level. This is either an indication for a p-type semiconductor if a band gap is assumed, or an indication for the fact that the oxide layer on Pt has semimetallic character.

3.1.2. *Gold* XPS investigations of the anodic oxide formed on Au led to the conclusion that gold in the oxide has the oxidation state +3. The actual composition of the film, however, is debatable. While Kim et al. /39/ assumed an $Au(OH)_3$, Dickinson et al. /40/ favoured an $Au_2O_3 \cdot H_2O$ layer, while Peuckert et al. /35/ believed the oxide to be AuOOH. In order to demonstrate the analysis procedure, data from the author´s lab are shown in Fig. 6 for the anodic film formation on Au in 1 M $HClO_4$. The Au 4f level shows the presence of oxidized Au at 1.5 V and 1.8 V. The chemical shift of 1.7 eV (Eb ($Au_{metal}$) = 84.0 eV, Eb ($Au_{oxide}$) = 85.7 eV) corresponds to Au (III) /16,39/. The O1s level, whose intensity increases significantly after oxide formation, gives further information. At a potential of 1.5 V two oxygen species can be clearly distinguished, with binding energies of 530.2 eV and 531.8 eV respectively. These levels can be attributed to O and OH in AuOOH /35/, Although, as stated above, the assignment of O1s binding

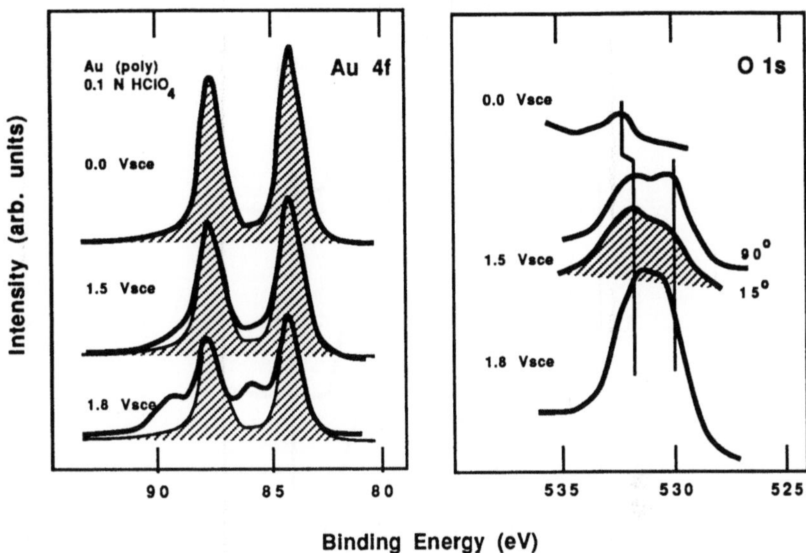


Figure 6. XP spectra of the Au 4f and O 1s levels for a polycrystalline Au electrode after electrochemical treatment at different potentials in 0.1 M $HClO_4$ for 2 min. Shaded spectra in Au 4f represent a clean Au surface. The O 1s spectra were recorded with 20 fold sensitivity at 90° (15° for shaded O1s spectrum).

energies is sometimes ambiguous. Interference with oxyanions from the electrolyte or contamination during transfer has always to be considered. The O 1s peak at 531.8 eV may well be due to $ClO_4^-$ species ( $E_b$ = 532.2 /16/), unless a total absence of Cl (or S in the case of $H_2SO_4$) has been verified. A rather interesting feature in Fig. 6 is the fact that the relative peak heights of the O 1s spectrum for the electrode emersed at 1.5 V are angle dependent. The high binding energy peak is increased at low emission angles, which demonstrates that this species is concentrated at the surface. While this observation would be easy to explain in terms of adsorbed $ClO_4^-$ counter ions, for AuOOH one has to assume a preferential orientation of the OH groups towards the surface. In agreement with the finding by Peuckert, the presence of $H_2O$ in the oxide film can be excluded on the basis of the O 1s spectra.

In addition to the chemical composition of the oxide layer, its electronic properties are of importance for the understanding of the electrochemical reactions proceeding on a thin oxide layer. It has been demonstrated that the anodic oxide has n-type semiconducting properties with a band gap of 1.75 eV /37/. This is nicely reflected in the UPS spectra of a thin oxide layer on Au grown at 1.7 Vsce (0.1 mA/cm$^2$) in 0.1M $HClO_4$ (Fig. 7). While the Fermi level is clearly visible for the clean sample, after oxidation the density of electronic states at $E_F$ is clearly zero. The top of the valence band occurs at a binding energy of 1.6 eV, which, together with the assumption of an n-type semiconductor, indicates that the band gap should be slightly larger, in accordance with other data. The three emission maxima appearing in the spectrum of the oxidized sample are attributed to $ClO_4^-$ counter ions still adsorbed on the surface of the electrode /27/, since they disappear after rinsing the sample with ultra pure water. Assuming that the rinsed sample reflects the electronic density of states of the oxide grown on Au, the maximum in optical absorption

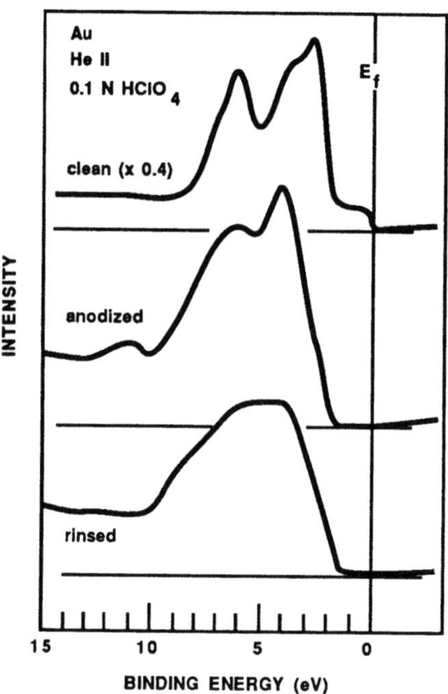


Figure 7. UP spectra of a clean Au electrode, after oxidation with a current density of 0.1 mA/cm$^2$, and after rinsing the oxidized sample with ultra-pure water. After ref. /38/

determined by McIntyre and Kolb /41/ at 4 eV may be attributed to the transition from the maximum of the valence band (O2p) to the conduction band. Similar UPS data were obtained for Pt /19,38/, Ru /38/, Ir /27/, and RuIr alloys /38/, demonstrating the usefulness of this technique for electrode surfaces.

3.1.3. *Ruthenium* In contrast to Au and Pt, oxide formation and reduction on Ru is not reflected in the cyclic voltammogram by clear anodic oxidation and cathodic reduction peaks, respectively. These structures are rather smeared out, and the actual onset of oxide formation is difficult to determine /42/. XPS analysis of the oxide formation on ruthenium has been performed in the author´s lab /43-45/, demonstrating thick oxide formation as soon as the electrode potential is higher than the oxygen evolution potential. Fig. 8 shows the binding energies of the Ru 3d5/2 and the O1s level as a function of electrode potential. The inflection point for the Ru signal occurs at about 1.25 $V_{sce}$, while that of the O1s level is found at 1.15 Vsce. The binding energy of 281 eV for the Ru3d level for potentials > 1.2 $V_{sce}$ indicates the formation of Ru (IV) as $RuO_2$ or $RuO_2 \cdot H_2O$, this assignment being made on the basis of the XPS study performed by Kim and Winograd on ruthenium-oxygen surfaces /46/. From the peak width and the shape of the O1s level it was concluded that a thick hydrous $RuO_2$ layer is formed. The presence of an oxide layer already at potentials cathodic of 1.1 Vsce could be verified by angle-dependent XPS measurements, which showed increasing Ru3d binding energies with decreasing emission angle, i.e. with decreasing sampling depth /43/.

Due to the extreme surface sensitivity of UPS, the initial stages of oxide formation

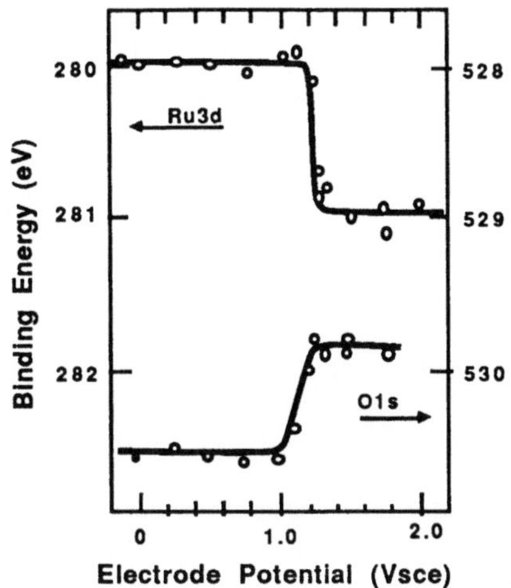


Figure 8. Binding energies of Ru 3d5/2 and O 1s levels of a bulk ruthenium electrode in 0.5M $H_2SO_4$. Different potentials were applied for 15 minutes. After ref. /43/.

on Ru can be directly observed in the valence band spectra. Fig. 9 shows the UP spectra recorded after emersion of a polycrystalline Ru electrode at potentials between 0.0 V and 0.8 $V_{sce}$. While the spectrum of the electrode emersed at 0.0 V already differs from the spectrum of the clean sample - probably as a consequence of $OH^-$ or counter ion adsorption - the development of the $t_{2g}$ band at 0.8 eV below $E_F$ is clearly visible for potentials above

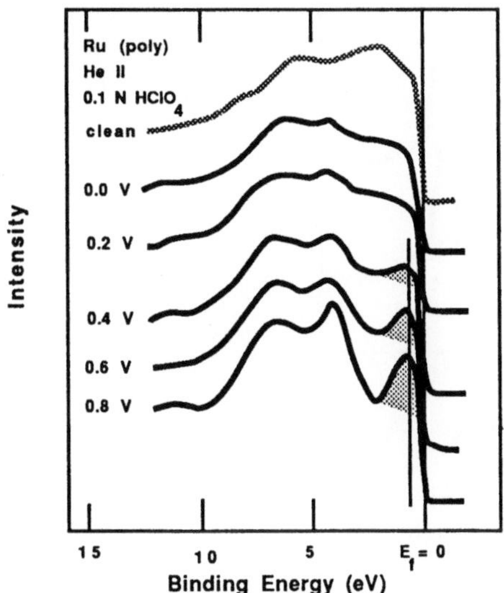


Figure 9. UP spectra of a Ru electrode emersed at different potentials from 0.1M $HClO_4$. Emersion potential vs. sce is shown beside each curve. The spectrum of the clean electrode is shown for comparison. After ref. /38/

0.4 Vsce. The $t_{2g}$ band is characteristic for Ru compound formation, with Ru most likely in the tetravalent state. The additional structures developing around 4.0 eV and 6.5 eV can be attributed to $ClO_4^-$ counter ions. The presence of an adsorbed hydroxide or oxide layer between 0.3 and 0.8 V has been suggested by Vukovic et al. /47/. These oxide layers are reversibly reduced at 0.1 V.

Ruthenium-oxide based electrodes, usually known as DSA (dimensionally stable anode) electrodes, are the best catalysts for anodic chlorine and oxygen evolution, and have therefore been the subject of several XPS investigations. Augustynski et al./48/ have investigated DSA-type $RuO_2$ electrodes prepared by thermal decomposition on Ti. After electrolysis in NaCl they found Cl species on the surface, enrichment in Ti and, besides $RuO_2$, some Ru (VI). They already proposed the importance of this higher ruthenium oxide, most probably $RuO_3$, for the catalytic performance and stability of the material. Kötz et al. /44/ have confirmed the presence of $RuO_3$ on the surface of thermally prepared $RuO_2$ electrodes, and have shown that this species is also formed after annealing of reactively sputtered pure $RuO_2$ electrodes at temperatures above 400° C. The importance of this species for the stability of $RuO_2$ electrodes during $O_2$ evolution in acid was reflected by the significant stability increase of those electrodes annealed above 400° C. Fig. 10 shows the XP spectra of the Ru 3d level of reactively sputtered $RuO_2$ on Ti before and after annealing at 450° C. After annealing the higher oxide is clearly evidenced by additional structures shifted by 1.6 eV, from 280.8 eV to 282.4 eV. A binding energy of 282.5 eV for the Ru 3d level was attributed to $RuO_3$ by Kim and Winograd /46/.

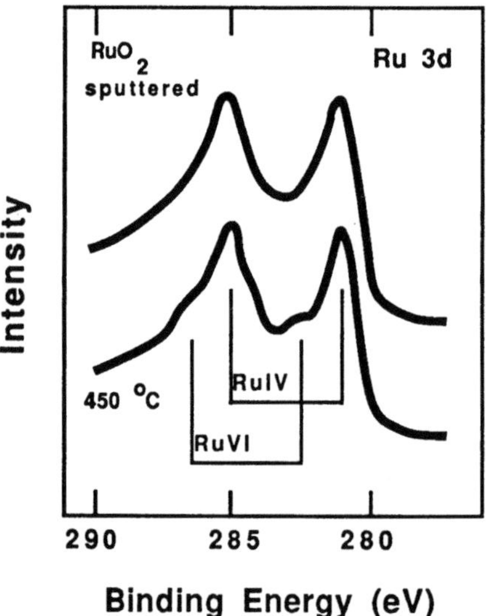


Figure 10. XP spectra of the Ru3d level of reactively sputtered $RuO_2$ on a titanium substrate. The two spectra were recorded with the sample as prepared and after annealing at 450° C for 1 h. After ref. /44/.

The results of several XPS and reflection spectroscopy studies performed on Ru and $RuO_2$ electrodes have led to a model for $O_2$ evolution on Ru and $RuO_2$ proposed by Kötz et al.

/43/, which is mainly based on the presence and relative stability of $RuO_3$. Recent XPS investigations by O'Grady et al. /49/ have shown that $RuO_3$ is also present on the surface of $RuO_2$ single crystal surfaces, the relevance of this high oxidation state for the catalytic reaction being emphasized again. These authors have analysed the O1s level of various $RuO_2$ single crystal surfaces.

Very recently XPS has been applied to ruthenium oxide electrodes after use as a cathode for hydrogen evolution in acid electrolytes /50/. It could be shown that the oxide is not reduced to the metal, even after prolonged electrolysis, and that the specific behaviour of the oxide towards underpotential deposition was maintained during $H_2$ evolution.

3.1.4 *Iridium* The electrochemical properties of iridium and iridium oxide have been studied by many groups, both because of its good catalytic performance for $O_2$ evolution and of its electrochromic effect, which was thought to be of technical importance for display applications. XPS has been used to identify the chemical composition of the oxides, the growth mechanism, and the origin of the coloration - bleaching process /44,51-53/. Thick, electrochemically formed oxides on Ir will be treated in 3.2.

If the electrode potential for a clean metallic Ir electrode is gradually made anodic, the formation of a thin oxide layer can be monitored by XPS. In order to evaluate the oxide thickness as a function of potential semi quantitatively, XPS data have to be taken at a small electron emission angle (higher surface sensitivity), and the measured signal has to be deconvoluted. The results of such a study by Kötz et al. /44/ are reproduced in Fig. 11. A significant potential dependence of the oxide thickness is observed, indicating corrosion of the oxide at potentials above the $O_2$ evolution potential. However, even at a potential of

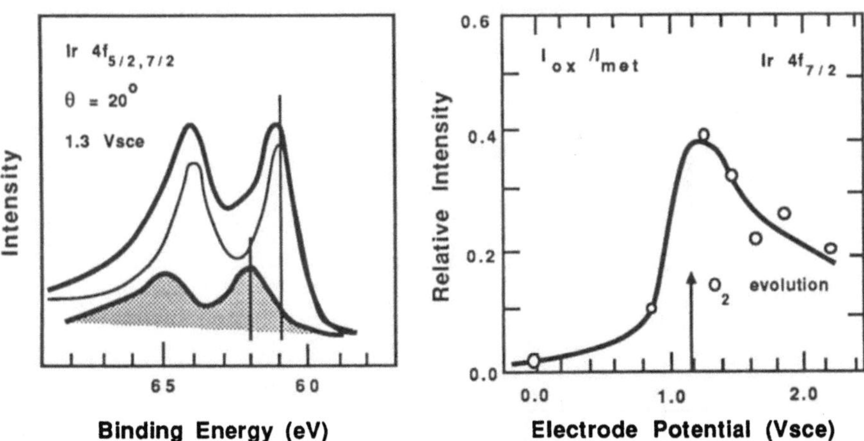


Figure 11. Left: XP spectra of the Ir4f levels of an iridium electrode after polarization at 1.3 Vsce in 0.5M $H_2SO_4$ for 15 min. The thin line indicates the contribution of the metal and the shaded area that of the oxide layer. Right: Relative intensity of the oxide over the intensity of the substrate Ir as a function of electrode potential. The maximum corresponds to roughly 1 ML (3 Å). After ref. /44/.

1.3 Vsce, where the oxide thickness is maximum, the XP spectrum of the Ir4f levels is dominated by the substrate metal. Deconvolution of the spectrum reveals a chemical shift of 1.2 eV, from $E_b = 60.8$ eV for the metallic state to 62 eV. According to Kim et al. /39/ this binding energy can be attributed to $IrO_2$. Similar results were obtained by Hall et al.

/52/. Augustynski et al. /51/ concluded, on the basis of an Ir4f and O1s level analysis, that the oxide layer formed is $Ir(OH)_4$. Peuckert has determined the composition of the adlayer as $IrO(OH)_2$ /54/.

As can be seen in Fig. 11, oxide thickness increases rapidly just before $O_2$ evolution starts. The actual onset of oxide (hydroxide) layer formation is difficult to detect from XPS data. UPS data again give some more insight, as a consequence of its higher surface sensitivity. Spectra similar to those shown in Fig. 9 could be obtained for Ir by Kötz et al. /27/. From these spectra it could be concluded from the appearance of the $t_{2g}$ band at 1.5 eV below the Fermi level, that oxidation processes already occur at 0.5 Vsce. In addition, it could be demonstrated that the oxide formed during the first anodic scan is not reduced in a consecutive cathodic scan.

Finally, it should be mentioned that several investigations were performed using XPS and AES on thermally prepared ruthenium or iridium based oxides in order to characterize the electrode surface after preparation, before any electrochemical experiments /48,55,56/. These studies revealed significant deviations of the surface composition from the nominal composition of the starting materials.

## 3.2. ELECTROCHROMISM

A change in optical absorption, i.e. colour, of a compound occurring upon application of an electrical field is called electrochromism, this effect being observed for electrochemically formed oxides on tungsten and iridium. In order to produce an effect visible to the naked eye, i.e. with potential display applications, the electrochromic oxide layer has to be thick. Thick oxide layers can be grown on iridium by repetitive triangular potential cycling, or with a square wave potential modulation with upper and lower potential limits close to the

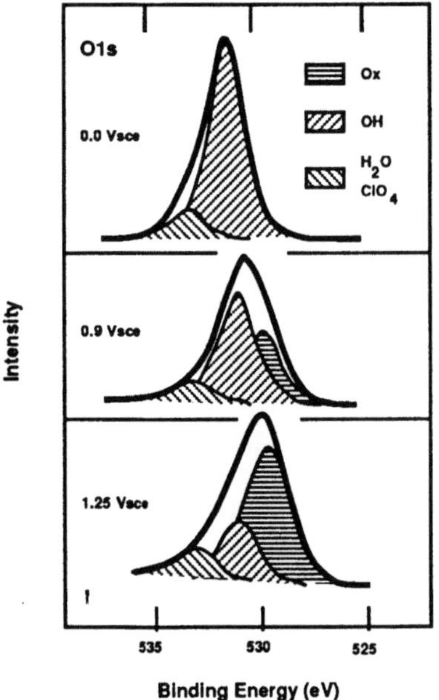


Figure 12. O1s level of thick iridium oxide films emersed at 0.0 V, 0.9 V, and 1.25 Vsce in 0.5M $H_2SO_4$. After Kötz et al. /53/.

$O_2$ evolution and $H_2$ evolution potentials, respectively. Optimization of this process was studied by Burke et al. /57/.

Formation of the thick oxide is reflected in the cyclic voltammogram by an increasing pseudo-capacitive current at potentials above 0.5 Vsce in 0.5M $H_2SO_4$ /53/. A reversible current peak at 0.7 Vsce corresponds to the change in coloration from the bleached to the coloured stage. At 1.15 Vsce, close to the onset of $O_2$ evolution, a second peak can be observed. XPS analysis of these films is straightforward, since the thickness of the layer under investigation exceeds the information depth of XPS. Fig. 12 shows the O1s level of a thick oxide film on an Ir electrode emersed at three potentials. At 0.0 Vsce the film is in the bleached state while at 0.9 Vsce and 1.25 Vsce the film is coloured. The respective spectra of the $Ir4f_{7/2}$ level exhibit a constant binding energy of 62.4 eV, indicating a higher oxidation state of Ir. Assignment of the iridium 4f 5/2 and 7/2 binding energy is rather difficult, due to the lack of reference samples.

While the O1s level in Fig. 12 at 0.0 Vsce could be fitted by one major component with a binding energy of 531.2 eV (OH), at more anodic potentials an increased peak width was observed, in addition to a shift of the maximum to 530.5 eV (0.9 Vsce) and 529.8 eV (1.25 Vsce), respectively. At 1.25 Vsce the measured XP spectrum could be fitted by an oxide (529.6 eV) and a hydroxide component with an intensity ratio of 2:1 /53/. The third component (see Fig. 12) with a binding energy of 533.1 eV, can be attributed to either adsorbed water or $SO_4^{2-}$ counter ions. It should also be mentioned, that the observed Ir/O ratio stays constant at roughly 3. Thus, these XPS results support the following anodic deprotonation mechanism

$$Ir(OH)_3 \rightarrow IrO(OH)_2 + H^+ + e^- \rightarrow IrO_2(OH) + 2H^+ + 2e^- \quad (5)$$

which had already been postulated by several authors on the basis of non-XPS investigations /58/.

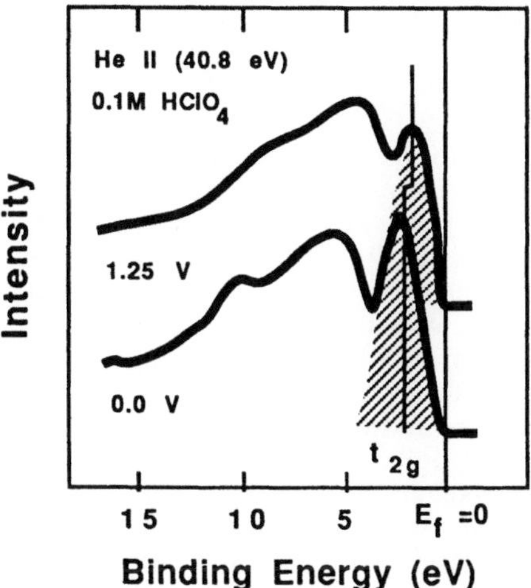


Figure 13. UPS valence band spectra of anodic iridium oxide films in the coloured (1.25 V) and bleached (0.0 V) state. After emersion the electrodes were rinsed with water for counter ion removal. After Kötz and Neff /27/.

Further insight into the electrochromism of iridium oxides could be gained from XPS and UPS valence band spectra. As is evident in Fig. 13, a significant shift of the $t_{2g}$ band of the oxide to higher binding energies (1.6 eV -> 2.5 eV) could be observed when the film was reduced from the coloured (1.25 V) to the bleached (0.0 V) state. There is a parallel decrease of the density of electronic states at the Fermi level: while it is finite in the coloured film, where $E_F$ lies inside the upper $t_{2g}$ band, it becomes zero for the bleached film, where $E_F$ falls within the empty gap between the completely filled $t_{2g}$ band and the empty $e_g$ band /59/. This observation readily explains the main characteristic of the electrochromic effect, since as a consequence of the change in the density of states at $E_F$ the bleached film has low conductivity, while the coloured film is a good conductor. Furthermore, optical transitions to the Fermi level, which are possible in the colored state, are rendered impossible in the bleached state because no transitions to the completely filled $t_{2g}$ band are possible.

Modulated reflectance spectra recently obtained on anodic iridium oxide films by Gutierrez et al. /60/ exhibit a shift of the main absorption structure to lower photon energies (longer wavelength) with increasing electrode potential. If one assumes optical transitions from lower filled states to the Fermi level to be responsible for this absorption, the energetic shift can be explained qualitatively by the lowering of the Fermi level due to withdrawal of electrons from the $t_{2g}$ band with increasing anodic potential.

This very same model would explain the coloration of a $WO_3$ film upon reduction and its bleaching by oxidation, which is the opposite sequence as compared to Ir. For oxidized$WO_3$ the $t_{2g}$ band is empty and the Fermi level is in a band gap. After reduction and proton insertion /58/ one electron moves to the $t_{2g}$ band, introducing a finite electronic density at the Fermi level and making additional optical transitions to the Fermi level possible.

If the difference of the two spectra in Fig. 13 is calculated, it becomes evident that the change in the O2p region (5 eV to 12 eV) can be attributed to additional OH bond formation in the film at 0.0 V /27/. OH bond formation is a result of proton insertion and valence state reduction from 4+ to 3+ upon bleaching (see eq. (5)).

## 3.3. UNDERPOTENTIAL DEPOSITION

XPS analysis was applied as early as 1977 to the underpotential deposition of Cu and Ag on polycrystalline platinum electrodes by Hammond and Winograd /6,7/. These UPD studies, which were extenden to Au(111) surfaces by Kolb et al. /29/ and to Pt by the author, are of interest for several reasons. Firstly, XPS may provide some information about the chemical state of the adatom on the metal surface, which is of importance for the understanding of the UPD phenomenon and its catalytic relevance. Secondly, UPD allows the controlled deposition of submonolayer amounts of a metal onto a substrate, thus enabling an easy test of the sensitivity of the XPS setup. Finally, if the valence band is inspected, the photo emission data give some additional information to the well known in-situ reflectance measurements of UPD systems.

The early results obtained by Hammond and Winograd demonstrated that the amount of electrochemically deposited metal, as determined by the charge passed, correlates well with the intensity of XPS signal of the metal deposited. In addition, it became clear upon inspection of the chemical shifts that the metal (Cu, Ag) is not oxidized during the transfer process. On the contrary, a chemical shift was consistently observed indicating an oxidation state "lower" than that of the metal: for Ag, as well as for Cu, binding energies smaller than the binding energy of the metal were found /6,7/. The results obtained by Kolb et al. /29/ and in the author´s lab /61/ fully confirmed these observations.

Fig. 14 shows the XP spectra of Ag deposited at monolayer and sub monolayer amounts onto a polycrystalline Pt electrode. Holding the potential for a certain time at -0.1 V results in deposition of roughly 2 ML whose binding energy corresponds to that of the bulk metal. At lower coverages the binding energy shifts gradually to lower values, reaching 386.8 eV at 1.0 V, which is about 1 eV lower than the bulk value. At a potential of 1.0 V the coverage of 0.1 ML is close to the detection limit of our system, estimated to be 5% of a monolayer for Ag.

Hamond and Winograd found a constant binding energy for the underpotential deposited Ag and Cu on Pt. A constant binding energy was also claimed by Kolb et al. for Cu on Au(111) /29/. The results presented in Fig. 14 and 15, however, indicate a shift of $E_b$, which gradually approaches the bulk value with increasing coverage. Fig. 15 shows a plot of the Ag3d and Cu2p core levels as a function of electrode potential in the UPD regime. Different symbols represent different series of measurement with new electrodes and electrolyte. Although the scattering of data points would hide any fine structure in these curves, the general shift towards lower kinetic energy (higher binding energy) with coverage is evident. Qualitatively the same effect was observed by Hammond and Winograd /6,7/, and recently by Shek et al. /62/, for evaporated metal films on polycrystalline Pt and on Pt (111), respectively. Shek et al. observed a shift of -0.6 eV with respect to the bulk value for

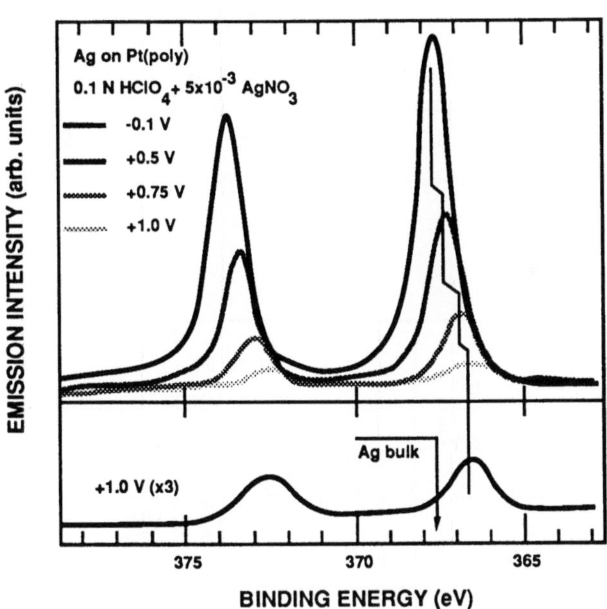


Figure 14. Ag3d core level spectra of silver underpotential deposited onto a polycrystalline Pt electrode at different electrode potentials. 0.1M $HClO_4$ + $10^{-3}$M $AgNO_3$. After /61/.

a Cu coverage of 0.3 ML. Unless further experimental evidence is provided one has to assume that the shift observed for Cu and Ag underpotential deposited on Pt is consistent with in-situ UHV measurements on evaporated layers. The still tentative explanations given by Shek et al. are based on electron rearrangement within the Cu constituent, a change in final state relaxation energy, and surface versus bulk effects. These authors exclude electron transfer between Pt and Cu as the origin of the decrease in binding energies relative to the bulk value.

It should be mentioned, however, that the substrate surface for the evaporated films and the electrochemically deposited layers are different sine on the surface of the Pt electrode one has to assume the presence of a certain amount of counter ions providing the applied potential difference. The fact that counter ions are preserved on the surface is

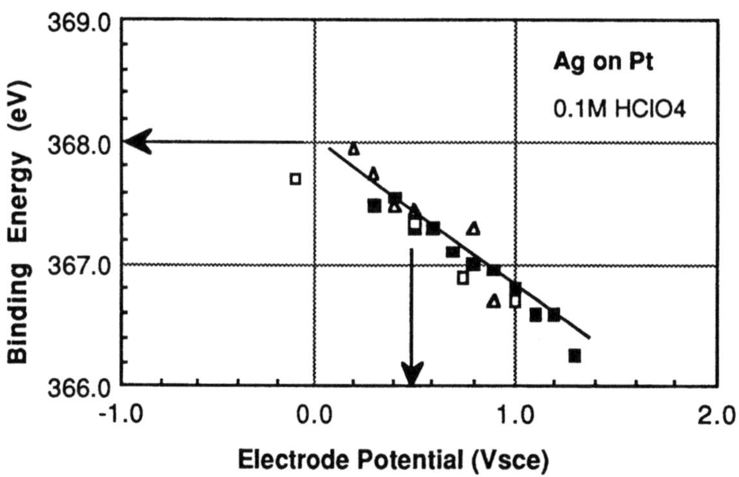


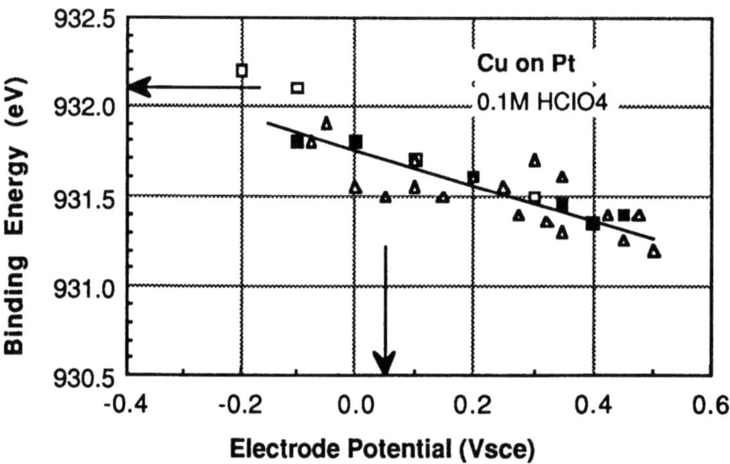


Figure 15. XPS core level shifts for Ag (top) and Cu (bottom) deposited onto Pt. The arrows indicate kinetic energies and bulk deposition potentials of the bulk metals. Different symbols represent different measurements with new electrodes and electrolyte.

verified by work function measurements, which exhibit a 1:1 correlation with polarization potential even during underpotential deposition (see 3.6.). Thus, the work function change during electrodeposition of the first Cu monolayer is roughly 0.5 V, while the work function changes by 1.2 eV for the evaporated Cu mono atomic layer on Pt.

Recent measurements on the UPD system Cu on ruthenium exhibit a chemical shift of roughly -0.5 eV with respect to the bulk value, although the electochemical UPD shift in this case is only 0.15 V /50,61/. Further investigations of the underpotential deposition phenomenon are needed in order to understand the origin of the negative chemical shift of the adsorbate. These investigations should be performed preferably on single crystal electrodes, where sharp desorption peaks allow a separarion between the influence of coverage and potential. Additional information should be gained from simultaneous LEED measurements.

Recent UPS data obained for the UPD systems Cu on Pt and Ag on Pt in the author's lab /61/ show, in addition to the shift of the Cu2p core level a simultaneous shift of the Cu3d state at 2.5 eV below the Fermi level, which is in good agreement with what was observed by Shek et al. /62/ for Cu evaporated on Pt. Fig. 16 shows the UP spectra for three different coverages of Cu on Pt(poly) and also the spectra for the clean Pt substrate and a bulk Cu sample. The development of the Cu 3d level, with a binding energy of 2.2 eV at 0.3 V, 2.35 eV at 0.1 V and 2.6 eV at -0.2 V, is clearly evidenced. The position for the bulk sample was determined to be 2.7 eV. Additional structures observed for the electrode emersed at -0.2 V are due to $ClO_4^-$ species at the surface. These

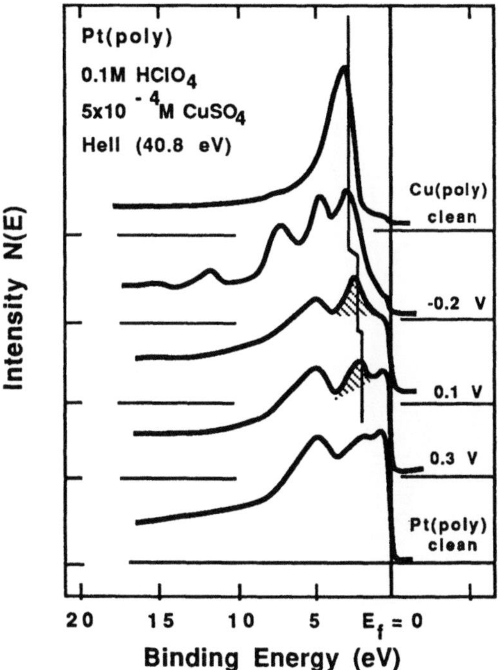


Figure 16. UP spectra for Cu underpotential deposited on Pt(poly). The electrode potentials of 0.3V, 0.1V and -0.2 Vsce correspond to coverages of 0.6, 1.0 and 2.3 ML, respectively. After /61/.

species can be removed by rinsing the sample with $H_2O$, but simultaneously a significant reduction of the work function is observed. On the basis of these UPS data previously determined optical absorption spectra for Cu on Pt can now be assigned. For Cu on Pt a coverage-dependent optical absorption maximum was found: it appeared at 2.3 eV for 0.55 ML and at 2.5 eV for 0.93 ML /63/. Thus, it can be deduced that this optical transition occurs from the Cu3d level to the Fermi level, since the binding energy (which is the difference between a given energy level and the Fermi level) of the Cued level increases with coverage. Similarly, good coverage-dependent UPS spectra could be measured for the system Ag on Pt, where a shift of the Ag 4d valence band was also observed. Here $ClO_4^-$ counter ion adsorption did not show up in the spectra /61/.

## 3.4. CORROSION

In principle, corrosion studies by means of XPS do not differ from, for instance, investigations of oxide layer growth as presented in 3.1. Therefore this paragraph will be kept relatively short with respect to the huge amount of corrosion studies available in the literature. A detailed discussion is beyond the scope of this chapter, and for further reading references 64 and 65 are recommended. Examples and references given in the following are certainly incomplete.

It is primarily the aim of corrosion studies to determine the chemical nature of the (passivating) oxide film formed during corrosion (passivation) in different solutions. Using ion sputtering, the composition of the film can be monitored as a function of thickness.

Perhaps the most studied corrosion system is the passivation of iron or stainless steel. Several in-situ and ex-situ spectroscopies have been applied so far: impedance, ellipsometry, reflectance spectroscopy, Raman scattering, SIMS, AES, STEM, Mössbauer spectroscopy, and, last but not least, XPS, to resolve the question as to whether the

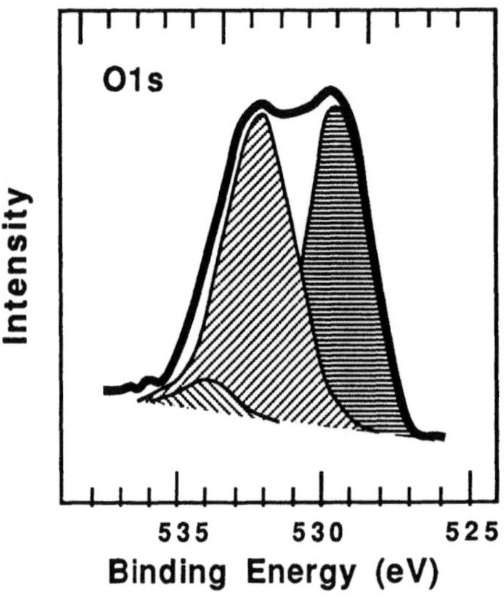


Figure 17. Oxygen 1s level of the passive layer on iron grown in borate buffer pH 8.4 at 0.65 Vrhe. After ref. /66/.

passivating film contains water or OH. The presence of OH has been confirmed by several groups /66-68/, who consistently found that OH is concentrated at the surface of the passivating film. Fig. 17 shows the O1s level of a passive film grown on iron in borate buffer /66/, clearly indicating $OH^-$ and $O^{2-}$ species at binding energies of 532 eV and 529.4 eV, respectively. After sputtering, the peak with higher binding energy ($OH^-$) disappears rapidly, which shows that the major portion of the passive film is free of OH. The same pattern was also found in iron-chromium alloys /69/ and in stainless steel /68/.

A recent study of the corrosion of stainless steel in chloride containing electrolytes at open circuit potentials performed by Brüesch et al./68/ has revealed that the passivating film contains chromium oxide. The concentration of this oxide has a maximum within the film, the height and position of this maximum being dependent on the chromium content of the steel. Significant changes of the film properties were observed at a critical Cr concentration of 12%. Brüesch et al. /68/, who used the electrochemical preparation cell described in Fig. 4, also found an enrichment of the outer surface layer in water or OH. On the basis of this study a phenomenological model has been developed by Brüesch et al. /70/ for the transition from corrosion to passivation in binary alloys. If corrosion in chlorine containing electrolytes is to be investigated, XPS can be used to determine the chlorine content of the passivating layer /67/.

As already mentioned, ex-situ XPS measurements are usually criticized because of the uncertainty with respect to possible changes occurring during transfer and exposure to the UHV. Kuroda et al. /71/ have observed changes in the diffraction spots of a passive film on iron when performing electron diffraction studies in a hydrated, and subsequently dehydrated, environment. Changes in the structure, however, are not necessarily related to changes in elemental composition.

### 3.5. NONMETALLIC ELECTRODES

3.5.1. *Graphite* Graphite is a well known electrode material, which is of significant relevance from a technical as well as a scientific point of view. Surface properties of carbon or graphite electrodes are mainly determined by surface functional groups produced on the surface by different treatment processes. By means of XPS the nature of these surface groups can be determined on the basis of a combined analysis of the O1s and C1s levels. This method, however, is not unequivocal, because chemically different surface groups may have similar binding energies.

Sherwood et al. /72/ have investigated the surface of carbon fibers by means of XPS. Significant changes in the C1s spectrum were observed as a function of the chemical and electrochemical oxidation procedures applied.

Very recently Sundberg et al. /73/ have investigated the correlation between the performance of glassy carbon electrodes with respect to $O_2$ reduction and the nature of the surface groups produced by anodic oxidation as characterized by means of XPS. In agreement with Sherwood et al. three shifted C1s peaks were found, which were attributed to (i) carbon in phenol, alcohol or ether ( C-OH, $\Delta E=1.2$ eV), (ii) carbonyl or quinone groups (C=O, $\Delta E=2.6$ eV) and (iii) carboxylic type groups (O=C-OH, $\Delta E=4.2$ eV). After electrochemical reduction, the XPS C1s peak assigned to carbonyl groups disappeared completely /73/. Investigations performed in the author´s lab for glassy carbon in 1M $HClO_4$ confirm the above observations.

3.5.2. *Semiconductor electrodes.* Semiconducting electrode materials have gained increasing attention in recent years, mainly in connection with photoelectrochemical reactions. Solar conversion efficiencies of up to 12% have been achieved with photo electrochemical devices where surface modification of the electrode plays a crucial role. Surface modification may be intentionally produced by especially designed chemical treatments, or be the consequence of some unwanted corrosion proccesses. In any case, knowledge of the composition and depth profile of such surface layers is a necessary prerequisite for the understanding and optimization of the underlying electrochemical and chemical processes.

Solomun et al. /74/ have recently investigated the surface stoichiometric changes of n-GaAs after anodic treatment. As a result of their XPS analysis they found a significant increase in As, and a depletion of Ga, in the surface layer. The surface species were identified as As(III) and Ga(III) oxides. During cathodic polarization the perturbation of the surface composition produced anodically was retained.

With ternary semiconductors like $CuInSe_2$ and $CuInS_2$ analysis by XPS has revealed significant surface changes after chemical treatment. Etching of $CuInS_2$ reduces the Cu and In content of the surface while those of sulfur and oxygen increase /75/. The composition of the surface layer formed after etching could be correlated to the photoelectrochemical performance. Monitoring of the Ru 3d level allowed some insight into the effects of a $RuCl_3$ etching procedure; in combination with other analytical techniques the binding energy of 281.7 eV for $Ru3d_{5/2}$ was interpreted as being due to the formation of some Ru-S-O structure.

Surface modifications on $CuInSe_2$ produced in-situ in an $I_2$ - $I^-$ - $Cu^+$ - HI electrolyte were characterized by XPS, and a substantial depletion in $In^{3+}$, paralleled by an enrichment in $I^-$ and reduction of selenium to $Se^0$, was observed. The composition of the surface layer changed gradually into $CuInSe_2$ over a distance of several hundred Å /76/. For the same system it was demonstrated by means of UPS that the surface layer has p-type semiconducting properties, and was characterized as p-$CuISe_3$ with patches of $Se^0$ /77/.

## 3.6. WORK FUNCTION MEASUREMENTS AND DOUBLE LAYER STUDIES

So far we have discussed the application of XPS and UPS to the investigation of electrode surfaces which had been modified by some faradaic reaction (oxide formation, corrosion, metal deposition). It is also very tempting to gain new information about the composition of the double layer on a metal electrode by means of XPS. There is, however, a question which has to be answered first. Is it possible to remove the electrochemical double layer from the electrolyte and transfer it to the UHV system intact? A major step towards answering this question was given by Kolb et al. /29/, who investigated the amount of anions on the surface of an emersed electrode under UHV conditions by means of XPS as a function of the emersion potential. In order to achieve high sensitivity, ions with relatively high XPS scattering cross sections are to be preferred like $Cs^+$ ($Cs_2SO_4$) and $Br^-$ (CsBr) or $I^-$ (CsI). Kolb et al. found increasing amounts of anions on the surface of a gold electrode for increasing potentials anodic of the pzc, and of cations for potentials cathodic of the pzc. This result, which is in agreement with the simple electrostatic model of the double layer, indicates the possibility of a tranfer of the intact electrical double layer. Further discussion of these experiments is given in the chapter by D.M. Kolb.

During XPS studies of counter ions on the surfaces of emersed noble metal electrodes an interesting additional effect was observed by Kolb et al. /29/, and later on reconfirmed by D´Agostino and Hansen /18/ and Kötz et al /19/: the binding energy of the counter ion on the electrode surface after emersion was a function of the electrode potential during emersion. This effect was called "electrochemical shift" by Hansen /18/ and appears to occur with several counter ions. Fig. 18 shows the data obtained by Hansen et al. /18/.

It is most surprising that over a wide potential range a 1:1 correlation could be observed. Up to now this phenomenon is not understood. Possible explanations are based on the change of the electrostatic potential, or on the fact that counterions reside at a certain distance from the electrode surface, so that the proper reference level should be the vacuum level rather than the Fermi level. The latter point is supported by the fact that the work function change also exhibits the 1:1 correlation with the electrode potential.

For surface science studies of the solid/gas interface in UHV it has been demonstrated several times that the work function of the surface is extremely sensitive towards changes

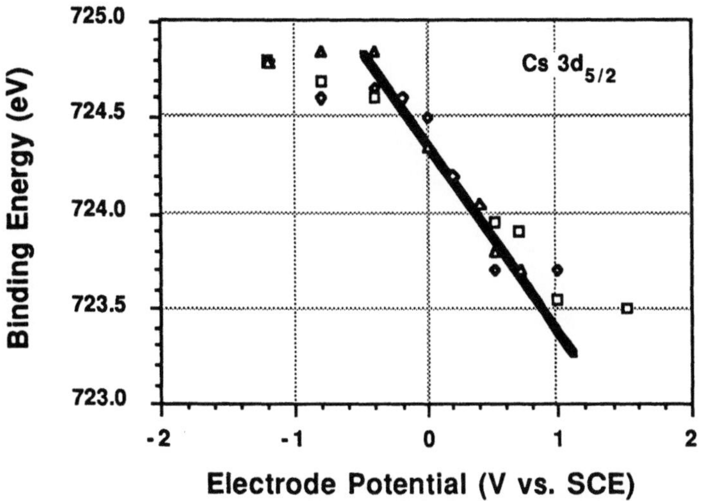


Figure 18. Binding energy of the $Cs3d_{5/2}$ level of Cs counter ions on the surface of an Au electrode as a function of the emersion potential. After /18/.

caused by adsorption and surface layer formation or rearrangement. Work function measurements in UHV are performed mainly by one of two methods, the vibrating capacitor (Kelvin probe) or by photoemission. The latter approach is usually realized with UPS. As has been mentioned above, the cutoff of the secondary electrons determines the minimum energy an electron has to have in order to overcome the surface barrier, i.e. the work function.

Work function measurements on emersed electrodes were first performed by Hansen et al. /28/ using a vertical rotating disc electrode, and measuring the work function with a Kelvin probe shortly above the electrolyte level. A linear relation between the electrode potential and the work function with a slope of 1 was found in a wide potential range. This result again indicates that the electrical double layer is retained after emersion at ambient pressure. In the author´s lab the experiments were carried a step further and the work function of an emersed electrode was determined under UHV conditions after emersion and transfer using UPS /19,26/. Again an 1:1 correlation between electrode potential and change in work function was found. Identical results were obtained on Au /19,26/, Pt /19/ and Ir /53/. While the determination of relative work function changes by means of the vibrating capacitor methode is easy, for an absolute measurement the work function of the Kelvin probe has to be known. On the other hand, if the photoelectric work function is determined the position of the Fermi level and the cutoff energy has to be measured accurately, which is possible only within the resolution of the instrument.

Fig. 19 shows the absolute work functions of three different electrodes after emersion and transfer to the UHV system as a function of the emersion potential. An 1:1 correlation is clearly evidenced within the scatter of data points which correspond to different experimental runs for the Au and Pt electrode. What is even more surprising is the fact that the three metal electrodes exhibit the same workfunction for the same electrode potential, although the work functions of the bare metals are significantly different.

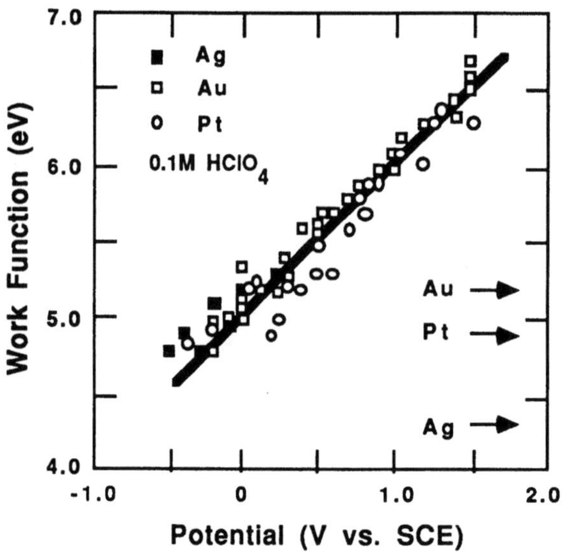


Figure 19. Absolute photoelectric work function for Au, Pt, and Ag electrodes emersed at different electrode potentials and transferred to the UHV system. Arrows indicate the work function of the clean polycrystalline electrode materials. After Kötz et al. /19/.

This result can only be understood if a quantitative preservation of the double layer constituents is assumed. According to Trasatti /78/ the work function of an electrode in solution $\phi_S$ is given by

$$\phi_S = \phi + ed\chi - eg_S(dip) + eg(ion) + \mu_{es} \tag{6}$$

$\phi_S$ in solution is equal to the work function of the bare metal $\phi$ with respect to the reference level in solution $\mu_{es}$ (chemical potential of electrons in the solution). $\phi$ is modified, due to the contact to the liquid phase, by the three terms $ed\chi$, $eg_S(dip)$ and $eg(ion)$. These terms describe the change in work function due to the contact with water, the contribution of oriented dipoles, and the contribution of counter charges, respectively. There is an 1:1 correlation between $\phi_S$ and the electrode potential E measured with respect to a reference electrode /78/, i.e.

$$E = \phi_S/e + \text{const.} \tag{7}$$

Provided all constituents of the double layer are preserved one should obtain a linear relation between $\phi$ after emersion and the electrode potential during emersion, which is indeed the case /19/. Unfortunately the scatter of data points is too big in order to detemine minor changes in slope, or steps, which would allow further insight into minor changes of the components of eq. (6). Therefore more experiments are needed, preferably on single crystal electrodes, with a higher accuracy in work function determination.

From Fig. 19 it is possible to determine the absolute work function of the reference electrode, which was found to be 4.85 ± 0.15 eV /19/. This result is relatively high when compared to the generally accepted value of 4.5 eV /28/, but it shows that at least the electric double layer is preserved. Absolutework function measurements have been performed and discussed in several publications /79/.

The question as to whether $H_2O$, which is also a component of the electrochemical double layer, is preserved has to await more reliable experimental data. Loss of $H_2O$ during transfer may shift the $\phi$ vs. potential curve by a small, probably constant, amount. On the basis of UPS /19,26/, XPS /19/, and AES /80/ measurements it was concluded that no significant amount of water is present after emersion, while other investigatiors interpreted XPS /13/ or TDS /35/ results as indicating the presence of $H_2O$.

Finally one important aspect, which nicely demonstrates the uniqueness of the electrochemical interphase, should be mentioned. In electrochemistry the work function of the electrode surface is determined only by the electrode potential and the respective double layer, and not by the electrode material or by modifications of the electrode surface. If there

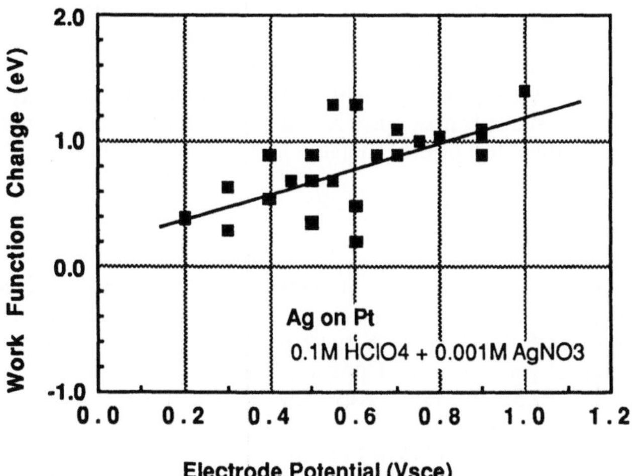


Figure 20. Change of the work function of a Pt electrode as a function of electrode potential in $Ag^+$ -containing electrolyte. Work function change is given with respect to the clean Pt surface. Deposition of 1ML is completed at 0.5 Vsce. After /61/.

is potential control there is a defined work function, independently of the surface properties. This is nicely reflected by the fact that the 1:1 correlation between $\phi$ and the electrode potential is maintained even after oxide formation at potentials above 1.0 Vsce for Au and 0.7 Vsce for Pt (Fig. 17). The very same behavior is more drastically seen in Fig.

20, where the work function of a Pt electrode is measured after emersion and transfer as a function of the electrode potential in $Ag^+$-containing electrolyte. Independently of the deposited amount of Ag, the work function is only determined by the electrode potential.

At the solid/vacuum interface oxygen adsorption or, especially, incorporation with oxide formation would show up as a drastic change in work function. In the case of Ag deposition onto Pt the work function of the sample should change from that of Pt to that of Ag after deposition of about one monolayer. As is evident in Fig. 20, however, at the electrochemical interface the work function of Pt is established ($\Delta\phi = 0$) at 0.0 Vsce, where the Ag coverage is > 1ML. The results of Fig. 20 indicate again, that at least the electrical double layer can be removed and transfered intact to the UHV system. The severe reduction in work function which would be expected for deposition of 1ML of Ag is compensated for by increased counter ion adsorption. Removal of these counter ions by rinsing the electrode with water results in a reduction of the work function /61/.

## 4. Concluding Remarks

The above examples on XPS and also UPS investigations have demonstrated the merits and problems of these techniques, both in general and when applied to electrode surface studies. Contradictory results obtained by different investigators applying XPS to nominally identical systems indicate that interpretation of the data is not always clear-cut. XPS and UPS data should always be substantiated by complementary in-situ or ex-situ techniques.

Provided high quality transfer systems are developed and applied, XPS measurements of binding energies of emersed counter ions (electrochemical shift), UPD adsorbates, and other components of the double layer ($H_2O$) will provide new insight into the chemistry of the interface. Valence band spectroscopy by means of UPS may become a valuable addition to optical in-situ spectroscopies sinc it allows a direct determination of the sample work function as a critical test of the emersion and transfer process.

## 5. References.


1. K. Siegbahn, C. Nordling, A. Fahlman, R. Nordberg, K. Hamrin, J. Hedman, G. Johansson, T. Bergmark, S.E. Karlsson, I. Lindgren, B. Lindberg, *ESCA - Atomic, Molecular and Solid State Structure Studied by Means of Electron Spectroscopy,* Almquist and Wiksells, Uppsala, 1967.
2. C.N. Berglund and W.E. Spicer, Phys. Rev. 136, 1030A (1964).
3. W.E. Spicer, in *Chemistry and Physics of Solid Sufaces IV*, edts. R. Vanselow and R. Howe, Springer Verlag, Berlin (1982) , p. 1.
4. B. Feuerbacher, B. Fitton and R.W. Willis, in *Photoemission and the Electronic Properties of Surfaces*, B. Feuerbacher, B. Fitton and R.R. Willis, Eds.,John Wiley & Sons, (1978), p. 1.
5. K.S. Kim, N. Winograd and R.E. Davis, J. Am. Chem. Soc. 93 (1971) 6269.
6. J.S. Hammond and N. Winograd, J. Electroanal. Chem., 80 (1977) 123.
7. J.S. Hammond and N. Winograd, J. Electrochem. Soc., 124 (1977) 826.
8. E. Yeager, W.E. O'Grady, M.Y.C. Woo and P. Hagans, J. Electrochem. Soc.,125 (1978) 346.
9. A.T. Hubbard, R.M. Ishikawa and J. Kalekaau, J. Electroanal. Chem., 86 (1978) 271.
10. R.N. Ross, J.Electroanal. Chem.,76 (1977) 139.

11. R.O. Ansell, T. Dickinson, A.F. Povey and P.M.A. Sherwood, J. Electroanal. Chem., 98 (1979) 69.
12. H. Neff, W. Foditsch and R. Kötz, J. Electron. Spec. Rel. Phen. 33 (1984) 171.
13. S. Haupt, U. Collisi, H.D. Speckmann and H.H. Strehblow, J. Electroanal. Chem., 194 (1985) 179.
14. *Photoemission in Solids I & II,* M. Cardona and L. Ley, eds., Topics in Applied Physics, Vol. 26 & 27, Springer Verlag, Berlin 1978.
15. *Photoemission and the Electronic Properties of Sufaces*, B. Feuerbacher, B. Fitton, R.F. Willis edts., John Wiley & Sons 1978.
16. Perkin-Elmer Corporation, *Handbook of X-Ray Photoelectron Spectroscopy*, Eden Prairie, Minnesota, 1977.
17. M. Cardona and L. Ley, in ref /14/, pp. 82 and 83.
18. A.T. D´Agostino and W.N. Hansen, Surf. Sci., 165 (1986) 268.
19. R. Kötz, H. Neff and K. Müller, J. Electroanal. Chem., 215 (1986) 331.
20. J.C. Fuggle, in *Handbook of X-ray and Ultraviolet Photoelectron Spectroscopy*, D. Brigg, ed.,Heyden & Son Ltd. London, 1977, p. 295.
21. K.S. Kim, W.E. Baitinger, J.W. Army and N. Winograd, J. Electron. Spec. Rel. Phen., 5 (1974) 351.
22. M. Pijolat and G. Hollinger, Surf. Sci., 105 (1981) 114.
23. T.A. Carlson and G.E. McGuire, J. Electron Spectrosc., 1 (1972/73) 161.
24. VG Scientific, ESCALAB 200-X, Data sheet, 1987.
25. G.K. Wertheim and P.H. Citrin in ref. /14/, p. 197.
26. H. Neff and R. Kötz, J. Electroanal. Chem., 151 (1983) 305.
27. R. Kötz and H. Neff, Surf. Sci., 160 (1985) 517.
28. W.N. Hansen and D.M. Kolb, J. Electroanal. Chem., 100 (1979) 493.
29. D.M. Kolb, D.L. Rath, R. Wille and W.N Hansen, Ber. Bunsenges. Phys. Chem. 87 (1983) 1108.
30. D. Scherson, S. Krause and D.M. Kolb, J. Electroanal. Chem., 176 (1984) 363.
31. M. Peuckert and H. Ibach, Surf. Sci., 136 (1984) 319.
32. A.M. Sukhotin, M.N. Shlepakov, Y.P. Kostikov and V.S. Strykanov, Elektrokhimiya, 18 (1982) 285.
33. W.E O´Grady, M.Y.C. Woo, P.L. Hagans and E. Yeager, J. Vac. Sci. Technol., 14 (1977) 365.
34. M. Peuckert, J. Electroanal. Chem., 185 (1985) 379.
35. M. Peuckert, F.P. Coenen and H.P. Bonzel, Surf. Sci., 141 (1984) 515.
36. E. Rach and J. Heitbaum, Electrochimica Acta, 31 (1986) 477.
37. J.W. Schultze and L. Elfenthal, J. Electroanal. Chem., 204 (1986) 153.
38. R. Kötz, to be published.
39. K.S. Kim, C.D. Sell and N. Winograd in "Electrocatalysis", Electrochem. Soc., M.W. Breiter, ed., (1974) 242.
40. T. Dickinson, A.F. Povey and P.M.A. Sherwood, J. Chem. Soc. Faraday Trans. I 71 (1975) 298.
41. J.D.E McIntyre and D.M. Kolb, Symp. Faraday Soc., 4 (1970) 99.
42. S.Hadzi-Jordanov, H. Angerstein-Kozlowska, M. Vukovic and B.E. Conway, J. Electrochem. Soc., 125 (1978) 1471.
43. R. Kötz, H.J. Lewerenz and S. Stucki, J. Electrochem. Soc., 130 (1983) 825.
44. R. Kötz, H.J. Lewerenz, P. Brüesch and S. Stucki, J. Electroanal. Chem., 150 (1983) 209.
45. R. Kötz amd S. Stucki, J. Electrochem. Soc., 132 (1985) 103.
46. K.S. Kim and N. Winograd, J. Catalysis, 35 (1974) 66.
47. M. Vukovic, H. Angerstein-Kozlowska and B.E. Conway, J. Appl. Electrochem., 12 (1982) 193.

48. J. Augustynski, L. Balsenc and J. Hinden, J. Electrochem. Soc., 125 (1978) 1093.
49. Lj. Atanasoska, W.E. O'Grady, R.T. Atanasoski and F.H. Pollak, Surf. Sci., in press.
50. R. Kötz and S. Stucki, J. Appl. Electrochem., 17 (1987) 1190.
51. J. Augustynski, M. Koudelka, J. Sanchez and B.E. Conway, J. Electroanal. Chem., 160 (1984) 233.
52. H.Y. Hall, P.M.A. Sherwood, J. Chem. Soc., Faraday Trans. I, 80 (1984) 135.
53. R. Kötz, H. Neff and S. Stucki, J. Electrochem. Soc., 131 (1984) 72.
54. M. Peuckert, Surf. Sci.,144 (1984) 451.
55. A. DeBattisti, R. Brina, G. Gavelli, A. Benedetti and G. Fagherazzi, J. Electroanal. Chem., 200 (1985) 93.
56. A. DeBattisti, G. Lodi, . Cappadonia, C. Battaglin and R. Kötz, J. Electrochem. Soc., to be published.
57. D.N. Buckley, L.D. Burke and J.K. Mulcahy, J. Chem. Soc. Faraday Trans. I, 72 (1976) 1896.
58. J.D.E McIntyre, S. Basu, W.F. Peck, W.L. Brown, and W.M. Augustyniak, Phys. Rev. B, 25 (1982) 7242.
59. L.F. Mattheiss, Phys. Rev. B13 (1976) 2433.
60. C. Gutierrez, M. Sanchez, J.I. Pena, C. Martinez, and M.A. Martinez, J. Electrochem. Soc., 134 (1987) 2119.
61. R. Kötz and H. Neff, to be published.
62. M.L. Shek, P.M. Stefan, I. Lindau, and W.E. Spicer, Phys. Rev B27 (1983) 7277
63. D.M. Kolb and R. Kötz, Surf. Sci. 64 (1977) 689.
64. P.M.A. Sherwood, Chem. Soc. Rev. 14 (1985) 1.
65. P.M.A. Sherwood, in *Contemporary Topics in Analytical and Clinical Chemistry*, Vol. 4, D.M. Hercules et al., eds., Plenum (1982) 205.
66. S.C. Tjong and E. Yeager, J. Electrochem. Soc. 128 (1981) 2251.
67. O.J Murphy, J. O'M Bockris, T.E. Pou, L.L. Tongson and M.D. Monkowski, J. Electrochem. Soc. 130 (1983) 1792.
68. P. Brüesch, K. Müller, A. Atrens and H. Neff, Appl. Phys. A38 (1985) 1.
69. S.C. Tjong, R.W. Hoffman and E.B. Yeager, J. Electrochem. Soc. 129 (1982) 1662.
70. P. Brüesch, K. Müller and H.R. Zeller, Surf. Sci. 169 (1986) L327.
71. K. Kuroda, B.D. Cahan, Gh. Nazri, E. Yeager and T.E. Mitchell, J. Electrochem. Soc. 129 (1982) 2163.
72. C. Kozlowski and P.M.A. Sherwood, J. Chem. Soc. Faraday Trans. I, 80 (1984) 2099.
73. K.M. Sundberg, Lj. Anastasoska, R. Anastasoski and W.H. Smyrl, J. Electroanal. Chem. 220 (1987) 161.
74. T. Solomun, R. McIntyre, W. Richtering and H. Gerischer, Surf. Sci. 169 (1986) 414.
75. H. Goslowsky, H.M. Kühne, H. Neff, R. Kötz and H.J. Lewerenz, Surf. Sci. 149 (1985) 191.
76. K.J. Bachmann, S. Menezes, R. Kötz, M. Fearheily and H.J. Lewerenz, Surf. Sci. 138 (1984) 475.
77. H.J. Lewerenz and R. Kötz, J. Appl. Phys. 60 (1986) 1430.
78. S. Trasatti, in *Comprehensive Treatise of Electrochemistry*, Vol. 1, J. O'M Bockris, B.E. Conway and E. Yeager eds., Plenum Press, New York, 1980, Ch.2, p.45
79. W.N. Hansen and G.J. Hansen, Phys. Rev. A36 (1987) 1396; Phys. Rev. Lett. 59 (1987) 1049; and references therein.
80. D. Aberdam, S. Traore, RT. Durand and R. Faure, Surf. Sci. 180 (1987) 319.

# RUTHERFORD BACKSCATTERING SPECTROSCOPY OF ELECTRODE SURFACES

R. KÖTZ
*Asea Brown Boveri*
*Corporate Research*
*CH-5405 Baden*
*Switzerland*

ABSTRACT. Rutherford Backscattering Spectroscopy (RBS), a standard interface analytical tool, has been applied only scarcely to the investigation of electrodes. After a short introduction to the basics of RBS most of the few studies, which used RBS for electrode surface analysis are reviewed, and the information provided by RBS is discussed. Recent in-situ RBS investigations of the electrode/electrolyte interface are presented, and problems incountered during this studies are discussed.

## 1. Introduction

Rutherford Backscattering Spectroscopy (RBS) is a well established tool for interface analysis in semiconductor device technology, especially for investigation of interdiffusion profiles and implantation profiles. For the analysis of electrodes this ex-situ technique has been applied rather scarcely, which may in part be due to the fact that RBS systems are expensive and not always available. The advantage of this technique is its capability of non destructive, quantitative elemental analysis without standards, and of depth profiling. A detailed description of the technique, its fundamental aspects, experimental details and evaluation procedures has been given by Chu et al. /1,2/ and by Mayer and Poate /3/.

In the following the kind of information available with RBS will be described together with its fundamental aspects. The possibility of in-situ RBS measurements will be discussed, and examples will be given where ex-situ and in-situ RBS have been applied for the investigation of electrode surfaces.

## 2. Fundamental Aspects of RBS

The basic arrangement of a RBS experiment is shown in Fig. 1. A beam of accelerated low mass ions ($He^+$, $He^{++}$ or $H^+$) with an energy between 500 keV and 5 MeV is directed onto the sample. Most of the projectiles are thermalized within a bulk sample or, in the case of a thin foil, are transmitted. Some of the ions are backscattered after collisions with target atoms and are energetically analyzed at the detector, which gives the number of backscattered projectiles with a certain kinetic energy per time interval. The RBS spectrum is a plot of detector counts versus energy of backscattered ions.



*C. Gutiérrez and C. Melendres (eds.),*
*Spectroscopic and Diffraction Techniques in Interfacial Electrochemistry,* 439–448.

## 2.1. INFORMATION PROVIDED BY RBS

2.1.1. *Elemental Specificy* Backscattering of the ions from the target surface occurs as a consequence of collisions between the ion and a nucleus of the sample. Conservation of momentum and energy leads to the kinematic equation

$$E_1 = E_0 K \qquad (1)$$

with

$$\sqrt{K} = (M_1 \cos\theta + \sqrt{(M_2^2 - M_1^2 \sin^2\theta)}) / (M_1 + M_2) \qquad (2)$$

where $E_0$ and $E_1$ are the energies of the incident and of the backscattered projectile, $M_1$ and $M_2$ correspond to the masses of the projectile and of the target atom, respectively, K is the kinematic factor, and $\theta$ is the scattering angle (usually about $170^o$) measued relative to the undeflected transmitted beam. Different masses of nuclei, i.e. different elements, can thus be identified by the energy of the backscattered particle.

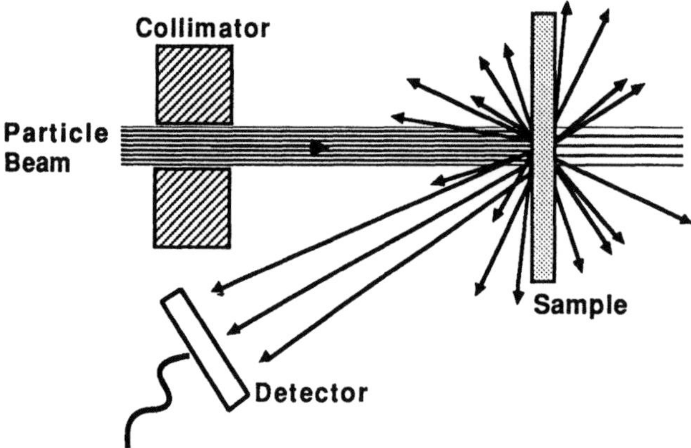


Figure 1. Basic arrangement of a Rutherford Backscattering experiment.

The energy of the scattered particle depends on the mass of the scattering nucleus at the surface and the angle of detection. Heavy elements appear at the high energy side of a RBS spectrum, light elements at the low energy end (see Fig. 2). Kinematic factors are tabulated for the elements as a function of the angle $\theta$ /2/.

2.1.2. *Depth Profiling* A particle traveling through a solid loses energy due to inelastic collisions. To a good approximation in thin films energy loss is directly proportional to thickness. The reduction in energy $\Delta E$ after having traversed a depth d is given by equation (3)

$$\Delta E = d\,[S] = N\,d\,[\varepsilon] \qquad (3)$$

where [S] is the energy loss factor, [$\varepsilon$] the stopping power, and N the atomic density (atoms/$cm^3$) of the material. The energy loss factor [S] can be calculated as

$$[S] = K\, dE/dx(E_0) + (1/\cos\theta)\, dE/dx(E_1) \tag{4}$$

where $E_0$ and $E_1$ are the energies of the incident and emerging particles, respectively. The energy loss factor is a function of the detection angle. In order to increase the depth sensitivity the scattered projectiles should be detected at a grazing angle, thus increasing the effective layer thickness. The energy loss factors are also tabulated /2/ for the different elements and energies. The highest values are in the order of 150 eV/Å.

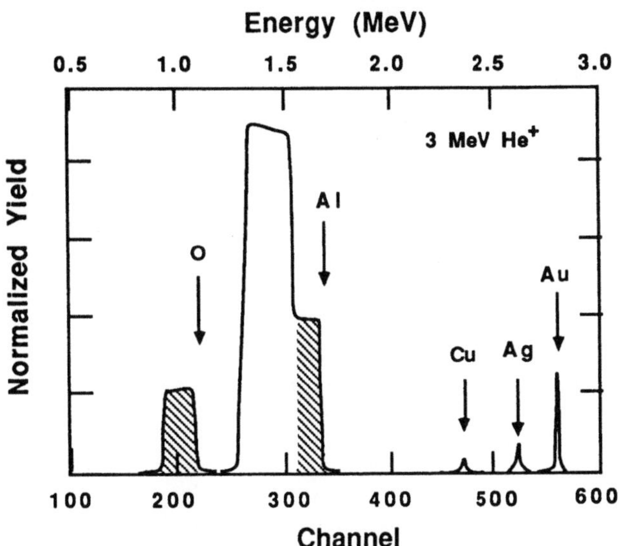


Figure 2. Simulated RBS spectrum for three monolayers of Au, Ag and Cu on top of an oxidized aluminum film. Thicknesses of Al and Al oxide are 5000 Å each. The surface energies of the different elements are indicated. The shaded areas indicate those parts of the spectrum originating from the oxide. Incident energy of $He^+$ ions is 3.0 MeV. Calculated with RUMP /4/.

A simulated RBS spectrum of an aluminum foil with a thickness of 5000 Å and an oxide layer with the same thickness on top is shown in Fig. 2. Monoatomic layers (thickness 3Å) of Cu, Ag and Au are deposited on top of the oxide. The different kinematic factors are clearly evidenced by the energetic positions of the elements. Au, with a kinematic factor of 0.92, appears at the right end of the energetic scale, while O with a factor of 0.36 appears at roughly 1.1 MeV. The depth profiling capability is evidenced by the width of the Al signal between 1.3 and 1.7 MeV and by the width of the oxygen signal originating from the oxide layer.

2.1.3. *Elemental Composition* In addition to what is in the sample (2.1.1.) and where (2.1.2.), RBS allows the determination of the composition of the target. The number of counts of backscattered and detected particles is related to the number of incident particles (beam current), the differential Rutherford scattering cross section, the acceptance angle of the detector and, of course, the density (atoms/cm$^3$) of target atoms N of the respective element. In a rather crude approximation (5-10% accuracy) the relative amount of two elements in a layer (e.g. Al and O in the oxide layer of Fig. 2) is given by

$$N_O / N_{Al} = H_O / H_{Al} (Z_{Al} / Z_O)^2 \quad (5)$$

where H is the height of the respective signal at the surface. In (5) equal stopping power factors are assumed for both elements in the oxide, and it is further assumed that the Rutherford scattering cross section is proportional to the square of the atomic number. For the simulated spectrum in Fig. 2 the calculated value is

$$N_O / N_{Al} = 1.1 / 1.9 (169 / 64) = 1.54$$

which is in relative good agreement with the composition of $Al_2O_3$. For a more accurate determination, more complicated calculations have to be performed /2/ or available simulation programs /4/ have to be used for fitting the experimental to calculated data. In the case of very thin films (e.g Au, Ag, and Cu in Fig.2) the total amount of atoms is proportional to the integral area under the peak.

## 2.2. EXPERIMENTAL CONSIDERATIONS

In the first place, RBS is an ex-situ technique, since a high vacuum in the order of $10^{-5}$ - $10^{-6}$ torr is required. In this range pumping is no problem, and a relatively fast sample turn over can be achieved.

Compared to other techniques like XPS, AES, or SIMS, depth resolution is rather limited. With a typical detector resolution of 15 keV (semiconductor detector) and an energy loss factor of, at the most, 150 eV/Å, a depth resolution of 100 Å is calculated, which decreases with decreasing [S]. Depth resolution may be improved by tilting the sample, by reducing the projectile energy, and by using an electrostatic analyzer (see 3.2.). A further limitation to depth resolution is energy straggling: due to statistical scattering events, the energy of the initially monochromatic particle beam becomes more and more spread out. For films with a thickness of some thousand Å straggling and detector resolution become comparable, this point becoming important for in-situ RBS arrangements (2.3.) Although depth resolution is rather poor in RBS, one should keep in mind that this resolution does not require sputtering for a sampling depth of several microns, while for techniques like XPS, AES or SIMS heavy sputtering is necessary for these thicknesses. RBS is non destructive, since sputtering does not occur because the nuclear stopping power is negligible as a consequence of the high energy of the projectiles.

Elemental resolution is also determined by the resolution of the detector. In addition, it depends on wether one wants to separate heavy or light elements. So separation of Pt and Ir is not possible because the kinematic factors differ only by 0.1 %, and the difference in energy will be only 3keV for a 3MeV projectile energy. On the other hand, O and N would be separated by 150 keV, which is ten times the detector resolution.

Elemental sensitivity increases with increasing atomic number (see Fig. 2). Therefore one has to reach a compromise between resolution and sensitivity. In general, it is always advantageous to try to detect heavy elements on a light substrate.

The energy of the backscattered ion contains mass resolution as well as depth resolution, so that it might be difficult in some cases to distiguish between a heavy atom located at a certain depth and a light atom on the surface. One way of solving this problem is tilting of the sample: the energy of the ions backscatterd by surface atoms should stay constant, while the energy of ions backscatered by the atoms beneath the surface will change.

Finally, an important feature of RBS when applied to single crystals should be mentioned. If a single crystalline target is properly aligned with respect to the direction of the projectile beam, channeling will occur. In this case the ion beam penetrates the crystal in an "open" direction, and the fraction of backscatterd particles will be reduced significantly as compared to a random direction. This method allows the detection of amorphous layers on single crystalline substrates /2,3/.

## 2.3. IN - SITU RBS

An electrochemical cell for in-situ RBS investigations of electrode surfaces has been developed in the author's laboratory /5/ and improved by M. Ciappa /6/. The key element is a thin silicon membrane with a thickness of around 1μm which separates the liquid phase from the vacuum of the beam line (FIg. 3). This membrane is etched into a silicon (100) wafer by means of standard etching techniques /7/. The metal electrode is evaporated onto the back of the Si wafer including the window. The projectile beam is focused onto this window, which has a typical size of 2mm x 2mm. The projectiles travel through the Si membrane and reach the electrode/electrolyte interface, from where backscattered particles carry information about the interfacial composition. The electrolyte volume needed for the cell is roughly 7 ml. Potential control is achieved by a standard three electrode arrangement.

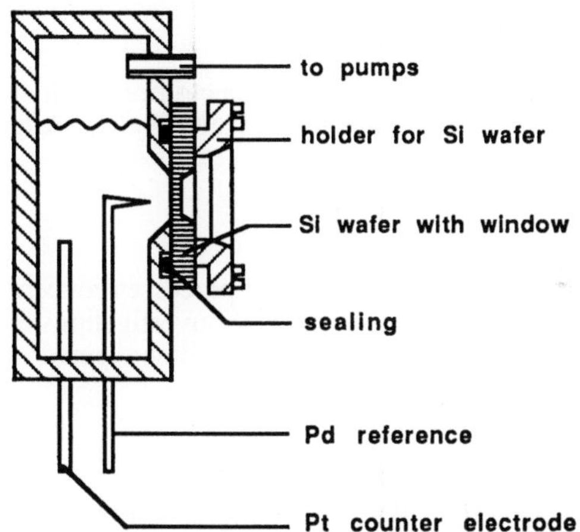


Figure 3. Sketch of the electrochemical cell for in-situ RBS measurements. After /5/.

The resulting RBS spectrum of the membrane - electrode - electrolyte system is plotted in Fig. 4. The electrode was a 1200 Å thick iridium layer, and the electrolyte was 0.5M $H_2SO_4$. The first in-situ RBS spectra were recorded for Cu deposited on an Ir electrode /5/. In 3.1 new in-situ results for oxide formation on Ta will be presented /6/.

The possibility of performing an in-situ, instead of an ex-situ experiment is a considerable advantage, but four problems arise:

(i) The fact that the projectiles have to pass the Si membrane before reaching the electrode/electrolyte interface reduces depth resolution significantly due to additional straggling. (ii) Deformation of the membrane as a consequence of the pressure difference of 20 mbar introduces stress and may lead to cracks in the evaporated film electrode. (iii) Investigation of bulk electrodes is not possible; and (iv) Thermalization of projectiles in the electrolyte within a distance of several µm from the electrode may lead to uncontrolled chemical (radical) reactions. This aspects have been investigated by M. Ciappa in some detail /6/.

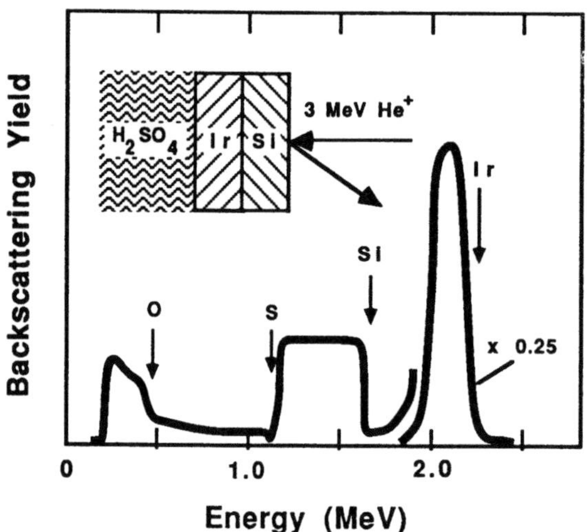


Figure 4 3 MeV in-situ Rutherford backscattering spectrum of an Ir electrode (0.12 µm) supported by a Si membrane (1.3 µm) in contact with 0.5M $H_2SO_4$. After /5/

## 3. RBS of Electrodes

The majority of RBS investigations of electrochemical processes concerns oxide formation on metal and metal alloy substrates /5,8-18/. A few RBS investigations on metal deposition /5,19,20/ and on fuel cell catalysts /21,22/ have also been performed. All of these investigations will be briefly mentioned in the following.

### 3.1. OXIDES AND OXIDE FORMATION

3.1.1. *Ruthenium and Iridium oxides* Backscattering spectroscopy can first of all be used to control the composition of "as prepared" electrodes before investigation of their electrochemical behavior. Such experiments were performed by Battaglin et al. /8/ for thermally prepared $RuO_2$ and $IrO_2$ DSA-type electrodes on Ti substrates. RBS showed a significant deficiency of oxygen for the ruthenium oxide, and an excess of O for the iridium oxide. The stoichiometries were determined to be $RuO_{1.3}$ and $IrO_{2.5}$, respectively. In addition, an interface region between the oxide and the Ti substrate was observed, with a thickness of about 0.2 µm.

Recently RBS was used to reconfirm XPS depth profile analysis of $RuO_2/TiO_2$ ruthenium-based mixed oxides supported on Ti /9/. The ruthenium deficiency observed previously by XPS could be confirmed by means of RBS.

Proton insertion into thick anodic iridium oxide films (AIROF) exhibiting electrochromism was studied by McIntyre et al. /10/ using RBS and nuclear reactions. The stoichiometry of the oxide in the bleached state was determined to be $HIrO_2 \cdot 3H_2O$, and $IrO_2 \cdot 2H_2O$ for the colored state. In addition, the insertion of Li in non aqueous electrolytes could be observed by means of nuclear reactions. RBS of the AIROF also revealed a gradual change of the film stoichiometry along the depth of the film. At the metal/oxide interface the film appeared to be less hydrated, its composition being $IrO_2$ and $HIrO_2$ for the colored and bleached state, respectively.

In the same study McIntyre et al. /10/ investigated the electrochromic effect in $WO_3$, and found the same proton insertion reaction for the reduction of the film. For tungsten, however, coloration occurs during reduction, and bleaching during oxidation. This is consistent with the $t_{2g}$ band filling observed by UPS of AIROFs after reduction. For tungsten the $t_{2g}$ band is empty in the oxidized state and occupied by one electron in the reduced state, thus making possible additional optical transitions to the Fermi level upon reduction. (see chapter by R. Kötz on XPS and UPS).

3.1.2. *Aluminum* Brown and Mackintosh /11,12/ have investigated the behavior of implanted atoms during the anodic oxidation of aluminum. As a model example the system Xe in Al was tested (Fig. 5).

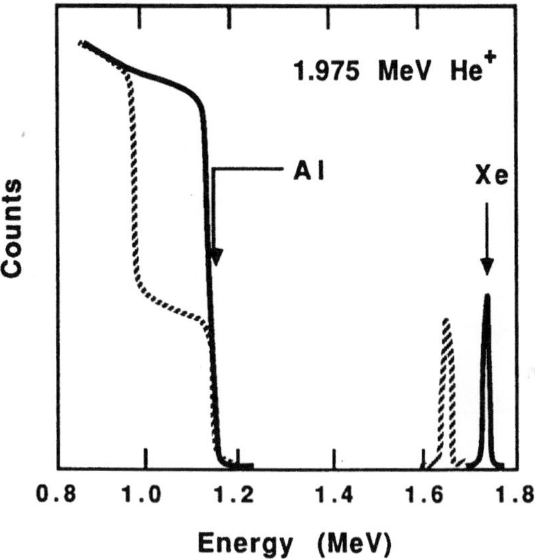


Figure 5. RBS spectrum of Xe implanted into Al before (——) and after (- - - -) anodic oxidation. After /11/.

After formation of the oxide in ammonium pentaborate at $2mA/cm^2$ the Xe implanted region moves from the surface into the oxide /11/, from which it is concluded that aluminum atoms, as well as oxygen atoms, are mobile during the oxidation. The direction of movement of the implanted metal ions depends on their initial position: ions implanted in a preformed oxide move outwards, while ions implanted into the metal beneath a natural air-formed oxide move inwards, during anodic oxidation.

The movement of cations during the anodic oxidation of superimposed metal layers was investigated by Perriere et al. /13/. The order of cations is preserved during the oxidation of Ta-Nb and Al-Nb systems, while it is reversed for Nb-Ta and Nb-Al. After oxidation of an Al layer on a Nb substrate, Nb moves outwards forming a $Nb_2O_5$ layer on top of the $Al_2O_3$ layer.

Oxide formation on aluminum was investigated by McCune and Russo /14/, Hurlen et al., /15/ and by Skeldon et al /16/. In these studies the oxide composition, incorporation of anions, and the oxide thickness were determined. The composition of the oxide formed after 20h in weakly alkaline solutions was determined to be $Al_2O_3 \cdot 1/2\ H_2O$ /15/. The inhibition effect of phosphate towards the water-Al reaction is clearly evidenced in the RBS spectra, which also showed that phosphorus is built into the surface layer /14/.

Strehblow et al. /17/ have investigated the anodic oxidation of the copper / aluminum alloys used for metallization of silicon integrated circuits. The RBS results clearly indicate that the resulting anodic oxide film is free of copper, which accumulates instead at the metal oxide interface within a layer of 650 Å thickness. The oxide layer grows linearly with time and potential, and the breakdown of the anodic oxide is related to the copper accumulation.

3.1.3. *Tantalum* Oxide growth on Ta-Si alloys has been investigated by Silverman and Schwarz /18/, who found that the anodic oxide layers can be divided into an outer and inner part. The outer layer is significantly enriched in Ta and is almost devoid of Si, while the inner layer is enriched in Si. The authors concluded that the Ta ions are mobile and move to the surface during oxide growth. The composition of the oxides was found to be stoichiometric $Ta_2O_5$ and $SiO_2$. For pure Ta films as well as Ta/Si films a linear relation between film thickness and anodization potential was found, with growth rates in 0.01% citric acid of 17.5 Å/V and 20 Å/V for Ta and Ta/Si, respectively.

In-situ RBS has been applied recently for the investigation of Ta in 0.5M $H_2SO_4$ /6/. The results were compared with those obtained ex-situ, and good agreement between the two methods was obtained. Fig. 6 shows the in-situ spectra of a Ta film electrode supported on the above mentioned Si membrane before and after anodization at 140 V in $H_2SO_4$. The energetic positions of the respective interfaces are indicated. The reduction in Ta thickness at the expense of Ta oxide growth is nicely reflected. In addition, the change in the oxygen edge due to oxide growth is also visible.

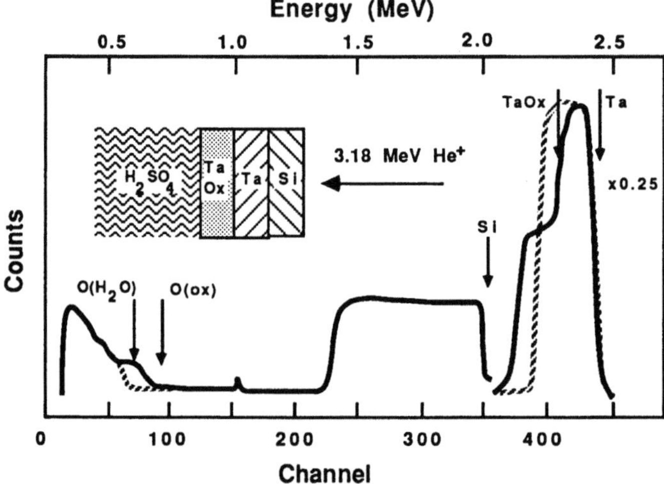


Figure 6. In-situ RBS spectrum of a Ta electrode (2300Å) oxidized at 140 V in 0.5M $H_2SO_4$ (full line) and of the fresh electrode (dashed line). Respective interfaces are indicated by arrows. Oxide thickness: 2400Å, Si membrane thickness: 13000Å. After Ciappa /6/.

The above mentioned problem of straggling can be seen when comparing the resolution of the surface Si edge at channel 350 and the Ta/electrolyte interface at channel 390 for the fresh sample. Detailed evaluation of the data /6/ showed that the resolution of 15 keV was reduced to 45 keV. Accurate determination of the oxide thickness as a function of the potential for the in-situ and ex-situ cases gave a growth rate of 16.1 Å/V for the oxide and a reduction in Ta thickness by 6.1 Å/V, which is in good agreement with literature data.

As can be seen in Fig. 6 the composition of the anodic oxide changes along its depth, with a decreasing O/Ta ratio towards the metal oxide interface. The O/Ta ratio was determined to be 2.5 at the metal/oxide interface, and 3.0 at the electrolyte/oxide interface /6/.

### 3.2. METAL DEPOSITION

The feasibility of in-situ RBS was first demonstrated for Cu deposition on Ir /5/. No relevant data concerning the electrochemical process were obtained in this test study.

Properties of electroless plated gold contacts on CdTe substrates were investigated by Mancini et al. /19/. It was found that an electrochemical reaction takes place between the gold chloride solution and the substrate, as a consequence of which the interface becomes depleted in cadmium. Reduction in Cd content influences the electrical properties significantly by formation of a p layer on an n-type sample.

Underpotential deposition of Pb on Ag was investigated by Frerichs et al. /20/. Using 0.5 MeV He ions and an electrostatic analyzer they claimed a depth resolution of 22 Å. The experiment clearly demonstrated that Pb atoms on the surface of the electrode can be removed quantitatively from the electrolyte and transfered to a vacuum chamber for ex-situ analysis. A nearly 1:1 correlation was obtained between the electrochemically determined number of adatoms and the quantitative RBS results.

### 3.3. FUEL CELL CATALYSTS

Finally, an interesting study concerning the distribution of Pt in fuel cell catalysts by means of RBS should be mentioned. Hyde et al. /21,22/ used 2.5 MeV protons for the analysis of electrodes from phosphoric acid fuel cells before and after use. In order to measure the Pt distribution over the active layer the electrode, cut at an angle, was examined by a focused (5µm spot) proton beam. It was shown that commercially available electrodes have a rather even distribution of Pt, and suffer a small loss of Pt even after a prolonged operation time. In the case of a Pt/V catalyst leaching of the vanadium component was observed.

## 4. Concluding Remarks

RBS is a powerful technique for non destructive, quantitative elemental analysis of surfaces, with a sampling depth of several µm. When used as an ex-situ technique for electrode surface analysis all of the known advantages and disadvantages apply. In order to make it an in-situ technique one has to pay the price of reduced depth resolution and of being restricted to thin film electrodes. It has still to be demonstrated that in-situ RBS, besides the advantage of on-line monitoring, provides more accurate information about the electrochemical interphase than ex-situ RBS.

Typical applications of RBS in electrochemistry are thick oxide formation, metal deposition, and monitoring of substrate-film interactions. Use of RBS for double layer studies is unlikely, but perhaps possible if depth resolution is increased, or if channelling is used on single crystal electrodes.

## 5. References.


1. W.K. Chu, J.W. Mayer, M.A. Nicolet T.M. Buck, G. Amsel, and F. Eisen, Thin Solid Films, 17 (1973) 1-41
2. W.K. Chu, J.W. Mayer and M.A. Nicolet, *Backscattering Spectrometry*, Academic Press, New York, 1978
3. J.W. Mayer and J.M. Poate, in *Thin Films - Interdiffusion and Reactions*, J.M. Poate, K.N. Tu and J.W. Mayer, eds.,The Electrochem. Soc., John Wiley & Sons, New York, 1978
4. L.R. Doolittle, Nuclear Instruments and Methods, B9 (1985) 344. The RUMP program is available from L. Doolittle, Materials Science and Engineering, Bard Hall, Cornell University, Ithaca, New York 14853, (607) 256-5031
5. R. Kötz, J. Gobrecht, S. Stucki and R. Pixley, Electrochim. Acta, 31 (1986) 169
6. M. Ciappa, Thesis, Physics Inst. University Zürich, 1986
7. K.E. Bean, IEEE Trans. Electron. Dev. ED25 (1978) 1185
8. C. Battaglin, A. Carnera, P. Mazzoldi, G. Lodi, P. Bonora, A. Daghetti and S. Trasatti, J. Electroanal. Chem. 135 (1982) 313
9. A. DeBattisti, G. Lodi, M. Cappadonia, C. Battaglin and R. Kötz, J. Electrochem. Soc., to be published.
10. J.D.E. McIntyre, S. Basu, W.F. Peck, W.L. Brown, and W.M. Augustyniak, Phys. Rev. B25 (1982) 7242
11. F. Brown and W.D. Mackintosh, J. Electrochem. Soc. 120 (1973) 1096
12. W.D. Mackintosh, F. Brown and H.H. Plattner, J. Electrochem. Soc. 121 (1974) 1281
13. J. Perriere, J. Siejka and S. Rigo, Corrosion Sci. 20 (1980) 91
14. R.C. McCune and C.J. Russo, J. Electrochem. Soc. 128 (1981) 555
15. T. Hurlen, A.T. Haug and G. Salomonsen, Electrochim. Acta 29 (1984) 1161
16. P. Skeldon, K. Shimizu, G.E. Thompson and G.C. Wood, Thin Solid Films, 123 (1985) 127
17. H.H. Strehblow, C.M. Melliar-Smith, and W.M. Augustyniak, in *Thin Film Phenomena - Interfaces and Interaction*, J.E.E. Baglin and J.M. Poate, eds., Proc. Vol. 78-2, The Electrochem. Soc., 1978
18. P.J. Silverman and N. Schwartz, J. Electrochem. Soc. 121 (1974) 550
19. A.M. Mancini, A. Quirini, L. Vasanelli, E. Perillo, E. Rosato, G. Spadaccini and E. Barbarino, J. Appl. Phys. 53 (1982) 5785
20. H.P. Frerichs, S. Kalbitzer, F.J. Demond, D.L. Rath and D.M. Kolb, Surf. Sci. 105 (1981) L271
21. C.J. Maggiore, P.J. Hyde and S. Srinivasan, Extended Abstracts, The Electrochem. Soc. Spring Meeting, Montreal, Canada, 1982
22. P.J. Hyde, C.J. Maggiore and S. Srinivasan, J. Electroanal. Chem. 168 (1984) 383

# ELECTROCHEMICAL APPLICATIONS OF SCANNING TUNNELING MICROSCOPY

A.J. Arvia
Instituto de Investigaciones Fisicoquímicas Teóricas y Aplicadas (INIFTA), Facultad de Ciencias Exactas, UNLP
Sucursal 4, Casilla de Correo 16
(1900) La Plata - Argentina

**ABSTRACT**. A brief description of the principles of the Scanning Tunneling Microscopy (STM) is presented, including the basic theory, experimental mounting and operation conditions. Applications of the new technique to characterize at the nm scale the topography of polycrystalline, single crystal, and amorphous metal surfaces are presented. Attention is paid to preferentially oriented and highly-dispersed metals of particular interest in electrochemistry. Surface imaging of graphite and semiconductors at the atomic scale are shown. The applications of STM to the study of surface configuration of different adsorbates is also included. The recent development of STM mounted in the electrochemical cell as an in situ technique to determine the topography of metal electrodes immersed in the electrolyte solution is described. The scanning tunneling spectroscopies are also briefly considered. Finally, the present state-of-the-art and the perspectives of STM in electrochemistry are discussed.

## 1. THE SCANNING TUNNELING MICROSCOPY

The Scanning Tunneling Microscopy (STM) has been recently developed by Binnig and Rohrer at the IBM Zürich Research Laboratory. This new imaging technique based on the electron tunneling effect allows one to obtain surface topography of solids (metals and semiconductors) at the atomic level [1,2].

It is worthwhile mentioning that the electron tunneling effect became extremely important a few decades ago to explain new facts such as the superconductivity phenomena of certain materials, the phonon spectra, and the electron-phonon coupling intensity in junctions [3-5]. The use of the electron tunneling effect as phonon spectrometer was also extended to semiconductors and to non-conducting oxides. Vibrational spectra and occasionally electronic transition spectra of inorganic and organic adsorbates can also be obtained [4,5]. Electron tunneling is also useful for plasmon excitation to investigate visible light emission due to metal electron energy loss.



*C. Gutiérrez and C. Melendres (eds.),*
*Spectroscopic and Diffraction Techniques in Interfacial Electrochemistry*, 449–469.

The application of the electron tunneling effect in STM has yielded for the first time topographic as well as spectroscopic data at the atomic level under non-destructive conditions of the specimen. Therefore, STM opens a revolutionary perspective in surface chemistry and physics, since it provides direct information on structural and reactivity properties of materials, most of which were derived until now from indirect measurements. STM is an outstanding technique with great potentialities and versatility for fundamental research in surface chemistry and physics, particularly of semiconductors and metals [6-12]. Recent investigations in molecular biology reveal STM to be a powerful complement of transmission electron microscopy [13,14]. Attempts are being made to extend the STM to in situ measurements, this possibility being particularly attractive for electrochemical work [15].

Very recently, Binnig, Quate and Gerber [16,17] proposed a new instrument, the atomic force microscope, for measuring the interaction forces between tip and sample surface, both on conducting and insulator materials [18].

## 2. PRINCIPLES OF STM

The basis of STM is the electron tunneling effect, which is a purely quantum phenomenon. Let us consider two plane metallic electrode surfaces 1 and 2 separated by the distance $\ell$, in vacuo (Fig. 1). If $\ell$ is sufficiently small, one obtains a finite probability for electron transfer from 1 to 2 through the gap $\ell$.

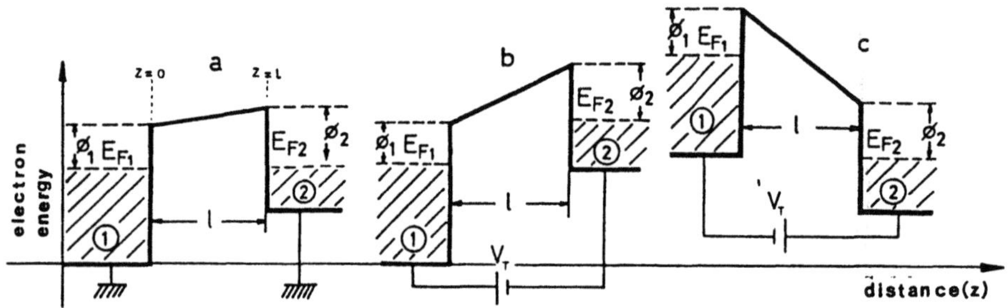


**Figure 1.** Energy diagrams for two metals 1 and 2 separated by the distance $\ell$ in vacuo at T = 0 K. $\phi_1$ and $\phi_2$ are the work functions of metals 1 and 2, respectively. a) Both metals grounded. b) Electron energy of metal 2 is modified with respect to that of metal 1 (elastic electron tunneling). c) Reverse situation to that described in b) (no electron tunneling).

When a small potential difference, $V_T$, is applied between metals 1 and 2, $i_T$, the tunneling current is given by the expression:

$$i_T = \alpha V_T \exp(-\beta\sqrt{\phi}\ell) \qquad (1)$$

where $\alpha$ and $\beta$ are proportionality constants, and $\phi = 1/2(\phi_1 + \phi_2)$ (Fig. 1,b). According to equation (1), $i_T$ depends exponentially on $\ell$ and on $\phi$. For $\phi \simeq 4$ eV, a change of 0.1 nm in $\ell$ produces an order of magnitude change in $i_T$. This dramatic change is the working principle of STM.

The experimental arrangement consists of the specimen surface, 2, and the metallic tip, 1, with a radius of curvature as small as possible. The tip is mounted on a piezoelectric device driven in the directions x, y, z by means of three voltages $V_x$, $V_y$, $V_z$ applied through a servomechanism which allows to work at constant $i_T$. In this way it is possible to shift the tip over the specimen surface (x,y) at a certain distance from it, by keeping the tunneling current constant, and to scan the surface in a raster pattern. Hence, by plotting the voltage $V_z$, which commands the vertical position of the tip with respect to the surface, against the voltages $V_x$ and $V_y$ which control the lateral scanning, a tridimensional image of the specimen surface can be obtained, as well as some information on the structure of the specimen surface through the work function, since this appears in expression (1).

However, a more realistic interpretation of the STM images can be advanced by means of the tridimensional theory of Tersoff and Hamman [20], who claimed that the electron tunneling current is related to a convolution of the density of electronic states (DOS) of the specimen surface and the tip. In this way, by modelling the tip as a spherical potential surface and assigning to it an s-type wave function, the conductance, $\sigma$, associated with the electron tunneling current, for small applied potentials appears as a function of $r_t$, the position of the centre of curvature of the tip, and the Fermi energy level of the specimen surface:

$$\sigma = f(r_t, E_F). \qquad (2)$$

The electron state density, $\rho(r_t, E_F)$, of the specimen surface at the Fermi level measured at the surface point opposite to the tip is given by the expression:

$$\rho(r_t, E_F) = \sum_{\nu} |\psi_\nu (r_t)|^2 \, \delta(eV_T - E_F) \tag{3}$$

where $V_T$ is the potential applied between the tip and the specimen surface. Thence, the STM image represents essentially a constant electronic density contour which for metallic surfaces mainly reflects the surface atomic structure.

The theory also provides an estimation of $\Delta$, the lateral resolution of the microscope, which is:

$$\Delta \simeq [2(R + \ell)]^{1/2} \tag{4}$$

where R is the tip radius and $\ell$ is the gap distance between metals 1 and 2. For $R = \ell = 0.5$ nm, one obtains $\Delta \simeq 0.2$–$0.3$ nm. At present the values of $\Delta$ are about 0.6 nm for metals, 0.3 nm for semiconductors, and 0.15 nm for graphite.

It is also possible to obtain local values of $\phi$, and also mapping $\phi$ for the surface, by modulating $\ell$ at the oscillation frequency $\omega$ and detecting only the modulated fraction of the tunneling current, $(i_T)_{mod}$, by means of a lock-in amplifier:

$$(i_T)_{mod} = \alpha V_T \exp(-\beta \ell_o \sqrt{\phi}) \tag{5}$$

where $\ell_o$ is the amplitude of the applied oscillation. The only unknown in eq. (1) is $\phi$, which can therefore be calculated.

## 3. EXPERIMENTAL MOUNTING

The STM essentially involves three different parts, namely, the STM unit including the mechanical mounting, the control electronics, and the data acquisition and image processing system (Fig. 2).

The fundamental mechanical devices are the proper STM unit and the vibration-free mounting. The STM unit consists of three piezoelectric ceramics arranged as a tripod whose vertex is placed in the tip (Fig. 3). The latter can be made from a tungsten wire ended in a fine tip usually prepared by electrochemical etching. The tip can be first roughly approximated to the surface to a distance smaller than 1 micron by means of an electrostatic-step-motor, and then its position controlled by potentials $V_x$, $V_y$, $V_z$, applied to the piezoceramics. The STM unit can be operated under either in an ultrahigh vacuum chamber or in contact with the atmosphere, as well as either at very low or room temperature, depending on the operating conditions required. Vibrations can be eliminated by filtering, using different mechanical and electrical damping devices.

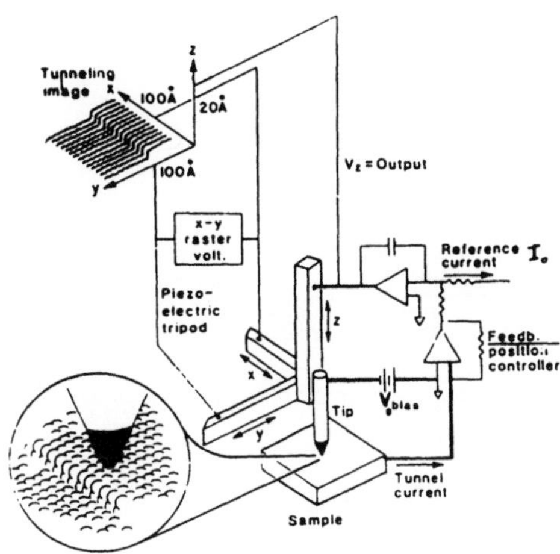


**Figure 2.** Scheme of an electron tunneling microscope (from ref. [21]).

The electronics of STM consists of a control unit made of a logarithmic amplifier, a filter, and a proportional derivating-integrating unit which keeps constant the tunneling current coming from the specimen, previously converted into a potential signal.

The programmed amplification of the signal out of the control unit operates the movement of the z-piezoceramic. The same signal goes to an analogic recorder and, slightly modified, is fed to the computer through an analogic-digital converter.

The software consists in data acquisition in real time and its further analysis. Finally, through image processing it is possible to obtain a noise-filtered real image, as well as the corresponding depth level profiles.

## 4. APPLICATIONS OF STM

STM has been already employed to study mostly ex-situ, the topography of metals, semiconductors and adsorbed molecules. Besides, the local electronic structure can be obtained by electron tunneling spectroscopy (see section 9). Some examples of these applications are presented in the following paragraphs.

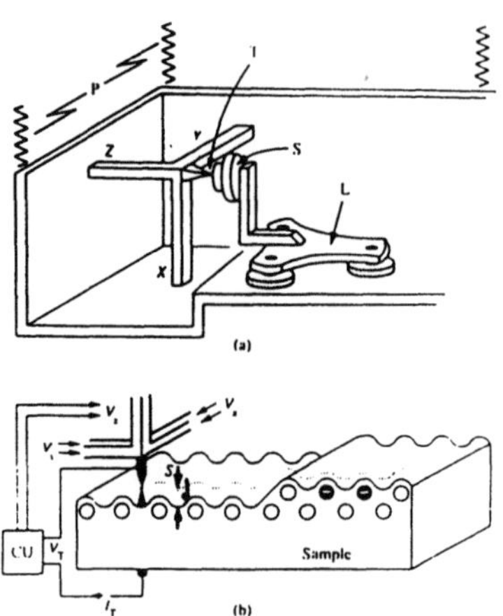


**Figure 3**. Principle of STM imaging (from ref. [21]).

## 4.1. Metals

STM has provided excellent imaging of polycrystalline and single crystal metal surfaces, amorphous metal surfaces, and evaporated metal films. It has also allowed the direct study of surface reconstruction processes and the influence of different adsorbates on those processes. Most of these experiments were made ex-situ under ultrahigh vacuum conditions, and included surface characterization by means of Auger spectroscopy and LEED.

The surface topography of a gold (100) single crystal (Fig. 4) shows large terraces separated by monoatomic steps, and corrugations on the terraces which are due to surface reconstruction [21,22]. The corrugation details reveal different types of reconstruction processes. Similarly, the surface of a platinum (100) single crystal shows an irreversible reconstruction after heating at $T > 420$ K into a 5x25 structure (hexagonal), which after annealing at $T > 1100$ K changes into an inverse hexagonal structure. No reconstruction occurs if a molecule such as CO, NO or ethylene is adsorbed on the metal surface.

STM has also been successfully employed to study the microtopography of preferentially oriented platinum surfaces [9,23] (Fig. 5). In this case it is illustrative to describe first the topography of a mirror-polished polycrystalline platinum surface (blank). It shows an inhomogeneous surface topography, with absence of any preferred crystallographic orientation. The r.m.s. average surface roughness is calculated as 11 nm, as expected for very flat,

**Figure 4.** STM topography of Au(100) single crystal. Flat terraces separated by monoatomic steps are seen. Divisions on the axis are 0.5 nm apart (from ref. [21]).

mechanically polished metal specimens. The STM image after 2 h electrochemical faceting [for experimental conditions see ref. [24]] reveals a surface smoother than that of the blank, with a few steps of variable length (x-axis) and of ~10 nm height (z-axis). In addition, the treated surface shows the formation of clusters, which in most cases are well localized. At this stage, the average height and width of clusters are estimated to be ~10 nm and ~50 nm. In certain regions the clusters show a definite orientation, forming ridges. After a longer electrochemical treatment the ridges develop into large plateaux and grooves. An illustrative example of the development of parallel ridges of ~5-10 nm height is shown in Fig. 5a.

Other surface regions of the same specimen show a dramatic development of smooth, square plaquettes of relatively large area (30 $nm^2$), terminating in pronounced grooves. Ridge-like structure in two directions at nearly 90° are observed, related to the symmetry of the (100) surface . This ridge-like topography is similar to that reported for Au(100) single crystals in air at atmospheric pressure [21,22].

Similar STM studies on stepped nickel surfaces in the presence of hydrogen revealed a p(2x2)2H superlattice [25]. The surface corrugation at terraces is about 0.1 nm and increases as the step border is approached.

### 4.2. The graphite surface

Graphite (HOPG, highly ordered pyrolitic graphite) is an important material for calibration purposes in STM (Fig. 6). The hexagonal lattice which is characteristic of graphite has been very clearly observed by STM imaging. The interpretation of the images is facilitated as the electronic structure of graphite is now very

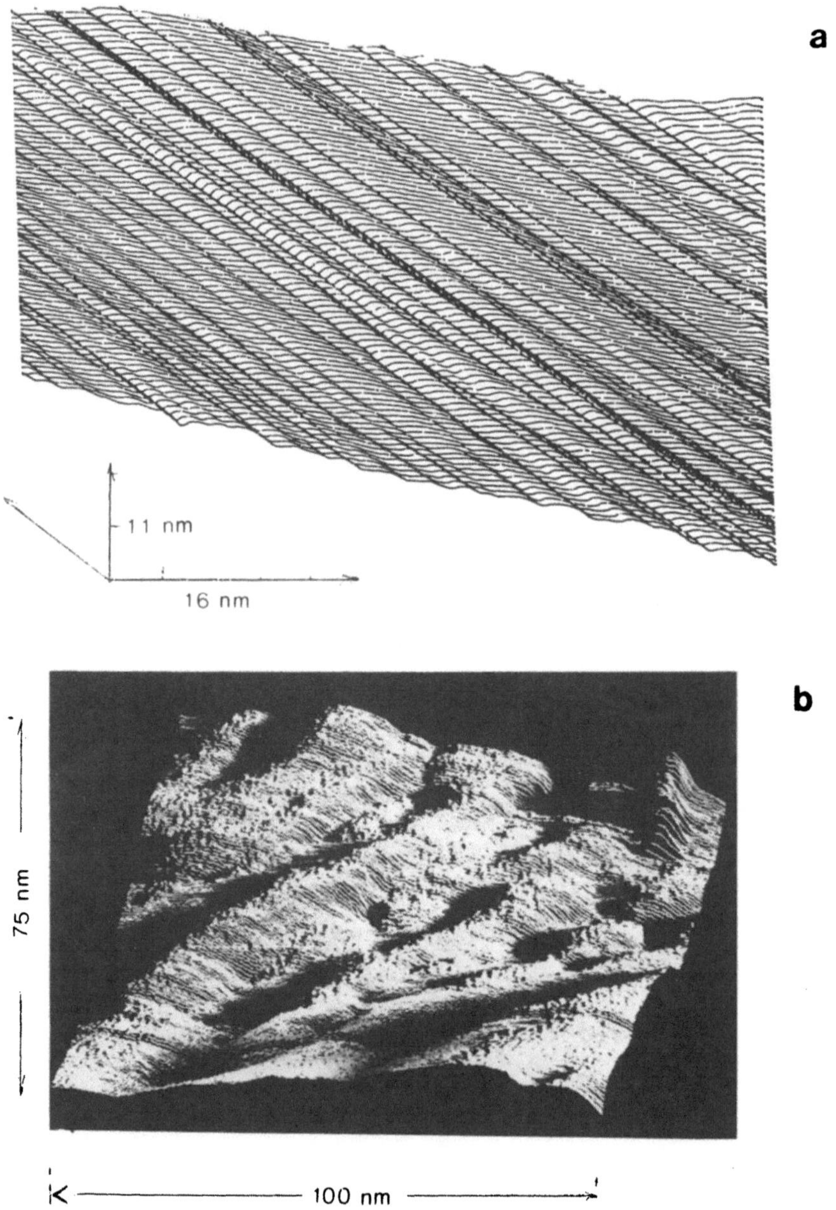


**Figure 5.** STM imaging of platinum (a) and gold (b) surfaces which had been subjected to electrochemical faceting (from ref. [23]).

well-established. The topographic image of graphite exhibits a clear periodicity, which corresponds to a hexagonal lattice where the maxima of the density of states belong to the centres of the hexagons. This produces a corrugation of 0.35 nm along the direction of one hexagon side. The STM image of graphite reveals a strong interaction between specimen surface and tip [21].

Graphite offers the possibility to observe the formation of lamellar compounds, such as those produced through reaction with ferric chloride solution. The image of the resulting surface exhibits large regions where the graphite structure remains unaltered, while others, also relatively large, corresponding to the lamellar compound [26].

The graphite interlamellar distance can be measured through STM imaging with good accuracy, and it can be taken as a reference length in STM image analysis [27].

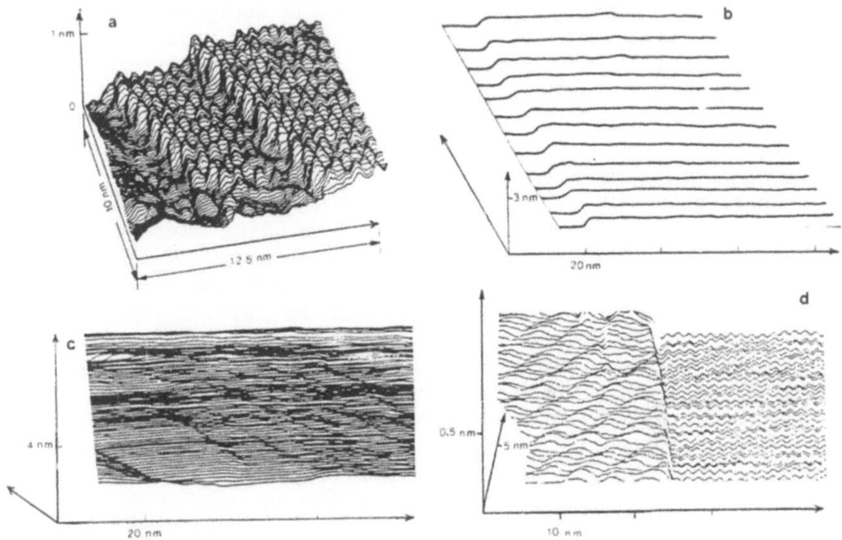


**Figure 6.** STM imaging. a) Si(111) showing 7x7 terraces due to atomic reconstruction separated by atomic steps. b) Exfoliated pyrolitic graphite; a 0.67 nm step can be noticed, coinciding with the reticular constant 2c =0.6708 nm. c) Au(100) single crystal surface treated in ultra high vacuum. Stepped 5x20 reconstruction is seen. d) Pt(100); two terraces separated by a 0.2 nm atomic step are observed.

### 4.3. Semiconductors

Probably the most impacting result coming out from STM in 1983 was the (7x7) reconstruction of Si(111) reported by Binnig and Rohrer [28]

(Fig. 7), which gave a conclusive answer to a controversy on the nature of that process. The lattice comprises 12 adatoms, and one can observe deep valleys at the corners of the unit cell, which is 2.7 nm wide. The presence of the adatoms cancels a certain number of residual bonds, thereby lowering the surface energy to a minimum value. For a stepped Si(7x7) the reconstruction effect continues just to the step border. Other silicon surfaces have also been successfully investigated by STM [29].

## 5. STM and surface microstructure of rough surfaces

Surface disorder and random roughness can play a determining role in the physicochemical properties of surfaces and interfaces. For polycrystalline, amorphous, or inhomogeneous surfaces, the experimental methods available with sufficiently high lateral resolution and vertical sensitivity are limited. It has been demonstrated that STM is a very powerful technique for studying random roughness at surfaces [30]. In this respect a good correlation between STM and LEED of massive single crystals, and STM and direct TEM lattice images of very thin layers, have been established.

Catalytic activity and surface roughness are of the outmost importance in heterogeneous catalysis, including electrocatalysis. The term roughness usually implies the existence of both macropores (macroroughness), which to a great extent are responsible for additional diffusional relaxation, and micropores (microroughness), which concern the effective catalytic area [32]. Despite the close relationship between microroughness and catalytic activity, many real systems involve complex macro- and micropore structures, which makes the direct determination of microroughness a difficult task. A new approach to overcome this drawback is foreseen by using metal surfaces which offer a large catalytic activity, a negligible micropore diffusional relaxation, and a distribution of active sites very close to that of the starting material [33]. This is the case, among others, of platinum electrodes in acid solution, which have been subjected to a relatively fast square wave potential cycling over a range such that a hydrous metal oxide multilayer is formed and immediately afterwards electroreduced, to yield a metal overlayer with a substantial increase in active area. In the case of platinum the new surface maintains the same electrochemical characteristics as the mirror-polished platinum surfaces for different processes, such as the electroadsorption /electrodesorption of hydrogen and oxygen at the atomic monolayer level [33]. The topography of this type of electrodes has been obtained at the nanometre scale by means of the STM. The untreated specimens refer to either the platinum surface as received (virgin specimen) or after it has been melted into a bead, quenched in water, and cleaned in acid, or after it has been subjected to several repetitive triangular potential scans at 0.2 V/s in the 0 to 1.5 V range (blank). The STM patterns of the blank are independent of the thermal treatment, shape and size of the specimen. One observes that the surface of the blank is much smoother than that of the virgin specimen, which means that the voltammetric cycling at 0.2 V/s has already caused smoothing of the surface after about 100 potential scans between 0 and 1.5 V.

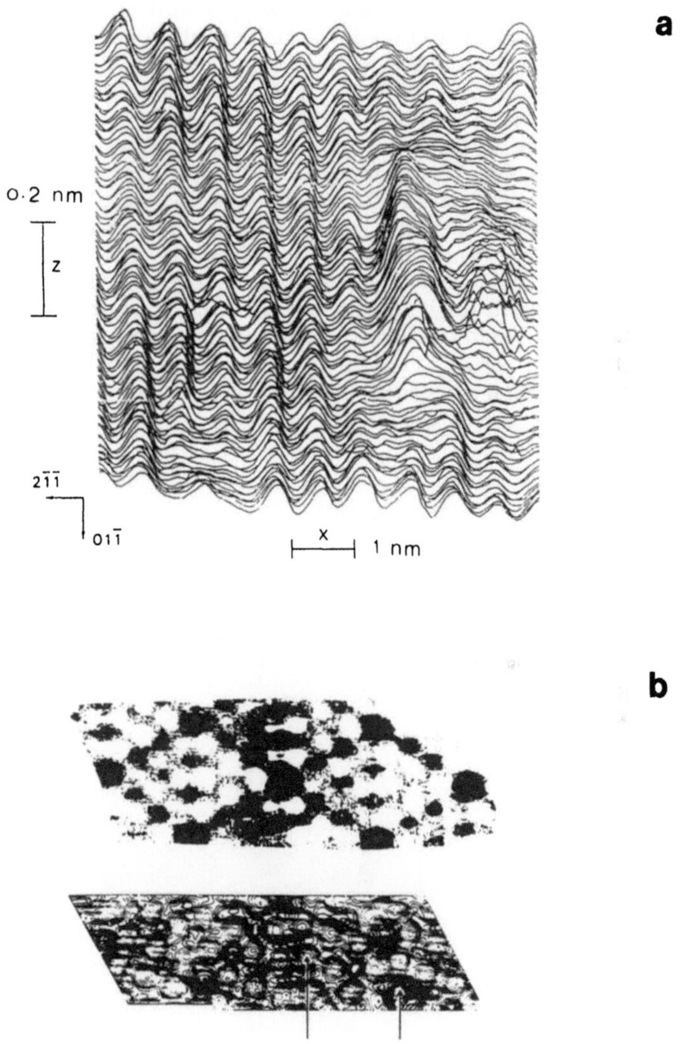


**Figure 7.** a) STM topography of Si(111) after 2x1 reconstruction. b) Image of Si(111) 7x7 reconstructed cell (from ref. [21]).

Platinum and gold electrodes were subjected to repetitive square wave potential sweeps (RSWPS treatment) to yield 40, 60, and 100-fold increases in the roughness factor as measured through the voltammetric charge. STM patterns of these specimens obtained by scanning different surface regions under various magnifications exhibit an inhomogeneous topography; an illustrative picture is presented in Fig. 8. The inhomogeneity is presumably related to different domains in the hydrous platinum oxide layer structure produced by the RSWPS treatment, which reflects the polycrystalline characteristics of the starting material. The non-homogeneous topography of treated platinum is characterized by a pebble-like structure (Fig. 8, region b) which appears either as high or low domes localized in domains randomly distributed over the surface. The transition from low to high dome structures occurs rather abruptly. After a 40x increase in voltammetric charge the average radius of the domes is ~10 nm (Fig. 8, region a). The top of each dome is relatively flat, with corrugations not higher than 0.1 nm. The prevalence of this morphology suggests that the structure of activated platinum electrodes results from the piling up of small nearly "rounded" platinum crystallites produced through the slow electroreduction of the hydrous oxide layer grown by the RSWPS treatment.

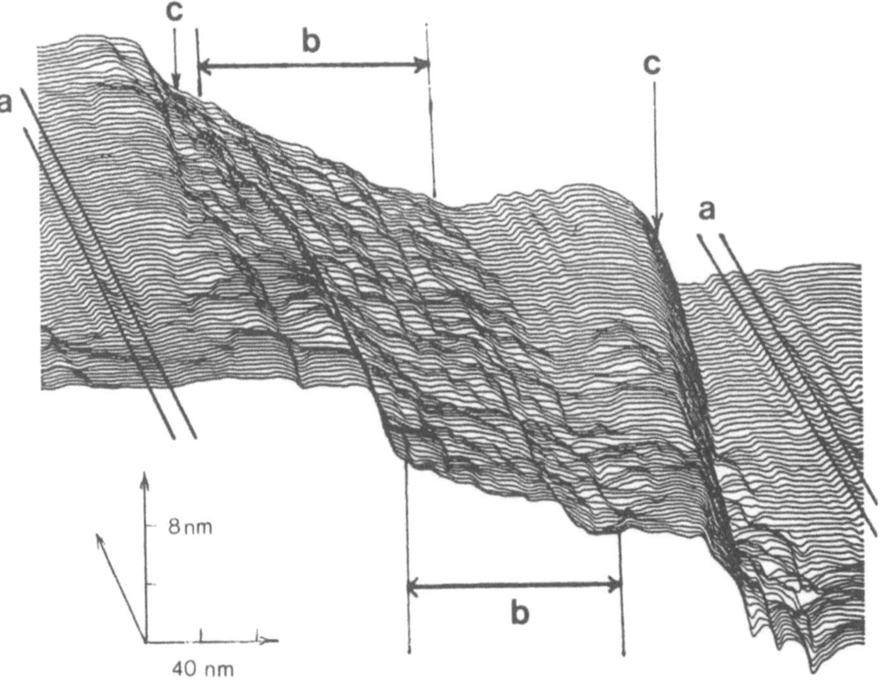


**Figure 8.** STM imaging of a platinum surface with a roughness factor of about 80. Three different topographies can be distinguished (a,b,c). Regions of type b are responsible for the roughness increase.

The structure from data analysis and processing for the case of gold clearly shows channels (white regions) surrounding the different domes. It appears that this randomly piled pebble-like structure penetrates into the bulk of the electroreduced metal overlayer, and is responsible for the change in catalytic activity.

In conclusion, the STM patterns furnish for the first time direct information about the microtopography of active metal electrodes measured ex-situ at the nm scale. They also indicate a correlation between the activation of the electrode and the change in surface topography.

## 6. AMORPHOUS ALLOYS

STM images of the amorphous alloys G16, containing nickel and cobalt as major components [34], present two characteristic regions: (i) very rough zones with high corrugations but showing, however, aligned structures, and (ii) flat regions of reduced size (100 nm) forming a clearly stepped structure. It is interesting to note that the two regions are observed on both blank (Fig. 9a) and treated electrodes, that is, electrodes which have been used as anodes at 0.2 $A/cm^2$ for 5 h in 1 M KOH (Fig. 9b). The boundaries between these two regions are sharp and well-defined as can be seen on Fig. 9b.

The sample is amorphous in the bulk and the crystalline phase is confined to the passivating spinel-like layer. The similarity for the STM images of treated and untreated samples indicates that the spinel-like layer is not changed by the electrochemical treatment. Small crystalline structures in an amorphous matrix have also been reported for amorphous palladium-copper-silicon alloys [35] although the crystallites were larger, and their number smaller, than in G16. The small size of those crystallites makes it impossible to see clearly developed X-ray diffraction patterns for the so-called amorphous metals. The STM lattice fringe spacing indicates the existence of regions with regular periodic atomic lattice domains. The microcrystalline domains resulting from STM real-time two-dimensional color level fluctuates considerably, as one should expect for a system where the atomic lattices of the domains are deformed [36]. For G16 the size of the crystallites is consistent with the crystallographic parameter of cubic face centred nickel spinel, which is 0.8112 nm [37].

The analysis of STM images of amorphous metals of electrocatalytic interest suggests that their catalytic activity is related to a combined effect of the surface compounds and the microcrystalline structure of the surface layer. This can explain the particle size effect in catalytic reactions [38,39].

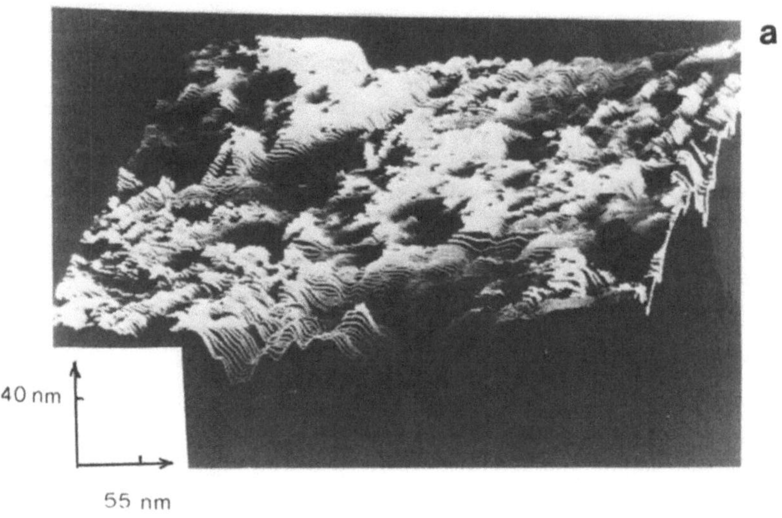


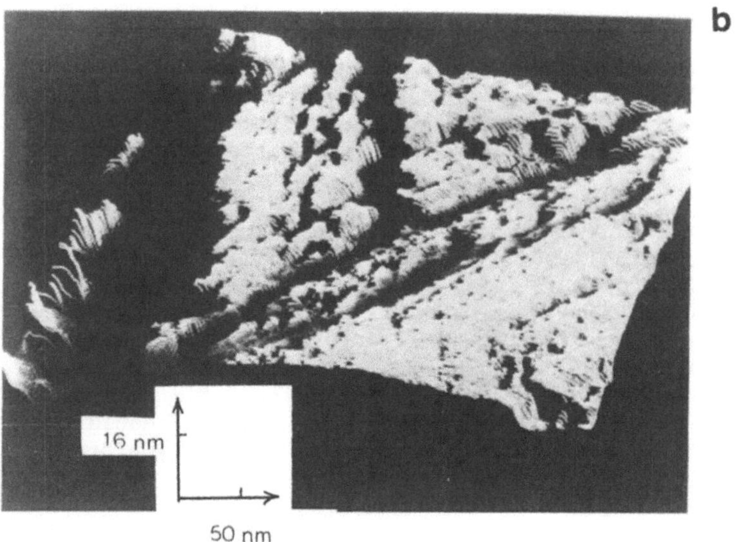


**Figure 9.** STM topography at different resolutions of G16 amorphous alloy (major components: Co and Ni). a) G16, blank. b) G16 treated.

## 7. STM OF ADSORBED SPECIES

The STM study of adsorbed species is particularly relevant in the physics of surfaces, in heterogeneous chemical processes, and in biology.

Let us consider as a first example the STM images of a copper phthalocyanine molecule adsorbed on silver [21]. This is a plane molecule containing four lobes, which appears as a disc- or cone-shaped molecule of about 1.2 nm diameter. Images often change in successive STM scans, especially on flat substrate areas, a fact which is consistent with surface diffusion.

STM images of barium stearate and diacetylenic acid deposited by the Langmuir-Blodgett technique on highly doped silicon substrates reveal a rough heterogeneous surface topography. A hilly landscape at the nm scale is derived from STM image inspection [40].

Fragments of DNA molecules deposited on graphite previously treated to render it hydrophilic can be observed through STM. These preliminary experiments aimed to distinguish the helix structure of the molecule.

Living systems such as the bacteriophage $\phi$ 29 have been observed through STM at room temperature and pressure by using a graphite structure [14]. The dimension and shape of the microorganism as deduced from this technique correlate with those already established by other experimental procedures. Similar studies were extended to biological membranes [41].

Very recently, changes resulting from the pinning of individual organic molecules to the graphite followed by STM have been reported [42]. This opens a clearly optimistic perspective towards molecular manipulation using scanning tunneling microscopes, and the possibilities of atomic scale engineering [43].

## 8. STM IN THE ELECTROCHEMICAL CELL

Very recently STM was developed as an in situ technique to determine the topography at the nm scale of different electrodes immersed in electrolyte solutions [15,44,45]. Since both the tip and the substrate are immersed in the electrolyte solution , electrochemical reactions (faradaic processes) take place both at the tip and the substrate. Therefore, the current is the sum of the tunneling and the faradaic currents. In order to separate the two contributions it is necessary to have an independent control of the STM unit, and try to bring to a minimum the faradaic current. The control of the STM unit can be achieved as in the case of vacuum electron tunneling, because for the immersed tip the same type of relationship among $i_T$, $\phi$ and $\ell$ is obeyed (equation 1). In general, the faradaic current must be kept lower than ±10 pA for a tip area immersed in solution of about $10^{-5}$ $cm^2$. This can be obtained by using a drop of electrolyte filling the gap between the tip and the working electrode (surface specimen). The useful potential window for in situ STM electrochemical work can be determined through the measurement of the polarization curve for each electrochemical system.

Different tip materials have been used for in situ STM electrochemical experiments, such as tungsten, platinum, platinum-iridium, tantalum, and glass covered platinum. In these cases the single tube piezoscanner STM design simplifies the cell mounting and operation control. The tip scan rates are in the range 20-80 nm/s.

In situ images of a graphite surface before and after electroplating with gold were obtained without removing the sample from the electroplating solution. Likewise, images from a preliminary investigation of electrochemical corrosion were obtained at 5 sec intervals [15]. STM studies of HOPG electrodes in 0.1 M NaCl were made with the main purpose of uncoupling of faradaic and tunneling currents in the presence of a redox couple, $[Fe(CN)_6]^{3-}/[Fe(CN)_6]^{4-}$, through the measurement of current/potential curves as a function of tip-substrate distance [46].

The electrodeposition of platinum on the basal plane of HOPG in acidic solution was followed by in situ STM. The initial formation of platinum nuclei can be observed as well as islands of platinum [47]. In situ STM was also used at the atomic level for platinum single crystals in acid electrolytes. The images exhibit flat domains which develop a clear corrugation after voltammetric cycling [48]. This corrugation is to some extent comparable to that described before for platinum electrodes with preferentially oriented surfaces.

In situ STM adsorption of lead on Ag(100) single crystals by applying a negative potential to the tip furnishes stable images, where no changes of the substrate topography can be observed after one adsorption cycle. This is not the case, however, for the lead phase, which develops steps and induces the recrystallization of the substrate, presumably due to lead dissolution into silver [49].

Very interesting in situ topographies of gold single crystals at the atomic level in acid electrolytes were recently shown, exhibiting flat domains and monoatomic steps [50,51]. A single adsorption/desorption cycle for chloride ion produces a considerable distortion of the surface structure.

Recently, a novel electrochemical technique to form a preferentially etched pattern on a semiconductor surface in solution was advanced by employing a modified version of STM, called scanning electrochemical and tunneling microscope (SETM) [52]. With the SETM the tip can be held at a longer distance and then, through faradaic processes it is possible to produce modifications of the electrode surface. This technique has been used for fine line etching of n-GaAs.

## 9. SCANNING TUNNELING SPECTROSCOPY (STS)

Over the last 20 years different electron tunneling spectroscopies (ETS) have been applied to many systems. In ETS vibrations are eliminated by using a metal-insulator-metal (MIM) configuration. This field has been recently reviewed by Walmsley [53], which gives the following classification of ETSs in increasing energy range:

i) superconducting energy gap spectroscopy (0-5 meV)
ii) electrode phonon spectroscopy (5-50 meV)
iii) barrier and semiconductor phonon spectroscopy (30-100 meV)
iv) molecular vibration spectroscopy (30-500 meV)
v) molecular electronic spectroscopy (0.5-2.0 eV)
vi) electron plasmon spectroscopy (1.5-5.0 eV).

Now that the problem of vibrations has been eliminated, it is possible to obtain with the STM the ET spectrum of any given point of the surface. If the tip-substrate distance is held constant (by means of e.g. a sample-and-hold system), the current $i_T$ that flows under an applied potential difference, $V_T$, is given by:

$$i_T(V_T) \simeq \int_0^{eV_T} \rho_T(E-eV_T)\rho_S(E,r_t)dE \tag{6}$$

where $\rho_T$ and $\rho_S$ refer to the density of electronic states for the tip and the specimen, respectively, and $r_t$ denotes the position of the tip. Now, by changing the applied potential from $V_T$ to $V_T$+V one obtains:

$$i_T(V_T+V) \simeq \int_0^{e(V_T+V)} \rho_T(E-eV_T+eV)\rho_S(E,r_t)dE \tag{7}$$

The difference between equations (6) and (7) is due to the contribution of those electronic states comprised between $eV_T$ and $e(V_T+V)$.

Tunneling spectra are recorded by superimposing a small ac voltage on top of the fixed dc voltage, and measuring the modulation of $i_T$ by phase-sensitive detection. For fixed tip position and distance the $di_T/dV$ vs V profile yields the electronic structure of that point. Fixing the voltage at a given peak of $di_T/dV$ it is possible to obtain an image $di_T(x,y)/dV$ which allows to resolve spacially the electronic structure of the surface (see e.g. [54]).

## 10. SOME CONSIDERATIONS RELATED TO STM DATA

The use of STM has at present some limitations which will be gradually overcome in the next few years. These limitations are related to: i) the high resolution of STM; ii) the STM unit itself and tip topography; and iii) the tip-metal sample interactions, which include screening effects.

Except when features are sought, in general, the high resolution of STM requires a systematic sampling of the surface area and a number of images sufficiently large to assure a representative description of the real topography of the sample.

The ideal tip should be constituted by a single atom, but the actual configuration of the tip is far from the ideal situation. The SEM micrograph of a good STM tungsten tip shows a tip radius of about 0.1 μm. The tip is then modelled as the sum of a number of atomic tips, so that the tip response is that of the different tips at the atomic level. Therefore, this implies that the structure of the tip itself can produce shadowed STM images due to the fact that along the scanning of the specimen surface the distance of each atomic tip to the surface changes and, accordingly, the atomic tip closest to the surface provides the largest electron tunneling contribution. This situation changes continuously along the scanning and depends on the topography of the tip at the atomic scale (multiple tip effect).

The reproducibility and stability of the STM images also depend on the mechanical stability of the STM unit and piezoelectric drifts caused by the electric field, and the temperature.

The electron tunneling current under ideal conditions concerns the passage through a vacuum gap of distance $\ell$ between the tip and the substrate (normal tunneling). However, various STM images described in the preceding sections involve the existence of a molecule (adsorbed species) occupying a certain space between the tip and the substrate. In these cases the topography resulting from the scanning tip run at a constant $\ell$ may involve a tunneling current with electron transfer through the adsorbate. The probability of this type of event depends on the electric diagram for the tunneling barrier consisting of the tip, the metal sample, and the molecule adsorbed on the surface. A more complex situation arises when the adsorbed molecule, or some other species, contributes through electron transfer to the total electron tunneling current. It has been suggested that forces partly due to polarization induced by the effective electric field, perturb the tip-adsorbate-substrate structure during STM measurements [55].

On the other hand, the tip-substrate electronic wave function interactions, either attractive or repulsive, lead to deformations (compression or expansion) of the STM images. In addition, STM image deformations can be caused by the existence of discrete positive and negative charge centers (cations and anions) regularly distributed at the sample surface. This effect can be accounted for by running two STM images changing the tip potential either from positive to negative or viceversa. The sum of these images offers the entire topography of the solid at atomic level. The STM topographies of GaAs and graphite are typical examples of the application of this procedure.

## 11. PERSPECTIVE OF STM IN ELECTROCHEMISTRY

It is clear that the development of non-destructive surface techniques sensitive to the atomic level and applicable in-situ are of the outmost importance for solving a great number of intriguing problems in surface physical chemistry. Possible applications of STM in heterogeneous catalysis and electrochemistry can be briefly summarized as follows [56].

(i) Determination of local surface topography at the atomic level and surface spectroscopy of a new type.

(ii) Imaging data which can be related to the double layer structure at interphases, both under equilibrium and non-equilibrium conditions. The possible coupling of STM to other surface techniques becomes particularly interesting here.

(iii) Study of adsorbate-substrate interactions.

(iv) Restructuring and ageing processes at surfaces.

Among specific electrochemical problems which can be investigated through the application of STM one can mention the following ones [52].

(i) Determination of roughness characteristics of solid electrodes.

(ii) Imaging of metal surface topographies at the early stages of attack in aggressive media.

(iii) Morphology of passive layers, possible imaging of 2D and 3D nucleation and growth processes, structural influence of water content, and ageing effects.

(iv) Adsorption of corrosion inhibitors on different surfaces.

(v) Formation and structure of organic films at electrodes.

In conclusion, there is a promising future for STM in all fields of heterogeneous catalysis and electrochemistry. Future developments, particularly those involving possible in-situ measurements and coupling with other surface techniques, will extend the capability of STM.

## ACKNOWLEDGMENT

The author thanks Prof. N. García and Prof. A.M. Baró, Departamento de Física Fundamental, Universidad Autónoma de Madrid, Spain, for their interesting and cooperative efforts in a joint program on the application of STM to electrochemical problems.

## REFERENCES

[1] G. Binnig and H. Rohrer, Helv.Phys.Lett., **55**, 726 (1982); Surf.Sci., **126**, 236 (1983); Sci.Amer., **253**, 50 (1985); Les Prix Nobel en 1986, pp. 85-111, The Nobel Foundation, Stockholm, Sweden (1987).

[2] G. Binnig, H. Rohrer, Ch. Gerber and W. Weibel, Appl.Phys.Lett.,

**40**, 178 (1982); Phys.Rev.Lett., **50**, 120 (1983).

[3] R.M. Eisberg, Fundamentals of Modern Physics, J. Wiley, New York (1967); S. Gasiorowicz, Quantum Physics, J. Wiley, New York (1974).

[4] R.C. Jaklevic and J. Lambe, Phys.Rev.Lett., **17**, 1139 (1966).

[5] J. Lambe and R.C. Jaklevic, Phys.Rev., **165**, 821 (1968).

[6] G. Binnig and H. Rohrer, Surf.Sci., **131**, 1379 (1983).

[7] W. Hösler, Ber.Bunsenges.Phys.Chem., **90**, 205 (1986).

[8] J.A. Golovchenko, Science, **232**, 48 (1986).

[9] J. Gómez, L. Vázquez, A.M. Baró, N. García, C.L. Perdriel, W.E. Triaca and A.J. Arvia, Nature , **323**, 612 (1986).

[10] L. Vázquez, J.M. Gómez Rodríguez, J. Gómez Herrero, A.M. Baró, N. García, J.C. Canullo and A.J. Arvia, Surf.Sci., **181**, 98 (1987).

[11] L. Vázquez, J. Gómez, A.M. Baró, N. García, M.L. Marcos, J. González-Velasco, J.M. Vara, A.J. Arvia, J. Presa, A. García and M. Aguilar, J.Am.Chem.Soc., **109**, 1730 (1987).

[12] R. Miranda, N. García, A.M. Baró, R. García, J.L. Peña and H. Rohrer, Appl.Phys.Lett., **47**, 367 (1985).

[13] N. García, A.M. Baró, R. Miranda, H. Rohrer, Ch. Gerber, R. García Cantú and J.L. Peña, Metrología, **21**, 566 (1985).

[14] A.M. Baró, R. Miranda, J. Alaman, N. García, G. Binnig, H. Rohrer, Ch. Gerber and J.L. Carrascosa, Nature, **315**, 253 (1985).

[15] B. Drawke, R. Sonnesfeld, J. Schneir and P.K. Hansma, Surf.Sci., **181**, 92 (1987).

[16] G. Binnig, C.F. Quate and Ch. Gerber, Phys.Rev.Lett., **56**, 930 (1986).

[17] G. Binnig, Ch. Gerber, E. Stoll, T.R. Albrecht and C.F. Quate, Surf.Sci., **189/190**, 1 (1987).

[18] H. Heinzelmann, P. Grütter, E. Meyer, H. Hidber, L. Rosenthaler, M. Ringger and H.J. Güntherodt, Surf.Sci., **189/190**, 29 (1987).

[19] E.L. Wolf, Principles of Electron Tunneling Spectroscopy, Oxford University Press, New York (1985).

[20] J.Tersoff and D.R. Hamman, Phys.Rev., **B31**, 2 (1985).

[21] J. Klein, S. Gauthier, S. Rousset and W. Sacks, J.Microsc.-Spectrosc..Electron., **12**, 403 (1987).

[22] G. Binnig, H. Rohrer, Ch. Gerber and E. Weibel, Surf.Sci., **134**, 319 (1983).

[23] A.J. Arvia, Actividad, **1**, 3 (1987).

[24] A.J. Arvia, J.C. Canullo, E. Custidiano, C.L. Perdriel and W.E. Triaca, Electrochim.Acta, **31**, 1359 (1986).

[25] G.F.A. van de Walle, H. van Kempen, P. Wyder and C.J. Flipse, Surf.Sci., **181**, 27 (1987).

[26] J.M. Soler, A.M. Baró, N. García and H. Rohrer, Phys.Rev.Lett., **57**, **444** (1986).

[27] H.J. Mamin, E. Ganz, D. Abraham, R. Thomson and J. Clarke, J.Phys.Rev., **B34**, 9015 (1987).

[28] G. Binnig, H. Rohrer, F. Salvan, Ch. Gerber and A.M. Baró, Surf.Sci., **157**, L373 (1985).

[29] R.S. Becker, J.A. Golovchenko, E.G. McRae and B.S. Swartzentruber, Phys.Rev.Lett., **55**, 2028 (1985); R.M. Feenstra, W.A. Thompson and

A.P. Fein, Phys.Rev.Lett., **56**, 608 (1986).

[30] J.K. Gimzewski and A. Humbert, IBM J.Res.Develop., **30**, 472 (1986).

[31] T. Kessler, A.M. Castro Luna, W.E. Triaca and A.J. Arvia, J.Appl.Electrochem., **16**, 693 (1986).

[32] J.O'M. Bockris and S. Srinivasan, Fuel Cells: Their Electrochemistry, McGraw-Hill, New York, Chapter 5, p. 230 (1969).

[33] A.C. Chialvo, W.E. Triaca and A.J. Arvia, J.Electroanal.Chem., **146**, 93 (1983); J.Electroanal.Chem., **171**, 303 (1984).

[34] G. Kreysa, J. Gómez, A.M. Baró and A.J. Arvia, in preparation.

[35] R. Wiesendanger, M. Ringger, L. Rosenthaler, H.R. Hidber, P. Oelhafen, H. Rudin and H.-J. Güntherodt, Surf.Sci., **181**, 46 (1987).

[36] Y. Hirotsu, M. Uchara and M. Ueno, J.Appl.Phys., **59**, 3081 (1986).

[37] P. Poix, Bull.Soc.Chim. France, 1085 (1965).

[38] O.M. Poltorak and V.S. Boronin, Russ.J.Phys.Chem., **40**, 1436 (1966).

[39] J.R. Anderson, Structure of Metallic Catalysts, Academic Press, New York (1975).

[40] H. Fuchs, W. Schrepp and H. Rohrer, Surf.Sci., **181**, 391 (1987).

[41] A. Stemmer, R. Reichelt, A. Engel, J.P. Rosenbusch, M. Ringger, H.R. Hidber and H.-J. Güntherodt, Surf.Sci., **181**, 394 (1987).

[42] J.S. Foster, J.E. Frommer and P.C. Arnett, Nature, **331**, 324 (1988).

[43] J.B. Pethica, Nature, **331**, 301 (1988).

[44] R. Sonnenfeld and P.K. Hansma, Science, **232**, 211 (1986).

[45] P. Lustenberger, H. Rohrer, R. Christoph and H. Siegenthaler, J.Electroanal.Chem., **243**, 225 (1988).

[46] C.L. Ni, L.D. Bell, C.P. Bankston and W.J. Kaiser, Abstract No. 1472, The Electrochemical Society, Spring Meeting, (1987).

[47] K. Itaya and S. Sugawara, Abstract No. 1412, The Electrochemical Society, Spring Meeting, (1987).

[48] K. Itaya and S. Sugawara, personal communication.

[49] R. Christoph, H. Siegenthaler, H. Wiese and H. Rohrer, 4th International Fischer Symposium on Electrochemical Phase Formation and Dissolution, Karlsruhe (1988).

[50] J. Wiechers, T. Twomey, D.M. Kolb and R.J. Behm., J.Electroanal.Chem.,**248**, 451 (1988).

[51] J. Wiechers, T. Twomey, D. Kolb and R.J. Behm, 4th International Fischer Symposium on Electrochemical Phase Formation and Dissolution, Karlsruhe (1988).

[52] C.W. Lin, F.R.F. Fan and A.J. Bard, J.Electrochem.Soc., **134**, 1038 (1987).

[53] D.G. Walmsley, Surf.Sci., **181**, 1 (1987).

[54] G.F.A. van de Walle, H. van Kempen, P. Wyder and C.J. Flipse, Surf.Sci., **181**, 27 (1987).

[55] J.K. Gimzewski, E. Stoll and R.R. Schlitter, Surf.Sci., **181**, 267 (1987).

[56] A.J. Arvia, Surf.Sci., **181**, 78 (1987).

## SUBJECT INDEX

MIX
Papier aus verantwortungsvollen Quellen
Paper from responsible sources
FSC® C105338

If you have any concerns about our products,
you can contact us on
**ProductSafety@springernature.com**

In case Publisher is established outside the EU,
the EU authorized representative is:
**Springer Nature Customer Service Center GmbH**
**Europaplatz 3, 69115 Heidelberg, Germany**

Printed by Libri Plureos GmbH
in Hamburg, Germany